Circuit Theory

Continuous and Discrete-Time Systems, Elements of Network Synthesis

C.P. KURIAKOSE

Formerly Professor of Electrical Engineering, Engineering College, Trivandrum and Pro-Vice Chancellor and Dean of Engineering Faculty, Cochin University of Science and Technology

Currently, Visiting Professor of Electrical Engineering, Government Model Engineering College, Cochin

PHI Learning Private Limited

New Delhi-110001

2011

Rs. 350.00

CIRCUIT THEORY: Continuous and Discrete-Time Systems, Elements of Network Synthesis
C.P. Kuriakose

ISBN-978-81-203-2643-9

The export rights of this book are vested solely with the publisher.

Third Printing ... ... October, 2011

Published by Asoke K. Ghosh, PHI Learning Private Limited, M-97, Connaught Circus, New Delhi-110001 and Printed by Mudrak, 30-A, Patparganj, Delhi-110091.

To the loving memory of my wife

Sosamma

Contents

Part 2

Linear Time-invariant (LTI) Systems

Part 3
Advanced Topics in Network Analysis

Part 4
Elements of Network Synthesis

List of Tables

List of Tables

Foreword

Circuit Theory which lays the foundation for any course in Electronics has been introduced by authors taking different approaches. Some treat this as a branch of physics supported by theorems enabling analysis of circuits, while others have taken an engineering point of view leading to a set of formulae to be remembered during one's career. In the latter case there is a loss of rigour of the underlying principles or basics. This happens more so when Circuit Theory is taught through "computer-based" assignments.

This book has circumvented these pitfalls. Professor C.P. Kuriakose who has brought in his more than half a century of teaching experience—at the university for full time students and at professional society courses for employed students—has undertaken, and successfully completed, an immensely difficult task, that of producing a palatable text on Circuit Theory. He has thoughtfully selected the appropriate topics for inclusion in this work, and has utilized to great advantage his unquestioned flair for simple, easy to understand exposition of relatively complex subjects. The result is a book which will serve as a textbook for engineering students. Professor Kuriakose strongly believes that practice fosters perfection, especially if it is circuit analysis. A variety of illustrative examples, solved problems and exercises carefully guide the student from basics of electricity to the heart of circuit (network) theory which is supported by the mathematical tools of transforms. The inclusion of a chapter on PSpice and MATLAB is sure to whet the interest of the reader for further exploration of the subject—especially the advanced topics.

The author has also taken care to make it "self-contained" for self-study by employed students of AMIE, and also a "quick-refresher" book of circuit designers.

I believe it will benefit many readers and be a fine reference.

V. Chunder
Director
NPOL, Cochin

Preface

Circuit Theory and Field Theory together form the foundation of nearly all curricula in the discipline of Electrical and Electronics Engineering. Several books are available on these subjects for engineering students in Universities (and Polytechnics) for use in the second and subsequent years; however some of them are too elementary and a few of them are too advanced, and there is a felt need for a medium level, concise and reasonably priced textbook which covers in full the syllabi of Circuit Theory for undergraduate courses at an optimum level of presentation, at the same time incorporating introductory concepts in advanced topics. It is assumed that the reader would have studied basic Physics and Electrical engineering principles. The book is presented in four parts:

1 Fundamental concepts in electrical engineering
2 Linear time-invariant systems
3 Advanced topics in network analysis
4 Elements of network synthesis

There are in all nineteen chapters, and each chapter begins with an introductory statement and incorporates a number of illustrative examples, selected solved problems and a set of chapter-end exercises (with hints and answers). The book should, therefore, serve as textbook for regular degree courses in Electric Circuit Theory, Network Analysis and Synthesis and Linear Circuits and Systems; and a reference book for postgraduate studies. The book will prove to be a source book for self-study by candidates preparing for AMIE, IETE, GATE, UPSC Engineering Service examination and other competitive examinations.

Equal emphasis is given to both the analytical and the numerical aspects of the subject. Besides the worked examples, solved problems and chapter-end exercises, the book contains a set of review problems (with hints and answers). The reader is advised to fill up the missing steps and work his way through the solution. This may perhaps be a little inconvenient but will definitely pay rich dividends.

Part 1 is mainly an upward revision of principles of Physics and Electrical Engineering, which will be very useful particularly for self-study or reference.

Part 2 covers the main topics in the analysis of Linear time-invariant systems. Every effort has been made to cover the undergraduate syllabi of Circuit Theory of several Indian Universities, in spite of the wide disparity in the syllabi of different Institutions.

An attempt has been made to highlight simple and elegant solutions to many of the network problems facilitated by the use of *transform methods* (particularly Laplace transformation) as against classical methods. In many textbooks transform methods are mentioned in passing and that too towards the end, thereby denying the reader the advantages of using the transforms in a large number of problems in such books. Part 2 is organised in ten chapters. They are:

Chapter 6—Techniques of circuit analysis, Branch current method, Mesh current method, Node voltage method, Chapter 7—Network topology, Chapter 8—Differential equations—Solution by classical methods, Chapter 9—Laplace transforms in network analysis, Chapter 10—Steady state sinusoidal response of higher order networks, Network theorems for ac networks, Chapter 11—Terminal characteristics of networks, Chapter 12—Network functions, Chapter 13—Wave filters—Frequency selective circuits—Passive and active filters, Chapter 14—Fourier series and Fourier transforms. The impact of computers upon instruction in electric circuits is without parallel in Electrical engineering education. Chapter 15 serves as an introduction to computer aided circuit analysis.

Part 3 is devoted to certain topics which are gaining more and more prominence in undergraduate syllabi. They are Chapter 16—State variable analysis, Chapter 17— Discrete systems and *z*-transforms, Chapter 18—Non-linear systems.

The concluding Part 4 contains Chapter 19 which introduces elements of Network synthesis.

Finally, a set of Review Questions covering all the chapters are added. The appendices cover Matrices and Determinants, Complex variables, Operational amplifiers, and Coupled circuits and the transformer. Suggestions for the improvement in the text are welcome.

C.P. Kuriakose

Acknowledgements

I am greatly privileged to get the book released with a Foreword from eminent scientist and educationist, V. Chunder, who is the Director of the prestigious National Physical and Oceanographic Laboratory, Cochin.

The book is dedicated to my erstwhile colleagues and students who have made possible a satisfying and rewarding teaching career for me. Many have helped with the preparation of this book. I am indebted to Prof. Dr. Paulose Jacob, Head of Computer Science department, Cochin University of Science and Technology, Prof. Dr. Varghese Paul, Head of IT department, and Prof. V.P. Devasiah, Head of Electronics department, Model Engineering College, Cochin for their very valuable suggestions from time to time. I am grateful to Ms. Ann Jacob who was my colleague at Old Dominion University (USA) twinning programme held at Rajagiri College, Cochin, Mr. Eldho Baby who has specialized in computer aided network analysis for his active support (Chapter 15 is his contribution), Ms. S. Deepthi who has specialized in networks, for her invaluable assistance throughout and Mr. Syam V. Krishnan, a specialist in computer simulation, for his assistance. I want to express my gratitude to all at Prentice-Hall of India. Finally I wish to thank my children for their continuing support and encouragement, and for whom all this effort was happily made, the youngest of them Maya is now a software enthusiast and has been of considerable help to me.

C.P. Kuriakose

Part 1

Fundamental Concepts in Electrical Engineering

Circuit Variables

1

CHAPTER CONTENTS

Electrical Engineers' world is truly an amazing one comprising systems as diverse as power, communication, control, computers and signal processing. The analysis as well as the synthesis and design of such systems are facilitated by the techniques of **Electric Circuit Theory**.

An electrical ***network*** is a system consisting of ***elements*** such as resistors, capacitors, inductors, and energy sources. Simple networks are often referred to as electric ***circuits***. An electric ***circuit*** is a mathematical model that approximates the behaviour of a simple network such as a lamp connected to a voltage source through a switch. But several lamps connected in parallel to a source may be treated as a network. Several such networks will constitute a system. However very often these terms are used interchangeably.

Electric circuit theory is a special case of electromagnetic field theory, the study of static and moving electric charges. But the application of the general theory is cumbersome and requires the use of advanced mathematics. For this reason field theory is generally reserved for senior classes of undergraduate courses. This book is devoted primarily to network analysis and the use of the techniques developed, in the solution of engineering problems. However synthesis of elementary circuits is briefly covered in the final chapter.

1.1 CIRCUIT CONCEPTS

Consider a circuit/network represented by the 'box' in Fig. 1.1. Given the network S and the input $x(t)$ to determine the output $y(t)$ is the ***analysis*** problem. Given the input $x(t)$ and the desired output $y(t)$ to design the network is the ***synthesis*** problem.

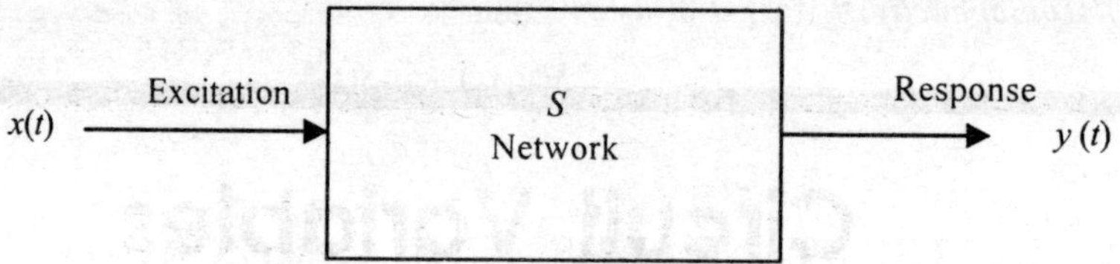

Fig. 1.1

Characteristics of electrical networks: A ***continuous-time*** network is characterized by input $x(t)$ and output $y(t)$ which are functions of the continuous variable t; whereas in a ***discrete-time*** network they vary only at discrete values of time. Thus input equals $x(n)$ and output equals $y(n)$ where $n = 1, 2, 3, \ldots$ are discrete points on the time axis. Continuous-time network are described by *differential equations*, and discrete-time circuits by *difference equations*.

A constant (***time-invariant***) network has a response, the shape of which is dependent only on the shape of the excitation and not on the time of application of the excitation. Thus, if

$$S\{x(t)\} = y(t) \tag{1.1}$$

$$S\{x(t - t_0)\} = y(t - t_0) \tag{1.2}$$

a time-shift t_0 in the input produces an identical time-shift in the output.

A ***lumped*** element has physical dimensions which do not affect its describing equation. That is to say electrical effects happen simultaneously throughout the system. If d is the largest dimension of the element and λ is the wavelength of the signal, a lumped element satisfies the condition

$$d << \lambda \tag{1.3}$$

where

$$\lambda = c/f \tag{1.3a}$$

in terms of the velocity of electromagnetic waves, c and frequency of the signal, f. In a ***distributed*** network variations with respect to the dimensions of the component are important. Total differential equation describes a lumped element network; on the other hand distributed network is described by the *partial differential equations*, involving space and time.

Principle of linearity: A circuit described by Eq. (1.1) is ***linear***, if

$$\begin{aligned} S(a_1x_1 + a_2x_2) &= a_1S(x_1) + a_2S(x_2) \\ &= a_1y_1 + a_2y_2 \end{aligned} \tag{1.4}$$

where a_1 and a_2 are constants.

Two corollaries result from the principle of linearity given by Eq. (1.4).

(i) Additivity: If $a_1 = a_2 = 1$ then

$$\begin{aligned} S(x_1 + x_2) &= S(x_1) + S(x_2) \\ &= y_1 + y_2 \end{aligned} \tag{1.5}$$

(ii) Homogencity: If $a_1 \neq 0,\ a_2 = 0$ then

$$S(a_1x_1) = a_1S(x_1)$$
$$= a_1y_1 \tag{1.6}$$

1.2 UNITS, STANDARDS AND DIMENSIONS

Measurement presupposes a well defined system of units. The SI units (abbreviation for Systeme Internationale d'unites) are used by most of the engineers throughout the world. The seven ***fundamental*** units and two ***supplementary*** units are tabulated in Table 1.1.

Table 1.1 Fundamental Units in SI

	Quantity	*SI unit*	*Unit symbol*
Fundamental units	Length	metre	m
	Mass	kilogram	kg
	Time	second	s
	Electric current	ampere	A
	Temperature	kelvin	K
	Luminous Intensity	candela	cd
	Amount of substance	mole	mol
Supplementary units	Plane angle	radian	rad
	Solid angle	steradian	sr

The definition of ***metre*** has undergone several changes from time to time, the latest definition being in terms of the velocity of light, replacing the ***krypton metre***. The ***ampere*** is now defined as *force per metre of two parallel straight conductors of infinite length and negligible cross-section at a distance of one metre apart in vacuum.* Definition of all other units will be found in Reference 1.

Derived units: All other quantities such as force, work, power, voltage, can be expressed in terms of the fundamental quantities, raised to various powers—these are called the ***dimensions*** of the derived quantity.

For example, Force = Mass × Acceleration

$$[F] = [M]\,[L]\,[T^{-2}]$$
$$[F] = [MLT^{-2}] \tag{1.7}$$

and the dimensions of force may be written as [1, 1, –2]

$$\text{Energy} = \text{Force} \times \text{Displacement}$$
$$[W] = [MLT^{-2}]\,[L]$$
$$[W] = [ML^2T^{-2}] \tag{1.8}$$

Derived units commonly used in network analysis are listed in Table 1.2.

Table 1.2 Derived Units

Quantity	*Quantity symbol*	*Unit*	*Dimension*	*Unit symbol*
Area	A	square metre	L^2	m^2
Volume	V	cubic metre	L^3	m^3
Velocity	v	metre per second	LT^{-1}	m/s
Force	F	newton	MLT^{-2}	N, (kg-m/s^2)
Pressure	P	pascal	$ML^{-1}T^{-2}$	Pa, (N/m^2)
Work, Energy, Heat	W	joule	ML^2T^{-2}	J, (Nm)
Power	p	watt	ML^2T^{-3}	W, (J/s)
Frequency	f	hertz	T^{-1}	Hz
Electric charge	Q	coulomb	TI	C, (As)
Electric potential	V, E	volt	$ML^2T^{-3}I^{-1}$	V, (J/C)
Resistance	R	ohm	$ML^2T^{-3}I^{-2}$	Ω, (V/A)
Capacitance	C	farad	$M^{-1}L^{-2}T^4I^2$	F
Inductance	L	henry	$ML^2T^{-2}I^{-2}$	H
Magnetic flux	ϕ	weber	$ML^2T^{-2}I^{-1}$	Wb
Permeability	μ	henry/metre	$MLT^{-2}I^{-2}$	H/m

1.3 CIRCUIT VARIABLES

1.3.1 Electric Current (*I*)

The ampere A, has been chosen as the fourth fundamental unit of the SI. All other electrical quantities may be defined in terms of the basic quantities of mass, length, time and current.

1.3.2 Electric Charge (*Q*)

An electric current consists of a flow of electric charge through a conducting medium and hence the time integral of current is charge. The unit of charge is ***coulomb***, defined as the charge transferred by a current of one ampere in one second. Conversely, the ampere may be defined as the flow of one coulomb of charge through a cross-section in one second.

$$i = dQ/dt$$

and the dimensions of charge are:

$$[Q] = [T]\,[I] \tag{1.9}$$

1.3.3 Potential Difference (*V*)

The unit of potential difference is volt which is defined in terms of energy required to move a charge from one point to another. The ***volt*** is that electric potential difference which exists between two points if a charge of 1 coulomb receives or delivers one joule of energy in moving between them. The dimensions of potential difference are:

$$[V] = [W]/[Q]$$

and from Eqs. (1.8) and (1.9),

$$[V] = [ML^2T^{-2}]/[TI]$$

hence,

$$[V] = [ML^2T^{-3}I^{-1}] \tag{1.10}$$

1.3.4 Electric Power and Energy

If charge is continuously transported from one point to another through a potential difference, energy W is transferred; and power p, the rate of doing work is:

$$p = dW/dt$$

but

$$v = dW/dq$$

and

$$i = dq/dt$$

hence

$$p = dW/dq \times dq/dt$$

$$p = vi \tag{1.11}$$

The total energy is:

$$W = \int p\, dt$$

$$W = \int vi\, dt \tag{1.12}$$

1.4 DIMENSIONAL ANALYSIS

Dimensional analysis is a powerful tool in circuit theory. The principle of ***dimensional homogeneity*** states that every term in an analytical equation must have the same dimensions. As an example, consider the familiar equation for the distance travelled by a particle moving with initial velocity u and acceleration a,

$$s = ut + \tfrac{1}{2}at^2$$

the dimensions of ut are $[LT^{-1}]\ [T] = [L]$

which is the dimension of distance s.

and the dimensions of $\tfrac{1}{2}at^2 = [LT^{-2}]\ [T^2] = [L]$

which again is the dimension of distance s.

The technique may also be used to deduce the *form* of an equation describing a physical phenomenon.

SUMMARY

- The scope of circuit theory includes analysis and synthesis.
- Electrical networks are classified as continuous-time networks or discrete-time networks.
- A constant network has a response which is dependent only on the shape of the excitation and not on the time of application of excitation.
- A lumped element circuit is governed by total differential equations whereas a distributed network is described by partial differential equations (including space and time).
- Principle of linearity involves two important concepts namely additivity and homogeneity.
- The International System of Units (SI) is summarized in Table 1.1 (base units) and Table 1.2 (derived units).
- Network analysis is based on the variables of current and voltage.
- Current is the rate of change of charge and has the unit of ampere ($i = dq/dt$).
- Voltage is energy per unit charge created by charge separation and has the unit of volt ($v = dW/dq$).
- Power is energy per unit of time and is equal to the product of terminal voltage and current and has the unit watt ($p = dW/dt = vi$).

Circuit Elements

2

CHAPTER CONTENTS

Electrical systems consist of interconnected components, most of which are non-linear but the analysis of such systems is facilitated by replacing the actual components by models having linear characteristics. The analysis of the resulting ideal circuits is the core of circuit theory.

There are three basic ***passive*** circuit elements, namely resistor, capacitor and inductor which either dissipate or store electrical energy; and two *active* circuit elements, namely voltage source and current source which supply electrical energy.

2.1 PASSIVE ELEMENTS

Passive elements are the resistors, inductors and capacitors.

2.1.1 The Resistor

A lumped time-invariant linear resistor is defined by its voltage-current relationship. In SI units the circuit symbol for ***resistance*** and the sign convention adopted are shown in Fig. 2.1. Note that the direction of current flow is opposite to that of the electrons which constitute it.

$$v = Ri \tag{2.1a}$$

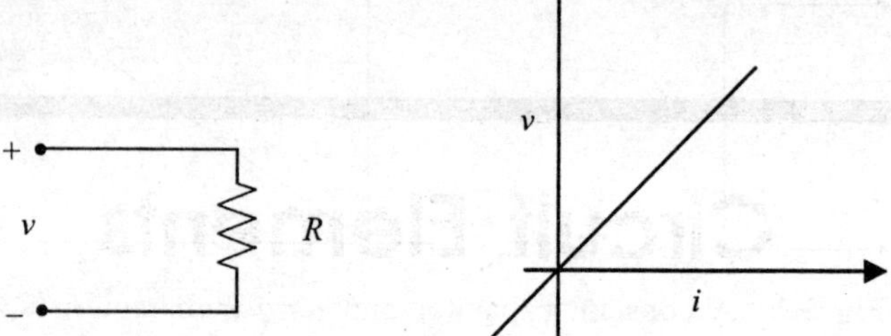

Fig. 2.1 A constant resistor and its v-i characteristics.

The current is proportional to the voltage and R is the constant of proportionality. The inverse relationship gives the current in terms of the voltage.

$$i = \frac{1}{R}v$$

or
$$i = Gv \tag{2.1b}$$

where G is the ***conductance*** measured in siemens S.

The output of a resistor depends only on the present input (and not on past values); such an element is called a **memory-less element**.

Resistors in series: The effective resistance between the terminals of a series combination of resistors is given by

$$R = R_1 + R_2 + R_3 + \cdots \qquad \text{(see section 3.4.1)}$$

Resistors in parallel: The effective resistance between the terminals of a parallel combination of resistors is given by

$$\frac{1}{R} = \frac{1}{R_1} + \frac{1}{R_2} + \frac{1}{R_3} + \cdots \qquad \text{(see section 3.4.2)}$$

Power dissipated by a resistor:

$$p = vi = i^2R = v^2G$$

2.1.2 The Capacitor

A constant lumped linear capacitor is defined by its charge-voltage relation,

$$q = Cv \tag{2.2}$$

where q is measured in coulombs and C is measured in farads.

The current is given by

$$\frac{dq}{dt} = i = C\frac{dv}{dt} \tag{2.2a}$$

which is the basic i-v relationship for the capacitor.

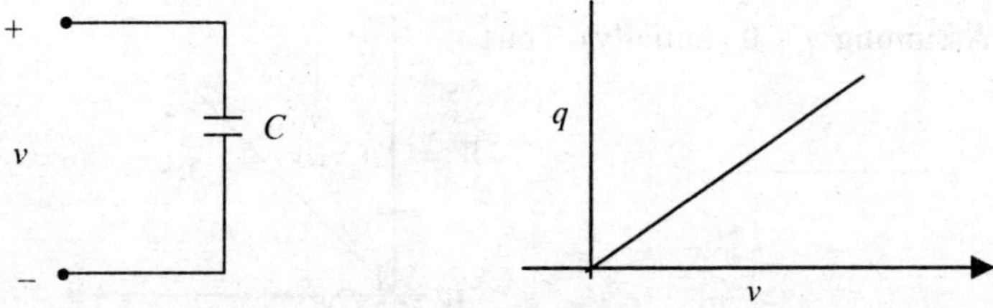

Fig 2.2 A constant capacitor and its *q-v* characteristics.

The inverse relationship is obtained as:

$$\frac{q(t)}{C} = v(t) = \frac{1}{C}\int_{-\infty}^{t} i(t)\,dt \tag{2.3}$$

which shows that the past values of the input *i* affect the output (unlike the resistor). Such an element is called a **dynamic element**. Rewriting Eq. (2.3) by expressing the integral as the sum of two integrals,

$$v(t) = \frac{1}{C}\int_{-\infty}^{0} i(t)dt + \frac{1}{C}\int_{0}^{t} i(t)\,dt$$

$$= \frac{1}{C}q(0) + \frac{1}{C}\int_{0}^{t} i(t)\,dt$$

where $q(0)$ is the initial charge on the capacitor. Thus,

$$v(t) = v(0) + \frac{1}{C}\int_{0}^{t} i(t)\,dt \tag{2.4}$$

where $v(0)$ is the initial voltage across the capacitor.

Energy stored in the capacitor:

$$W = \int_{-\infty}^{t} vi\,dt$$

$$= \int_{-\infty}^{t} vC\left(\frac{dv}{dt}\right)dt$$

$$= \int_{-\infty}^{v} Cv\,dv$$

(Note the change in the variable of integration from *t* to *v*)

Assuming $v = 0$ (initially). Thus,

$$W = \int_{-\infty}^{v} Cv\,dv$$

$$W = \frac{1}{2}Cv^2 \tag{2.5}$$

Capacitance of a parallel plate capacitor: It is given by

$$C = \varepsilon A/d$$

where, A is the area of the plates, d is the distance between the plates, ε is the absolute permittivity of the medium ($\varepsilon_0\, \varepsilon_r$), ε_0 is the permittivity of free space (8.854×10^{-12} F/m) and ε_r is the relative permittivity (or dielectric constant) of the medium.

Capacitors in parallel: The equivalent capacitance of capacitors in parallel is:

$$C = C_1 + C_2 + C_3 + \cdots \qquad \text{(see section 3.4.3)}$$

Capacitors in series: The effective capacitance of capacitors in series is:

$$\frac{1}{C} = \frac{1}{C_1} + \frac{1}{C_2} + \frac{1}{C_3} + \cdots \qquad \text{(see section 3.4.4)}$$

EXAMPLE 2.1 In the circuit of Fig. 2.3, capacitor C_1 (4 F) is charged to a potential of 12 V and the capacitor C_2 (2 F) is uncharged. Determine the energy stored in the circuit. Also determine the potential on each capacitance and the total energy stored after the switch has been closed.

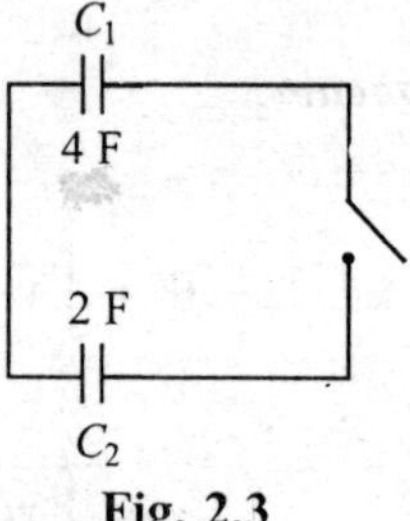

Fig. 2.3

SOLUTION:

Before closing the switch,

$$\text{Energy stored} = \frac{1}{2}C_1V^2$$

$$= \frac{1}{2} \times 4 \times 12^2 = 288 \text{ J} \qquad \textbf{Ans.}$$

and $\qquad$ Charge $Q = CV = 4 \times 12 = 48$ C

After closing the switch,

$$\text{Charge } Q = (C_1 + C_2)\, V$$

where V is the common terminal voltage.

$$48 = (4 + 2)\, V$$

$$V = 8 \text{ V} \qquad \textbf{Ans.}$$

and the energy stored after closing the switch

$$= \frac{1}{2} CV^2$$

$$= \frac{1}{2} (4 + 2)\, 8^2 = 192 \text{ J} \qquad \textbf{Ans.}$$

2.1.3 The Inductor

A constant lumped linear inductor is defined by its flux-current relation.

$$\phi = Li(t) \tag{2.6}$$

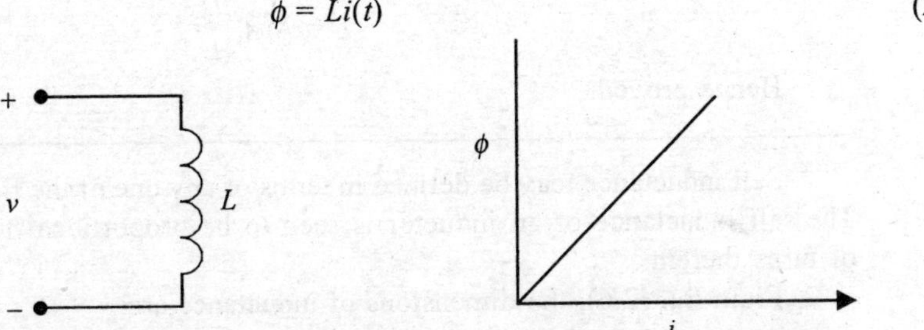

Fig. 2.4 A constant linear inductor and its flux-current characteristics.

In SI units, flux ϕ is measured in webers Wb and inductance L in henry H.

The most usual form of an inductor is a solenoid. The voltage *induced* across an inductor of N turns is, by Faraday's law of electromagnetic induction,

$$e = -N \frac{d\phi}{dt} \tag{2.7}$$

$$\text{Applied voltage } v = N \frac{d\phi}{dt}$$

$$= N \left(\frac{d\phi}{di} \right) \left(\frac{di}{dt} \right)$$

$$v = L \frac{di}{dt} \tag{2.8}$$

where

$$L = N \left(\frac{d\phi}{di} \right) \tag{2.9}$$

is the ***coefficient of self induction*** (or ***self inductance***) measured in henries.

EXAMPLE 2.2 Show that the self inductance in an air cored inductor is:

$$L = N^2 A\mu_0/l$$

SOLUTION:

The flux produced is given by

Flux = m.m.f / Reluctance

$$\phi = \frac{NI}{l/(A\mu)}$$

where,

ϕ is the flux in webers, I is the magnetizing current, l is the length of magnetic circuit, A is the area of cross-section, μ_0 is the permeability of free space = $4\pi \times 10^{-7}$ H/m, and N is the number of turns.

Here, $$\frac{d\phi}{di} = NA\frac{\mu_0}{l}$$

and from Eq. (2.9),

$$L = N^2 A\frac{\mu_0}{l} \tag{2.10}$$

Hence proved.

Self inductance may be defined in terms of any one of the Eqs. (2.8), (2.9) or (2.10). The self inductance of an inductor is seen to be proportional to the square of number of turns therein.

From Eq. (2.6), the dimensions of inductance are:

$$[L] = [\phi] / [I]$$

and using Eq. (2.10), we get

$$= \frac{[ML^2T^{-2}I^{-1}]}{[I]}$$

$$[L] = [ML^2T^{-2}I^{-2}]$$

The inverse *i*-*v* relationships is obtained by the integration of Eq. (2.8).

$$\begin{aligned} i &= \frac{1}{L}\int_{-\infty}^{t} v\,dt \\ &= \frac{1}{L}\int_{-\infty}^{0} v\,dt + \frac{1}{L}\int_{0}^{t} v\,dt \\ &= i(0) + \frac{1}{L}\int_{0}^{t} v\,dt \end{aligned} \tag{2.11}$$

where $i(0)$ is the initial current in the inductor.

Total energy supplied to the inductor: It is given by

$$W = \int_{-\infty}^{t} v\,i\,dt$$

$$= \int_{-\infty}^{t} L\left(\frac{di}{dt}\right) i\,dt$$

$$= \int_{-\infty}^{i} L\,i\,di$$

(Note the change in the variable of integration from t to i.)

Assuming that $i = 0$ (initially)

$$W = \int_{0}^{i} L\,i\,di$$

$$W = \frac{1}{2} L\,i^2 \qquad \text{J} \tag{2.12}$$

Note that the energy stored in an inductor is dependent only on current.

Inductors in series: The effective inductance of a series combination is:

$$L = L_1 + L_2 + L_3 + \cdots \qquad \text{(see section 3.4.5)}$$

Inductors in parallel: The effective inductance of a parallel combination of inductors is given by

$$\frac{1}{L} = \frac{1}{L_1} + \frac{1}{L_2} + \frac{1}{L_3} + \cdots \qquad \text{(see section 3.4.6)}$$

Principle of duality: The voltage-current equations for inductance and capacitance are:

$$v = L\frac{di}{dt} \qquad \text{or} \qquad i = \frac{1}{L}\int_{-\infty}^{t} v\,dt$$

and

$$i = C\frac{dv}{dt} \qquad \text{or} \qquad v = \frac{1}{C}\int_{-\infty}^{t} i\,dt$$

Capacitance and inductance are said to be dual quantities. The principle of duality will be considered in detail in Section 6.4.

Mutual inductance: The mutual inductance between two coils which are magnetically coupled is given by

$$M = K\sqrt{L_1 L_2} \qquad \text{(see Section 3.4)}$$

2.2 ACTIVE ELEMENTS

2.2.1 Voltage Sources and Current Sources

An electrical source is a device that is capable of converting energy in non-electrical form into electrical form and vice versa. An ***ideal voltage source*** is a circuit element that maintains a prescribed voltage across its terminals irrespective of the current flowing between these terminals. Similarly an ***ideal current source*** is a circuit element that maintains a prescribed current through its terminals irrespective of the voltage across these terminals.

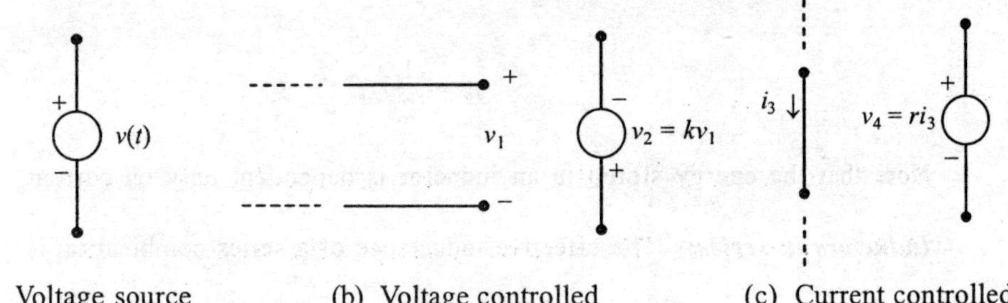

(a) Voltage source (symbol) (b) Voltage controlled voltage source (c) Current controlled voltage source

Fig. 2.5

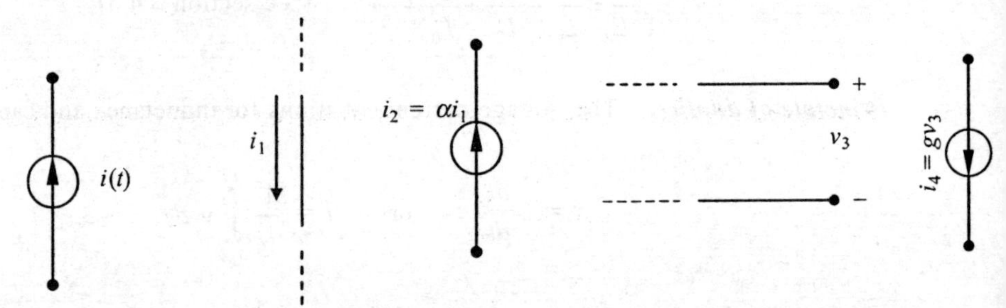

(a) Current source (symbol) (b) Current controlled current source (c) Voltage controlled current source

Fig. 2.6

Ideal voltage and current sources may be either independent sources or dependent sources. An ***independent source*** in a circuit does not depend upon voltage or current elsewhere in the circuit; whereas a ***dependent source*** establishes a voltage or current whose value depends on the value of a voltage or current elsewhere in the circuit. The circuit symbols for the ideal independent sources as well as those for ideal dependent sources are shown in Fig. 2.5 and Fig. 2.6.

2.2.2 Representation of Physical Sources

Voltage sources: In many physical sources the output voltage does not remain constant with current but it decreases almost linearly with current. A car battery, for example, has this type of characteristic. Such a characteristic is represented by an ideal source in series with an ideal resistance as shown in Fig. 2.7(a).

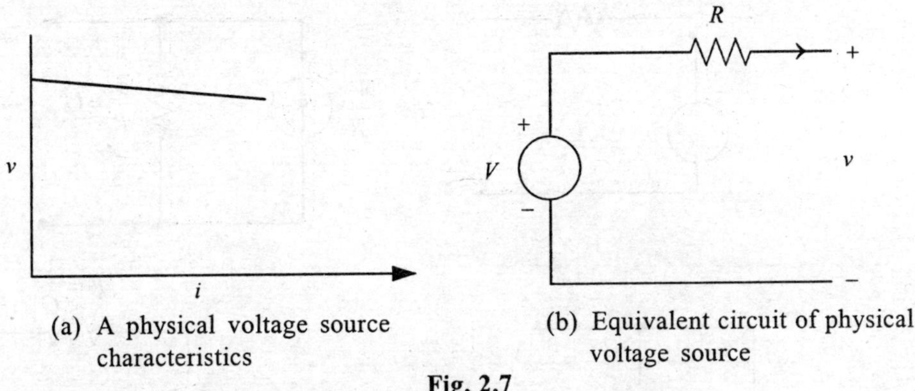

(a) A physical voltage source characteristics

(b) Equivalent circuit of physical voltage source

Fig. 2.7

The volt-ampere equation of this branch is:

$$v = V - iR \tag{2.13}$$

Current sources: Many of the current sources have output characteristics similar to that shown in Fig. 2.8(a). Such a characteristic is the simplest way of representing a physical current source. The physical source is taken as the parallel combination of an ideal current source and ideal conductance as shown in Fig. 2.8(b).

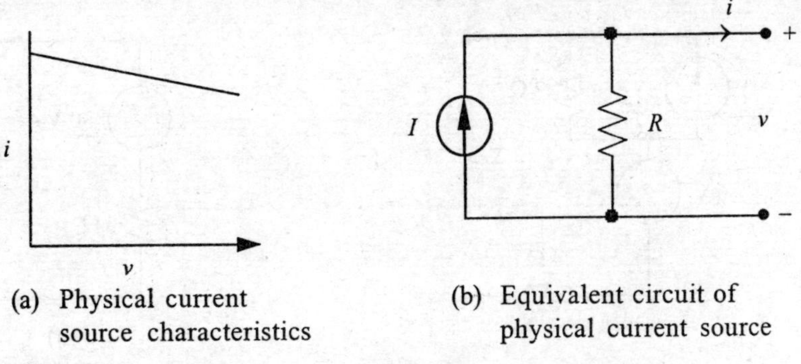

(a) Physical current source characteristics

(b) Equivalent circuit of physical current source

Fig. 2.8

The volt-ampere equation is

$$i = I - vG \tag{2.14}$$

Source transformation (equivalence): It can be shown that the source branches in Fig. 2.9 have the same volt-ampere equation and, therefore, equivalent with respect to an external network. The conditions for this are:

$$R = 1/G \quad \text{and} \quad IR = V$$

Thus, voltage source branch consisting of a 6 V source in series with a resistance of 2 Ω may be replaced by a current source consisting of a 3 A source in parallel with a 2 Ω resistance.

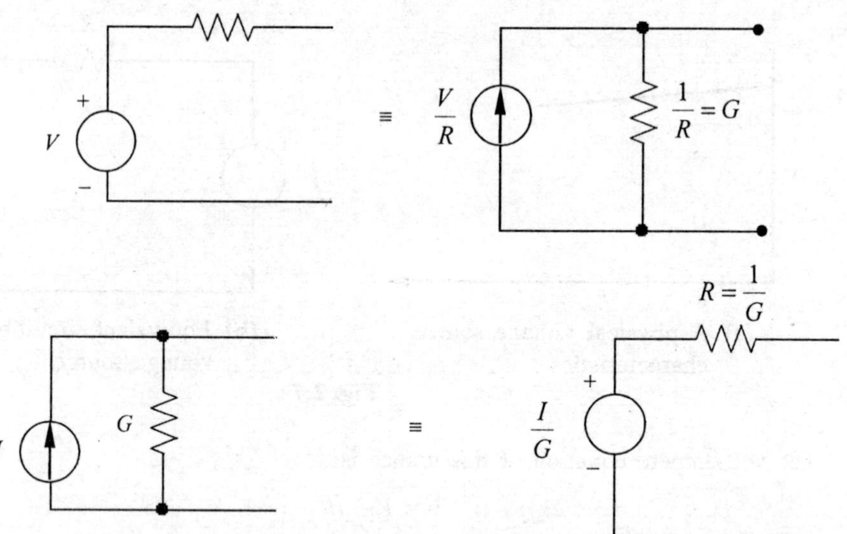

Fig. 2.9 Source equivalence.

Compound sources: If a network contains several sources it is often possible to reduce them to a single source branch. For example, consider the network of Fig. 2.10, which has two voltage sources.

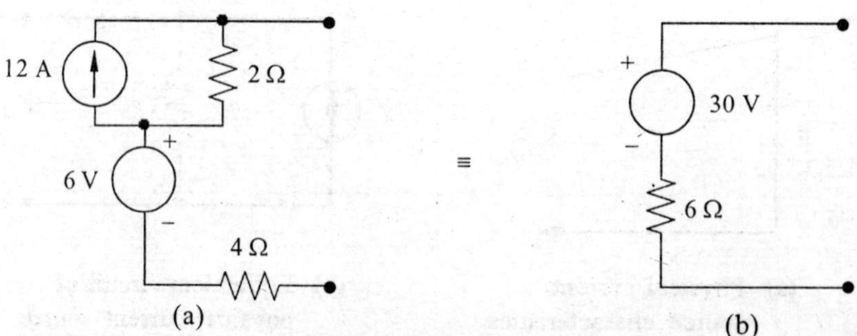

Fig. 2.10 A compound source and its equivalent.

The source branch consists of a current source and a voltage source. The 12 A current source and 2 Ω resistance are replaced by a 24 V voltage source in series with a 2 Ω resistance. The two voltage sources are then added to give 30 V and the two series resistances added to give a 6 Ω resistance. The equivalent source branch is shown in Fig. 2.10(b).

Note: Examples and problems relating to network elements will be included in Chapter 3 on network laws.

SUMMARY

- Electric circuits consist of elements which are classified as passive or active.
- The passive elements are the resistors, inductors and capacitors. The characteristics of the ideal elements are shown in Section 2.1. Series and parallel connections of resistive, capactitive and inductive circuits are also included. Mutual inductance between coils which are magnetically coupled is described in Section 3.4.
- The active elements are voltage sources and current sources. Ideal energy sources may be either independent sources or dependent sources.
- Source transformation and compound sources are also included.

Solved Problems

P2.1 Find the value of R in the circuit shown in Fig. P2.1

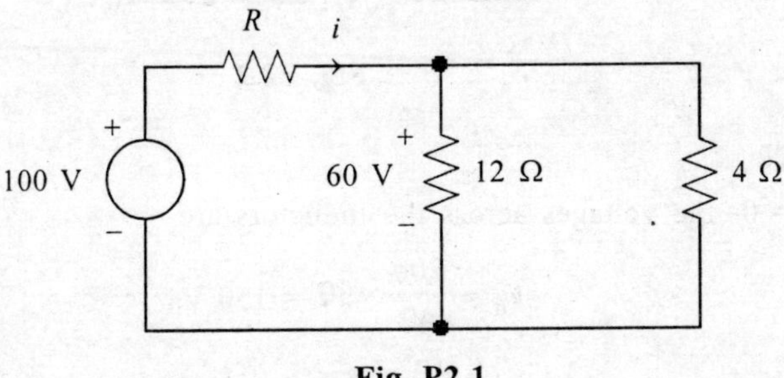

Fig. P2.1

SOLUTION:

Current in the 12 Ω resistance is:

$$\frac{60}{12} = 5\text{ A}$$

Total current from the source is given by

$$\frac{4}{(12+4)} i = 5 \quad \text{[using current division in a parallel circuit]}$$

which gives $\qquad i = 20$ A

R is given by

$$20\,R = 100 - 60$$

or

$$R = 2\ \Omega \qquad \textbf{Ans.}$$

P2.2 In the circuit of Fig. P2.2 the inductors and the capacitors do not contain any stored energy. If the switch is closed at time $t = 0$, find the voltages across the capacitors and the inductors at (a) $t = 0$ (b) $t = \infty$.

Note that with a sudden change in voltage will cause an inductance to act as an open circuit and a capacitance as a short circuit; at $t = \infty$ capacitances act as open circuit and inductances act as short circuit.

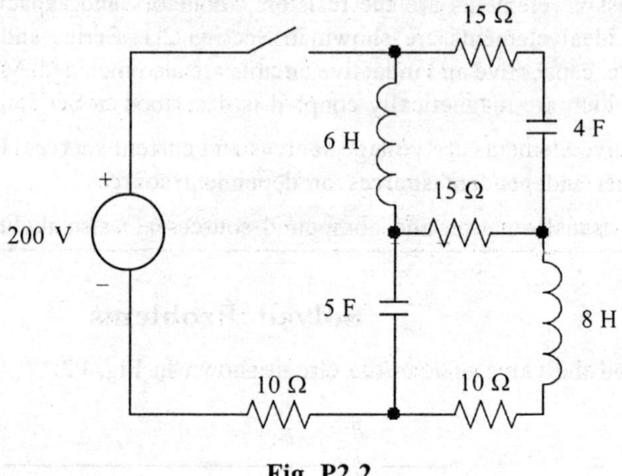

Fig. P2.2

SOLUTION:

***(a)* At *t* =** 0 the voltages across the inductors are

$$V_6 = \frac{200}{40} \times 30 = 150\ \text{V}$$

$$V_8 = \frac{200}{40} \times 15 = 75\ \text{V} \qquad \textbf{Ans.}$$

***(b)* At *t* =** ∞ the voltages across the capacitors are

$$V_4 = \frac{200}{35} \times 15 = 85.71\ \text{V}$$

$$V_5 = \frac{200}{35} \times 25 = 142.85\ \text{V} \qquad \textbf{Ans.}$$

Exercise

E2.1 Determine the current distribution in the circuit of Fig. E2.1

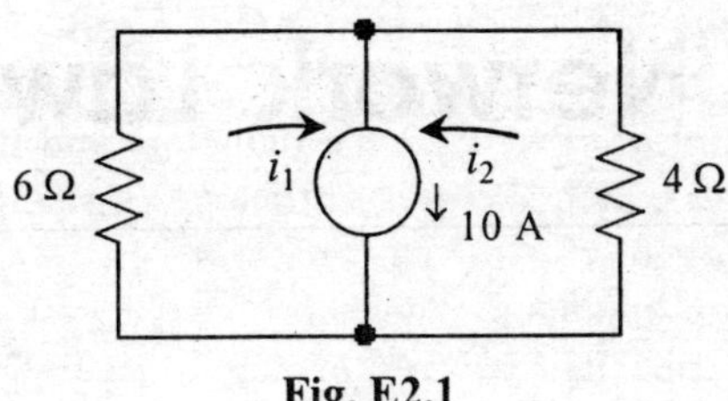

Fig. E2.1

[**Hint:** Use current division in a parallel circuit.]

[*Ans.* $i_1 = 4$ A, $i_2 = 6$ A]

Note: Other exercises will be added in Chapter 3, where the network laws are introduced.

Network Laws

3

CHAPTER CONTENTS

There are three basic laws which govern electric circuits/networks. They are:

- Ohm's law
- Kirchoff's current law
- Kirchoff's voltage law

The behaviour of electric circuits carrying currents can be predicted from the postulates contained in these laws.

Any electrical network comprises ***branches***, ***junctions*** (nodes) and ***meshes***; and there exists a one to one correspondence between these and the three network laws.

- Ohm's law governs the *v-i* relationship in a branch and hence is termed ***branch law***.
- Kirchoff's current law refers to current at any junction and hence is termed ***junction law***.
- Kirchoff's voltage law refers to voltage around a mesh and hence is termed ***mesh law***.

The laws which are applicable to the steady state condition for direct current networks are described in following sections.

3.1 BRANCH LAW (OHM'S LAW)

In 1827 G.S Ohm discovered that a metallic conductor, maintained at a fixed temperature, will pass a current directly proportional to the applied voltage; this volt-ampere relationship in Eq. (3.1) is known as **Ohm's law** for ideal resistance element.

$$v = Ri \tag{3.1}$$

In SI units v is the voltage across the conductor in volts, i is the current through the conductor in amperes and R is defined as the resistance of the element in ohms. (See Fig. 3.2)

The law may be stated in an alternate form in terms of conductance G (inverse of resistance) measured in siemens (S).

$$i = Gv \tag{3.2}$$

The electron theory of metallic conduction may be used to explain Ohm's law. (See Reference 1)

3.2 JUNCTION LAW (KIRCHHOFF'S CURRENT LAW—KCL)

KCL relates to current at any junction in a network. According to the principle of conservation of charge and because the junction cannot store charge, the sum of the currents entering the junction must equal the sum of the currents leaving the junction. Thus, the algebraic sum of all the currents at a junction is zero.

$$\sum i = 0 \tag{3.3}$$

In Fig. 3.1, the KCL may be written in the form,

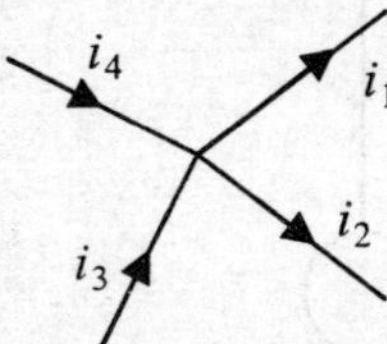

Fig. 3.1

$$i_1 + i_2 - i_3 - i_4 = 0$$

or

$$i_1 + i_2 = i_3 + i_4$$

3.3 MESH LAW (KIRCHHOFF'S VOLTAGE LAW—KVL)

The mesh law relates to the potential difference existing around any closed path and is a consequence of the principle of *conservation of energy*. In such a closed path the total work done in taking a unit charge around the path must be zero. As the voltage is a measure of the energy lost or gained by a unit charge, the algebraic sum of the voltages around any closed path must be zero.

$$\sum v - \sum Ri = 0 \tag{3.4a}$$

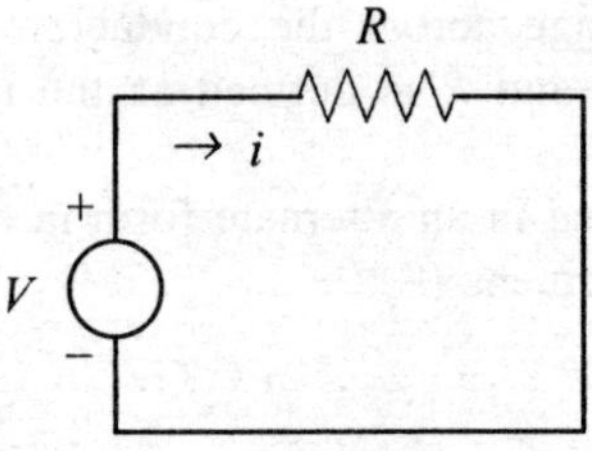

Fig. 3.2

Equation (3.4a) represents the total voltage method and

$$\sum v = \sum Ri \tag{3.4b}$$

is the equated voltage method.

EXAMPLE 3.1 For the network shown in Fig. 3.3, determine the supply current and the source e.m.f.

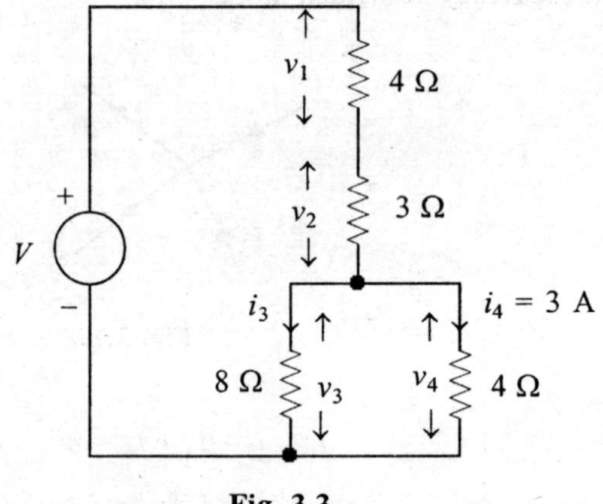

Fig. 3.3

SOLUTION:

Since R_3 and R_4 are in parallel,

$$\begin{aligned} v_3 &= i_4R_4 = 3 \times 4 \\ &= 12 \text{ V} = i_3R_3 \\ &= i_3 \times 8 \end{aligned}$$

or $$i_3 = 1.5 \text{ A}$$

By Kirchoff's first law,

$$i = i_3 + i_4$$
$$= 1.5 + 3 = 4.5 \text{ A}$$

Also $$v_1 = iR_1$$
$$= 4.5 \times 4 = 18 \text{ V}$$
$$v_2 = iR_2$$
$$= 4.5 \times 3 = 13.5 \text{ V}$$

By Kirchoff's second law

$$V = v_1 + v_2 + v_3$$
$$= 18 + 13.5 + 12 = 43.5 \text{ V} \quad \textbf{Ans.}$$

EXAMPLE 3.2 In the circuit shown in Fig. 3.4, find the current i_1 and the voltage v.

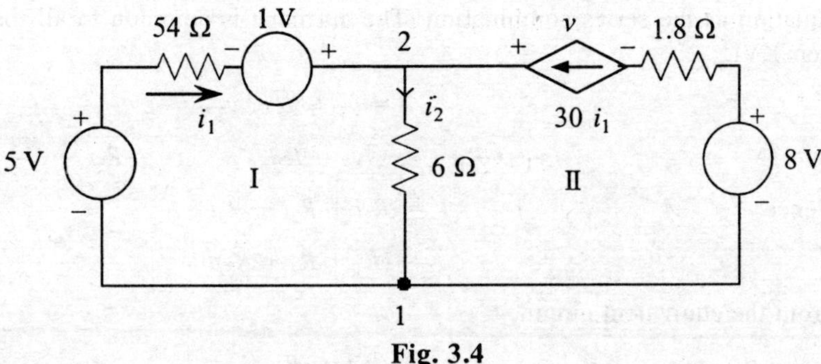

Fig. 3.4

SOLUTION:

Applying KCL at node 2,

$$i_2 = i_1 + 30i_1 = 31i_1$$

For mesh I,

$$5 = 54i_1 - 1 + 6i_2$$

Solving,

$$i_1 = 25 \text{ mA} \quad \textbf{Ans.}$$

For mesh II,

$$8 + v - 1.8 \times 30i_1 - 6i_2 = 0$$

Substituting the values of i_1 and i_2, we get

$$v = -2 \text{ V} \quad \textbf{Ans.}$$

3.4 APPLICATION OF NETWORK LAWS TO SIMPLE DC NETWORKS

3.4.1 Resistors in Series

A series combination of resistors is shown in Fig. 3.5(a).

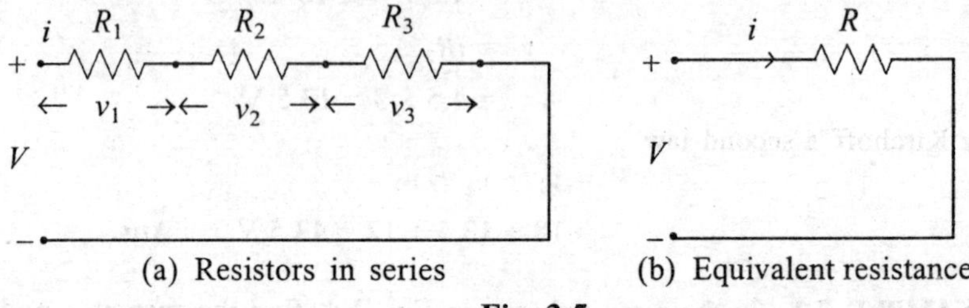

(a) Resistors in series (b) Equivalent resistance

Fig. 3.5

It is required to find a single equivalent resistance R which will have the same volt-ampere equation as the series combination. The current i is common to all the resistances, and from KVL,

$$V = v_1 + v_2 + v_3$$

But, $\quad v_1 = R_1 i \qquad v_2 = R_2 i \qquad v_3 = R_3 i \qquad$ (3.5a)

Hence $\quad V = R_1 i + R_2 i + R_3 i$

$$= (R_1 + R_2 + R_3)i \qquad (3.5b)$$

From the equivalent circuit,

$$V = Ri$$

Hence $\quad Ri = (R_1 + R_2 + R_3)i$

and the equivalent resistance of the series combination is:

$$R = R_1 + R_2 + R_3 \qquad (3.6)$$

Potential distribution in a series circuit: From Eqs. (3.5a) and (3.5b),

$$\frac{v_1}{V} = \frac{R_1 i}{(R_1 + R_2 + R_3)i}$$

$$= \frac{R_1}{R_1 + R_2 + R_3}$$

Hence $\quad v_1 = \dfrac{R_1 V}{R_1 + R_2 + R_3} \qquad$ (3.7a)

Similarly, $\quad v_2 = \dfrac{R_2 V}{R_1 + R_2 + R_3} \qquad$ (3.7b) and $$v_3 = \frac{R_3 V}{R_1 + R_2 + R_3} \tag{3.7c}$$

It is seen that in the series combination of resistors the voltage across each resistor is proportional to its resistance that is to say the total voltage is divided in proportion to the individual resistance.

3.4.2 Resistors in Parallel

In the parallel connection shown in Fig. 3.6 the voltage is common to all the resistors. From KVL,

$$i = i_1 + i_2 + i_3$$

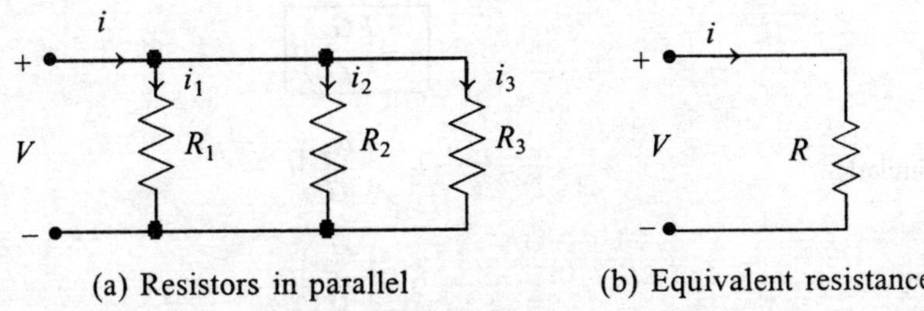

(a) Resistors in parallel (b) Equivalent resistance

Fig. 3.6

or $$i = \frac{V}{R_1} + \frac{V}{R_2} + \frac{V}{R_3}$$

$$= \left(\frac{1}{R_1} + \frac{1}{R_2} + \frac{1}{R_3}\right) V$$

From the equivalent circuit,

$$i = \left(\frac{1}{R}\right) V \tag{3.8}$$

where R is the equivalent resistance of the parallel combination.

$$\left(\frac{1}{R}\right) V = \left(\frac{1}{R_1} + \frac{1}{R_2} + \frac{1}{R_3}\right) V \tag{3.9}$$

It follows $$\frac{1}{R} = \frac{1}{R_1} + \frac{1}{R_2} + \frac{1}{R_3}$$

and $$R = \frac{R_1 R_2 R_3}{R_1 R_2 + R_2 R_3 + R_3 R_1}$$

With only two resistors in parallel,

$$\boxed{R = \frac{R_1 R_2}{R_1 + R_2}}$$

Current distribution in a parallel circuit: In terms of conductance, Eq. (3.9) may be written as:

$$GV = G_1V + G_2V + G_3V$$

and
$$\frac{i_1}{i} = \frac{G_1V}{G_1V + G_2V + G_3V}$$

where $G = G_1 + G_2 + G_3$, the equivalent conductance of the parallel combination.

or
$$\boxed{i_1 = \left(\frac{G_1}{G}\right)i} \tag{3.10}$$

Similarly,
$$i_2 = \left(\frac{G_2}{G}\right)i$$

and
$$i_3 = \left(\frac{G_3}{G}\right)i$$

Thus, the current in any parallel branch is proportional to its conductance (which shows the advantage of using conductances for parallel circuit calculations).

With two resistors in parallel the current in branch 1 with resistance R_1 is, from Eq. (3.10),

$$i_1 = \frac{\dfrac{1}{R_1}}{\left(\dfrac{1}{R_1} + \dfrac{1}{R_2}\right)}i$$

or
$$i_1 = \frac{R_2 i}{R_1 + R_2} \tag{3.11}$$

The extension of the expression for current division with three or more resistances in parallel is cumbersome, but in terms of conductances it is readily obtained from Eq. (3.10).

EXAMPLE 3.3 Determine the equivalent resistance between the points 1 and 2 of the circuit shown in Fig. 3.7.

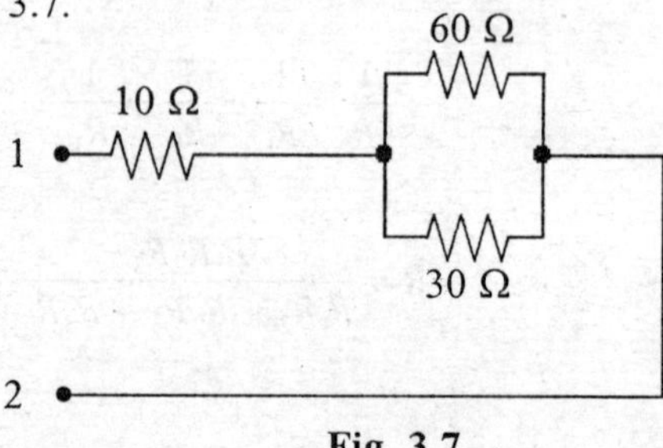

Fig. 3.7

SOLUTION:
The equivalent resistance between points 1and 2 is:

$$R_{12} = 10 + (60 \parallel 30)$$
$$= 30\,\Omega \quad \textbf{Ans.}$$

EXAMPLE 3.4 Determine the resistance between terminals 1 and 2 of the circuit shown in Fig. 3.8.

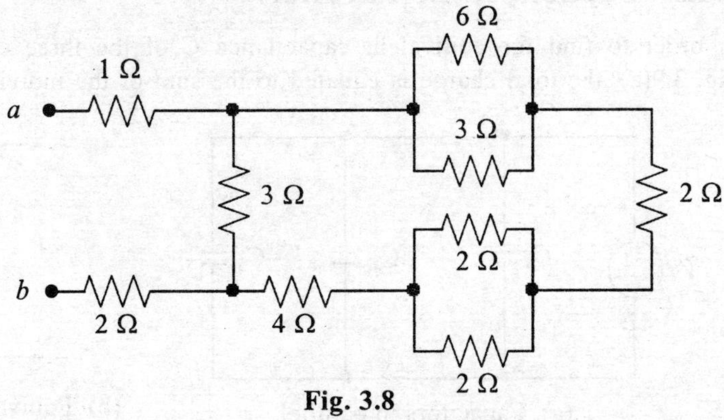

Fig. 3.8

SOLUTION:
The equivalent resistance between points a and b is:

$$R_{ab} = 1 + [3 \parallel \{(6 \parallel 3) + 2 + (2 \parallel 2) + 4\}] + 2$$
$$= 1 + \{3 \parallel (2 + 2 + 1 + 4)\} + 2$$
$$= 1 + 9/4 + 2$$
$$R_{ab} = 5.25\,\Omega \quad \textbf{Ans.}$$

Power dissipated by a resistor:

$$p = vi$$
$$= i^2 R = \frac{v^2}{R}$$
$$= v^2 G = \frac{i^2}{G}$$

EXAMPLE 3.5 Three resistors of 4 Ω, 6 Ω and 8 Ω are connected in series to a 36 V source. What is the potential across the 6 Ω resistor. What is the power dissipated by the 8 Ω resistor.

SOLUTION:
The current drawn from the source

$$i = \frac{36}{(4+6+8)} = 2\text{ A}$$

Potential across the 6 Ω resistor is given by

$$v = Ri = 6 \times 2 = 12 \text{ V} \qquad \textbf{Ans.}$$

Power dissipated by the 8 Ω resistor is given as:

$$p = i^2R = 2^2 \times 8 = 32 \text{ W} \qquad \textbf{Ans.}$$

3.4.3 Capacitors in Parallel

In order to find the equivalent capacitance C of the three capacitors in parallel in Fig. 3.9(a), the total charge is equated to the sum of the individual charges.

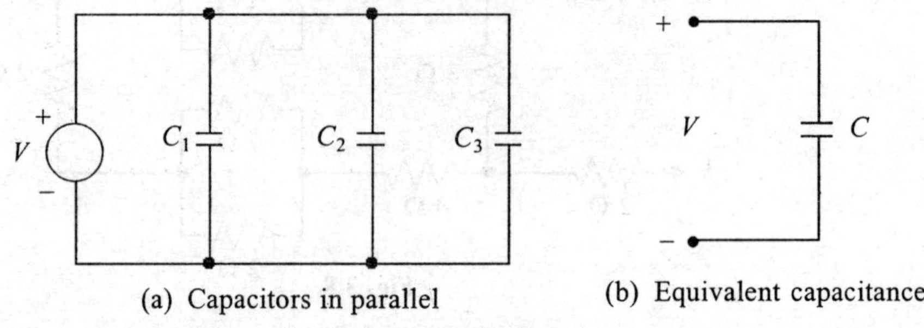

(a) Capacitors in parallel (b) Equivalent capacitance

Fig. 3.9

Thus,

$$\begin{aligned} q &= q_1 + q_2 + q_3 \\ &= VC_1 + VC_2 + VC_3 \\ &= V(C_1 + C_2 + C_3) \\ &= VC \end{aligned}$$

Thus, the effective capacitance of the parallel combination shown in Fig. 3.9(b) is:

$$\boxed{C = C_1 + C_2 + C_3} \qquad (3.12)$$

3.4.4 Capacitors in Series

In the series connection of capacitors all of them have a charge of Q on their plates, hence,

$$\begin{aligned} V &= v_1 + v_2 + v_3 \\ &= \frac{q}{C_1} + \frac{q}{C_2} + \frac{q}{C_3} \\ &= q\left(\frac{1}{C_1} + \frac{1}{C_2} + \frac{1}{C_3}\right) \\ &= \frac{q}{C} \end{aligned}$$

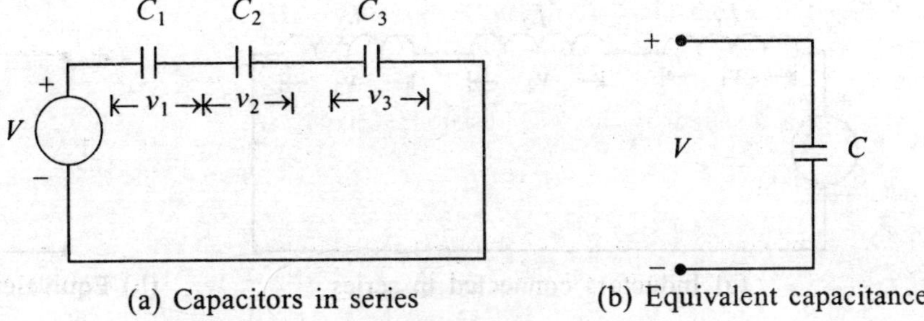

(a) Capacitors in series (b) Equivalent capacitance

Fig. 3.10

Thus, the effective capacitance of the series combination shown in Fig 3.10(b) is:

$$\boxed{\frac{1}{C} = \frac{1}{C_1} + \frac{1}{C_2} + \frac{1}{C_3}} \tag{3.13}$$

EXAMPLE 3.6 In Fig. 3.11 find the value of C_1 so that the equivalent capacitance $C = 2$ F

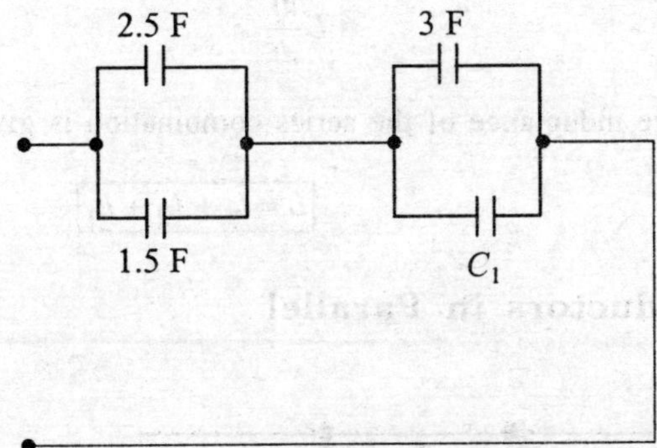

Fig 3.11 The equivalent capacitance.

SOLUTION:

$$C = 2 = \frac{(2.5 + 1.5)(3 + C_1)}{(2.5 + 1.5) + (3 + C_1)}$$

or

$$\frac{4(3 + C_1)}{7 + C_1} = 2$$

which gives

$$C_1 = 1 \text{ F} \qquad \textbf{Ans.}$$

3.4.5 Inductors in Series

Figure 3.12(a) shows three inductors connected in series. Current through each inductance must be changing at the same rate since the current is common. Thus, by KVL,

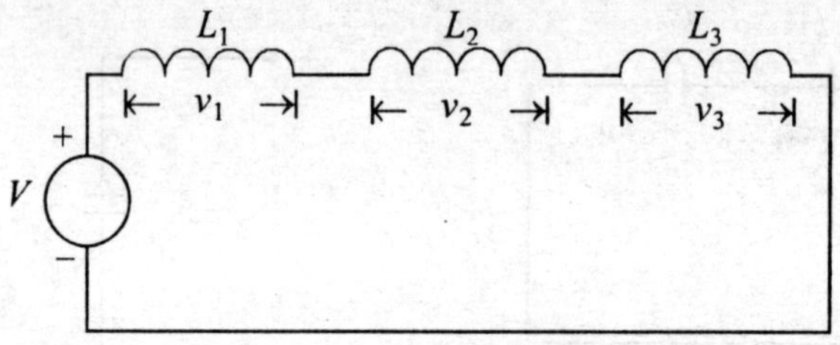

(a) Inductors connected in series

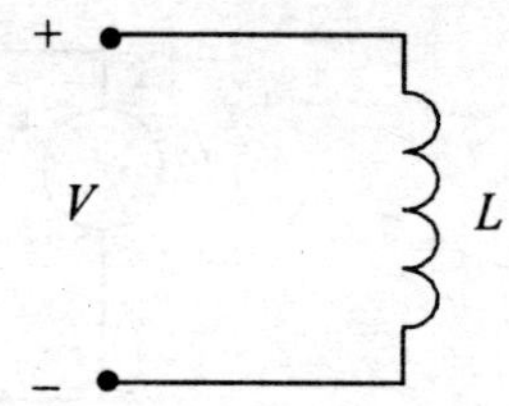

(b) Equivalent inductance

Fig. 3.12

$$
\begin{aligned}
V &= v_1 + v_2 + v_3 \\
&= L_1 \frac{di}{dt} + L_2 \frac{di}{dt} + L_3 \frac{di}{dt} \\
&= (L_1 + L_2 + L_3)\left(\frac{di}{dt}\right) \\
&= L \frac{di}{dt}
\end{aligned}
$$

The effective inductance of the series combination is given by

$$\boxed{L = L_1 + L_2 + L_3} \tag{3.14}$$

3.4.6 Inductors in Parallel

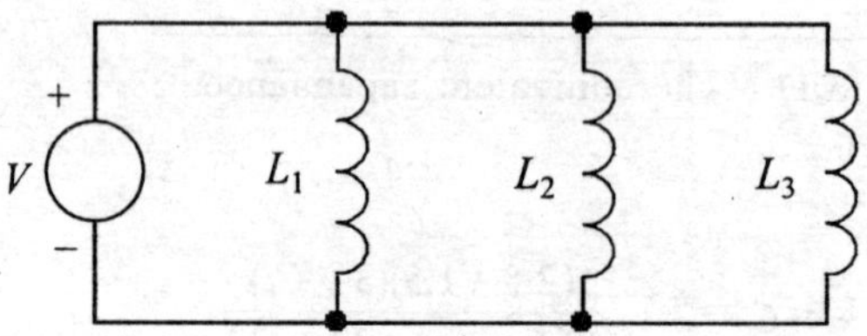

(a) Inductors connected in parallel

(b) Equivalent inductance

Fig. 3.13

The voltage across each inductor is same and

$$V = L_1 \frac{di}{dt} = L_2 \frac{di}{dt} = L_3 \frac{di}{dt}$$

But, $$i = i_1 + i_2 + i_3$$

Hence,

$$\frac{di}{dt} = \frac{di_1}{dt} + \frac{di_2}{dt} + \frac{di_3}{dt} \quad \text{(Note this step.)}$$

$$= \frac{V}{L_1} + \frac{V}{L_2} + \frac{V}{L_3}$$

$$= V\left(\frac{1}{L_1} + \frac{1}{L_2} + \frac{1}{L_3}\right)$$

$$= \frac{V}{L}$$

Thus, the effective inductance of the parallel combination shown in Fig. 3.13(b) is given by

$$\boxed{\frac{1}{L} = \frac{1}{L_1} + \frac{1}{L_2} + \frac{1}{L_3}} \tag{3.15}$$

EXAMPLE 3.7 Determine the effective inductance of the network shown in Fig. 3.14. Given $L_1 = 2$ H, $L_2 = 2$ H, $L_3 = 4$ H, $L_4 = 2$ H, $L_5 = 4$ H.

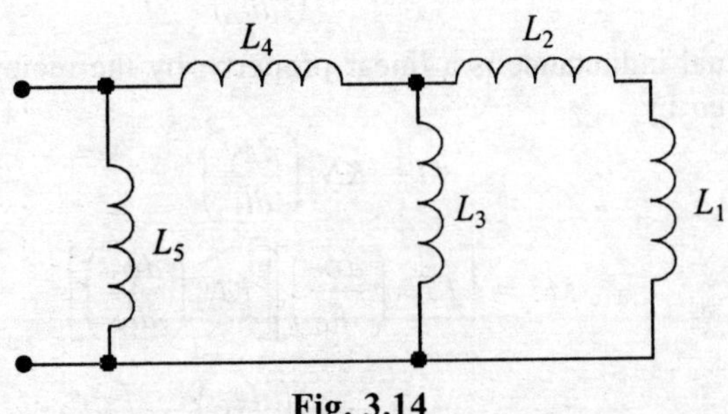

Fig. 3.14

SOLUTION:
The given circuit (Fig. 3.14) reduces as shown in Figs. 3.14(a), 3.14(b), 3.14(c).

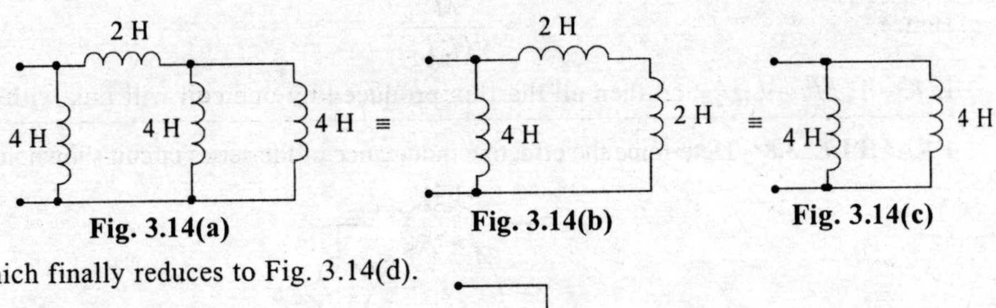

Fig. 3.14(a) **Fig. 3.14(b)** **Fig. 3.14(c)**

which finally reduces to Fig. 3.14(d).

2 H

Fig. 3.14(d)

$$L = 2\text{ H} \qquad \textbf{Ans.}$$

Mutual inductance: Consider two coils which are magnetically coupled. The voltage induced in coil II can be related to a time-varying current in coil I by a parameter known as **mutual inductance** M. The self inductance of the two coils are L_1 and L_2 henries, respectively. A current i_1 in coil I produces a flux ϕ_1 in coil I and a flux ϕ_m $(= K\phi_1)$ in coil II, K being the ***coefficient of coupling*** (≤ 1). Like self inductance L, mutual inductor has the unit henry.

E.m.f. induced in coil II,

$$e = -\left(\frac{d\phi_m}{dt}\right)N_2$$

or

$$e = -K\left(\frac{d\phi_1}{dt}\right)N_2$$

$$e = -\left[KN_2\left(\frac{d\phi_1}{di_1}\right)\right]\left(\frac{di_1}{dt}\right)$$

$$e = -M\frac{di_1}{dt} \tag{3.16}$$

where

$$M = KN_2\left(\frac{d\phi_1}{di_1}\right) \tag{3.17a}$$

As mutual inductance is a linear property, by the reciprocity theorem (Chapter 4), it is also given by

$$M = KN_1\left(\frac{d\phi_2}{di_2}\right) \tag{3.17b}$$

Thus,

$$M^2 = \left[KN_2\left(\frac{d\phi_1}{di_1}\right)\right]\left[KN_1\left(\frac{d\phi_2}{di_2}\right)\right]$$

$$= K^2N_1\left(\frac{d\phi_1}{di_1}\right)\ N_2\left(\frac{d\phi_2}{di_2}\right)$$

$$= K^2L_1L_2$$

Hence,

$$K = \frac{M}{\sqrt{(L_1L_2)}} \tag{3.18}$$

If $K = 1$, $M^2 = L_1L_2$, and then all the flux produced by one coil will link with the other.

EXAMPLE 3.8 Determine the effective inductance of the series circuit shown in Fig. 3.15.

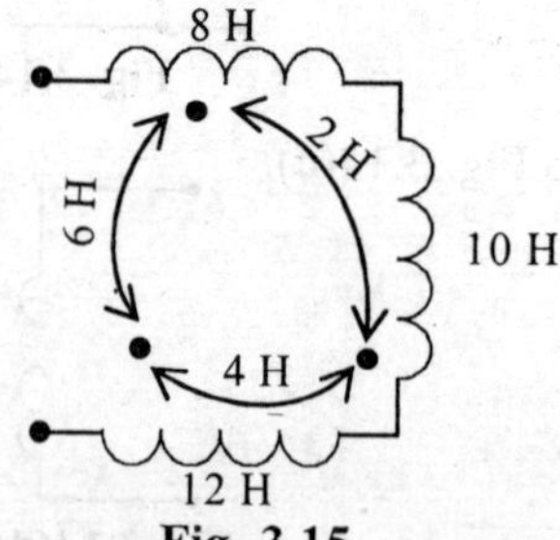

Fig. 3.15

For the dot convention in mutual inductance, see Reference 3.

SOLUTION:

The equivalent inductance

$$\begin{aligned} L_{eq} &= L_1 + L_2 + L_3 - 2M_{12} + 2M_{23} - 2M_{31} \\ &= 8 + 10 + 12 - 2 \times 2 + 2 \times 4 - 2 \times 6 \\ &= 22 \text{ H} \end{aligned}$$

SUMMARY

- Most network problems can be solved by applying Kirchoff's laws to produce simultaneous equations. Use of the network theorems to simplify the solution of networks are considered in Chapter 4.

Solved Problems

P3.1 Figure P3.1 shows a circuit with a dependent source. Find i_2 and v_2.

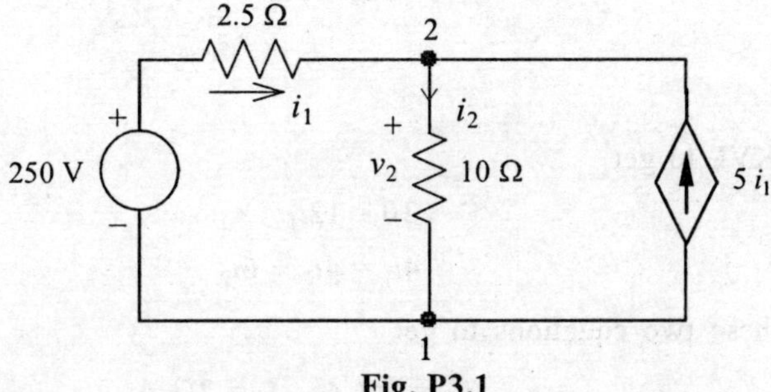

Fig. P3.1

SOLUTION:

The equation for the circuit containing the voltage source 250 V, the 2.5 Ω resistor and the 10 Ω resistor is:

$$250 = 2.5i_1 + 10i_2 \tag{1}$$

At node 2, KCL gives

$$i_2 = i_1 + 5i_1$$

or

$$i_2 = 6i_1$$

Substitute in the Eq. (1) to get

$$250 = 2.5i_1 + 10 \times 6i_1$$

or

$$i_1 = \frac{250}{62.5} = 4 \text{ A}$$

and

$$\begin{aligned} i_2 &= 6i_1 \\ &= 6 \times 4 \\ &= 24 \text{ A} \end{aligned}$$

The required voltage is:

$$v_2 = 10 \times 24$$
$$= 240 \text{ V}$$

P3.2 In the circuit of Fig. P3.2, find the voltage v_0 across the 6 Ω resistor. Show that the total power developed in the circuit equals total power dissipated.

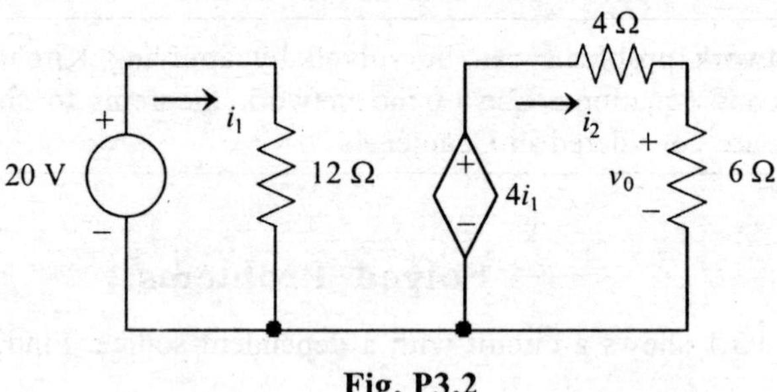

Fig. P3.2

SOLUTION:

Apply KVL to get

$$20 = 12i_1$$
$$4i_1 = 4i_2 + 6i_2$$

Solve these two equations to get

$$i_1 = 5/3 \text{ A}, \quad i_2 = 2/3 \text{ A}$$

Now, $\quad v_0 = i_2\, 6 = 4 \text{ V}$

Power delivered by the 20 V source,

$$p_1 = -20 \times 5/3 = (-33.33) \text{ W}$$

Power delivered to the dependent voltage source,

$$p_2 = 4i_1 \times -i_2$$
$$= 4 \times 5/3 \times -2/3 = (-4.44) \text{ W}$$

The power delivered to the resistors

$$= (5/3)^2\, 12 + (2/3)^2\, 4 + (2/3)^2 6$$
$$= 37.77 \text{ W}$$

Thus, total power developed equals total power dissipated.

P3.3 Three resistors of 4 Ω, 8 Ω and 16 Ω are connected in parallel to a 14 A source. Determine the current passing through the 4 Ω resistor. What is the power dissipated by the 8 Ω resistor. What is the source voltage.

SOLUTION:

The current passing through the 4 Ω resistor,

$$= \frac{G_4}{G_4 + G_8 + G_{16}} I$$

$$= \frac{1/4}{1/4 + 1/8 + 1/16} \times 14 = 8 \text{ A} \qquad \textbf{Ans.}$$

Power dissipated by the 4 Ω resistor

$$= I^2 R$$

$$= 8^2 \times 4 = 256 \text{ W} \qquad \textbf{Ans.}$$

Source voltage $= i_4 R_4$

$$= 8 \times 4 = 32 \text{ V} \qquad \textbf{Ans.}$$

P3.4 When the two coils are connected in series their effective inductance is found to be 5.0 H. When the connections to one of the coils are reversed the effective inductance is 3.0 H. Coefficient of coupling is 0.6. Calculate the self inductance of each coil and the mutual inductance.

$$L = L_1 + L_2 \pm 2M \tag{1}$$

$$= L_1 + L_2 \pm 2K\sqrt{L_1 L_2}$$

$$5 = L_1 + L_2 + 2K\sqrt{L_1 L_2}$$

and

$$3 = L_1 + L_2 - 2K\sqrt{L_1 L_2}$$

It follows

$$4 = L_1 + L_2$$

Hence

$$5 = 4 - L_2 + L_2 + 2 \times 0.6 \times \sqrt{(4 - L_2)L_2}$$

$$1 = 1.2\sqrt{(4 - L_2)L_2}$$

$$1 = 1.2^2 (4L_2 - L_2^2)$$

$$L_2^2 - 4L_2 + 0.694 = 0$$

Solving,

$$L_2 = 3.81 \text{ or } 0.18$$

Therefore,

$$L_1 = 0.18 \text{ or } 3.81 \qquad \textbf{Ans.}$$

and M is given by the Eq. (1)

$$5 = 4 + 2M$$

$$M = 0.5 \text{ H} \qquad \textbf{Ans.}$$

P3.5 Determine the effective capacitance of the network shown in Fig. P3.3.

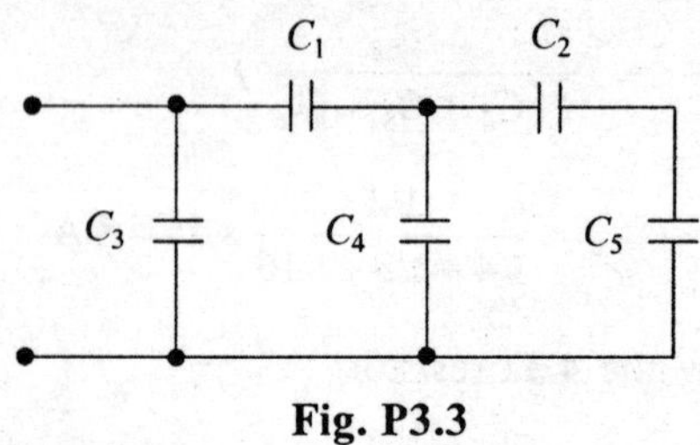

Fig. P3.3

Given $C_1 = 2$ F, $C_2 = 2$ F, $C_3 = 1$ F, $C_4 = 1$ F, and $C_5 = 2$ F.

SOLUTION:

The given circuit (Fig. P3.3) reduces as shown in Figs. P3.3(a), P3.3(b) and P3.3(c).

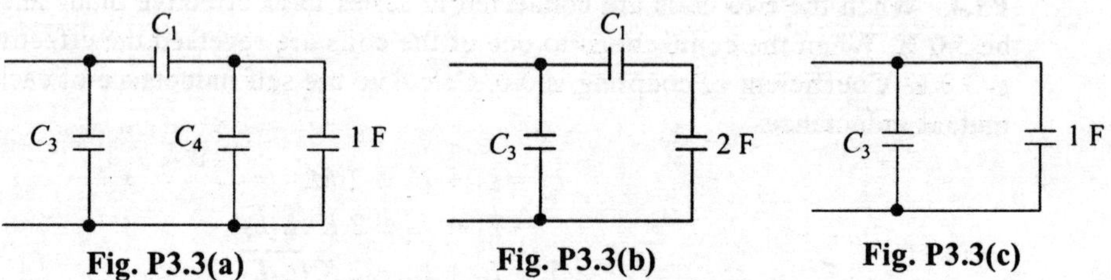

Fig. P3.3(a) **Fig. P3.3(b)** **Fig. P3.3(c)**

which reduces as shown in Fig. P3.3(d).

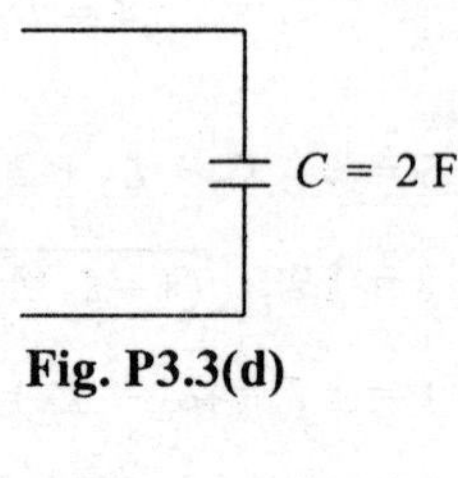

Fig. P3.3(d)

$$C = 2 \text{ F} \quad \textbf{Ans.}$$

P3.6 Determine the effective inductance of the network shown in Fig. P3.4.

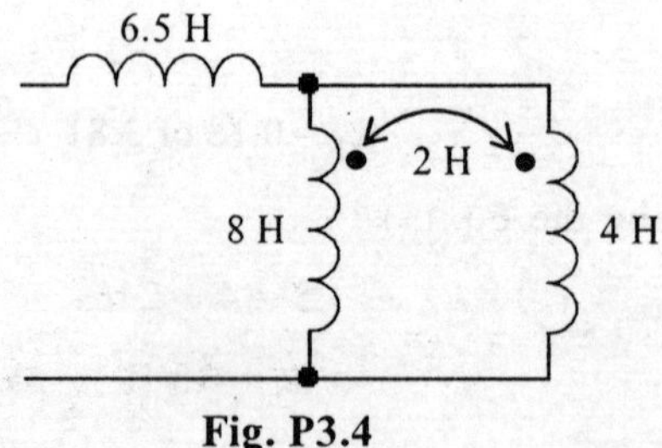

Fig. P3.4

SOLUTION:

The 8 H and 4 H coils can be replaced by a single coil having an inductance of

$$L = \frac{L_1L_2 - M^2}{L_1 + L_2 - 2M}$$

$$= \frac{8 \times 4 - 2^2}{8 + 4 - 2 \times 2}$$

$$= 3.5 \text{ H}$$

Hence the equivalent inductance of the circuit is obtained as:

$$L_{eq} = 6.5 + 3.5 = 10 \text{ H} \qquad \textbf{Ans.}$$

Note the polarity of the voltages produced by the mutual inductance as indicated by the dots.

Exercises

E3.1 Twelve resistors each of 1 Ω resistance are joined together as shown in Fig. E3.1. Determine the effective resistance between points *A* and *B* and the effective resistance between points *A* and *C*.

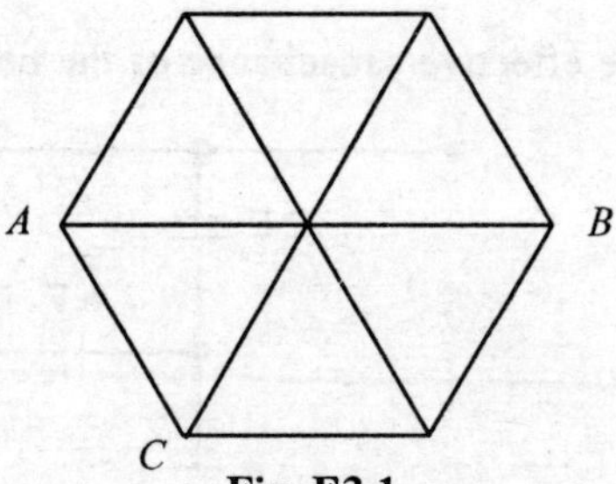

Fig. E3.1

[*Ans:* R_{AB} = 4/5 Ω, R_{AC} = 11/20 Ω]

E3.2 Twelve resistors each of 1 Ω resistance are joined in the form of a cube as shown in Fig. E3.2. Determine the effective resistance of the network between points *A* and *B* and between points *A* and *C*.

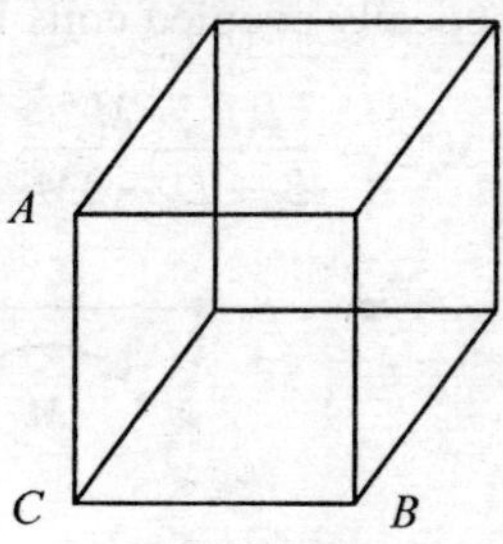

Fig. E3.2

[*Ans:* R_{AB} = 3/4, Ω, R_{AC} = 7/12 Ω]

E3.3 A coil consists of two similar sections wound on a common core. Each section has an inductance of 0.12 H. Calculate the inductance of the coil when the sections are connected in series.

[*Ans:* 0.48 H]

E3.4 Three inductors are connected as shown in Fig. E3.3. Given L_1 = 0.5 H, and L_2 = 0.6 H. Find L_3 such that the equivalent inductance is 0.7 H with respect to terminals *A* and *B*.

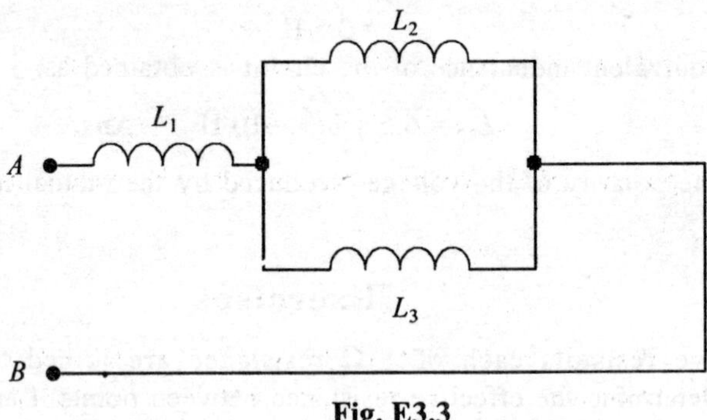

Fig. E3.3

[*Ans:* L_3 = 0.3 H]

E3.5 Determine the effective capacitance of the network shown in Fig. E3.4.

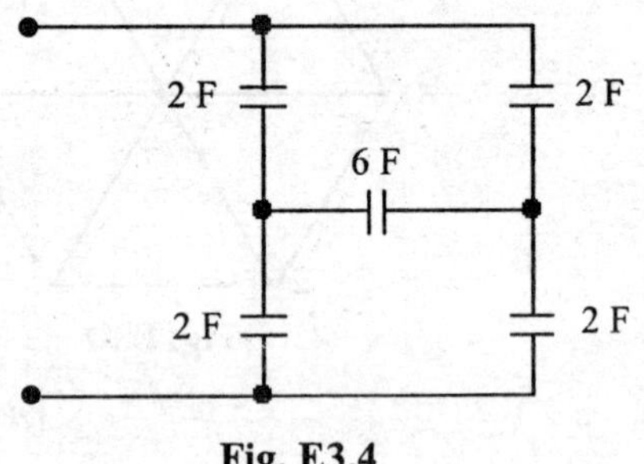

Fig. E3.4

[**Hint:** Note that the capacitances constitute a balanced bridge.]

[*Ans:* 2 F]

E3.6 Show that the magnetically coupled coils in Fig. E3.5 can be replaced by a single coil having an inductance of $L = \dfrac{L_1L_2 - M^2}{L_1 + L_2 - 2M}$

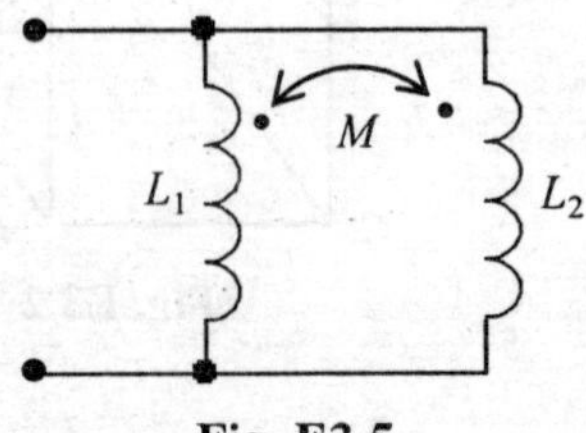

Fig. E3.5

Network Theorems

4

CHAPTER CONTENTS

Solving electrical networks for voltages and currents in them, may be reduced to the simplest forms if it can be assumed that the passive elements in the networks are linear and bilateral; in other words, independent of the magnitude and direction of the current. The behaviour of networks carrying steady direct current can be predicted by the postulates of Ohm's law and Kirchoff's laws, which are elaborated in textbooks in basic electrical engineering.

Concepts of impedance as a complex operator, and of electrical voltage and current phasors, enable the extension of the dc methods to determine the steady state conditions in networks containing sinusoidal driving voltages having the same frequency (see Chapter 10). They may be further extended into the transient state if instantaneous values of voltages and currents are used with inductance and capacitance, in place of impedance (see Chapter 10).

There are over 30 theorems available for simplifying the solution of network problems. These theorems can be derived from the three network laws (see Chapter 3), and may be divided into three categories:

- General network theorems
- Theorems for simplifying passive networks
- Theorems for simplifying active networks

4.1 GENERAL NETWORK THEOREMS

There are four general network theorems:

Superposition theorem: The superposition theorem is a direct consequence of the ***principle of linearity*** and it relates to the systems subject to simultaneous excitations from multiple sources. The theorem may be stated as *the total response of a linear system to a number of simultaneous different stimuli is the sum of the responses to the individual stimulus acting in the absence of the others.*

Reciprocity theorem: If a driving voltage in one branch of a network produces a current in any other branch, the same driving voltage in the second branch would produce the same current in the first.

Substitution theorem: A resistance R carrying a current i flowing in a given direction can be replaced by a driving voltage of value Ri acting in the opposite direction, and conversely.

Compensation theorem: If a network is modified by making a change ΔR in one of its branches, the resultant increment of current in any branch is equal to the current that would be produced by a driving voltage of value $-\Delta Ri$ acting in series with the modified branch where i is the previous current in the modified branch.

The superposition theorem is a general theorem applicable to all linear networks in the transient as well as in the steady state. It affords a useful method of solving two-source systems, which leads to the definition of *driving point and transfer impedances*. The other theorems follow from the application of the superposition theorem to steady state conditions. Proofs of these theorems are not given here as they can be found in Reference 1.

EXAMPLE 4.1 In the circuit of Fig. 4.1, find the current i in the 1 Ω resistance.

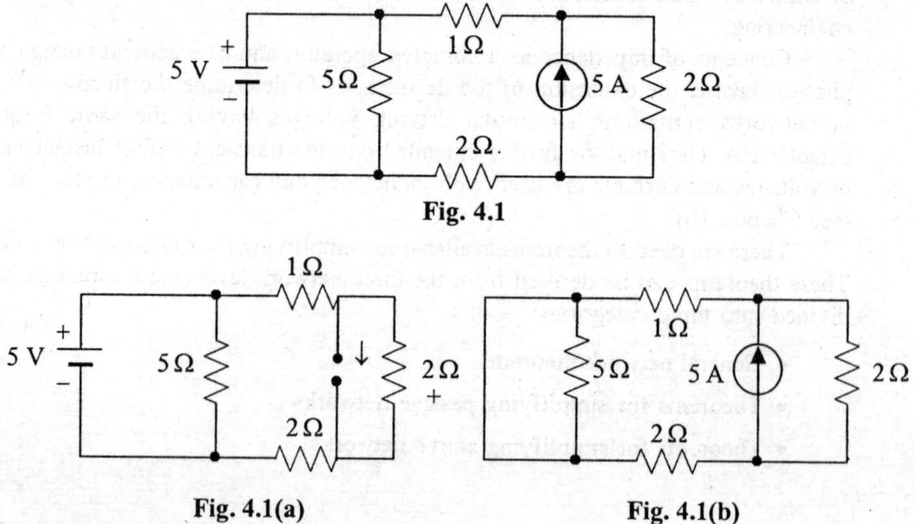

Fig. 4.1

Fig. 4.1(a)

Fig. 4.1(b)

SOLUTION:

The current in the 1 Ω resistance due to the voltage source acting alone [with current source open circuited as shown in Fig. 4.1(a)] is given by

$$\frac{5}{1+2+2} = 1 \text{ A}$$

The current in the 1 Ω resistance due to the current source acting alone with the voltage source short-circuited as shown in Fig. 4.1(b) is given by

$$-5 \times \frac{2}{5} = -2 \text{ A}$$

The total current in the 1 Ω resistance is:

$$1 - 2 = -1 \text{ A} \qquad \textbf{Ans.}$$

EXAMPLE 4.2 Show that the circuit shown in Fig. 4.2 is reciprocal.

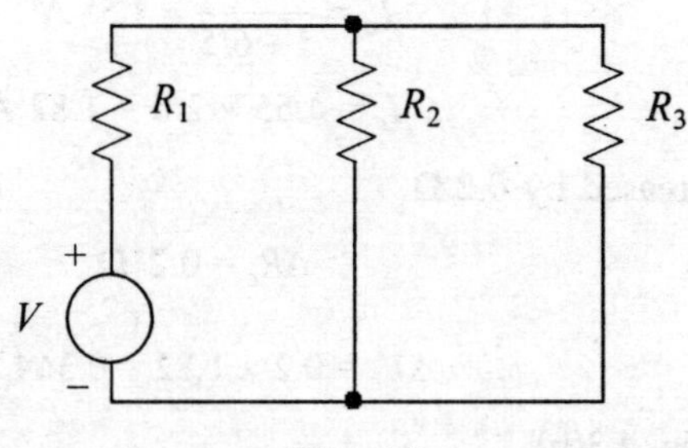

Fig. 4.2

SOLUTION:

$$I_1 = \frac{V}{R_1 + \dfrac{R_2 R_3}{R_2 + R_3}} = V \frac{R_2 + R_3}{R_1 R_2 + R_1 R_3 + R_2 R_3}$$

Therefore,

$$I_3 = I_1 \frac{R_2}{R_2 + R_3} = V \frac{R_2}{R_1 R_2 + R_1 R_3 + R_2 R_3}$$

The transfer impedance is:

$$R_{13} = \frac{V}{I_3} = \frac{R_1 R_2 + R_1 R_3 + R_2 R_3}{R_2}$$

To determine the transfer impedance R_{31}, the source voltage V is transferred to branch 3 and the resulting current in branch 1 is calculated. This involves only the interchange of R_1 and R_3 in the above expression. There is no change in its value and, therefore, R_{31} must be equal to R_{13}.

EXAMPLE 4.3 Consider the network of Fig. 4.3 with resistances of magnitude $R_a = 1\ \Omega$ $R_b = 2\ \Omega$ $R_c = 3\ \Omega$ and $V_s = 10$ V. Determine the current in R_c if its value is increased by 0.2 Ω. Use compensation theorem.

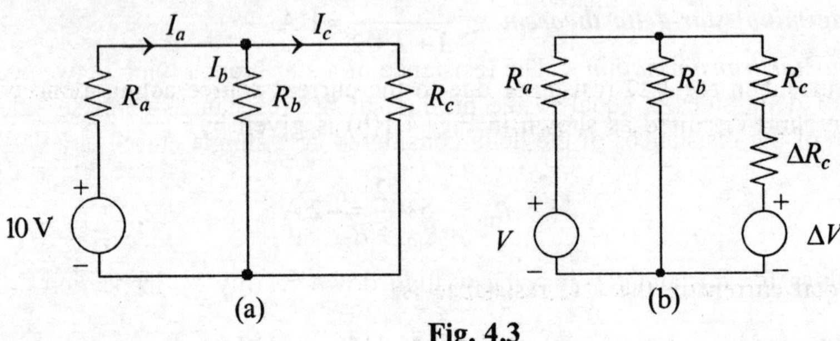

Fig. 4.3

SOLUTION:

$$I_a = \frac{10}{1 + 6/5} = 4.55 \text{ A}$$

$$I_c = 4.55 \times 2/5 = 1.82 \text{ A}$$

if now R_c is increased by 0.2 Ω

$$\Delta R_c = 0.2\ \Omega$$

and

$$\Delta V_c = 0.2 \times 1.82 = 0.364 \text{ V}$$

From circuit in Fig. 4.3(b),

$$\Delta I_c = -\frac{0.364}{3.2 + 2/3} = -0.094 \text{ A}$$

Thus, the current in R_c after its increase is given by

$$1.82 - 0.094 = 1.726 \text{ A} \qquad \textbf{Ans.}$$

This may be checked by using $R_c = 3.2\ \Omega$ in Fig. 4.3(a).

4.2 REDUCTION THEOREMS

4.2.1 Reduction of Passive Networks

In networks, two branches connected in ***series***, can be reduced into a single equivalent branch by adding the component resistances.

$$R = R_1 + R_2$$

Parallel branches may be reduced to a single branch by adding the component conductances.

$$G = G_1 + G_2$$

An unwanted junction of three branches can be eliminated by Kennelly's three-terminal star-delta transformation. The following theorems which can be derived from the basic network laws are applicable to linear bilateral networks in the steady state:

Three terminal-star-delta theorem

I. Delta-star transformation. The resistance of a star branch lying between and replacing a pair of delta sides is equal to the product of the resistances of these sides divided by the sum of the resistances of the delta considered as a simple closed circuit. That is to say

$$R_{10} = \frac{R_{12}R_{31}}{R_{12} + R_{23} + R_{31}} \tag{4.1}$$

The values for R_{20} and R_{30} may be written down readily by inspection.

II. Star-delta transformation. Referring to Fig. 4.4, the resistance of a delta side corresponding to and replacing any two star branches is given by Eq. (4.2).

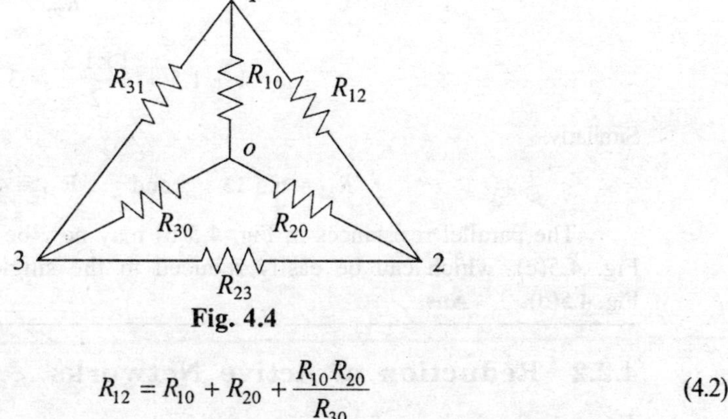

Fig. 4.4

$$R_{12} = R_{10} + R_{20} + \frac{R_{10}R_{20}}{R_{30}} \tag{4.2}$$

Note: The values of R_{23} and R_{31} may be written down readily by inspection. For proof of Eqs. (4.1) and (4.2), see Reference 1.

EXAMPLE 4.4 For the circuit shown in Fig. 4.5(a) find the equivalent resistance of the circuit between points A and B.

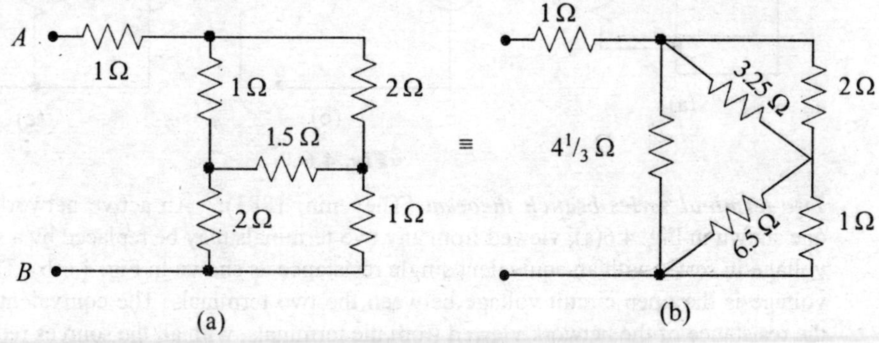

1 Ω
1.24 Ω
4.33 Ω
≡
2.419 Ω
0.866 Ω

(c) (d)

Fig. 4.5

SOLUTION:

The star network 10, 20 and 30 may be replaced by an equivalent delta as shown in Fig. 4.5(b).

$$R_{12} = R_{10} + R_{20} + \frac{R_{10}R_{20}}{R_{30}}$$

$$= 1 + 1.5 + \frac{1 \times 1.5}{2} = 3.25\ \Omega$$

Similarly,

$$R_{23} = 6.5\ \Omega \qquad \text{and} \qquad R_{31} = 4.33\ \Omega$$

The parallel resistances in Fig. 4.5(b) may now be combined to give the circuit in Fig. 4.5(c), which can be easily reduced to the single resistance of 2.419 Ω as in Fig. 4.5(d). **Ans.**

4.2.2 Reduction of Active Networks

Three useful theorems for active networks are:

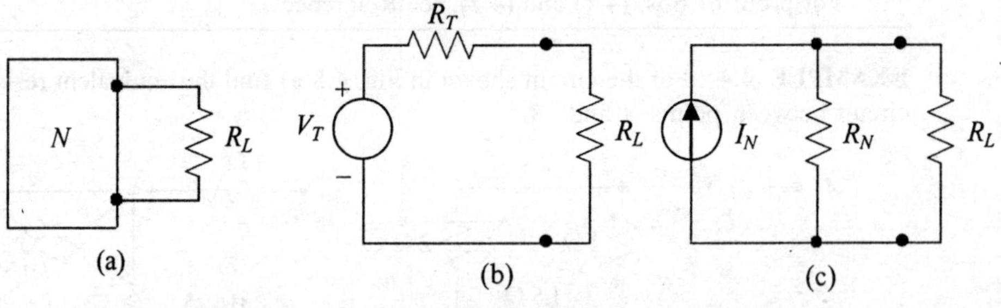

Fig. 4.6

Two-terminal series branch theorem (Thevenin, 1883): An active network such as the one shown in Fig. 4.6(a), viewed from any two terminals may be replaced by a single driving voltage in series with an equivalent single resistance as shown in Fig. 4.6(b). The equivalent voltage is the open circuit voltage between the two terminals. The equivalent resistance is the resistance of the network viewed from the terminals, with all the sources reduced to zero.

Two-terminal parallel branch theorem (Norton): An active network such as the one shown in Fig. 4.6(a), viewed from any two terminals may be replaced by a current source shunted by a resistance as shown in Fig. 4.6(c). The current source is the short circuit current between the terminals. Norton resistance is the same as Thevenin resistance.

Nodal voltage theorem (Millman): If the voltage between any node zero and a number of points 1, 2, 3, in a network are known then the voltage between node zero and a second node n to which these points are connected, is given by

$$V_{0n} = \frac{V_{01}G_1 + V_{02}G_2 + V_{03}G_3}{G_1 + G_2 + G_3} \tag{4.3}$$

as can be seen from Fig. 4.7.

where V_{0n} is the voltage between node zero and node n, and G is branch conductance.

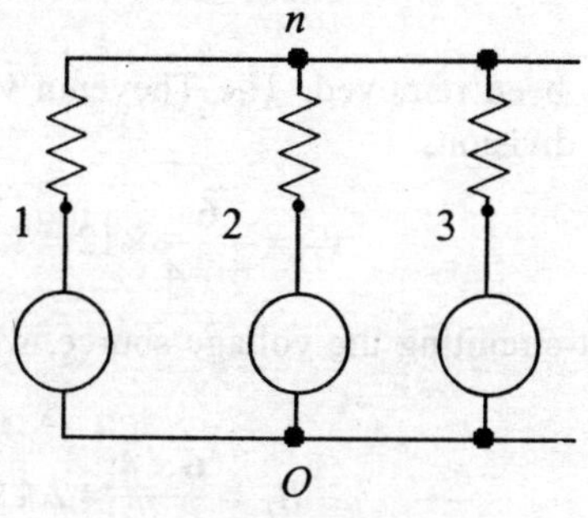

Fig. 4.7

The application of this theorem to a simple resistance circuit is given in Example 4.5.

EXAMPLE 4.5 Calculate the voltage of node n with reference to 0. Deduce the current in the 4 Ω resistance as shown in Fig. 4.8.

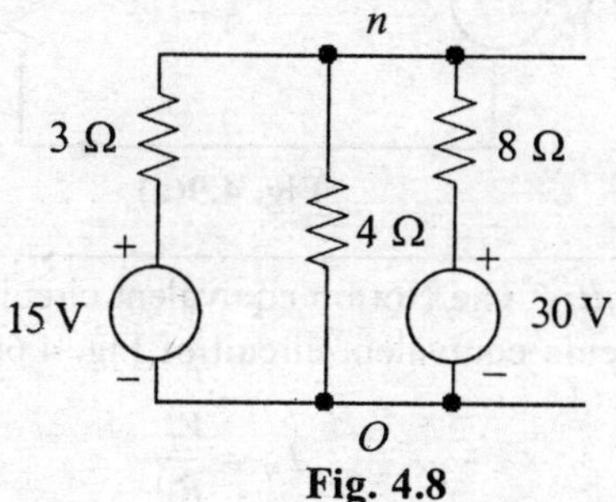

Fig. 4.8

SOLUTION:

From Millman's theorem,

$$v_1G_1 + v_2G_2 = v_{n0}(G_1 + G_2)$$

which gives

$$v_{n0} = 19.10 \text{ V} \quad \textbf{Ans.}$$

The current in the 4 Ω resistance $= \dfrac{v_{n0}}{3 \parallel 8 + 4} = 3.09$ A **Ans.**

EXAMPLE 4.6 Determine the Thevenin equivalent for the circuit shown in Fig. 4.9.

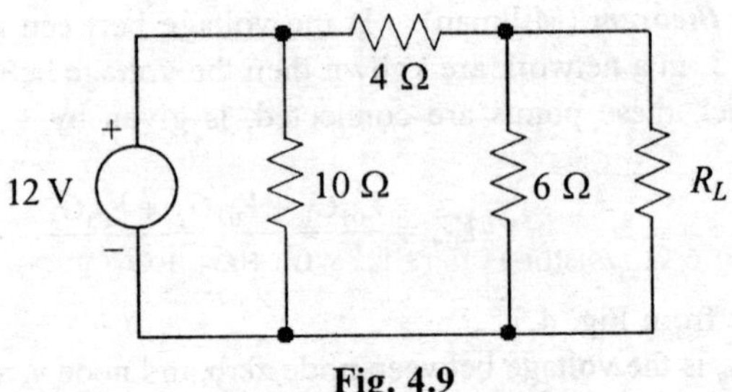

Fig. 4.9

SOLUTION:
Assume that R_L has been removed. The Thevenin voltage is the voltage across the 6 Ω resistor. By voltage division,

$$v_T = \frac{6}{6+4} \times 12 = 7.2 \text{ V}$$

R_T is found by short-circuiting the voltage source, which leaves the 6 Ω and 4 Ω resistors in parallel so that

$$R_T = \frac{6 \times 4}{6+4} = 2.4\ \Omega$$

The Thevenin equivalent circuit is:

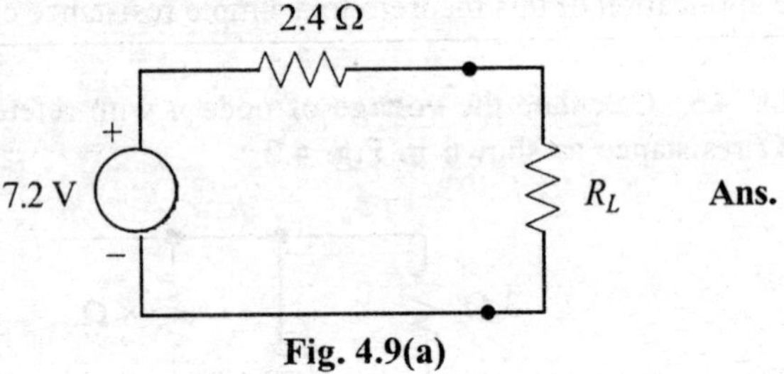

Fig. 4.9(a)

Norton equivalent circuit: The Norton equivalent circuit can also be obtained by source conversion of the Thevenin equivalent circuit of Fig. 4.6(b).

$$I_N = \frac{V_T}{R_T}$$

and $$R_N = R_T$$

EXAMPLE 4.7 For the circuit of Fig. 4.9, find the Norton equivalent circuit.

SOLUTION:
R_N is the same as R_T found earlier, i.e., $R_N = 2.4\ \Omega$

To find i_N, short circuit terminals AB as shown in Fig. 4.10.

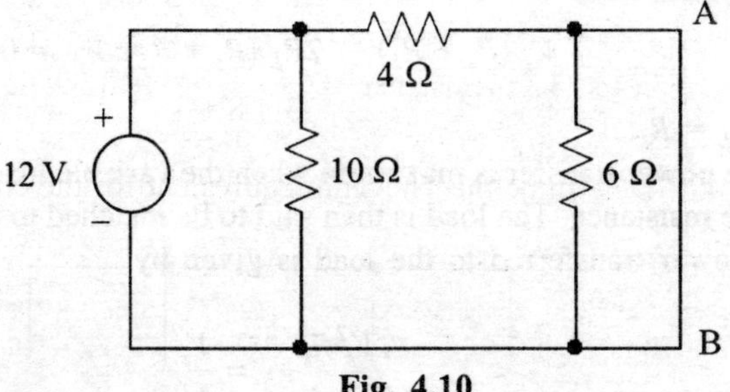

Fig. 4.10

Since the 6 Ω resistor is parallel with the short circuit, it can be ignored and I_N can be calculated as:

$$I_N = \frac{12}{10 \| 4} \times \frac{10}{14} = 3 \text{ A} \qquad \textbf{Ans.}$$

Note: For circuits with both dependent and independent sources, R_T cannot be directly found. In such cases one method is to determine V_T and I_N and then find R_T or R_N.

4.3 OTHER NETWORK THEOREMS

4.3.1 Maximum Power Transfer Theorem

The theorem states that in order to extract maximum power from a physical source into a resistive load, the internal resistance of the source must be equal to the load resistance.

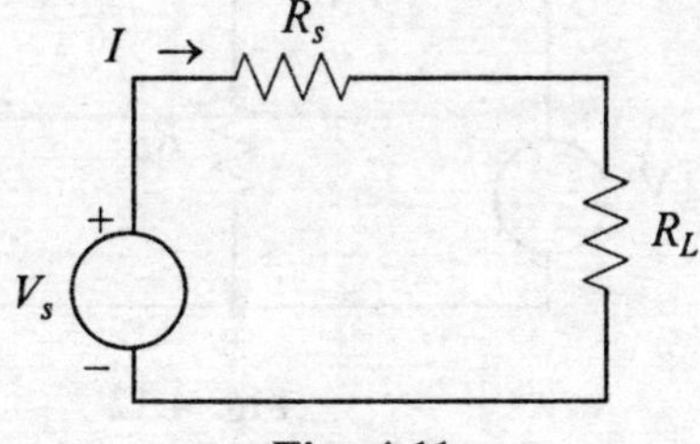

Fig. 4.11

The current into the load resistance is:

$$I = \frac{V_s}{R_L + R_s}$$

Power absorbed by the load resistance is:

$$P_L = I^2 R_L = \frac{V_s^2 R_L}{(R_L + R_s)^2}$$

For P_L to be maximum $\dfrac{dP_L}{dR_L} = 0$

or $$V_s^2 (R_s + R_L)^2 - 2R_L (R_s + R_L) V_s^2 = 0$$

which gives $R_L = R_s$.

Thus, the power transfer is maximum when the variable load resistance is equal to the constant source resistance. The load is then said to be matched to the source. The maximum value of the power transferred to the load is given by

$$\frac{V_s^2 R_s}{(R_s + R_s)^2} = \frac{V_s^2}{4R_s}$$

The total power supplied by the source equals

$$\frac{V_s^2}{(2R_s)^2}$$

Note that the source resistance absorbs power equal to the power in the load resistance. Thus, the maximum power that can be transferred to the load is only 50% of the source output.

The maximum power transfer theorem applicable to a.c. circuits is stated in Chapter 10.

EXAMPLE 4.8 In the circuit shown in Fig. 4.12 what value of R_L will absorb maximum power? What is the value of this maximum power?

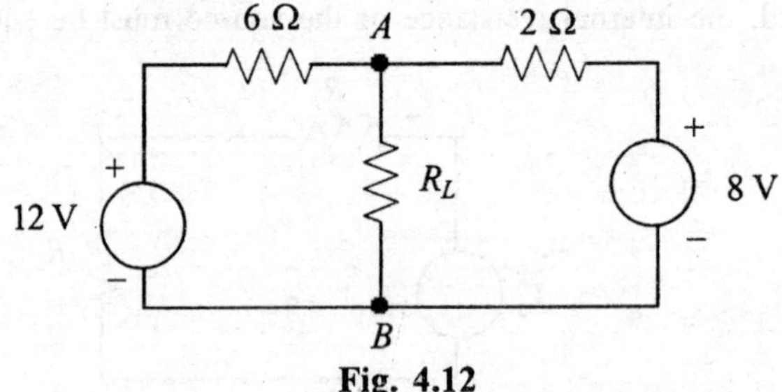

Fig. 4.12

SOLUTION:

The first step is to find the Thevenin equivalent of the given circuit with respect to terminals A and B.

The Thevenin voltage across AB is given by

$$V_T = 8 + \frac{(12-8)2}{8} = 9 \text{ V}$$

The Thevenin resistance is the resistance of the circuit as seen from AB with the voltage sources short-circuited.

$$R_T = \frac{6 \times 2}{8} = \frac{3}{2} \Omega$$

For maximum power transfer,
Load resistance AB = Source resistance = 3/2 Ω **Ans.**

The maximum power transferred is:

$$I_L^2 R_L = \left[\frac{9}{3/2+3/2}\right]^2 \times \frac{3}{2} = 13\frac{1}{2}\ \text{W} \quad \textbf{Ans.}$$

4.3.2 Tellegen's Theorem

Tellegen's theorem states that *the total power in a network is zero provided the branches in the network obey Kirchoff's laws.*

Tellegen's theorem is independent of the nature of the branches; they can be linear or non-linear, passive or active, constant or time varying, provided they satisfy Kirchoff's law. (For proof see Chapter 7.)

Since the instantaneous power $p(t)$ is, by definition, the rate of change of energy $w(t)$ at time t then Tellegen's theorem implies the principle of conservation of energy.

since $\Sigma\, p(t) = 0$ (4.4)
then $\Sigma\, w(t) = \text{Constant}$

4.4 SOURCE EQUIVALENCE

A voltage source consisting of a voltage V in series with a resistance R may be replaced by a current source consisting of a current V/R in parallel with a resistance R as shown in Fig. 4.13.

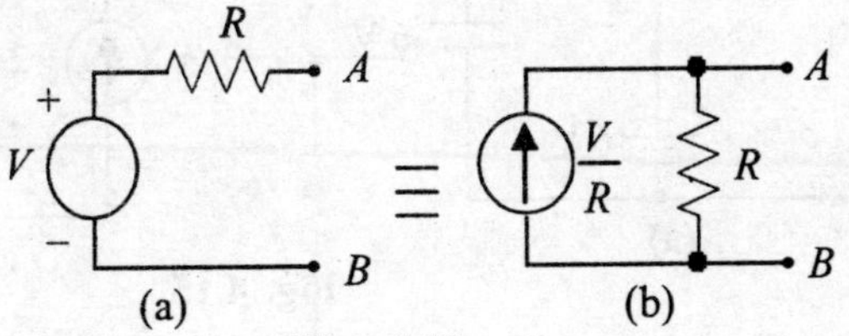

(a) (b)

Fig. 4.13

A current source with a resistance R in parallel may be replaced by a voltage source $I_s R$ in series with a resistance R as shown in Fig. 4.14.

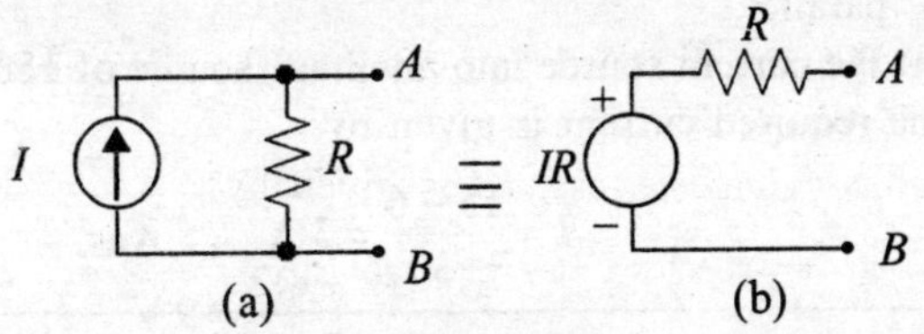

(a) (b)

Fig. 4.14

The relation between V and I in Fig. 4.13 that ensures equivalence with respect to nodes A and B is expressed as:

$$I = \frac{V}{R}$$

If a resistance R_L is connected between A and B in Fig. 4.13(a), the current in R_L is:

$$I_L = \frac{V}{R + R_L} \tag{4.5}$$

If the same resistance R_L is connected between A and B in Fig. 4.13(b) then

$$I_L = \frac{RI}{R + R_L} \tag{4.6}$$

If the two circuits are equivalent, the load current must be the same. Thus,

$$V = RI$$

or

$$I = \frac{V}{R}$$

Conversely the current source in Fig. 4.13(b) may be replaced by a voltage source in Fig. 4.13(a) and

$$V = RI$$

Compound sources: If a network contains both current sources and voltage sources, it is advantageous to reduce it to a network with only one type of sources. [see Example 4.9]

EXAMPLE 4.9 In Fig. 4.15, $I_s = 3$ A, find the value of the current I.

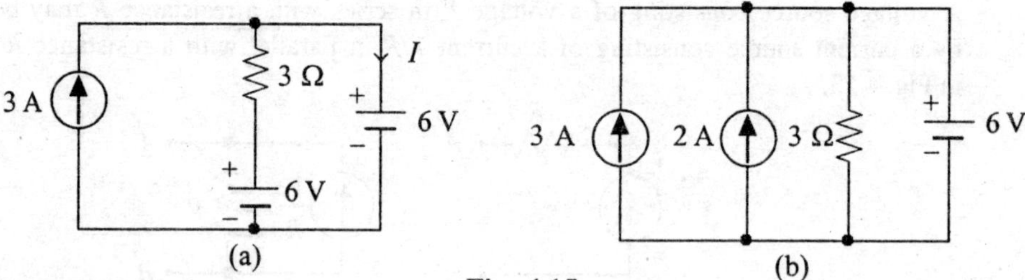

Fig. 4.15

SOLUTION:

Convert the voltage source with 3 Ω resistor in series to a current source with 3 Ω resistor in parallel as shown in Fig. 4.15(b).

Combine the two current sources in Fig. 4.15(b) to get a single current source and a resistance in parallel.

Reconvert the current source into a voltage source of 15 V in series with a resistance 3 Ω. Finally the required current is given by

$$I = \frac{15 - 6}{3} = 3 \text{ A} \qquad \textbf{Ans.}$$

SUMMARY

- Network theorems as applied to direct current networks have been considered in this chapter. Source equivalence, a very useful technique in network analysis has been illustrated with examples and solved problems. Their application to alternating current circuits will be taken up in Chapter 10.

Solved Problems

P4.1 Use superposition theorem to calculate the current in the network shown in Fig. P4.1.

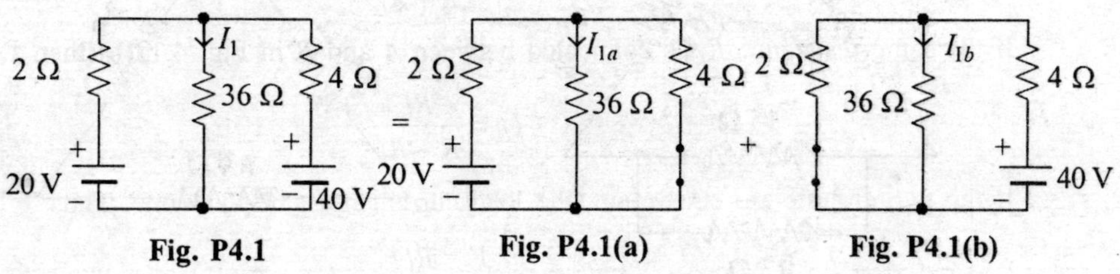

Fig. P4.1 **Fig. P4.1(a)** **Fig. P4.1(b)**

SOLUTION:

With 20 V acting alone the circuit reduces to Fig. P4.1(a) and

$$I_{1a} = \frac{20}{2 + \dfrac{36 \times 4}{36 + 4}} \times \frac{4}{36 + 4} = \frac{5}{14} \text{ A}$$

With the 40 V source acting alone the circuit reduces to Fig. P4.1(b) and the current is:

$$I_{1b} = \frac{40}{4 + \dfrac{2 \times 36}{2 + 36}} \times \frac{2}{36 + 2} = \frac{5}{14}$$

Total current

$$I_1 = I_{1a} + I_{1b}$$

$$= \frac{5}{14} + \frac{5}{14} = 0.714 \text{ A} \qquad \textbf{Ans.}$$

P4.2 In the network shown in Fig. P4.2(a), calculate the equivalent resistance between nodes A and B.

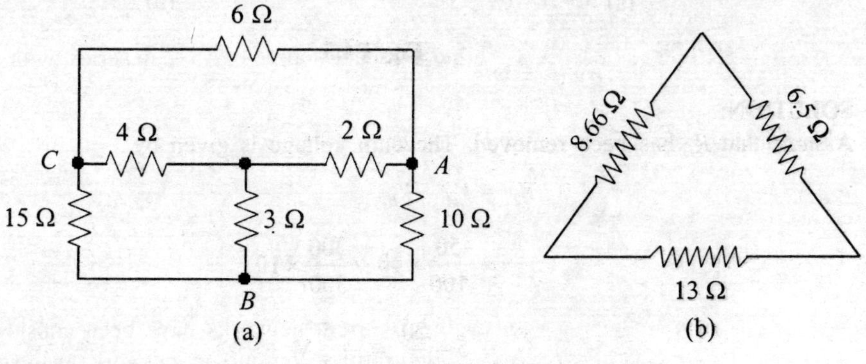

(a) (b)

(c)

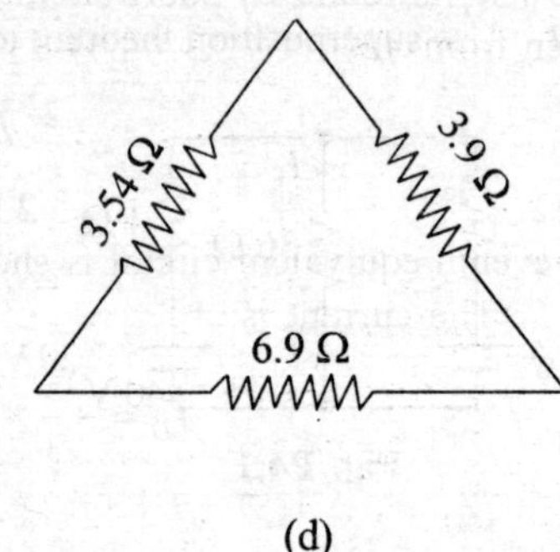

(d)

Fig. P4.2

SOLUTION:
Find the equivalent delta connected circuit as shown in Fig. P4.2(b) for the star-connected resistors 2 Ω, 3 Ω and 4 Ω.

From Fig. P4.2(b), the circuit of Fig. P4.2(a) can be arranged as in Fig. P4.2(c), which can be reduced to Fig. P4.2(d). The equivalent resistance between A and B is given as:

$$= 3.9 \parallel (3.54 + 6.9) = 2.84\ \Omega$$

P4.3 In the bridge circuit of Fig. P4.3(a) find the current through the detector D.

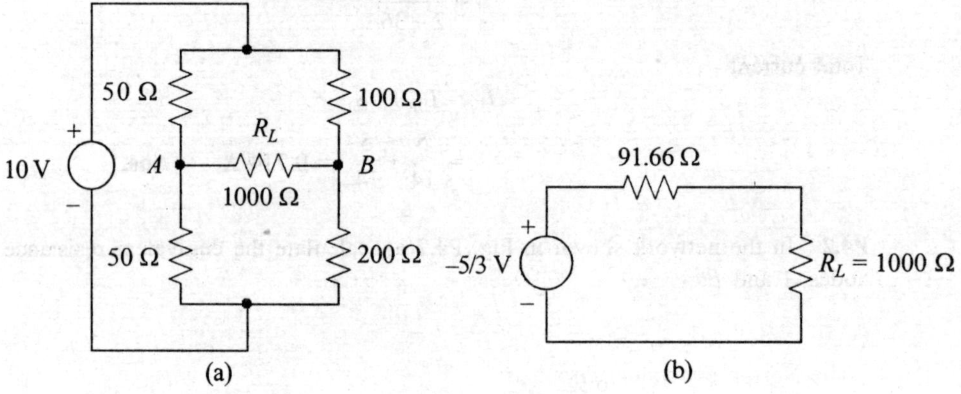

(a) (b)

Fig. P4.3

SOLUTION:
Assume that R_L has been removed. Thevenin voltage is given by

$$V_T = V_A - V_B$$

$$= \frac{50}{100} \times 10 - \frac{200}{300} \times 10$$

$$= 5 - \frac{20}{3} = -\frac{5}{3}\ \text{V}$$

B is at a higher potential compared to A.
Now R_T is found by short circuiting the voltage source and determining the resistance as seen from AB.

$$R_T = 50 \parallel 50 + 100 \parallel 200$$
$$= 91.66\ \Omega$$

Thevenin equivalent circuit is shown in Fig. P4.3(b).
The current is:

$$I_D = -\frac{5}{3} \times \frac{1}{91.66 + 1000} = -1.5\ \text{mA} \qquad \textbf{Ans.}$$

P4.4 In the circuit of Fig. P4.4(a) determine the Norton equivalent circuit.

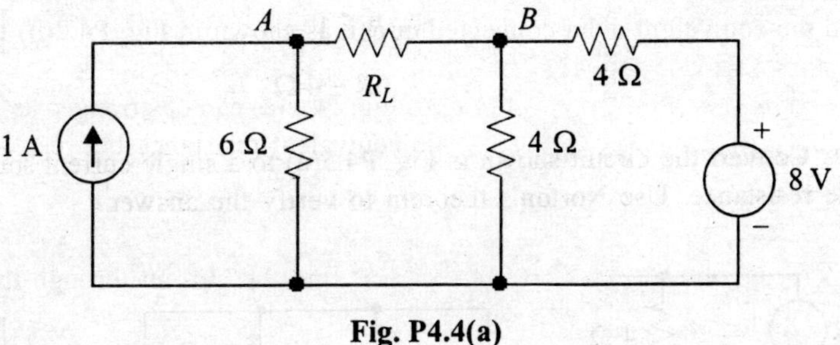

Fig. P4.4(a)

SOLUTION:

$$R_N = 6 + (4 \parallel 4) = 8\ \Omega$$

$$V_B = \frac{8}{8} \times 4 = 4\ \text{V}$$

Thus,

$$I_N = 1 \times \frac{6}{8} - \frac{4}{6+2} = \frac{1}{4}\ \text{A}$$

The Norton equivalent circuit is:

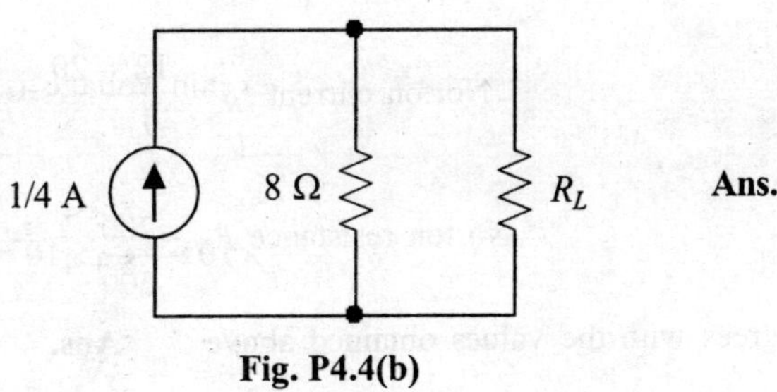

Ans.

Fig. P4.4(b)

P4.5 (a) Find the value of the load resistance R_L which will absorb maximum power in the circuit shown in Fig. P4.4(a). What is the value of the maximum power absorbed under these conditions? (c) Determine the resistance required in the current source which will cause the load resistance to absorb zero power. For maximum power to the load, $R_L = R_T$

(a)
$$R_T = 6 + 2 = 8\ \Omega$$

$$V_T = 6 \times 1 - 4 \times 2 = -2\ \text{V}$$

which gives the load current $I = -\dfrac{2}{8+8} = -\dfrac{1}{8}$ A

(b) Hence power in the load resistance = 1/8 W

(c) To absorb zero power, the potential difference across the load terminals must be zero. So if R is the source resistance,

$$R \times 1 = 4$$

or
$$R = 4\ \Omega$$

P4.6 Convert the circuit shown in Fig. P4.5(a) to a single current source in parallel with single resistance. Use Norton's theorem to verify the answer.

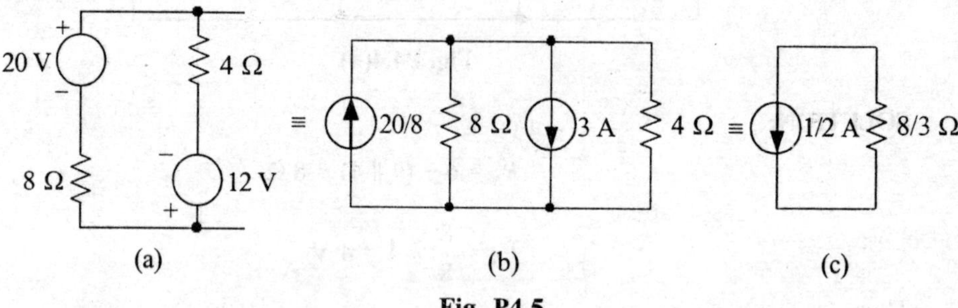

Fig. P4.5

SOLUTION:

Convert the two voltage sources into current sources as shown in Fig. P4.5(b). Combine the two current sources to obtain a single current source of value 0.5 A in parallel with resistance of 8/3 Ω as shown in Fig. P4.5(c).

Now,

$$\text{Norton current } I_N = \frac{12}{4} - \frac{20}{8} = \frac{1}{2}\ \text{A}$$

$$\text{Norton resistance } R_N = \frac{8 \times 4}{8 + 4} = \frac{8}{3}\ \Omega$$

which agrees with the values obtained above. **Ans.**

Exercises

E4.1 Find the voltage across the 1 Ω resistor in Fig. E4.1 by using superposition theorem.

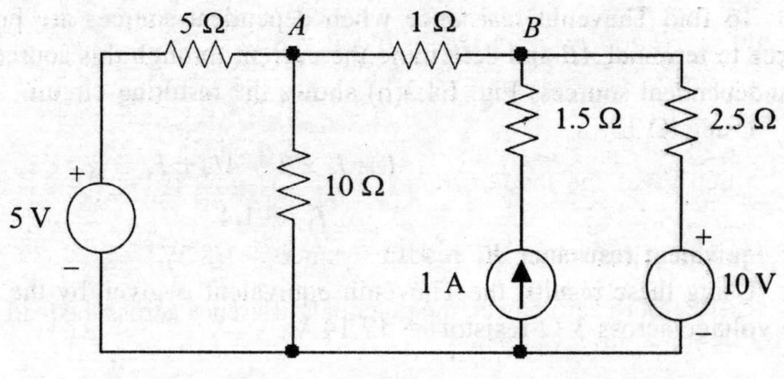

Fig. E4.1

[*Ans:* –1.075 V]

E4.2 Find the resistance across the terminals *A* and *B* of the network shown in Fig. E4.2.

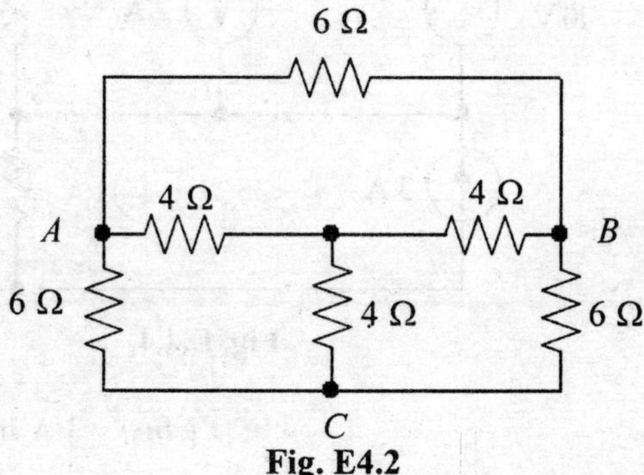

Fig. E4.2

[*Ans:* 8/3 Ω]

E4.3 Use Thevenin's theorem to find the voltage across the 3 Ω resistor in Fig. E4.3.

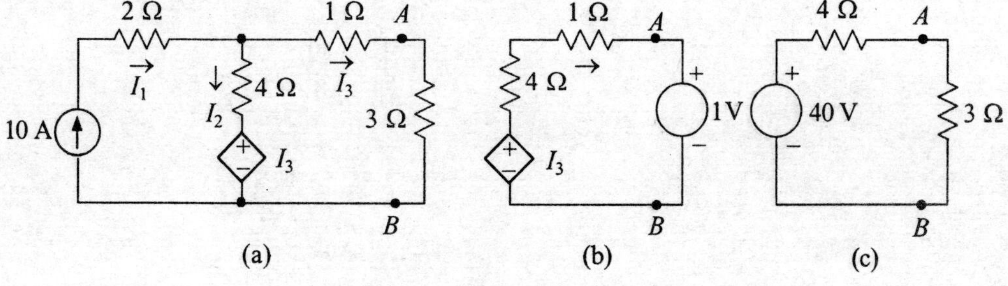

Fig. E4.3

[**Hint:** On removing the 3 Ω resistor, the current I_3 is zero and hence the dependent voltage source has no voltage across it. Voltage V_{AB} is just the voltage across the 4 Ω resistor, which gives $V_T = 40$ V.]

To find Thevenin resistance when dependent sources are present, apply a voltage source to terminal *AB* and determine the current through this source. (after the removal of all independent sources) Fig. E4.3(b) shows the resulting circuit.

Using KVL

$$1 = I_3 \times 1 - 4I_3 + I_3$$

or

$$I_3 = -1/4$$

The equivalent resistance $R_T = 4$ Ω

Using these results, the Thevenin equivalent is given by the circuit in Fig. E4.3(c). The voltage across 3 Ω resistor = 17.14 V.

E4.4 Convert the circuit shown in Fig. E4.4 to a single current source, in parallel with single resistance. Use Norton's theorem to verify the answer.

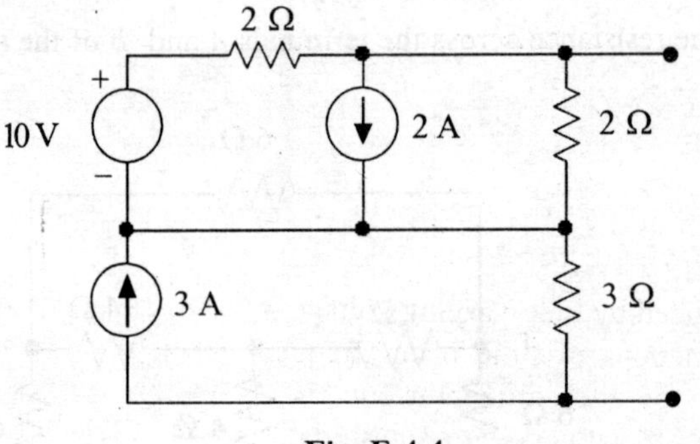

Fig. E.4.4

[*Ans:* 3 A in parallel with 4 Ω resistance.]

Alternating Current Circuit

5

CHAPTER CONTENTS

Circuits energized by time-varying voltage or current sources are considered in this chapter; sinusoidal functions of time occur frequently and naturally in many physical systems. Fundamentals of alternating current circuits are introduced followed by steady state response of RLC networks to sinusoidal voltages or currents .

5.1 FUNDAMENTALS OF ALTERNATING QUANTITIES

5.1.1 Generation of an Alternating Voltage

The principle of an alternator is illustrated in Fig. 5.1. A loop *AB* is carried by a spindle *DD* rotated at a constant speed in a uniform magnetic field produced by poles *NS*. The ends

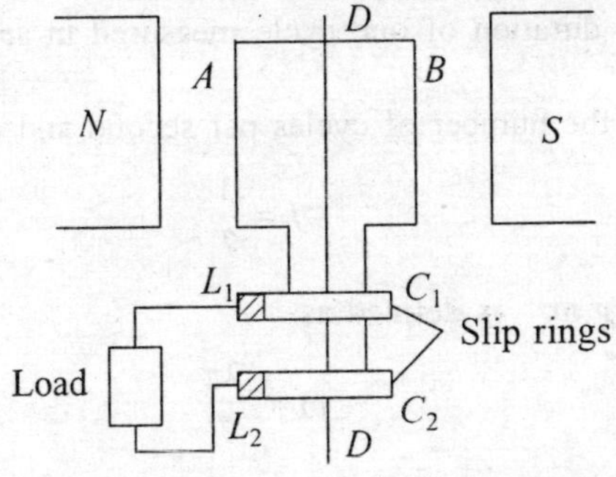

Fig. 5.1 A simple alternator.

of the loop are brought out to two slip rings C_1 and C_2 attached to but insulated from DD. Wearing on these are carbon brushes L_1 and L_2, which are connected to an external load.

Faraday's law for rotational e.m.f. gives

$$e = 2Blv \sin \theta \tag{5.1a}$$

$$= E_m \sin \theta \tag{5.1b}$$

where B is the flux density in tesla, l is the length in metres of coil sides, v is the peripheral velocity of the coil, θ is the angle between the plane of the coil and the normal to the flux direction.

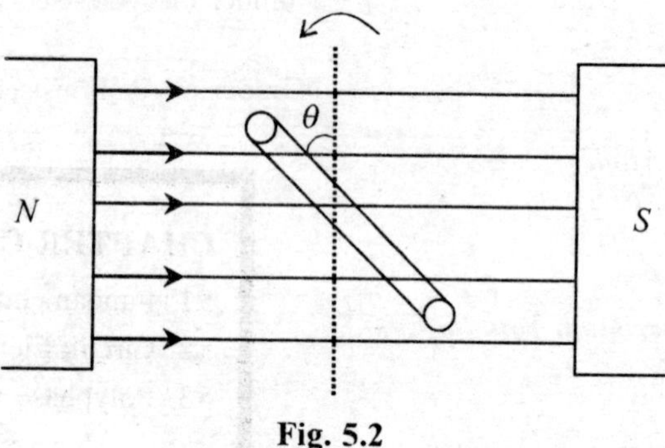

Fig. 5.2

Figure 5.2 shows clearly the variation of e.m.f. in the simple alternator. One cycle of e.m.f. is completed in the coil in one rotation.

The e.m.f. is maximum when $\theta = 90°$, and is given by

$$E_m = 2Blv \tag{5.2}$$

5.1.2 Definitions

Waveform: is the variation of a quantity such as voltage or current shown on a graph to a base of time or rotation.

Cycle: denotes each repetition of a variable quantity recurring at equal intervals.

Period T: is the duration of one cycle measured in seconds.

Frequency f: is the number of cycles per second and denoted by hertz (Hz), i.e.,

$$f = \frac{1}{T} \tag{5.3}$$

Angular frequency ω: is defined as:

$$\omega = \frac{2\pi}{T} \tag{5.4}$$

Instantaneous value: is the magnitude of a waveform at any instant. Instantaneous values are denoted by e, v, i.

EXAMPLE 5.1 In the simple generator shown in Fig. 5.1 the coil consists of 50 turns and is rotated at 3000 r.p.m. The magnetic field has a uniform density of 0.10 T. Mean area/turn is 50 cm^2. Calculate: (a) the frequency, (b) the period, (c) maximum value of the e.m.f. generated and (d) the value of the generated e.m.f. when the coil has rotated through 45°.

SOLUTION:

(a) Frequency

$$f = \text{Number of cycles/sec}$$

$$= \text{Number of rotations/sec} = \frac{3000}{60} = 50 \text{ Hz}$$

(b) Period

$$T = \frac{1}{50} = 0.02 \text{ s}$$

(c) Maximum value of e.m.f.

From Eq. (5.2),

$$E_m = 2Blv \text{ (for the two coil sides)}$$

$$= 2Bl.2\pi rn$$

where r is the radius of the coil.

$$E_m = 2\pi BA\, n$$

$$= 2\pi \times 0.1 \times 50 \times 10^{-4} \times \frac{3000}{60} \times 50 = 7.85 \text{ V}$$

(d) For θ = 45°

$$\sin 45 = 0.707$$

and $$e = 7.85 \times 0.707 = 5.55 \text{ V}$$

5.1.3 Representative Values of Alternating Quantities

Maximum value (peak value): is the maximum instantaneous value measured from its zero value. Maximum values have the symbols E_m, V_m and I_m.

Average value: of a waveform is given by the ratio of the area enclosed over a cycle to the base. In the case of symmetrical waveforms, such as sinusoids, this definition would give zero as the average value and hence the average is taken only over half a cycle. Thus, the average value of a sinusoidal voltage may be obtained by integrating the function over a half-period.

From Eq. 5.1(b),

$$E_{av} = \frac{1}{\pi}\int_0^{\pi} E_m \sin\theta \, d\theta$$

$$= \frac{E_m}{\pi}[-\cos\theta]_0^{\pi}$$

Thus,

$$E_{av} = \frac{2E_m}{\pi} = 0.637\,E_m \tag{5.5}$$

In terms of time,

$$E_{av} = \frac{1}{(T/2)}\int_0^{T/2} E_m \sin\omega t \, dt$$

$$= \left(\frac{2E_m}{T}\right)\int_0^{T/2} \sin\omega t \, dt = \frac{2E_m}{\pi} \tag{5.6}$$

Similarly, the average current is:

$$I_{av} = \frac{2I_m}{\pi} \tag{5.7}$$

Root mean square (r.m.s.) or effective value: The effective value of an alternating current is that value of direct current which when passed through a resistance produces the same heating effect as produced by the alternating current while passing through the same resistance for the same time.

If the sinusoidal current is passed through a resistance R, the instantaneous heating effect equals i^2R. The heat produced during a complete cycle is:

$$\int_0^{2\pi} i^2 R \, d\theta = I_m^2 R \int_0^{2\pi} \sin^2\theta \, d\theta$$

$$= \frac{I_m^2 R}{2}\int_0^{2\pi} (1-\cos 2\theta)\, d\theta$$

$$= \frac{I_m^2 R}{2} 2\pi = \pi I_m^2 R$$

The average heating effect over a period is, thus

$$= \frac{I_m^2 R\pi}{2\pi}$$

$$= \frac{1}{2} I_m^2 \, R \tag{5.8}$$

If now I is the value of direct current through the same resistance to produce the same heating effect

$$I^2R = \frac{1}{2} I_m^{\ 2} R$$

or $$I = I_m/\sqrt{2} = 0.707\, I_m \tag{5.9}$$

It can thus be seen that the effective value of an alternating current is the square root of the mean of the squares of the instantaneous values and hence termed as **r.m.s. value**.

The relationship derived for current is also applicable to voltage, and

$$E = E_m/\sqrt{2} \qquad \text{(for sine waves)}$$

Note that the symbols for r.m.s. values have no suffix attached to them.

Peak or crest factor K_P: is defined as *the ratio of the maximum value to the r.m.s. value of an alternating quantity.*

For a sine wave,

$$K_P = \frac{E_m}{E_m/\sqrt{2}} = 1.414 \tag{5.10}$$

Form factor K_f: is defined as *the ratio of the r.m.s. value to the average value.*

For a sine wave,

$$K_f = \frac{E_m/\sqrt{2}}{(2/\pi)E_m} = \frac{\pi}{2\sqrt{2}} = 1.11 \tag{5.11}$$

EXAMPLE 5.2 Calculate the form factor for the waveform shown in Fig. 5.3.

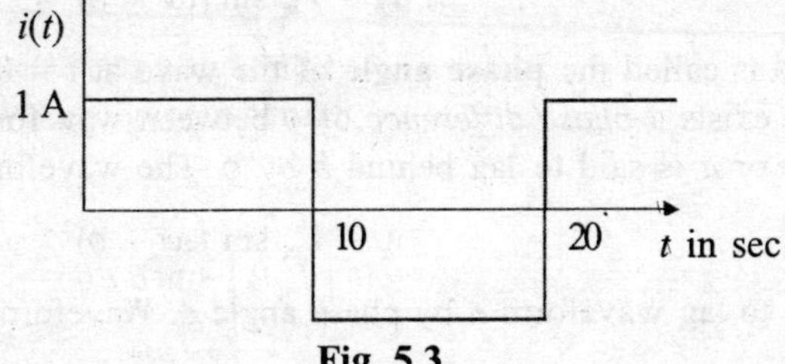

Fig. 5.3

SOLUTION:

By inspection

$$i_{av} = 1.0 \text{ A}$$

r.m.s. value is given by

$$I^2 = \frac{1}{T}\int_0^T i^2 dt = \frac{1}{10}\int_0^{10} 1^2\, dt = 1$$

$$I = 1 \text{ A}$$

$$K_f = \frac{I}{i_{av}} = 1$$

5.1.4 Average Value and rms Value of Non-sinusoidal Current and Voltage

The average and r.m.s. values of non-sinusoidal quantities may be determined either graphically or analytically.

5.1.5 Phase Angle and Phase Difference in Alternating Waveforms

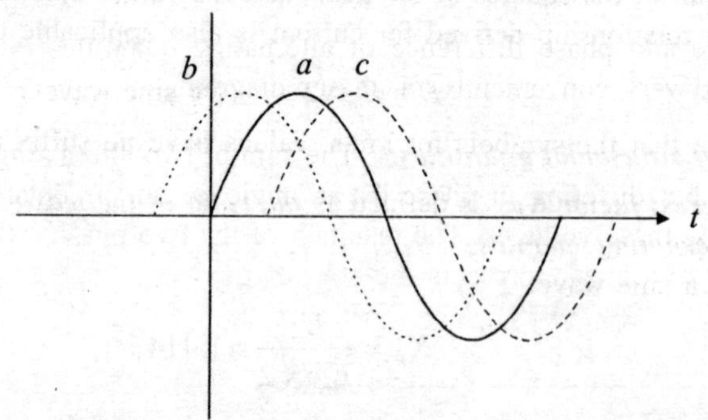

Fig. 5.4

In Fig. 5.4, waveform a may be written as:

$$v_a = V_m \sin \omega t$$

where ω is specified in radians per second. The wave form b may be written as:

$$v_b = V_m \sin (\omega t + \phi)$$

The angle ϕ is called the **phase angle** of the wave at $t = 0$.

There exists a *phase difference* of ϕ between waveforms a and b; and b is said to lead a by ϕ or a is said to lag behind b by ϕ. The waveform c may be written as:

$$v_c = V_m \sin (\omega t - \phi)$$

This is said to lag waveform a by phase angle ϕ. Waveform c is said to lag a by a phase angle ϕ.

5.1.6 Representation of Sinusoids by Phasors

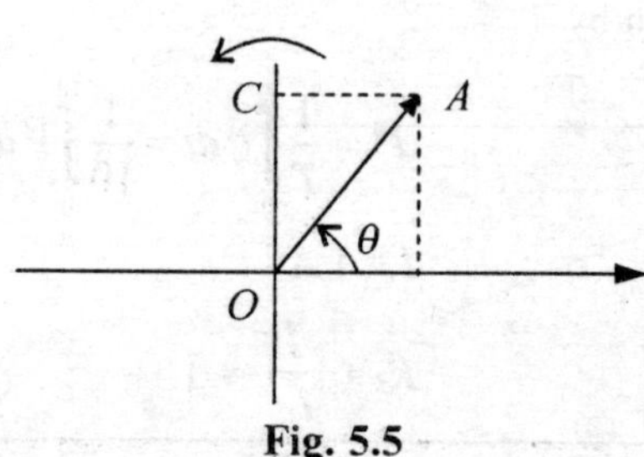

Fig. 5.5

Consider a directed line segment of magnitude equal to the maximum value of an alternating current, $OA = I_m$, rotated anticlockwise about point O at a uniform angular velocity of ω radians per second. When OA has rotated through an angle θ,

$$OC = OA \sin\theta = I_m \sin\theta = i,$$

Hence the projection of OA on the vertical axis represents the instantaneous value of the current.

If f is the frequency of current in hertz, then the angular velocity ω in radians per second is:

$$\omega = 2\pi f \tag{5.12}$$

Phase and phase difference of alternating quantities (voltage and current) can be represented very conveniently in phasor diagrams.

Addition of sinusoidal quantities: The sum of two alternating voltages, having the same frequency but differing in phase by an angle ϕ, can be obtained by adding the phasors representing these voltages. The resultant of the two phasors OA and OB in Fig. 5.6(a) is given by OC which represents the sum of the two voltages.

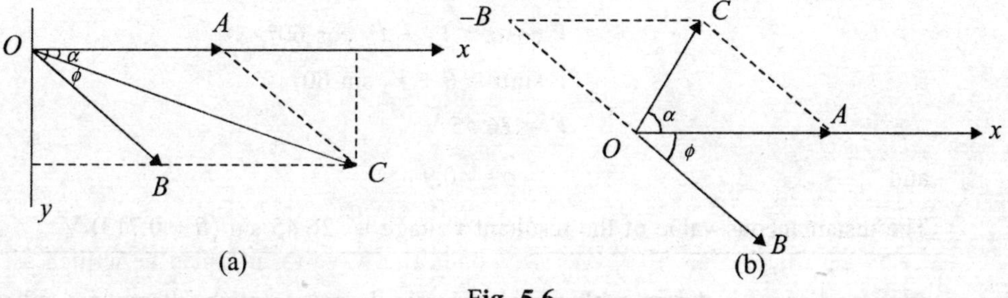

Fig. 5.6

Subtraction of sinusoidal quantities: The phasor difference ($OA – OB$) is shown as OC in Fig. 5.6(b).

EXAMPLE 5.3 The instantaneous values of two alternating voltages are represented by

$$v_1 = 30 \sin\theta$$

$$v_2 = 20 \sin(\theta - \pi/3)$$

Derive an expression for the instantaneous value of (i) the sum and (ii) the difference of these voltages.

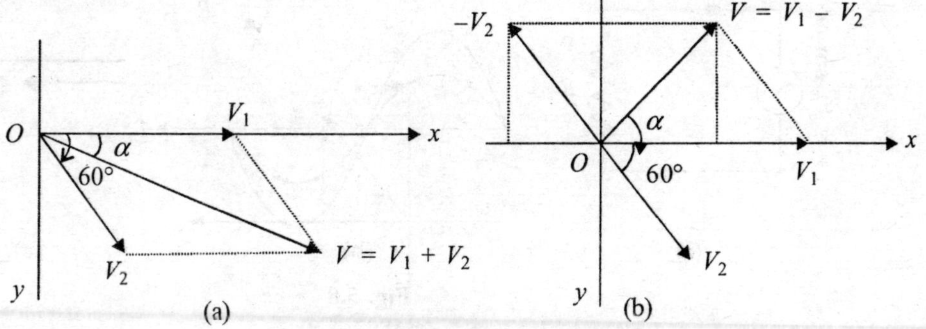

Fig. 5.7

SOLUTION:

(i) Sum of voltages: The phasor V_1 is taken as the reference and phasor V_2 is drawn 60° behind V_1. The diagonal V of the parallelogram drawn on V_1 and V_2 represents the phasor sum of V_1 and V_2. The magnitude and angle of V can be determined by resolving the phasors in the horizontal and vertical directions.

$$V \cos \alpha = V_1 + V_2 \cos 60° \tag{1}$$

$$V \sin \alpha = -(V_2 \sin 60° + 0) \tag{2}$$

From (1) and (2), we get

$$\text{Resultant } V = 43.6 \text{ V}$$

and $\alpha = -23.4°$

The instantaneous value of the resultant voltage is 43.6 sin (θ – 0.41) V. **Ans.**

(ii) Difference of voltages: The subtraction of phasors is shown in Fig. 5.7(b). Resolving horizontally

$$V \cos\alpha = V_1 + V_2 \cos 60°$$

$$V \sin\alpha = \theta + V_2 \sin 60°$$

$$V = 26.45 \text{ V}$$

and $\alpha = 40.9$

The instantaneous value of the resultant voltage is 26.45 sin (θ + 0.713) V. **Ans.**

Phasor diagrams drawn with r.m.s. values: In representing alternating voltages and currents by phasors it is assumed that their waveforms are sinusoidal, in which case the r.m.s. value is known to be 0.707 times the maximum value. Further ammeters and voltmeters are generally calibrated in r.m.s. values of currents and voltages. It is, therefore, the practice to draw phasor diagrams using r.m.s. values instead of maximum values.

5.2 CIRCUIT ELEMENTS IN FREQUENCY DOMAIN

5.2.1 Alternating Current in a Circuit with Resistance Only

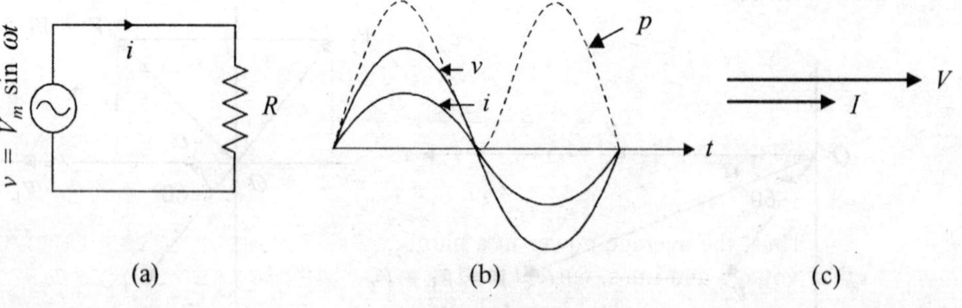

Fig. 5.8

In the circuit of Fig 5.8(a), the voltage at any instant is:

$$v = V_m \sin \omega t \tag{5.13a}$$

The current is given by

$$i = \frac{V_m}{R} \sin \omega t = I_m \sin \omega t \tag{5.13b}$$

where $I_m = \dfrac{V_m}{R}$

In terms of r.m.s. values

$$I = \frac{V}{R} \tag{5.13c}$$

which is Ohm's law for a resistive circuit.

Equations (5.13a) and (5.13b) show that the voltage and current are in phase. The voltage and current waveforms are shown in Fig. 5.8(b). The phasors representing the voltage and current are coincident as shown in Fig. 5.8(c).

Power in a resistive circuit: The instantaneous power is given by

$$p = vi = V_m \sin \omega t \, I_m \sin \omega t$$

$$= V_m I_m \sin^2 \omega t = \frac{1}{2} V_m I_m (1 - \cos 2\omega t) \tag{5.14}$$

Equation (5.14) represents a sinusoidal wave with a frequency double that of the component voltage and current waves, and displaced from the time axis as shown in Fig. 5.8(b).

The average power is obtained as:

$$P = \frac{1}{T}\int_0^T \frac{1}{2}(V_m I_m)[1 - \cos 2\omega t]\, dt$$

$$= \frac{V_m I_m}{2T}\left[t - \left(\frac{\sin 2\omega t}{2\omega}\right)\right]$$

$$= \frac{V_m I_m}{2} = \frac{V_m}{\sqrt{2}} \frac{I_m}{\sqrt{2}}$$

$$= VI \tag{5.15}$$

Thus, the average power in a purely resistive circuit is given by the product of the r.m.s. voltage and r.m.s. current; and is measured in watts.

5.2.2 Alternating Current in a Circuit Containing Inductance Only

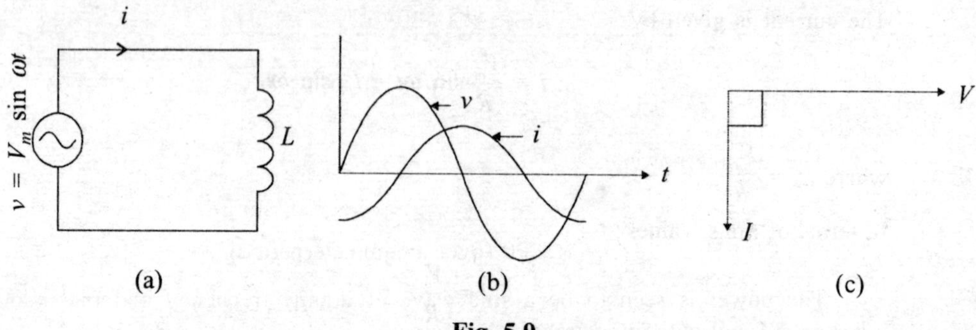

Fig. 5.9

In an inductor the applied voltage

$$v = L\frac{di}{dt}$$

or

$$di = \frac{v}{L}\,dt$$

If

$$v = V_m \sin \omega t, \tag{5.16}$$

Then

$$i = \int \frac{V_m}{L} \sin \omega t \, dt$$

$$= -\frac{V_m}{L\omega} \cos \omega t \tag{5.17a}$$

$$= -I_m \cos \omega t \tag{5.17b}$$

where $I_m = V_m / L\omega$

neglecting the constant of integration, which is significant only for transient conditions. In terms of r.m.s. values,

$$I = \frac{V}{L\omega} = \frac{V}{X_L} \tag{5.18}$$

where L is the inductance measured in henries and X_L (= $L\omega$) is the inductive reactance measured in ohms; this is Ohm's law for an inductive circuit. The voltage and current waveforms are shown in Fig. 5.9(b). The voltage and current phasors are shown in Fig. 5.9(c), which clearly indicates the 90° lag of the current with respect to the voltage in an inductive circuit.

Power in an inductive circuit: The instantaneous power is given by

$$p = vi$$

$$= (V_m \sin \omega t)(-I_m \cos \omega t)$$

$$= -\frac{1}{2} V_m I_m \sin 2\omega t \qquad (5.19)$$

The average power is:

$$P = -½ (V_m I_m) \text{ average of } \sin 2\omega t$$

$$= 0 \text{ (over a complete period)} \qquad (5.20)$$

The power is seen to be a sine wave of double frequency and maximum value $½V_m I_m$ = VI, but with an average value zero.

EXAMPLE 5.4 A resistance of 10 Ω is in series with a pure inductance of 62.8 mH and the circuit is connected to a 200 V, 50 Hz supply as shown in Fig. 5.10. Calculate the magnitude and phase angle of the current.

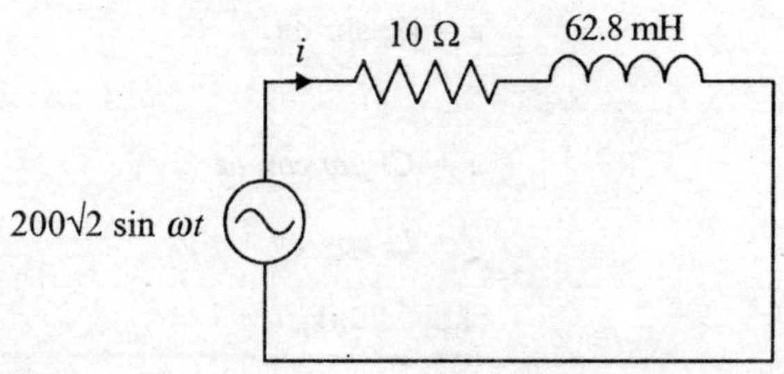

Fig. 5.10

SOLUTION:

$$\text{Reactance } X_L = 2\pi f L$$

$$= 2\pi \times 50 \times 62.8 \times 10^{-3} = 20\ \Omega$$

$$\text{Impedance } Z = (R^2 + X_L^2)^{½}$$

$$= (10^2 + 20^2)^{½} = 22.36\ \Omega$$

$$\text{Circuit current } I = V/Z$$

$$= 200/22.36 = 8.94 \text{ A} \qquad \textbf{Ans.}$$

$$\text{Phase angle } \phi = \tan^{-1}\left(\frac{X_L}{R}\right) = \tan^{-1}(20/10) = 63.43° \text{ lagging} \qquad \textbf{Ans.}$$

5.2.3 Alternating Current in a Circuit Containing Capacitance Only

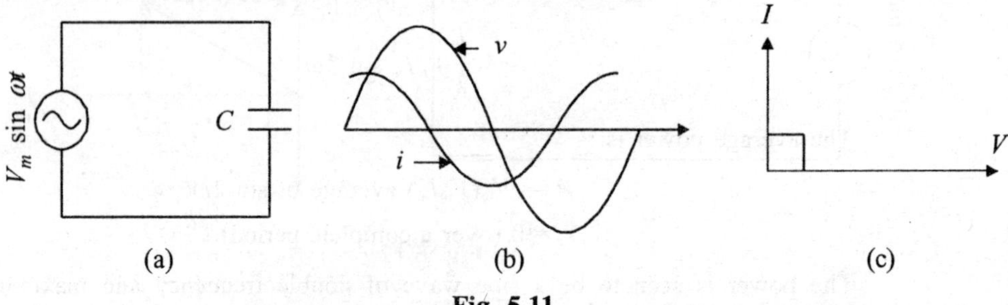

Fig. 5.11

In a capacitor, the current is given by

$$i = C\frac{dv}{dt}$$

If
$$v = V_m \sin \omega t \tag{5.21}$$

Then
$$i = CV_m\omega \cos \omega t \tag{5.22a}$$

$$= I_m \cos \omega t \tag{5.22b}$$

where
$$I_m = C\,\omega V_m \tag{5.23}$$

In terms of r.m.s. values, Eq. (5.23) becomes

$$I = \frac{V}{1/(C\omega)} = \frac{V}{X_C} \tag{5.24}$$

where C is the capacitance measured in farads and $X_C(1/C\omega)$ is the capacitive reactance in ohms; this is Ohm's law for the capacitive circuit.

Equations (5.21) and (5.22b) show that the current leads the voltage by 90° in a capacitance.

The voltage and current waveforms are shown in Fig. 5.11(b). The corresponding phasors are shown in Fig. 5.11(c), which clearly indicates the 90° lead of the current with reference to the voltage.

5.2.4 General Series Circuit

A circuit with resistance, inductance and capacitance all connected in series is shown in Fig. 5.12(a). This circuit is connected across an alternating source of voltage V and angular frequency ω.

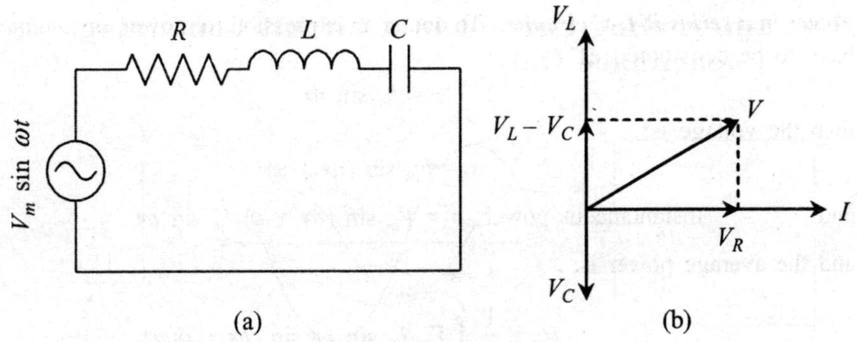

Fig. 5.12

Choosing current I as the reference phasor, the phasor diagram is shown in Fig. 5.12(b). The voltage drops in R, L and C, respectively, are:

$V_R = RI$ in phase with I

$V_L = L\omega I$ in quadrature (current lagging)

$V_C = I/(C\omega)$ in quadrature (current leading)

Since the phasor sum of the three component voltages must be equal to the applied voltage V (effective value),

$$\therefore \qquad V^2 = V_R^2 + (V_L - V_C)^2 \tag{5.25a}$$

$$= I^2[R^2 + (L\omega - 1/C\omega^2)] \tag{5.25b}$$

and

$$I = \frac{V}{[R^2 + (L\omega - 1/C\omega)^2]^{1/2}} \tag{5.26a}$$

$$= \frac{V}{[R^2 + (X_L - X_C)^2]^{1/2}} = \frac{V}{(R^2 + X^2)^{1/2}} = \frac{V}{Z} \tag{5.26b}$$

where X_L is inductive reactance, X_C is capacitive reactance, X is net reactance, Z is impedance. All of these are measured in ohms.

Phase angle between V and I is:

$$\phi = \tan^{-1} X/R \tag{5.27}$$

and power factor is:

$$\cos \phi = R/Z \tag{5.28}$$

If $X_L > X_C$, net reactance X is inductive and the current lags behind the applied voltage.

If $X_L = X_C$, the impedance $Z = R$ and the current $I = V/R$ in phase with V.

If $X_L < X_C$, the net reactance is capacitive and the current leads the applied voltage.

R–L circuits and R–C circuits may be considered as special cases of the general series circuit.

Power in a series R-L-C circuit: To obtain an expression for power, *instantaneous* values have to be considered. If

$$i = I_m \sin \omega t$$

then the voltage is:

$$v = V_m \sin (\omega t + \phi)$$

and Instantaneous power $p = V_m \sin (\omega t + \phi)\, I_m \sin \omega t$

and the average power is:

$$P = \frac{1}{T}\int_0^T V_m I_m \sin \omega t \sin (\omega t + \phi)\, dt$$

$$= \frac{V_m I_m}{2} \cos \phi$$

Since the average of sin ωt cos ωt (= ½ sin $2\omega t$) over a complete period is zero.

or $$P = VI \cos \phi$$

The factor by which the product volt-ampere is to be multiplied in order to obtain average power is called **power factor**. (= cos ϕ for sine waves). Average power is also known as **active power** or **true power**. Reactive power is defined as $VI \sin \phi$. Thus the impedance triangle of Fig. 5.10(b) represents power to a different scale.

EXAMPLE 5.5 A circuit having a resistance of 10 Ω, inductance of 0.12 H and a capacitance of 90 μF in series is connected across a 230 V 50 Hz supply. Calculate: (i) the impedance, (ii) the current, (iii) the voltages across *R*, *L* and *C* and (iv) the phase difference between the current and the supply voltage.

SOLUTION:

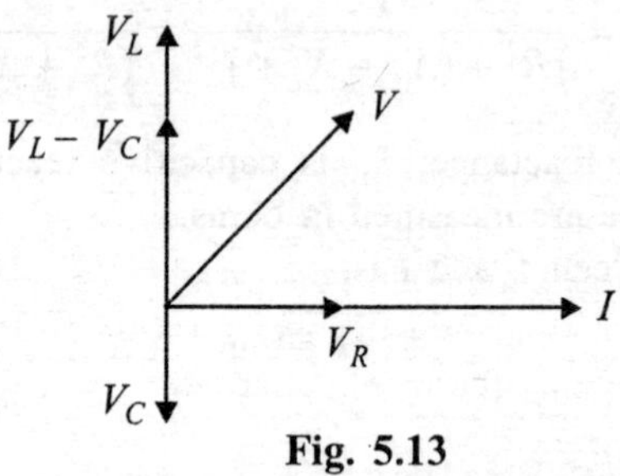

Fig. 5.13

The phasor diagram for the circuit is shown in Fig. 5.13.

(i) Impedance

$$\text{Impedance } Z = \sqrt{\left\{10^2 + \left(2\pi \times 50 \times 0.12 - \frac{10^6}{2\pi \times 50 \times 90}\right)^2\right\}}$$

$$= \sqrt{100 + (37.69 - 35.36)^2} = 10.26\ \Omega \qquad \textbf{Ans.}$$

(ii) *Current*

Current $I = V/Z = 230/10.26 = 22.42$ A **Ans.**

(iii) Voltage

Voltage across R $V_R = 10 \times 22.42 = 224.2$ V

Voltage across L $V_L = X_L I = 37.69 \times 22.42 = 845$ V

Voltage across C $V_C = X_C I = 35.36 \times 22.42 = 792.77$ V **Ans.**

(iv) Phase difference

The phase difference between current and supply voltage is

$$\cos^{-1}\frac{V_R}{V} = \cos^{-1}\frac{224.2}{230} = 12.89° \quad \textbf{Ans.}$$

5.2.5 General Parallel Circuit

A circuit with resistance R, inductance L and capacitance C all connected in parallel across an alternating source of voltage V (effective value) and angular frequency ω is shown in Fig. 5.14(a).

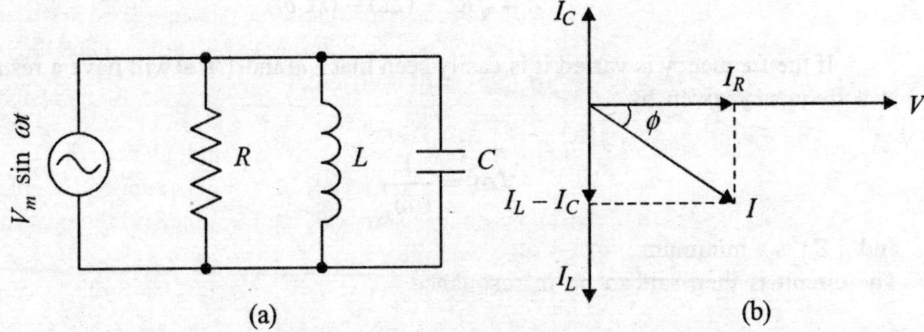

Fig. 5.14

The current in the three elements are:

$I_R = V/R$, in phase with the applied voltage V.

$I_L = V/L\omega$, lagging the applied voltage by 90°.

$I_C = VC\omega$, leading the applied voltage by 90°.

Choosing applied voltage V as the reference, the phasor diagram is as shown in Fig. 5.14(b). It follows that the resultant current is:

$$I^2 = I_R^2 + (I_L - I_C)^2$$

$$= V^2\left[\frac{1}{R^2} + \left(\frac{1}{L\omega} - C\omega\right)^2\right]$$

and

$$I = V\sqrt{\frac{1}{R^2} + \left(\frac{1}{L\omega} - C\omega\right)^2} = VY$$

where $Y = \sqrt{1/R^2 + (1/L\omega - C\omega)^2}$ is the admittance of the circuit measured in siemens.

If $I_C > I_L$, the resultant current I leads the applied voltage.

If $I_C = I_L$, the resultant current is in phase with the applied voltage.

and if $I_C < I_L$, the resultant current lags behind the applied voltage.

The parallel circuits with two elements only are special cases of the general parallel circuit.

5.2.6 Resonance in Alternating Current Circuits

In ac circuits the supply voltage more often operates with a varying frequency. The effect of such frequency variation in a circuit comes under the purview of *resonance*. There are applications in diverse fields such as communication systems (radio, television, telephones), power systems and control systems where the study of resonance becomes very relevant. The effect of frequency variation in an *R-L-C* circuit is introduced in the next section. A measure of its effect known as **Q factor** is also included.

Resonance in a series circuit: The *R-L-C* circuit in Fig. 5.12(a) has input impedance of

$$Z = \sqrt{R^2 + (L\omega - 1/C\omega)^2}$$

If the frequency is varied it is easily seen that $L\omega$ and $1/C\omega$ will have a resultant zero at a frequency given by

$$L\omega_0 = \frac{1}{C\omega_0} \tag{5.29}$$

and $|Z|$ is a minimum.

The circuit is then said to be in resonance.

Note that the capacitive reactance, inversely proportional to ω, is higher at low frequencies, while the inductive reactance, directly proportional to ω, is greater at the higher frequencies. Hence net reactance at frequencies below ω_0 is capacitive and at frequencies above ω_0 the circuit appears inductive.

Frequency response: The voltage transfer function (See Section 12.2) is plotted in Fig. 5.15. The points where the response is $1/\sqrt{2}$, the half power points, are at ω_{c1} and ω_{c2}.

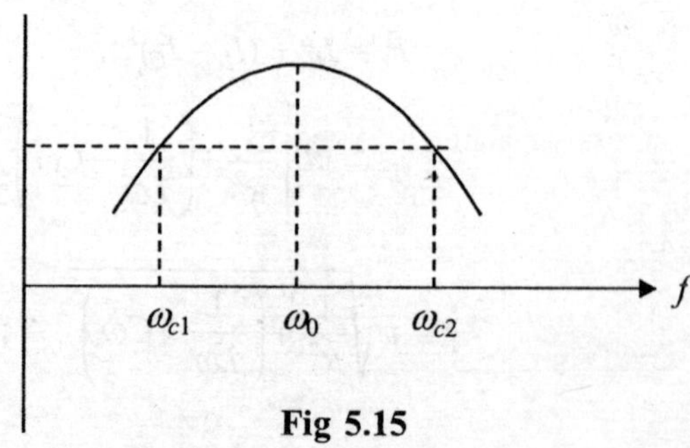

Fig 5.15

Bandwidth β is the difference between these two frequencies

or $$\boxed{\beta = \omega_{c2} - \omega_{c1}} \tag{5.30}$$

Quality factor for the series circuit is defined as:

$$\boxed{Q = \omega_0 L/R}$$

Half power frequencies can be expressed in terms of circuit elements or in terms of ω_0 and Q. The bandwidth can be shown to be

$$\beta = R/L \tag{5.31a}$$

$$= \omega_0/Q \quad \text{(see Problems)} \tag{5.31b}$$

Resonance in parallel circuit: In the parallel circuit of Fig. 5.14(a) resonance occurs if $I_L = I_C$, when the current is in phase with the applied voltage. This occurs when $1/L\omega = C\omega$ which leads to the resonance frequency $\omega_0 = 1/\sqrt{LC}$. It may be noted that the three elements R, L and C whether connected in series or parallel have the same values for resonant frequency.

EXAMPLE 5.6 A circuit having a resistance 5 Ω, inductance 0.4 H and a variable capacitance in series, is connected across a 230 V 50 Hz supply. Calculate: (i) the capacitance to give resonance, (ii) the voltages across the inductance and capacitançe and (iii) the Q factor of the circuit.

SOLUTION:

(i) Capacitance

At resonance $$2\pi f L = \frac{1}{2\pi f C}$$

∴ $$C = \frac{1}{(2\pi \times 50)^2 \times 0.4} = 25.3\ \mu\text{F} \quad \textbf{Ans.}$$

(ii) Voltage

$$I = \frac{V}{R} = \frac{230}{5} = 46\ \text{A}$$

which gives voltage across inductance

$$V_L = 2\pi \times 50 \times 0.4 \times 46 = 5780.53\ \text{V} \quad \textbf{Ans.}$$

And voltage across capacitor is given by

$$V_C = V_L = 5780.53\ \text{V}$$

(iii) Q factor

$$Q = \frac{2\pi \times 50 \times 0.4}{5} = 25.13 \quad \textbf{Ans.}$$

5.3 POLYPHASE CIRCUITS

The ac networks considered so far consisted only of single-phase circuits. In a single-phase network power has been shown to be time varying (see section 5.2.4). However, in a balanced polyphase network power remains constant (see section 5.3.3). It is desirable to have a steady flow of power from source to the load and hence polyphase circuits are used in the diverse fields of generation, transmission, distribution and utilization of electrical power. Three-phase circuits are the industry standard. Two-phase circuits are used in control systems.

5.3.1 Three-phase Systems

A simple three-phase generator, as shown in Fig. 5.16, has three coils which generate voltages of the same amplitude and frequency but with a phase difference of $2\pi/3$ with respect to each other.

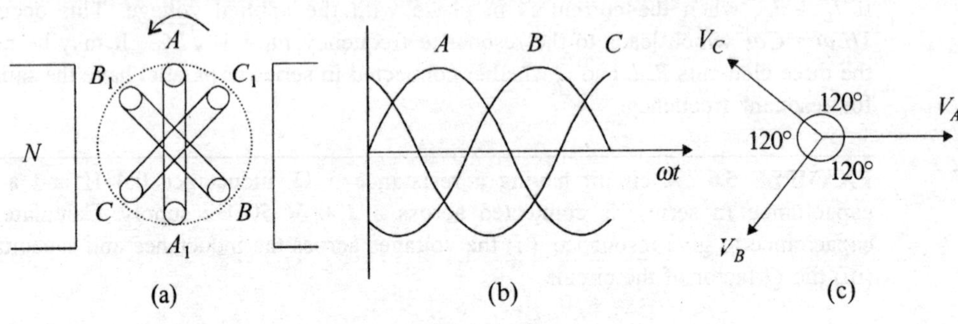

Fig. 5.16

If the phase voltages are:

$$v_A = V_m \sin \omega t$$
$$v_B = V_m \sin (\omega t - 2\pi/3)$$
$$v_C = V_m \sin (\omega t - 4\pi/3) = V_m \sin (\omega t + 2\pi/3)$$

The wave diagram of the source voltages is shown in Fig. 5.16(b); and the phasor diagram is shown in Fig. 5.16(c).

Counter clockwise rotation produces voltages in the coils in the sequence *ABC*. If the direction of rotation is changed the sequence becomes *CBA*. The letters *RYB* (red, yellow, blue) are often used in place of *ABC* and to denote the phase sequence accordingly. Slip rings required for connecting the source voltages to a stationary load are not shown in the Fig. 5.16(a).

5.3.2 Interconnection of Phases

Interconnection of the coils internally facilitates reduction in the number of external connections and consequently in the number of slip rings. In the *star connection* corresponding coil ends (A_1, B_1, C_1) are joined at a common point N which is referred as to the *neutral* and the free ends (A, B, C) are brought out to *lines* R, Y, B of the three-phase system.

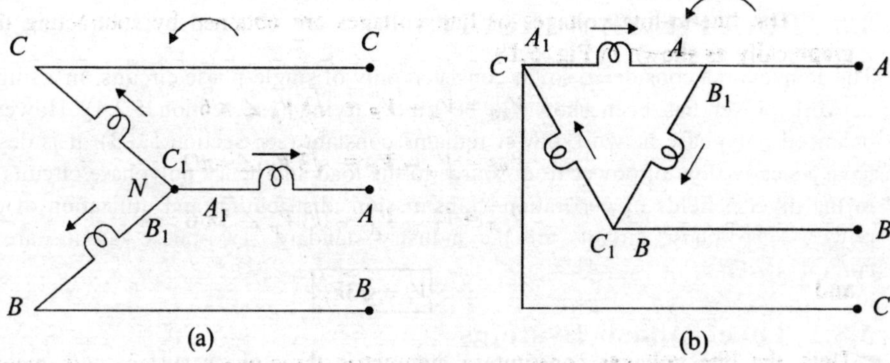

Fig. 5.17

If the neutral point is carried along with the lines, the system becomes three-phase four wire. The coils can be connected also as shown in Fig. 5.17(b) making a *delta-connected* three-phase system. A delta-connected generator can have a neutral point only through the use of a Δ-*Y* transformer.

A balanced three-phase system presupposes a symmetrical three-phase voltage source and equal load impedances; otherwise the system is unbalanced.

Symmetrical three-phase voltage source

1. Star connection. Any one voltage in a three-phase system fixes the angles of all other voltages.

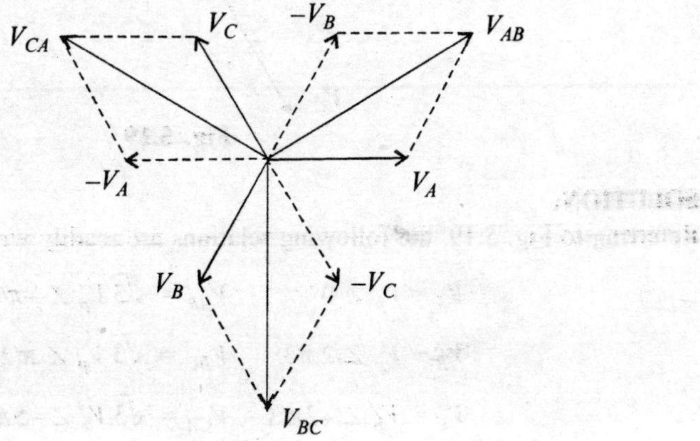

Fig. 5.18

In Fig. 5.18, V_A is chosen as the reference. The line to neutral or phase voltages are given by

$$V_A = V_p \angle 0°, \qquad V_B = V_p \angle -2\pi/3, \qquad V_C = V_p \angle 2\pi/3$$

where $V_p = V_m/\sqrt{2}$

The line-to-line voltages or line voltages are obtained by subtracting the phasors graphically as shown in Fig. 5.18.

$$V_{AB} = V_A - V_B = \sqrt{3}\, V_p \angle \pi/6$$

$$V_{BC} = V_B - V_C = \sqrt{3}\, V_p \angle -\pi/2$$

$$V_{CA} = V_C - V_A = \sqrt{3}\, V_p \angle 5\pi/6$$

and

$$\boxed{V_1 = \sqrt{3} V_P}$$

Thus, the line voltages constitute a symmetric three-phase system with magnitude $\sqrt{3}$ times.

EXAMPLE 5.7 Obtain the phase voltages and line voltages with *CBA* sequence.

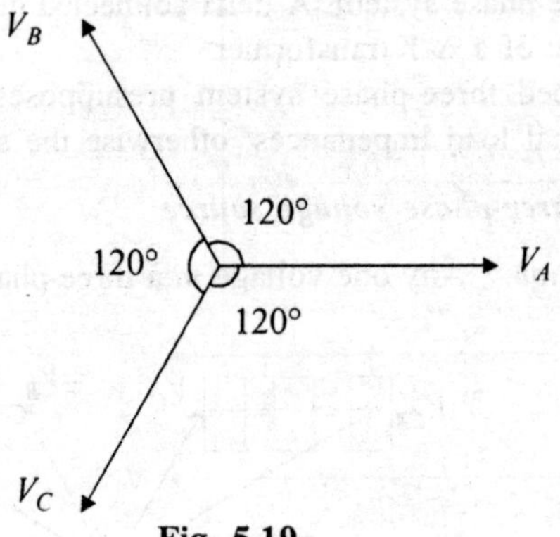

Fig. 5.19

SOLUTION:
Referring to Fig. 5.19, the following relations are readily written:

$$V_A = V_p \angle 0° \qquad V_{AB} = \sqrt{3}\, V_p \angle -\pi/6$$

$$V_B = V_p \angle 2\pi/3 \qquad V_{BC} = \sqrt{3}\, V_p \angle \pi/2$$

$$V_C = V_p \angle -2\pi/3 \qquad V_{CA} = \sqrt{3}\, V_p \angle -5\pi/6 \qquad \textbf{Ans.}$$

---

2. Delta connection. The three single-phase voltage sources may be connected also as shown in Fig. 5.17(b) which is the delta-connected system. The phase voltages are obviously equal to the respective line voltages or

$$\boxed{V_l = V_p}$$

Balanced three-phase currents

1. Star connection. Three identical impedances are connected to the symmetrical three-phase voltage source as shown in Fig. 5.20(a), which give rise to a balanced system. The impedances are taken as:

$$Z_a = Z_b = Z_c = |Z| \angle \phi$$

and the three-phase currents are:

$$I_a = V_p/|Z| \angle (0 - \phi)$$
$$I_b = V_p/|Z| \angle (-2\pi/3 - \phi)$$
$$I_c = V_p/|Z| \angle (2\pi/3 - \phi)$$

The phasor diagram of a balanced three-phase system is shown in Fig. 5.20(b). The line currents are obviously equal to the respective phase currents. The current in the neutral line is:

$$I_n = I_a + I_b + I_c$$

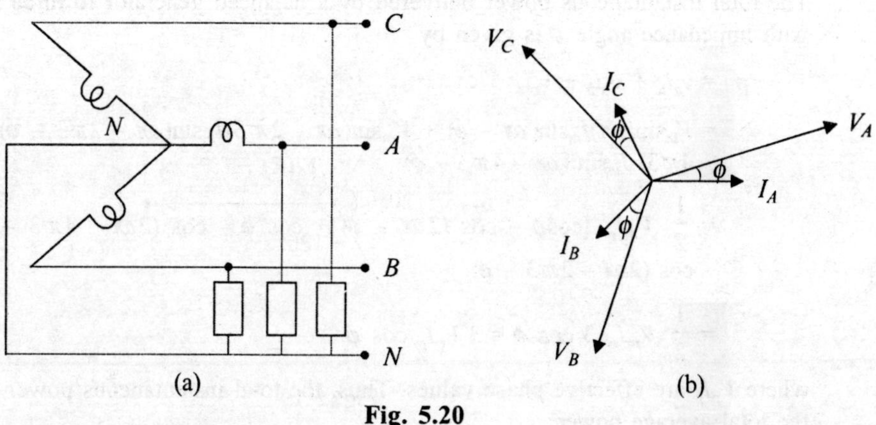

Fig. 5.20

which by adding the three phasors graphically is seen to be always zero and the neutral line may, therefore, be omitted if the system is balanced.

2. Delta connection.

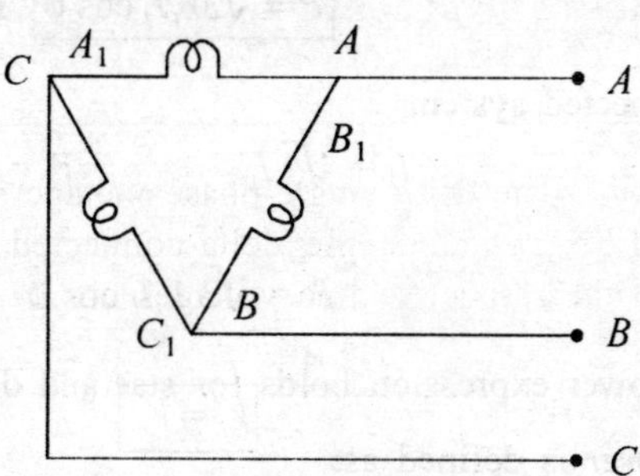

Fig. 5.21

The phase currents are:

$$I_{ab} = V_{ab}/Z_{ab} \quad I_{bc} = V_{bc}/Z_{bc} \quad I_{ca} = V_{ca}/Z_{ca}$$

The line current phasors are, therefore,

$$I_a = I_{ab} - I_{ca} = \sqrt{3}\ I_p \angle -\pi/6$$

$$I_b = I_{bc} - I_{ab} = \sqrt{3}\ I_p \angle \pi/2$$

$$I_c = I_{ca} - I_{bc} = \sqrt{3}\ I_p \angle -5\pi/6$$

Thus, in a balanced delta connection

$$\boxed{I_l = \sqrt{3} I_p}$$

5.3.3 Three-phase Power

The total instantaneous power delivered by a balanced generator to three identical loads with impedance angle ϕ is given by

$$\begin{aligned} p &= v_a i_a + v_b i_b + v_c i_c \\ &= V_m \sin(\omega t) I_m \sin(\omega t - \phi) + V_m \sin(\omega t - 2\pi/3) I_m \sin(\omega t - 2\pi/3 - \phi) + V_m \sin(\omega t - 4\pi/3)\, I_m \sin(\omega t - 4\pi/3 - \phi) \\ &= \frac{1}{2} V_m I_m\ [\cos\phi - \cos(2\omega t - \phi) + \cos\phi - \cos(2\omega t - 4\pi/3 - \phi) + \cos\phi - \cos(2\omega t - 2\pi/3 - \phi)] \\ &= \frac{1}{2} V_m I_m\, 3\cos\phi = 3\, V_p I_p \cos\phi \end{aligned} \tag{5.32}$$

where $V_p I_p$ are effective phase values. Thus, the total instantaneous power is the same as the total average power.

For the star-connected system,

$$I_l = I_p \qquad V_l = \sqrt{3}\, V_p$$

Hence from Eq. (5.32),

$$\boxed{P = \sqrt{3} V_l I_l \cos\phi}$$

For the delta-connected system,

$$I_l = \sqrt{3}\, I_p \qquad V_l = V_p$$

From Eq. (5.32),

$$P = \sqrt{3}\, V_l I_l \cos\phi$$

Thus, the same power expression holds for star and delta balanced systems.

Reactive power is defined as:

$$Q = \sqrt{3}\, V_l I_l \sin\phi$$

Apparent power is defined as:

$$S = \sqrt{3}\, V_l I_l$$

Summarizing

Y connection		*Δ connection*
$V_l = \sqrt{3}\, V_p$		$V_l = V_p$
$I_l = I_p$		$I_l = \sqrt{3}\, I_p$
	$P = \sqrt{3}\, V_l I_l \cos \phi$	

Note: Unbalanced delta-connected and star-connected loads are discussed in Chapter 10 where the complex notation is introduced.

EXAMPLE 5.8 Three non-reactive loads of 5 kW, 3 kW, and 2 kW are connected between the neutral and the *R*, *Y* and *B* phases, respectively of a three-phase four-wire system. The line voltage is 400 V. Find the current in each line conductor and in the neutral wire.

SOLUTION:
The first step is to determine the values of resistances in the star connected load. Since this is a four wire system, the phase voltage is $400/\sqrt{3}$. Hence

$$5 \times 10^3 = \frac{(400/\sqrt{3})^2}{R_1}$$

which gives

$$R_1 = 10.55\ \Omega$$

Similarly,

$$3 \times 10^3 = \frac{(400/\sqrt{3})^2}{R_2}$$

which gives

$$R_2 = 17.77\ \Omega$$

and

$$2 \times 10^3 = \frac{(400/\sqrt{3})^2}{R_3}$$

$$R_3 = 26.66\ \Omega$$

Hence, the phase currents are:

$$I_R = V/R_1 = \frac{400}{\sqrt{3} \times 10.55} = 21.89 \text{ A} \qquad \textbf{Ans.}$$

$$I_Y = V/R_2 = \frac{400}{\sqrt{3} \times 17.77} = 12.99 \text{ A} \qquad \textbf{Ans.}$$

$$I_B = V/R_3 = \frac{400}{\sqrt{3} \times 26.66} = 8.66 \text{ A} \qquad \textbf{Ans.}$$

$$I_N = 21.89 - \frac{1}{2} \times 12.99 - \frac{1}{2} \times 8.66 = 11.07 \text{ A} \qquad \textbf{Ans.}$$

Measurement of power

1. Star connected balanced load with the neutral point accessible. A wattmeter W is

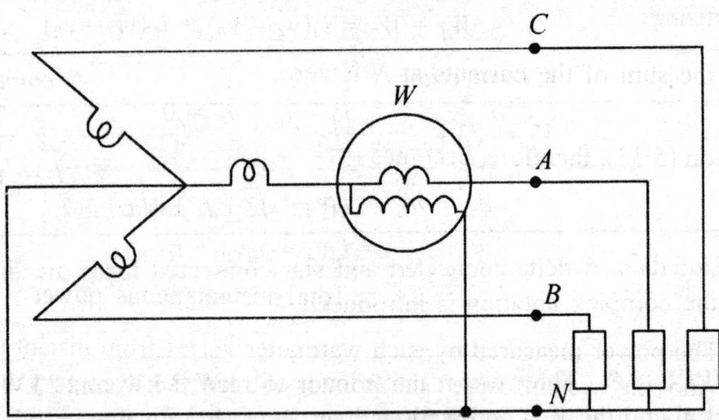

Fig. 5.22

connected with the current coil in series with one line, and the voltage coil between that line and the neutral point as shown in Fig. 5.22. The reading of the wattmeter gives the power per phase and, therefore, the total active power = 3 × wattmeter reading.

2. Two-wattmeter method—Balanced or unbalanced, star or delta connected system. Two wattmeters are connected as shown in Fig. 5.23 with current coils in series with any two lines, and the voltage coil connected between these lines and the third line.

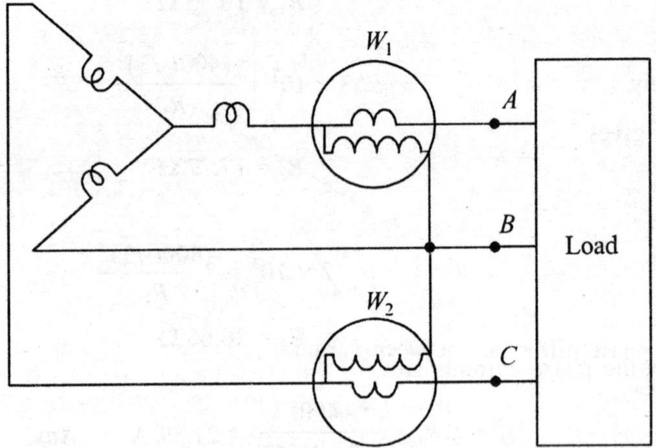

Fig. 5.23

The instantaneous current through the current coil of wattmeter W_1 is i_A and the instantaneous voltage across the potential coil is $(v_A - v_B)$. The instantaneous power measured by the wattmeter is:

$$W_1 = i_A\,(v_A - v_B)$$

Similarly, the instantaneous power measured by the second wattmeter is:

$$W_2 = i_C\,(v_C - v_B)$$

Hence the total power measured by the two wattmeters is given as:

$$W_1 + W_2 = i_A(v_A - v_B) + i_C(v_C - v_B) \tag{5.33}$$

Since the sum of the currents at N is zero,

$$i_A + i_B + i_C = 0$$

Equation (5.33), therefore, becomes

$$\begin{aligned} W_1 + W_2 &= i_A v_A + i_C v_C - v_B(i_A + i_C) \\ &= i_A v_A + i_B v_B + i_C v_C \\ &= \text{Total instantaneous power} \end{aligned}$$

The power measured by each wattmeter varies from instant to instant but the inertia of the moving system causes the pointer to read the average value of power. Hence the sum of the two-wattmeter readings gives the average value of the total power absorbed by the three phases.

Note that the method is valid for star or delta connection with balanced or unbalanced load.

Power-factor measurement. Consider three identical impedances $Z = |Z| \angle \phi$ star-connected to the system of Fig. 5.23. The corresponding vector diagram is shown in Fig. 5.24.

The r.m.s. current through the current coil of wattmeter W_1 is I_A. The potential difference across the voltage coil is V_{AB}.

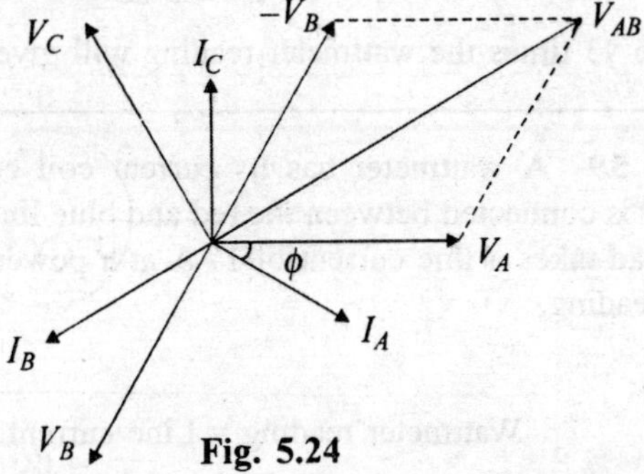

Fig. 5.24

The phase difference between I_A and V_{AB} is $(30° + \phi)$ and hence the wattmeter reads

$$W_1 = I_A V_{AB} \cos\,(30 + \phi) = I_l V_l \cos\,(\phi + 30°) \tag{5.34}$$

Similarly,
$$W_2 = I_l V_l \cos\,(\phi - 30°) \tag{5.35}$$

Now, Total power

$$\begin{aligned} P &= W_1 + W_2 \\ &= I_l V_l \cos\,(\phi + 30°) + I_l V_l \cos\,(\phi - 30°) \\ &= I_l V_l\, 2 \cos\phi \cos 30 = \sqrt{3}\; I_l V_l \cos\phi \end{aligned}$$

and $$\frac{W_1 - W_2}{W_1 + W_2} = \frac{\cos(\phi + 30) - \cos(\phi - 30)}{\cos(\phi + 30) + \cos(\phi - 30)} = \frac{2\sin\phi \sin 30}{2\cos\phi \cos 30} = \frac{\tan\phi}{\sqrt{3}}$$

or $$\tan\phi = \sqrt{3}\,\frac{W_1 - W_2}{W_1 + W_2} \tag{5.36}$$

The power factor cos ϕ is readily obtained from Eq. (5.36).

Measurement of reactive power kVAR. The method of obtaining the reactive power in a three-phase circuit is shown in Fig. 5.25.

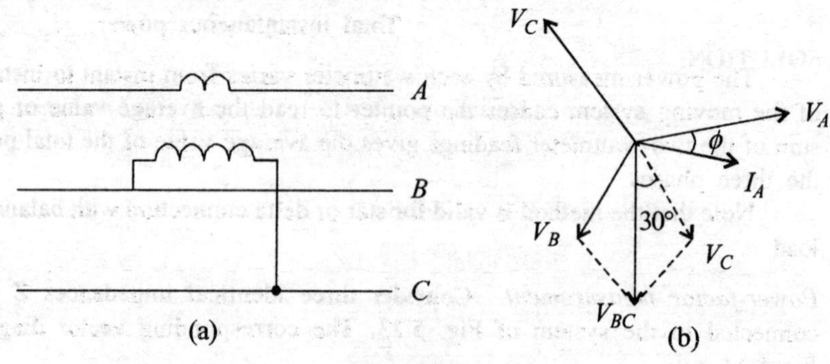

Fig. 5.25

$$\text{Wattmeter reading} = I_A V_{BC} \cos(90 - \phi) = I_l V_l \sin\phi$$

Hence √3 times the wattmeter reading will give the total reactive power.

EXAMPLE 5.9 A wattmeter has its current coil connected in the yellow line and its voltage coil is connected between the red and blue lines. The line voltage is 400 V and the balanced load takes a line current of 17 A at a power factor of 0.8 lagging. Calculate the wattmeter reading.

SOLUTION:

$$\text{Wattmeter reading} = \text{Line current} \times \text{Line voltage} \times \sin\phi$$

$$= 17 \times 400 \times 0.6 = 4080 \text{ W} \qquad \textbf{Ans.}$$

5.3.4 Two-phase System

A balanced two-phase generator has two voltage sources with voltages of the same amplitude and frequency but 90° out of phase.

EXAMPLE 5.10 A two-phase generator is represented in Fig. 5.26. If the loads are identical, say $Z \angle \theta$. Determine the currents and voltages.

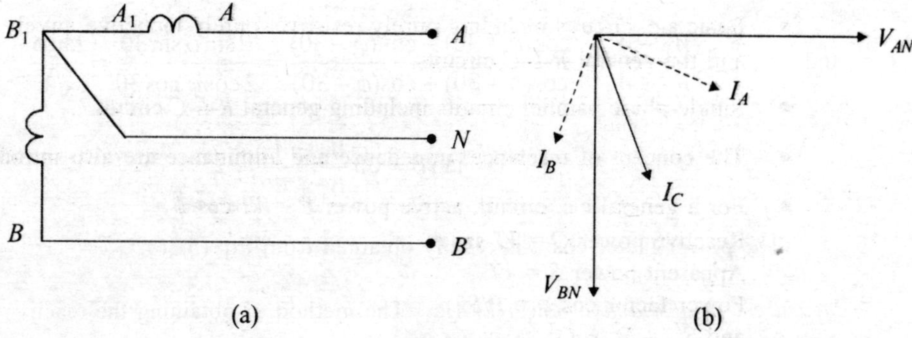

Fig. 5.26

SOLUTION:

$$v_A = V_m \sin \omega t \qquad i_A = I_m \sin(\omega t - \phi)$$

$$v_B = V_m \sin (\omega t - \pi/2) \qquad i_B = I_m \sin (\omega t - \pi/2 - \phi)$$

From the vector diagram of Fig. 5.26(b),

$$V_{AB} = V_{AN} - V_{BN} = \sqrt{2}V \angle 45°$$

$$I_C = I_A + I_B = \sqrt{2}I \angle -45° - \phi$$

Power in two-phase circuits: The instantaneous power delivered by the sources are:

$$p_a = V_m \sin \omega t I_m \sin (\omega t - \phi)$$

$$p_b = [-V_m \cos \omega t]\ [-I_m \cos(\omega t - \phi)]$$

The total instantaneous power is:

$$p = p_a + p_b$$

$$= \frac{V_m I_m}{2} [\{\cos\phi - \cos(2\omega t - \phi)\} + \cos\phi + \cos(2\omega t - \phi)\}]$$

$$= \frac{V_m I_m}{2} 2 \cos \phi = 2VI \cos \phi$$

Note that a phase voltage of V and a line voltage of $\sqrt{2}\,V$ are available and that the power flow is constant.

SUMMARY

The chapter covers

- Fundamentals of alternating quantities which include certain important definitions like maximum value, average value and r.m.s. value. The usefulness of phasor representation of alternating quantities has been brought out.

- Basic a.c. circuits including purely resistive, purely inductive, purely capacitive and the general *R-L-C* circuits.
- Single-phase parallel circuits including general *R-L-C* circuit.
- The concept of reactance, impedance and admittance are also introduced.
- For a general a.c. circuit, active power $P = VI \cos \phi$
  Reactive power $Q = VI \sin \phi$
  Apparent power $S = VI$
  Power factor $\cos \phi = P/S$
  and $S^2 = P^2 + Q^2$
- Resonance in series circuits, in parallel circuits and their significance. Concept of resonant frequency, bandwidth and quality factor *Q*.
- Finally three-phase circuits are introduced.

Complex notation, a very powerful technique in a.c. circuits will be applied to network problems in Chapter 10.

Solved Problems

P5.1 An alternating voltage has the equation $e = 280 \sin 314t$. Determine (i) the r.m.s. value, (ii) frequency and (iii) the instantaneous voltage when $t = 4$ ms.

SOLUTION:

(i) r.m.s. value

$$E_m = 280 \text{ V}$$

Hence r.m.s. value is:

$$V = 280/\sqrt{2} = 198 \text{ V} \qquad \textbf{Ans.}$$

(ii) Frequency

$$f = \frac{314}{2\pi} = 50 \text{ Hz} \qquad \textbf{Ans.}$$

(iii) Instantaneous voltage

When $t = 4$ ms,

$$e = 280 \sin (314 \times 4 \times 10^{-3} \times 180/\pi)^\circ$$

$$= 266.24 \text{ V} \qquad \textbf{Ans.}$$

P5.2 An alternating voltage has a periodic time $2T$. The voltage for a time $1/3\ T$ is 50 V, for $1/6\ T$, it is 20 V and zero for ½ T. Calculate the r.m.s. value and average value of this voltage.

SOLUTION:

$$I_{av} = \frac{50(T/3) + 20(T/6) + 0(T/2)}{T}$$

$$= 20 \text{ A} \quad \textbf{Ans.}$$

$$I^2 = \frac{50^2(T/3) + 20^2(T/6)}{T}$$

$$= 900$$

$$I = 30 \text{ A} \quad \textbf{Ans.}$$

P5.3 A capacitor 15.5 μF takes a current of 1.15 A when an alternating voltage of 230 V is applied to it. Calculate (i) the frequency of the applied voltage (ii) the resistance to be connected in series with the capacitor to reduce the current to 0.75 A and (iii) the phase angle of the resulting circuit.

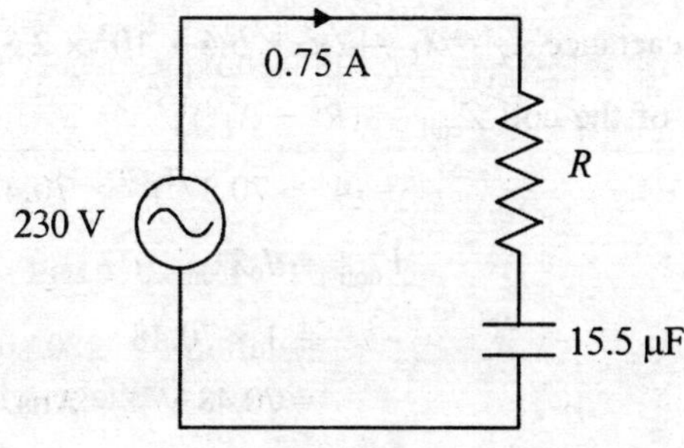

Fig. P5.1

SOLUTION:

(i) Frequency

$$X_C = \frac{230}{1.15} = 200\ \Omega$$

$$= \frac{1}{2\pi f C}$$

$$f = \frac{1}{2\pi \times 15.5 \times 10^{-6} \times 200} = 51.34 \text{ Hz} \quad \textbf{Ans.}$$

(ii) Resistance

$$Z = V/I = 230/0.75 = 306.66\ \Omega$$

$$= \sqrt{R^2 + X_C^2}$$

from which

$$R = 232.46\ \Omega \quad \textbf{Ans.}$$

(iii) Phase angle

$$\phi = \cos^{-1}(R/Z)$$

$$= \cos^{-1}(232.46/306.66)$$

$$= 40.71° \text{ leading} \quad \textbf{Ans.}$$

P5.4 A coil of resistance 4 Ω and inductance of 2 mH is connected in series with a capacitor of 0.4 μF. The circuit is connected to a 4 V variable frequency supply. Calculate: (i) the frequency at which resonance occurs, (ii) the voltage across the coil, (iii) the voltage across the capacitor and (iv) the Q factor of the circuit.

SOLUTION:

(i) Frequency

$$f_0 = \frac{1}{2\pi\sqrt{LC}} = \frac{1}{2\pi(2\times10^{-3}\times0.4\times10^{-6})^{1/2}} = 5.6\text{ kHz} \qquad \textbf{Ans.}$$

(ii) Voltage across coil

Current at resonance is given by

$$I_0 = V/R = 4/4 = 1\text{ A}$$

Inductive reactance $X_L = X_C = 2\pi \times 5.6 \times 10^3 \times 2 \times 10^{-3} = 70.37\ \Omega$

Impedance of the coil $Z_{coil} = (R^2 + X_L^2)^{1/2}$

$$= (4^2 + 70.37^2)^{1/2} = 70.48\ \Omega$$

$\therefore$

$$V_{coil} = I_0 Z_{coil}$$

$$= 1 \times 70.48$$

$$= 70.48\text{ V} \qquad \textbf{Ans.}$$

(iii) Voltage across capacitor

$$V_C = I_0 X_C$$

$$= 1 \times 70.37 = 70.37\text{ V} \qquad \textbf{Ans.}$$

(iv) Q factor

$$Q = \frac{\omega L}{R} = \frac{70.37}{4} = 17.59 \qquad \textbf{Ans.}$$

P5.5 A delta-connected load is shown in Fig. P5.2. Supply voltage is 400 V at 50 Hz. Calculate the phase currents.

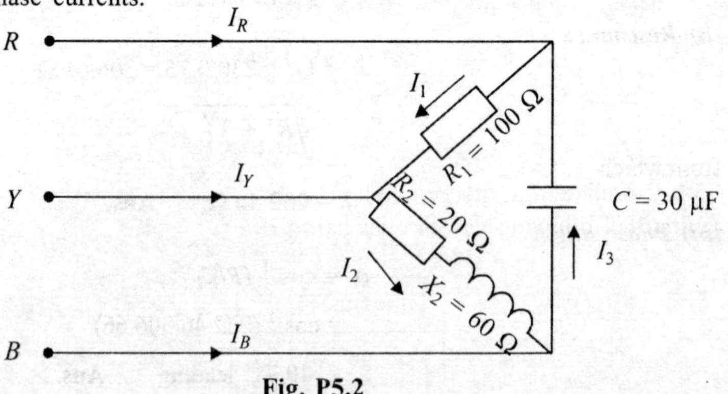

Fig. P5.2

SOLUTION:

Let i_1, i_2, i_3 be the phase currents in loads *RY*, *YB* and *BR*, respectively, assuming phase sequence *RYB*.

$$i_1 = \frac{400}{100} = 4 \text{ A in phase with } V_{RY}$$

$$i_2 = \frac{400}{(20^2 + 60^2)^{1/2}} = 6.3 \text{ A lags } V_{YB} \text{ by } \tan^{-1}\phi_2 \quad \textbf{Ans.}$$

where
$$\phi_2 = \tan^{-1}\frac{60}{20} = 71.56$$

Then
$$i_3 = 2\pi \times 50 \times 30 \times 10^{-6} \times 400$$
$$= 3.77 \text{ A leading } V_{BR} \text{ by } 90° \quad \textbf{Ans.}$$

P5.6 A star-connected load consists of three coils of resistance 2 Ω and reactance 3 Ω. The load is supplied at a line voltage of 400 V, 50 Hz. The total power in the load is measured by the two-wattmeter method. Calculate the separate readings of the wattmeters.

SOLUTION:

Check total power.

The line current I_l in the star-connected load $I_P = \dfrac{400/\sqrt{3}}{\sqrt{2^2 + 3^2}} = 64.05$ A

and
$$\phi = \tan^{-1}(3/2) = 56.31$$

Now, wattmeter reading $P_1 = V_l I_l \cos(30° + \phi)$

$$= 400 \times 64.05 \times \cos(30 + 56.31)$$

$$= 1.65 \text{ kW} \quad \textbf{Ans.}$$

and $P_2 = V_l I_l \cos(30° - \phi)$

$$= 400 \times 64.05 \times \cos(30 - 56.31) = 22.97 \text{ kW} \quad \textbf{Ans.}$$

Thus, Total power $= \sqrt{3}\, V_l I_l \cos\phi$

$= 24.63$ kW which is the sum of P_1 and P_2. **Ans.**

Exercises

E5.1 Calculate from first principles the r.m.s. value and form factor of an alternating current having the following values over half a cycle:

Time (ms)	0	1	2	3	4
Voltage (V)	0	200	200	200	0

[*Ans:* 163 V, Form factor = 1.09]

E5.2 A 100 Ω resistor and a 30 mH inductor and a 50 μF capacitor are connected in series across a supply voltage of 500 cos (500t + 30°) volts. Calculate the steady state current in the circuit.
[*Ans:* 13.13 A]

E5.3 A coil has resistance of 500 Ω and inductance 300 μH. Find the capacitance which when connected in parallel with the coil will produce resonance with a supply frequency of 1 MHz. If a second capacitor of 25 pF is connected in parallel with the first, find the frequency at which resonance will occur.
[*Ans:* Capacitance required for resonance at 1 MHz = 84.4 pF, with the second capacitor added in parallel with the first the resonant frequency = 0.88 MHz.]

E5.4 Derive the relationship between the line and phase currents for a balanced three-phase delta connected load. Three coils are connected in delta to a three-phase three- wire 400 V 50 Hz supply and take a line current of 6 A at 0.7 pf lagging. Calculate the resistance and inductance of the coils.
[*Ans:* R = 80.8 Ω, L = 0.26 H]

E5.5 In the balanced three-phase circuit shown in Fig. E5.1 the current coil of the wattmeter is connected in line A and the potential coil is connected across lines B and C. Show that the wattmeter reading multiplied by $\sqrt{3}$ equals the total reactive power taken by the load. Phase sequence is ABC.

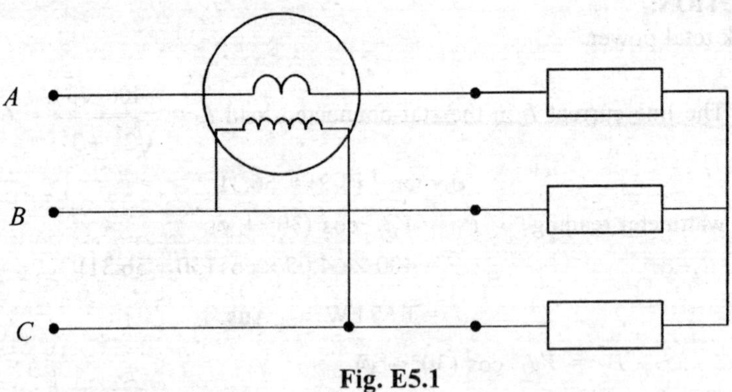

Fig. E5.1

[**Hint:** Use phasor diagram for the three-phase circuit.]

E5.6 The voltage from A to N in a balanced three phase circuit is 231 ∠–30°. If the phase sequence is positive. What is the value of V_{BC}.?
[**Hint:** Use vector diagram.]

[*Ans:* 400 ∠–120°]

E5.7 Two wattmeters read 240 W and 100 W when connected to measure the active power taken by a balanced three-phase load. Calculate the active power and the power factor of the load.

[*Ans:* 340 W, 0.814]

Part 2

Linear Time-invariant (LTI) Systems

Techniques of Circuit Analysis

6

CHAPTER CONTENTS

The object of circuit analysis is to determine the currents through and voltages across each branch of network, with given energy sources. If there are b branches in the network there will be $2b$ unknown variables. These will require $2b$ independent equations to be obtained from the basic laws governing electric circuits. However there are various techniques available for reducing the number of independent variables in a network.

If the network is a very simple one, it may be reduced by combining elements in series or parallel as in Chapter 5. More complicated network may be reduced by using reduction theorems such as star-delta transformation or Thevenin's theorem and its dual Norton's theorem.

When reduction methods fail, *network equations* are formed, which can then be solved for all the currents and voltages. Three methods are available; they are:

- Branch current method
- Mesh current method
- Node voltage method

6.1 BRANCH CURRENT METHOD

In order to reduce the number of independent current variables, and hence the number of equations to be solved, the current distribution itself is done making use of KCL. The method is illustrated in Example 6.1.

EXAMPLE 6.1 Solve the circuit shown in Fig. 6.1.

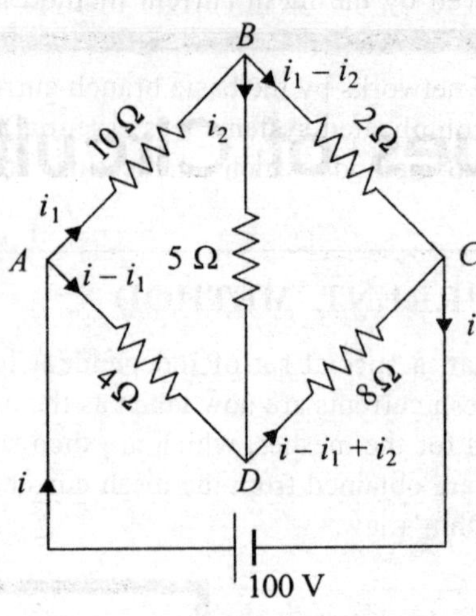

Fig. 6.1

SOLUTION:

In the closed circuit *ABDA*,

$$10i_1 + 5i_2 - 4(i - i_1) = 0$$

In the circuit *BCDB*,

$$2(i_1 - i_2) - 8(i - i_1 + i_2) - 5i_2 = 0$$

In the circuit *ADCA*,

$$4(i - i_1) + 8(i - i_1 + i_2) = 100$$

Rearranging the equations,

$$4i - 14i_1 - 5i_2 = 0$$
$$8i - 10i_1 + 15i_2 = 0$$
$$12i - 12i_1 + 8i_2 = 100$$

Solving these simultaneous equations (by Cramer's method),

$$i = 19.40 \text{ A} \qquad i_1 = 7.44 \text{ A} \qquad i_2 = -5.35 \text{ A}$$

The currents in the other branches can be obtained in terms of these three currents. Thus, all the currents are given by

$$i_{AB} = i_1 = 7.44 \text{ A} \qquad i_{BD} = i_2 = -5.35 \text{ A} \qquad i_{BC} = i_1 - i_2 = 12.79 \text{ A}$$

$$i_{AD} = i - i_1 = 11.96 \text{ A} \qquad i_{DC} = i - i_1 + i_2 = 6.61 \text{ A}$$

The voltages across each branch can be found, knowing their resistances (using Ohm's law). Thus,

$$v_{AB} = 74.4 \text{ V} \qquad v_{BD} = -26.75 \text{ V} \qquad v_{BC} = 25.58 \text{ V}$$

$$v_{AD} = 47.84 \text{ V} \qquad v_{DC} = 52.88 \text{ V}$$

This completes the solution. A check may be made by noting that $v_{AB} + v_{BC} \approx 100$ V. The same problem is solved by the mesh current method in Section 6.2.

The solution of networks by the basic branch current method is direct but laborious. When dealing with complicated systems it is essential to reduce the number of variables to the minimum. Two methods which enable this, are the mesh current and the node voltage techniques

6.2 MESH CURRENT METHOD

If a network is planar, a special set of independent loops, namely the meshes may be considered. These mesh currents are now taken as the unknown variables there after KVL equations are formed for the meshes, which are then solved to obtain the mesh currents. The branch currents are obtained from the mesh currents; and finally the branch voltages are obtained using Ohm's law.

EXAMPLE 6.2 Solve the circuit shown in Fig. 6.2 by the mesh analysis.

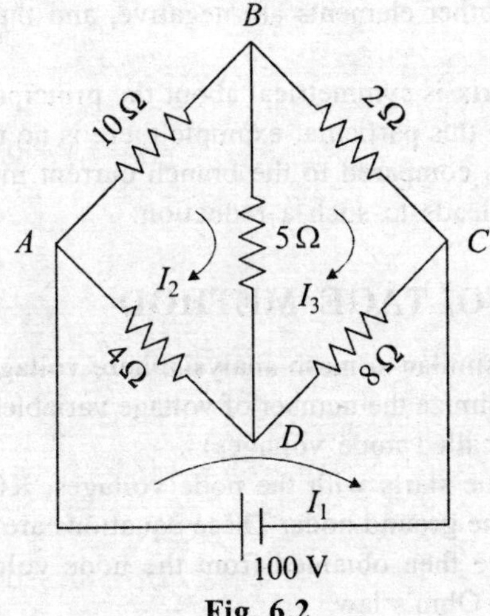

Fig. 6.2

SOLUTION:
The mesh equations are:

$$4(I_1 - I_2) + 8(I_1 - I_3) = 100$$
$$10I_2 + 5(I_2 - I_3) + 4(I_2 - I_1) = 0$$
$$2I_3 + 8(I_3 - I_1) + 5(I_3 - I_2) = 0$$

Rearranging,

$$12I_1 - 4I_2 - 8I_3 = 100$$
$$-4I_1 + 19I_2 - 5I_3 = 0$$
$$-8I_1 - 5I_2 + 15I_3 = 0$$

The three mesh currents are:

$$I_1 = 19.4 \text{ A} \qquad I_2 = 7.44 \text{ A} \qquad I_3 = 12.79 \text{ A}$$

The branch currents can be obtained as:

$$i_{AB} = I_2 = 7.44 \text{ A} \qquad i_{BD} = I_2 - I_3 = -5.35 \text{ A} \qquad i_{BC} = I_3 = 12.79 \text{ A}$$

$$i_{AD} = I_1 - I_2 = 11.96 \text{ A} \qquad i_{DC} = I_1 - I_3 = 6.61 \text{ A} \qquad \textbf{Ans.}$$

Note that the branch currents obtained are the same as those in the branch current method of the previous section hence the branch voltages are also the same a before.

In Example 6.2 the coefficient matrix of the equation can be written down by inspection of the three meshes in the circuit.

$$\begin{bmatrix} 12 & -4 & -8 \\ -4 & 19 & -5 \\ -8 & -5 & 15 \end{bmatrix}$$

Note: 1. All elements on the principal diagonal are positive and consist of the total resistances through which the mesh currents flow.
2. All the other elements are negative, and they are resistances common to two meshes.
3. The matrix is symmetrical about the principal diagonal.
4. While in this particular example there is no reduction in the number of current variables compared to the branch current method, generally the mesh current method leads to such a reduction.

6.3 NODE VOLTAGE METHOD

Nodal analysis is similar to mesh analysis, here voltages are used as unknown variables and in order to minimize the number of voltage variables use is made of n-1 node to datum voltages (usually called node voltages).

This technique starts with the node voltages, KCL equations are then written for the nodes except the ground node. These equations are solved for the node voltages. The branch voltages are then obtained from the node voltages. Finally the branch currents are obtained using Ohm's law.

EXAMPLE 6.3 Solve the circuit shown in Fig. 6.3 by nodal analysis.

SOLUTION:

C is selected as the ground node, leaving three independent nodes, but the voltage of node *A* is fixed by the voltage source, leaving only two independent nodes *B* and *D*.

Applying KCL at nodes *B* and *D*,

$$(V_B - 100)\frac{1}{10} + (V_B - V_D)\frac{1}{5} + V_B\frac{1}{2} = 0$$

$$(V_D - 100)\frac{1}{4} + (V_D - V_B)\frac{1}{5} + V_D\frac{1}{8} = 0$$

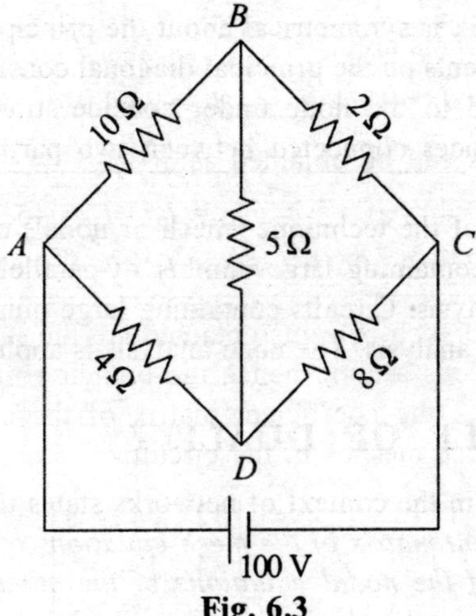

Fig. 6.3

Rearranging, we get

$$V_B\left(\frac{1}{10}+\frac{1}{5}+\frac{1}{2}\right)-V_D\frac{1}{5}=10 \tag{6.1a}$$

$$V_B\left(-\frac{1}{5}\right)+V_D\left(\frac{1}{4}+\frac{1}{5}+\frac{1}{8}\right)=25 \tag{6.1b}$$

Solving these simultaneous equations,

$$V_B = 25.59 \text{ V}, \qquad V_D = 52.38 \text{ V}$$

The branch voltages are found as follows:

$V_{AB} = V_A - V_B = 100 - 25.59 = 74.41$ V $\qquad V_{BD} = V_B - V_D = -26.79$ V

$V_{BC} = V_B - V_C = 25.59$ V $\qquad V_{AD} = V_A - V_D = 47.62$ V

$V_{DC} = V_D - V_C = 52.38$ V **Ans.**

The branch currents can be written down making use of Ohm's law. The values of the branch voltages and branch currents are seen to agree with those obtained from the two previous methods.

The coefficient matrix of current equations (6.1a and 6.1b) may be obtained directly from the given circuit

$$\begin{bmatrix} \frac{1}{10}+\frac{1}{5}+\frac{1}{2} & -\frac{1}{5} \\ -\frac{1}{5} & \frac{1}{4}+\frac{1}{5}+\frac{1}{8} \end{bmatrix}$$

Notes: 1. All the elements on the principal diagonals are positive.
2. All the other elements are negative.

3. The matrix is symmetrical about the principal diagonal.
4. The elements on the principal diagonal consists of the sum of the conductances connected to the node under consideration, the other elements consists of conductances connected between two particular nodes.

Note: The choice of the technique (mesh or nodal) depends on which results in *fewer* equations. Circuits containing large number of parallel branches and few nodes are best solved by nodal analysis. Circuits containing large number of nodes and few meshes are best solved by mesh analysis. The node analysis is applicable to non-planar networks also.

6.4 PRINCIPLE OF DUALITY

Principle of duality in the context of networks states that *two networks are dual to each other if the coefficient matrix of the mesh equations of one network is identical with the coefficient matrix of the nodal equations of the other.*

The elements in the dual networks obey the same law. The behaviour of the elements in one of the dual networks may be obtained from the other by interchanging voltages and currents; resistance and conductance form a pair of dual elements. If the resistance in a network has a value of 10 Ω, its volt-ampere equation is:

$$v = 10i$$

The element in the dual network has a conductance of 10 S and its volt-ampere equation is:

$$i = 10v$$

The power dissipated in a resistance of 10 Ω by a current of 2 A flowing through it is given as:

$$i^2R = 40 \text{ W}$$

In the dual network the conductance is 10 S and the voltage is 2 V, thus the power dissipated is

$$v^2G = 40 \text{ W}, \qquad \text{which is same as in the resistance.}$$

Again the volt-ampere equation for an inductance and a capacitance are:

$$v = L\frac{di}{dt} \qquad \text{and} \qquad i = C\frac{dv}{dt}$$

From the definition of duality the capacitance and inductance are dual quantities, thus the dual of a 3 F capacitance is a 3 H inductance.

Consider two networks, one a general series circuit and the other a general parallel circuit.

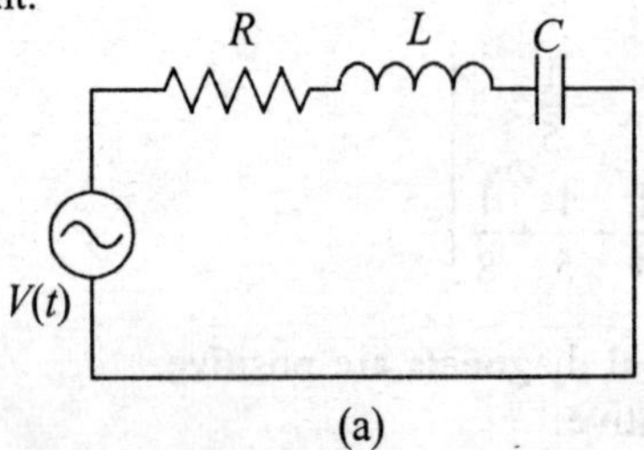

(a)

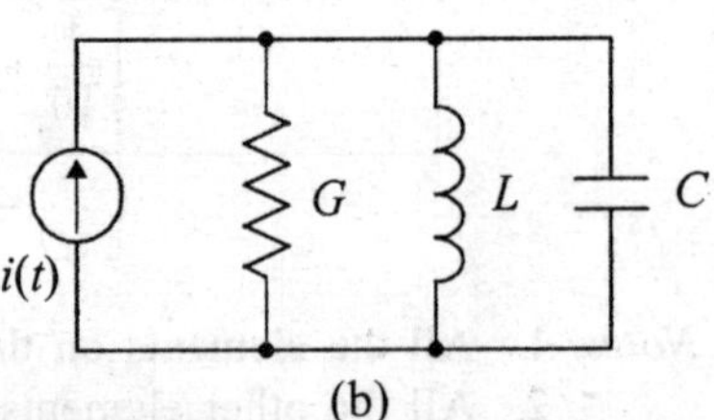

(b)

Fig. 6.4

The respective network equations are:

$$L\frac{di}{dt}+Ri+\frac{1}{C}\int i\,dt=v(t) \tag{6.2a}$$

$$C\frac{dv}{dt}+Gv+\frac{1}{L}\int v\,dt=i(t) \tag{6.2b}$$

It is seen that the following are analogous quantities:

Ri	and	Gv
$L\dfrac{di}{dt}$	and	$C\dfrac{dv}{dt}$
$\dfrac{1}{C}\int i\,dt$	and	$\dfrac{1}{L}\int v\,dt$

These lead to the pairs of dual quantities:

R	and	G
L	and	C
i	and	v
Loop	and	Node

A simple graphical construction may be followed for determining the dual of a given network.

EXAMPLE 6.4 Construct the dual of the given network in Fig. 6.5.

Procedure: Place a dot within each loop of the network. These dots correspond to independent nodes. Place a node external to the network; this is the datum node.

Connect the internal dots in the two loops, traversing only one element at a time. For each element traversed in the original network connect the dual element in the dual network.

Join the internal dots to the external dots crossing all external branches. The dual obtained is shown in Fig. 6.5(b).

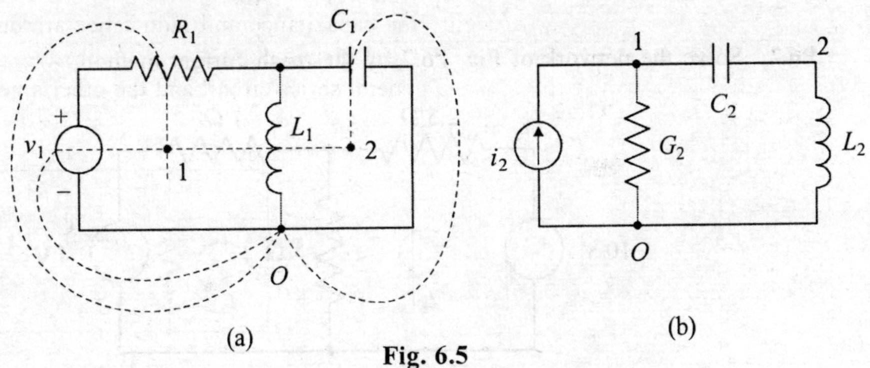

Fig. 6.5

SUMMARY

- Both mesh current and node voltage analysis are powerful tools in network analysis. If in a given circuit the number of meshes is less than the number of free nodes, mesh analysis may be used; if on the other hand the number of free nodes is less, node analysis is used. The node analysis is possible even if the circuit is non-planar. Topological basis for the methods presented here will be discussed in Chapter 7.
- The application of these techniques to a.c. circuit will be introduced in Chapter 10.
- The principle of duality can be used with great advantage in circuit analysis.

Solved Problems

P6.1 Use branch current in the network shown in Fig. P6.1 to find the current supplied by the 10 V source.

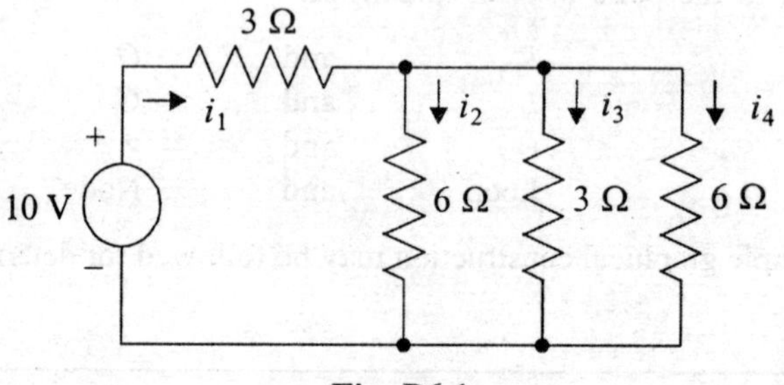

Fig. P6.1

SOLUTION:

$$i_2 \times 6 = i_3 \times 3$$
$$i_2 \times 6 = i_4 \times 6$$
$$10 = i_1 \times 3 + i_2 \times 6$$
$$i_1 = i_2 + i_3 + i_4$$

Solving for i_1, we get

$$i_1 = 20/9 \text{ A} \qquad \textbf{Ans.}$$

P6.2 Solve the network of Fig. P6.2 by the mesh current method.

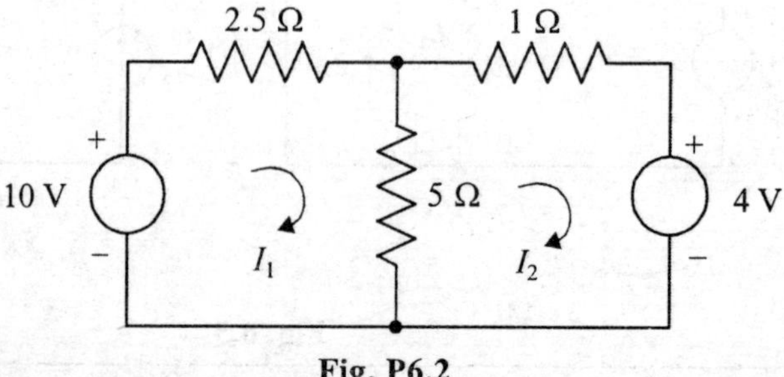

Fig. P6.2

SOLUTION:

Mesh currents are shown in the circuit diagram. Applying KVL to the two loops,

$$10 - 2.5I_1 - 5(I_1 - I_2) = 0$$

and
$$4 + 5(I_2 - I_1) + 1 \times I_2 = 0$$

Rearranging,

$$7.5I_1 - 5I_2 = 10$$

$$-5I_1 + 6I_2 = -4$$

Solving these two equations, we have

$$I_1 = 2\text{ A} \qquad I_2 = 1\text{ A} \qquad \textbf{Ans.}$$

The current in the 5 Ω resistor is obtained as:

$$I_3 = I_1 - I_2 = 1\text{ A} \qquad \textbf{Ans.}$$

P6.3 Use the mesh current method to determine the power dissipated in the 8 Ω resistor in the circuit shown in Fig. P6.3.

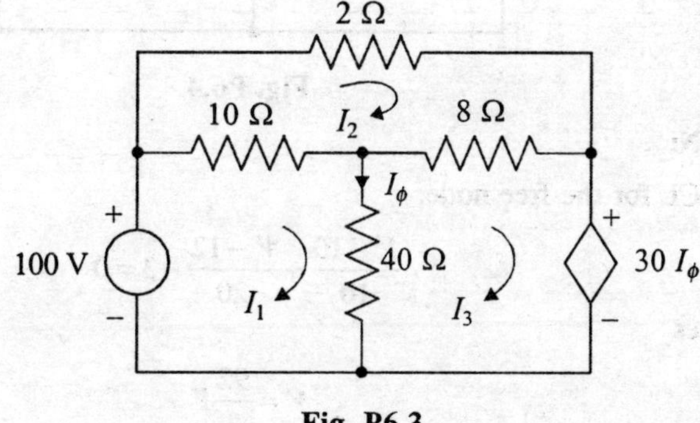

Fig. P6.3

SOLUTION:

The circuit has 6 branches and 4 nodes. Hence the number of independent meshes is given by

$$L = 6 - (4 - 1) = 3$$

Thus, three mesh currents are required to describe the circuit which are shown in Fig. P6.3.

$$100 = 10(I_1 - I_2) + 40(I_1 - I_3)$$

$$0 = 10(I_2 - I_1) + 2I_2 + 8(I_2 - I_3)$$

$$0 = 40(I_3 - I_1) + 8\,(I_3 - I_2) + 30I_\phi$$

where $I_\phi = I_1 - I_3$, which is the constraint imposed by the dependent source.

Solving these equations,

$$I_2 = 26 \text{ A}$$

$$I_3 = 28 \text{ A}$$

The current in the 8 Ω resistor = $I_3 - I_2 = 2$ A
and the power dissipated is:

$$(I_3 - I_2)^2 \times 8 = 32 \text{ W} \qquad \textbf{Ans.}$$

P6.4 Find the power delivered by the 3 A current source in the circuit shown in Fig. P6.4 using the node voltage method.

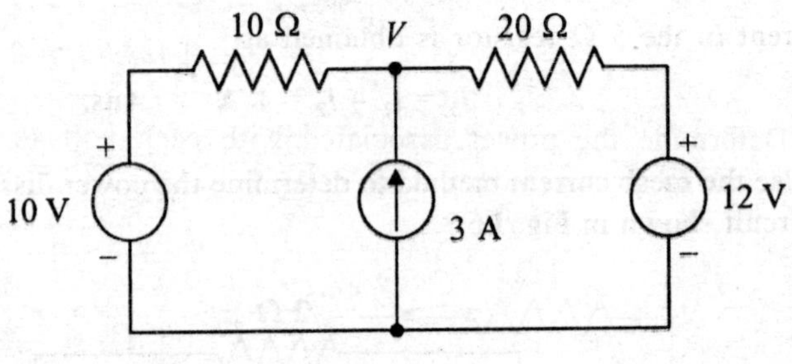

Fig. P6.4

SOLUTION:

Writing KCL for the free node,

$$\frac{V-10}{10} + \frac{V-12}{20} - 3 = 0$$

which gives

$$V = \frac{92}{3} V$$

Power delivered by the current source = 92/3 × 3 = 92 W **Ans.**

Exercises

E6.1 Find the dual of the network shown in Fig. E6.1.

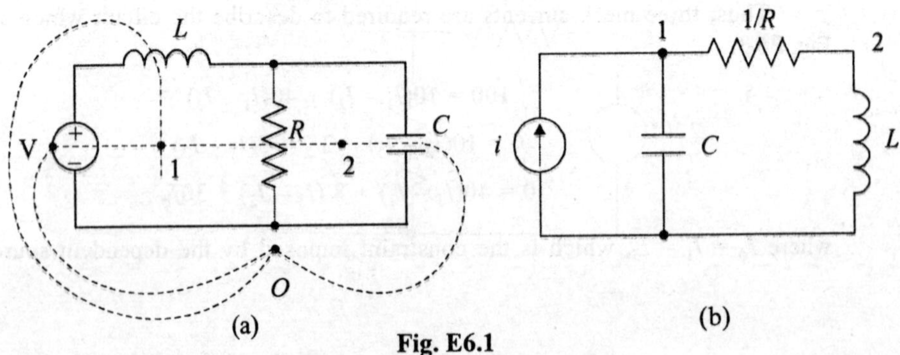

Fig. E6.1

E6.2 Solve the network in Fig. E6.2 by the branch current method.

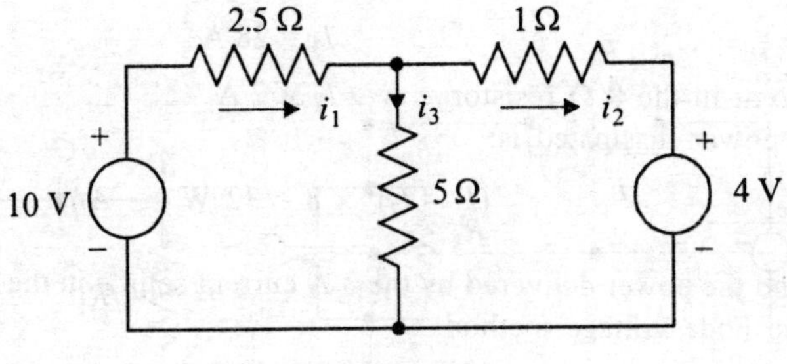

Fig. E6.2

[*Ans:* $i_1 = 2$ A, $i_2 = 1$ A, $i_3 = 1$ A]

E6.3 Determine the power associated with each voltage source in the circuit of Fig. E.6.3 using the mesh current method.

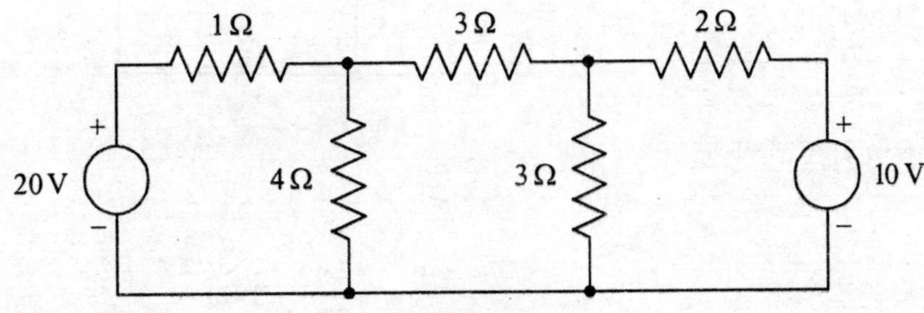

Fig. E6.3

[Hint: Write down the mesh current equations by inspection of the three meshes, as in Example 6.2. Solve the three mesh equations.] [*Ans:* $I_1 = 5.6$ A, $I_2 = 2$ A, $I_3 = -0.8$ A, Power associated with 20 V source = 112 W, Power associated with 10 V source = 8 W]

E6.4 Determine the current in the 0.5 Ω resistor of the circuit shown in Fig. E6.4 by the use of nodal analysis.

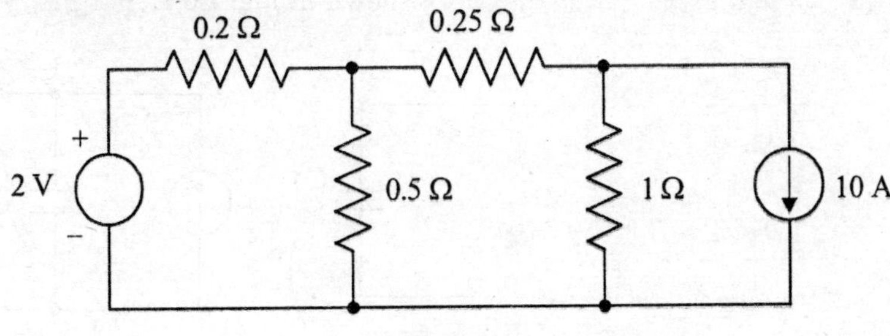

Fig. E6.4

[*Ans:* 0.513 A]

E6.5 Show that the two circuits in Fig. E6.5 are dual to each other.

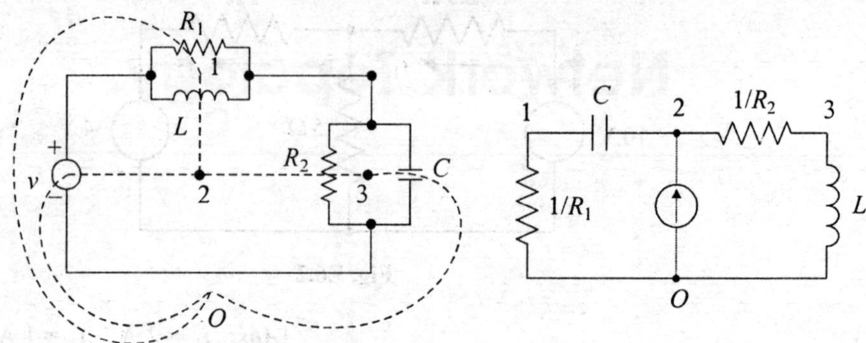

Fig. E6.5

Network Topology

7

CHAPTER CONTENTS

The chapter deals with the geometrical structure of electrical networks. Some of the related problems are systematic choice of network variables, and an efficient way of formulating mesh or node equations for complicated networks.

7.1 GRAPH OF A NETWORK

The ***graph*** of a network shows its geometrical structure. A network and its graph are shown in Fig. 7.1. Note that each element in the network has been replaced by a line segment in the graph. In a graph only interconnections of the segments are of interest (not their length or curvature).

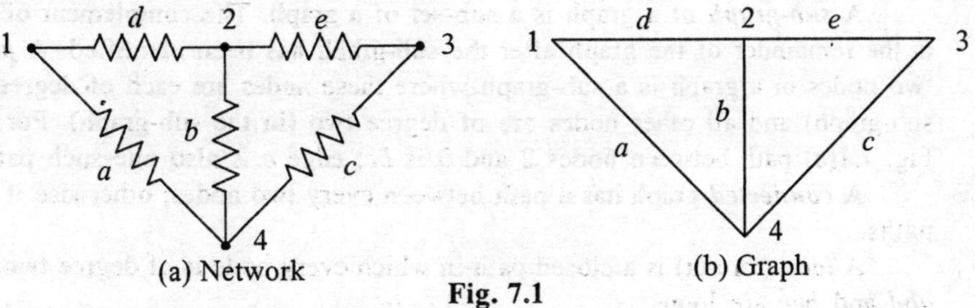

Fig. 7.1

A line segment together with its two end points is called an **edge**. The total number of edges in a graph is denoted by *e* (equal to 5 in Fig. 7.1).

End points can be joined to form a ***node***. The number of nodes in a graph is denoted by ***n*** (equal to 4 in Fig. 7.1). A graph is ***planar*** if it can be drawn on a plane surface so that no two edges intersect at a point which is not a node. The graph in Fig. 7.2(a) is planar but the graph in Fig. 7.2(b) is ***non-planar***.

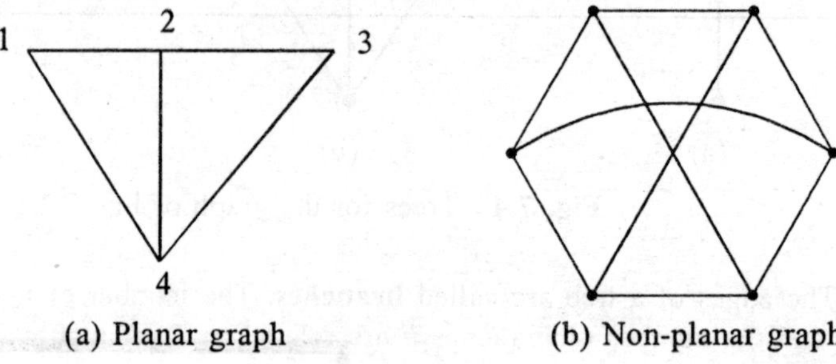

(a) Planar graph (b) Non-planar graph

Fig. 7.2

An edge and a node are incident to each other if the node is one of the two nodes of the edge. The ***degree*** of a node is the number of edges incident to this node; for example, the degree of node 2 in Fig. 7.1 is 3.

The two graphs in Fig. 7.3 appears to be different but they are topologically equivalent since the relationship of the edges and nodes is identical.

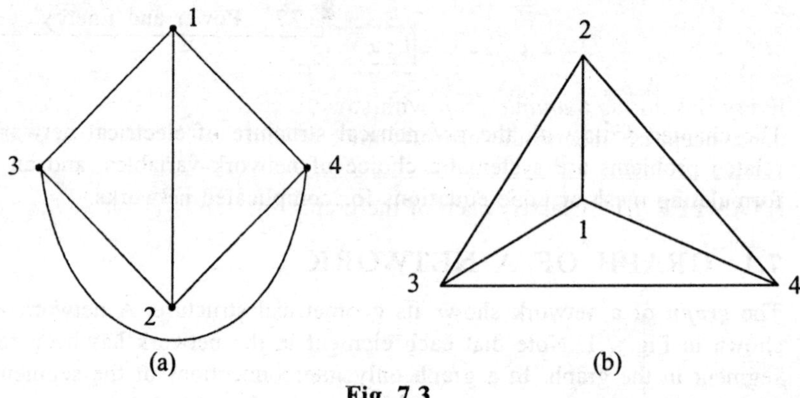

Fig. 7.3

A ***sub-graph*** of a graph is a sub-set of a graph. The complement of a sub-graph is the remainder of the graph after the sub-graph has been identified. A ***path*** between two nodes in a graph is a sub-graph where these nodes are each of degree one (in the sub-graph) and all other nodes are of degree two (in the sub-graph). For example, in Fig. 7.1(a) path between nodes 2 and 3 is *bc*; edge *e* is also one such path.

A ***connected*** graph has a path between every two nodes; otherwise it has separate paths.

A ***loop*** (circuit) is a closed path in which every node is of degree two. In Fig. 7.1, *abd* and *bce* are loops.

A *tree* of a connected graph is a connected sub-graph containing all the nodes but *no* loops. Figure 7.4 shows three of the trees of the graph in Fig. 7.1. A tree may be constructed by first putting in all the nodes and then connecting all of them without creating any loop.

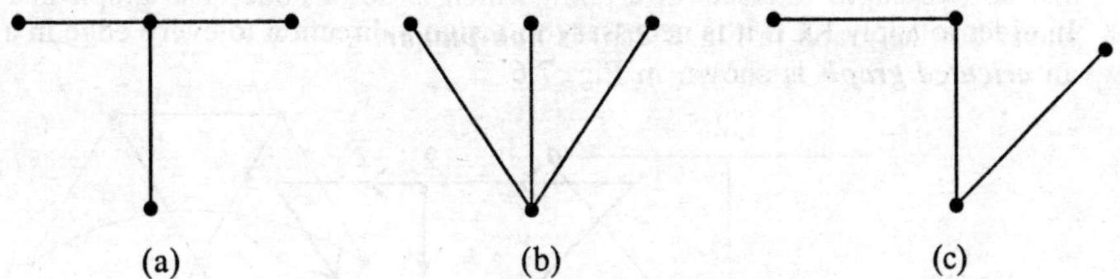

Fig. 7.4 Trees for the graph of Fig. 7.1.

The edges of a tree are called **branches**. The number of nodes n of a connected graph and the number of branches b are related by the equation:

$$b = n - 1 \tag{7.1}$$

The complement of a tree is called a **co-tree**. Its edges are called **links**. If they are l in number then

$$b + l = e \tag{7.2}$$

Since the tree and the co-tree form the original graph,

$$l = e - (n - 1) \tag{7.3}$$

or

$$\boxed{l = e - n + 1}$$

Every link forms a *unique loop* with some branches of the tree. Kirchhoff made use of the concept of trees in the proper choice of current variables in network analysis.

EXAMPLE 7.1 Classify each of the graphs in Fig. 7.5 as planar or non-planar.

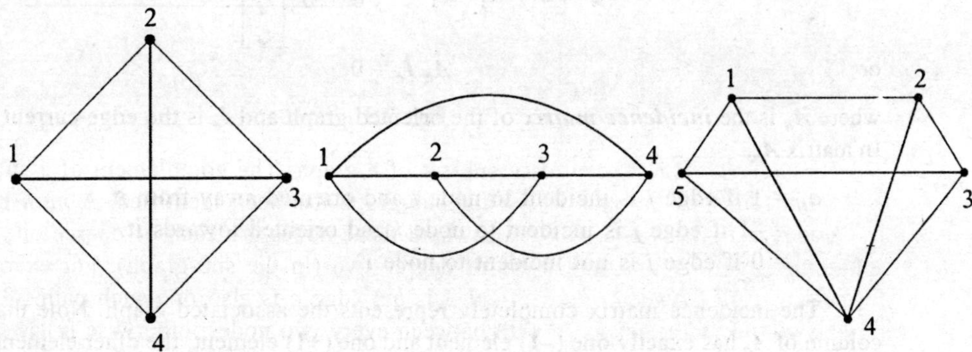

Fig. 7.5

SOLUTION:
All are planar. **Ans.**

7.2 INCIDENCE MATRIX *A*

In order to apply KCL it is necessary to assign a direction to every edge in a graph. Such an ***oriented graph*** is shown in Fig. 7.6.

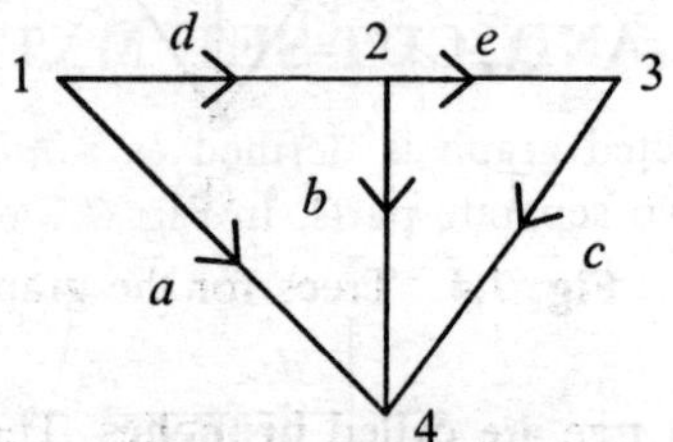

Fig. 7.6 Oriented graph.

KCL equation may now be written at each node,

Node 1, $i_a + i_d = 0$

Node 2, $i_b - i_d + i_e = 0$

Node 3, $i_c - i_e = 0$

Node 4, $-i_a - i_b - i_c = 0$

In matrix form these equations may be rewritten as:

$$\begin{array}{c} \text{Node} \\ 1 \\ 2 \\ 3 \\ 4 \end{array} \begin{array}{c} \text{Branch} \qquad\qquad \text{Link} \\ \begin{bmatrix} 1 & 0 & 0 & 1 & 0 \\ 0 & 1 & 0 & -1 & 1 \\ 0 & 0 & 1 & 0 & -1 \\ -1 & -1 & -1 & 0 & 0 \end{bmatrix} \end{array} \begin{bmatrix} i_a \\ i_b \\ i_c \\ \vdots \\ i_d \\ i_e \end{bmatrix} = 0$$

or $$A_a I_e = 0 \tag{7.4}$$

where A_a is the ***incidence matrix*** of the oriented graph and I_e is the edge current matrix. In matrix A_a,

a_{ij} = 1 if edge j is incident to node i and oriented away from it.
= –1 if edge j is incident to node i and oriented towards it.
= 0 if edge j is not incident to node i.

The incidence matrix completely represents the associated graph. Note that every column of A_a has exactly one (–1) element and one (+1) element; the other elements being zero. The rank of A_a is $(n - 1)$. See Appendix A for definition of rank.

Any one row preferably the last may be deleted from A_a without loss of information and the matrix now becomes A the ***reduced incidence matrix***.

$$A = \begin{bmatrix} 1 & 0 & 0 & 1 & 0 \\ 0 & 1 & 0 & -1 & 1 \\ 0 & 0 & 1 & 0 & -1 \end{bmatrix}$$

7.3 CUT-SETS AND CUT-SET MATRIX *B*

A cut-set of a connected graph is defined as a ***minimal*** set of edges whose deletion leaves the graph in two separate parts. In Fig. 7.7 edges *a* and *d* form a cut-set; so too *c* and *e*.

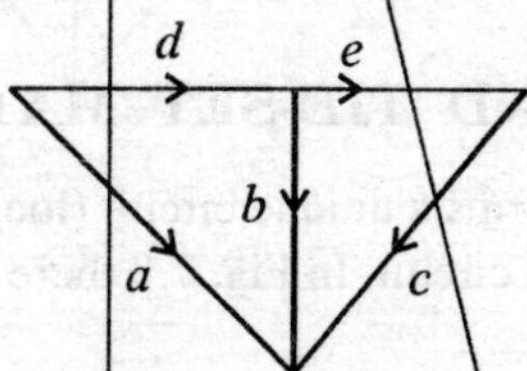

Fig. 7.7 Cut sets.

f-cut-set: If the cut-set contains only *one* branch of a tree of the graph, the set is called a **fundamental cut-set**. Orientation of each *f*-cut-set is defined with respect to the *tree* branch which defines it.

Cut-sets and KCL:

For the cut-set *ad* $\quad i_a + i_d = 0$
For the cut-set *bde* $\quad i_b - i_d + i_e = 0$
For the cut-set *ce* $\quad i_c - i_e = 0$

In matrix form these equations are:

$$\begin{bmatrix} 1 & 0 & 0 & 1 & 0 \\ 0 & 1 & 0 & -1 & 1 \\ 0 & 0 & 1 & 0 & -1 \end{bmatrix} \begin{bmatrix} i_a \\ i_b \\ i_c \\ \cdots \\ i_d \\ i_e \end{bmatrix} = 0$$

$$[U \;\vdots\; B_l][I_e] = 0$$

or
$$B\, I_e = 0 \tag{7.5}$$

where B is the *f*-cut-set matrix.

Note that the edge current vector is written as:

$$\boldsymbol{I_e} = \left[\frac{I_{br}}{I_{\text{link}}}\right]$$

The $\boldsymbol{B}$ matrix may be written *directly* from the oriented graph,

$$\begin{aligned} b_{ij} &= 1, \text{ if edge } j \text{ is in cut-set } i \text{ and their orientations agree.} \\ &= -1, \text{ if edge } j \text{ is in cut-set } i \text{ and their orientations do not agree.} \\ &= 0, \text{ if edge } j \text{ is not in cut-set } i. \end{aligned}$$

Note: From Eqs. (7.4) and (7.5) it is seen that matrices $\boldsymbol{A}$ and $\boldsymbol{B}$ in this case are equal; but in general they are only equivalent that is to say $\boldsymbol{B}$ may be obtained from $\boldsymbol{A}$ by elementary matrix transformations.

7.4 TIE-SETS AND TIE-SET MATRIX C (CIRCUIT MATRIX)

Every link of a co-tree forms a unique circuit (loop) with branches of the chosen tree. Such a circuit is a fundamental circuit. In Fig. 7.8 there are two links and hence two fundamental loops.

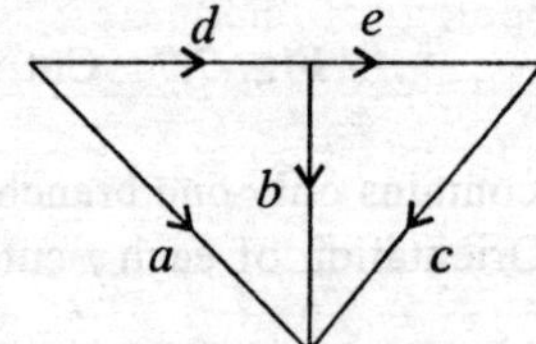

Fig. 7.8 Tie set.

abd (for link *d*)
bce (for link *e*)

The orientation for each fundamental circuit is chosen as that of its *link*, as in Fig. 7.7.

KVL for each fundamental circuit gives

$$\begin{aligned} v_d + v_b - v_a &= 0 \\ v_e + v_c - v_b &= 0 \end{aligned}$$

In matrix form

$$\begin{bmatrix} -1 & 1 & 0 & 1 & 0 \\ 0 & -1 & 1 & 0 & 1 \end{bmatrix} \begin{bmatrix} v_a \\ v_b \\ v_c \\ \cdots \\ v_d \\ v_e \end{bmatrix} = 0$$

$$[C_{br} \;\vdots\; U]\left[\frac{V_{br}}{V_{\text{link}}}\right] = 0$$

$$CV = 0 \tag{7.6}$$

The fundamental circuit matrix C is of order $(e - n + 1)$. Each element c_{ij} is given by

c_{ij} = 1, if edge j is in circuit i and their orientations agree.
= –1, if edge j is in circuit i and their orientations do not agree.
= 0, if edge j is not in circuit i.

7.5 TOPOLOGICAL RELATIONS

The matrices defined above are governed by the following topological relations:

1. $BC^T = 0$ $\quad CB^T = 0$ $\quad CA^T = 0$ (7.7)
2. $C_{br}{}^T + B_l = 0$ $\quad C_{br} = -B_l{}^T$ (7.8)
3. $B = A_{br}{}^{-1}A$ $\quad B_l = A_b{}^{-1}A_l$ (7.9)

Some of these relations are verified here in after.

EXAMPLE 7.2 Consider the oriented graph shown in Fig. 7.9. Form the A, B and C matrices and verify the topological relation $BC^T = 0$.

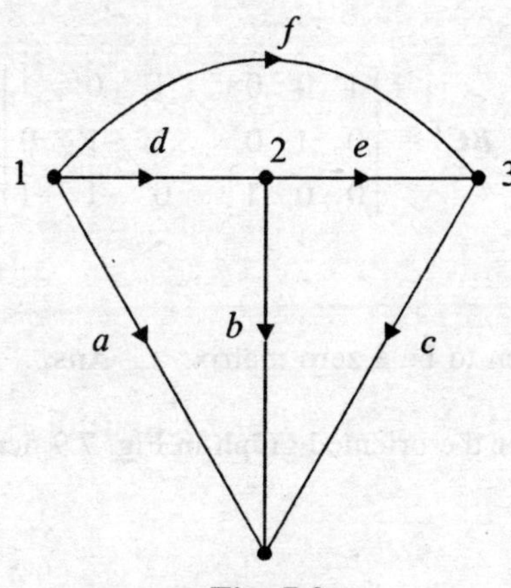

Fig. 7.9

SOLUTION:
The reduced incidence matrix is:

$$A = [A_{br} \mid A_l]$$

$$\begin{array}{c} \text{Node} \\ \\ A = \begin{array}{c} 1 \\ 2 \\ 3 \end{array} \end{array} \begin{array}{c} \text{Branch} \qquad\qquad \text{Link} \\ \begin{bmatrix} 1 & 0 & 0 & 1 & 0 & 1 \\ 1 & -1 & 0 & -1 & 1 & 0 \\ 0 & 0 & 1 & 0 & -1 & -1 \end{bmatrix} \end{array}$$

The cut-set matrix is:

$$B = [U \mid B_l]$$

$$\begin{array}{c} \\ \\ B = \end{array}\begin{array}{c} \\ a \\ b \\ c \end{array}\begin{array}{c} \text{Branch} \quad \text{Branch} \quad \text{Link} \\ \begin{bmatrix} 1 & 0 & 0 & 1 & 0 & 1 \\ 0 & 1 & 0 & 1 & -1 & 0 \\ 0 & 0 & 1 & 0 & -1 & -1 \end{bmatrix} \end{array}$$

The fundamental circuit matrix is:

$$C = [C_{br} \mid U]$$

$$\begin{array}{c} \\ \\ C = \end{array}\begin{array}{c} \\ d \\ e \\ f \end{array}\begin{array}{c} \text{Link} \quad \text{Branch} \quad \text{Link} \\ \begin{bmatrix} -1 & -1 & 0 & 1 & 0 & 0 \\ 0 & 1 & 1 & 0 & 1 & 0 \\ -1 & 0 & 1 & 0 & 0 & 1 \end{bmatrix} \end{array}$$

Hence
$$BC^T = \begin{bmatrix} 1 & 0 & 0 & 1 & 0 & 1 \\ 0 & 1 & 0 & 1 & -1 & 0 \\ 0 & 0 & 1 & 0 & -1 & -1 \end{bmatrix}\begin{bmatrix} -1 & 0 & -1 \\ -1 & 1 & 0 \\ 0 & 1 & 1 \\ 1 & 0 & 0 \\ 0 & 1 & 0 \\ 0 & 0 & 1 \end{bmatrix}$$

which is readily seen to be a zero matrix. **Ans.**

EXAMPLE 7.3 For the oriented graph in Fig. 7.9 derive B_l from the reduced incidence matrix A.

SOLUTION:

From Eq. (7.9),

$$B_l = A_{br}^{-1} A_l$$

Here

$$A_{br} = \begin{bmatrix} 1 & 0 & 0 \\ 0 & -1 & 0 \\ 0 & 0 & 1 \end{bmatrix}$$

Hence

$$A_{br}^{-1} = \begin{bmatrix} 1 & 0 & 0 \\ 0 & -1 & 0 \\ 0 & 0 & 1 \end{bmatrix}$$

and $$A_{br}^{-1} A_l = \begin{bmatrix} 1 & 0 & 0 \\ 0 & -1 & 0 \\ 0 & 0 & 1 \end{bmatrix} \begin{bmatrix} 1 & 0 & 1 \\ -1 & 1 & 0 \\ 0 & -1 & -1 \end{bmatrix}$$

$$= \begin{bmatrix} 1 & 0 & 1 \\ 1 & -1 & 0 \\ 0 & -1 & -1 \end{bmatrix} \quad \textbf{Ans.}$$

which is the B_l matrix.

Relationship between link currents and branch currents: The fundamental cut-set equations are:

$$BI_e = 0$$

or $$[U : B_l] \left[\frac{I_{br}}{I_{\text{link}}} \right]$$

Carrying out this multiplication

$$I_{br} = -B_l \, I_{\text{link}}$$

which relates branch currents and link currents.

7.6 LOOP AND NODE EQUATIONS

In terms of edge variables (currents and voltages) Kirchhoff's laws have been expressed as:

$$AI_e = 0 \tag{7.10}$$

$$CV_e = 0 \tag{7.11}$$

In order to relate the edge variables to mesh currents and node voltages (commonly used in network analysis), the following transformation of variables is introduced:

$$I_e = C^T I_m \tag{7.12}$$

$$V_e = A^T V_n \tag{7.13}$$

where I_m is a column matrix of the $(e - n + 1)$ loop or mesh currents.

$$I_m = \begin{bmatrix} i_1 \\ i_2 \\ \vdots \\ i_{e-n+1} \end{bmatrix}$$

and V_n is a column matrix of the $(n - 1)$ node-to-datum voltages.

$$V_n = \begin{bmatrix} v_1 \\ v_2 \\ \vdots \\ v_{n-1} \end{bmatrix}$$

Current transformation: Substituting Eq. (7.12) in Eq. (7.10), we get

$$AI_e = A(C^T I_m) = (AC^T)I_m = 0 \tag{7.14}$$

because $AC^T = 0$

Voltage transformation: From Eq. (7.13)

$$CV_e = C(A^T V_n) = (CA^T)V_n = 0 \tag{7.15}$$

because $CA^T = 0$

Derivation of loop and node equations: The Laplace transform method used in this derivation will be explained in Chapter 9.

The V-I relationship for an element is of the general form:

$$V_e(s) = \left[R_e + sL_e + \frac{1}{sC_e} \right] I_e(s) + E_e(s)$$

$$= Z_e(s)\, I_e(s) + E_e(s) \tag{7.16}$$

or $$I_e(s) = [G_e + 1/(sL_e) + sC_e]\; V_e(s) + J_e(s)$$

$$= Y_e(s)\, V_e(s) + J_e(s) \tag{7.17}$$

where $E_e(s)$ is the voltage source in an element and $J_e(s)$ is the current source across it.

These relations hold good for all elements of the network and in matrix form they become

$$V_e = Z_e I_e + E_e \tag{7.18}$$

$$I_e = Y_e V_e + J_e \tag{7.19}$$

Combining Eq. (7.18) with Eq. (7.11),

$$CV_e = C(Z_e I_e + E_e) = 0$$

which gives $$C\,Z_e I_e = -CE_e \tag{7.20}$$

or in terms of loop currents defined by Eq. (7.12),

$$CZ_e C^T I_m = -CE_e \tag{7.21}$$

$\boxed{CZ_e C^T = Z}$ is defined as the **loop impedance matrix**. (7.22)

and $-CE_e = E$ is the column matrix of the voltage sources around each loop.

On a dual basis the ***node equation*** may be derived from Eq. (7.19) and Eq. (7.10),

$$AI_e = A(Y_eV_e + J_e) = 0$$

Hence $$AY_eA^TV_n = -J_e \tag{7.23}$$

Thus $$AY_eA^T = Y, \text{ is the node admittance matrix} \tag{7.24}$$

and $-AJ_e = J$, is the column matrix of the current sources of each node.

EXAMPLE 7.4 Derive the node equations for the circuit of Fig. 7.10.

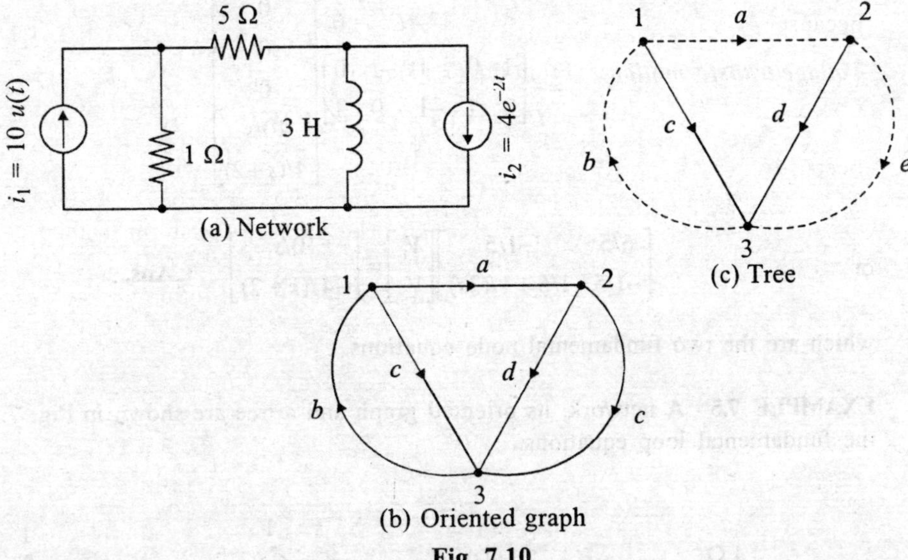

(a) Network

(b) Oriented graph

(c) Tree

Fig. 7.10

SOLUTION:
A network, its oriented graph and a tree are shown in Fig. 7.10. With node 3 grounded the reduced incidence matrix is:

$$\begin{array}{c} \\ 1 \\ 2 \end{array}\begin{array}{c} \begin{array}{ccccc} c & d & a & b & e \end{array} \\ \begin{bmatrix} 1 & 0 & 1 & -1 & 0 \\ 0 & 1 & -1 & 0 & 1 \end{bmatrix} \end{array}$$

The v-i relations for the edges are written by inspection:

$$\begin{bmatrix} I_c \\ I_d \\ I_a \\ I_b \\ I_e \end{bmatrix} = \begin{bmatrix} 1 & 0 & 0 & 0 & 0 \\ 0 & 1/3s & 0 & 0 & 0 \\ 0 & 0 & 1/5 & 0 & 0 \\ 0 & 0 & 0 & 0 & 0 \\ 0 & 0 & 0 & 0 & 0 \end{bmatrix} \begin{bmatrix} V_c \\ V_d \\ V_a \\ V_b \\ V_e \end{bmatrix} + \begin{bmatrix} 0 \\ 0 \\ 0 \\ 10/s \\ 4/(s+2) \end{bmatrix}$$

From Eq. (7.23) the node equations are:

$$\begin{bmatrix} 1 & 0 & 1 & -1 & 0 \\ 0 & 1 & -1 & 0 & 1 \end{bmatrix} \begin{bmatrix} 1 & 0 & 0 & 0 & 0 \\ 0 & 1/3s & 0 & 0 & 0 \\ 0 & 0 & 1/5 & 0 & 0 \\ 0 & 0 & 0 & 0 & 0 \\ 0 & 0 & 0 & 0 & 0 \end{bmatrix} \begin{bmatrix} 1 & 0 \\ 0 & 1 \\ 1 & -1 \\ -1 & 0 \\ 0 & 1 \end{bmatrix} V_n$$

$$= -\begin{bmatrix} 1 & 0 & 1 & -1 & 0 \\ 0 & 1 & -1 & 0 & 1 \end{bmatrix} \begin{bmatrix} 0 \\ 0 \\ 0 \\ 10/s \\ 4/(s+2) \end{bmatrix}$$

or

$$\begin{bmatrix} 6/5 & -1/5 \\ -1/5 & 1/5 + 1/(3s) \end{bmatrix} \begin{bmatrix} V_1 \\ V_2 \end{bmatrix} = \begin{bmatrix} 10/s \\ -4/(s+2) \end{bmatrix} \qquad \textbf{Ans.}$$

which are the two fundamental node equations.

EXAMPLE 7.5 A network, its oriented graph and a tree are shown in Fig. 7.11. Find the fundamental loop equations.

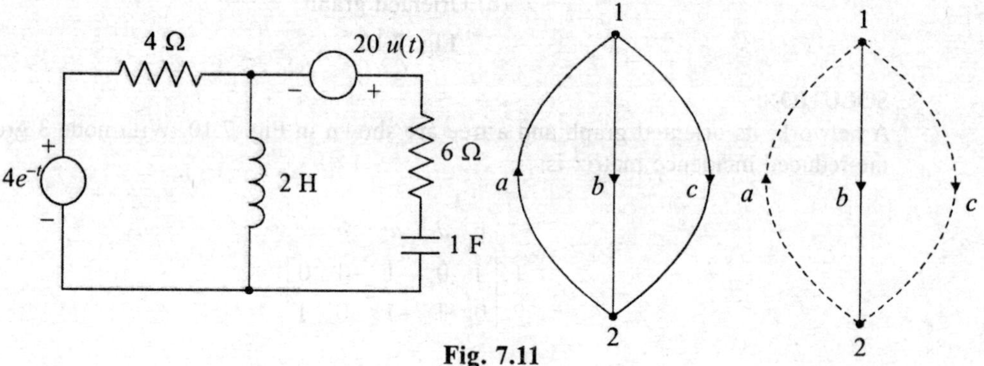

Fig. 7.11

SOLUTION:

Equation (7.6) gives the circuit matrix as:

$$C = \begin{bmatrix} 1 & 1 & 0 \\ -1 & 0 & 1 \end{bmatrix}$$

with columns labelled b, a, c.

The *v-i* relations for the edges are written by inspection:

$$\begin{bmatrix} V_b \\ V_a \\ V_c \end{bmatrix} = \begin{bmatrix} 2s & 0 & 0 \\ 0 & 4 & 0 \\ 0 & 0 & 6+1/s \end{bmatrix} \begin{bmatrix} I_b \\ I_a \\ I_c \end{bmatrix} + \begin{bmatrix} 0 \\ -4/(s+1) \\ -20/s \end{bmatrix}$$

The fundamental loop equations are: [from Eq. (7.2)]

$$\begin{bmatrix} 1 & 1 & 0 \\ -1 & 0 & 1 \end{bmatrix} \begin{bmatrix} 2s & 0 & 0 \\ 0 & 4 & 0 \\ 0 & 0 & 6+1/s \end{bmatrix} \begin{bmatrix} 1 & -1 \\ 1 & 0 \\ 0 & 1 \end{bmatrix} I_m$$

$$= -\begin{bmatrix} 1 & 1 & 0 \\ -1 & 0 & 1 \end{bmatrix} \begin{bmatrix} 0 \\ -4/(s+1) \\ -20/s \end{bmatrix}$$

or

$$\begin{bmatrix} 2s+4 & -2s \\ -2s & 2s+6+1/s \end{bmatrix} \begin{bmatrix} I_{m1} \\ I_{m2} \end{bmatrix}$$

$$= \begin{bmatrix} 4/(s+1) \\ 20/s \end{bmatrix} \qquad \textbf{Ans.}$$

7.7 POWER AND ENERGY

Tellegen's theorem: The instantaneous power delivered at time t to an oriented element r in a network is given by

$$p_r(t) = v_r(t)\, i_r(t)$$

The total instantaneous power in all the elements is given by

$$p(t) = v_1(t)i_1(t) + v_2(t)i_2(t) + \cdots + v_n(t)i_n(t) \tag{7.25}$$

Tellegen's theorem states that *the total instantaneous power in a network is zero, provided the elements obey Kirchhoff's laws.*

Express Eq. (7.25) as the scalar product of the column matrices $\boldsymbol{i}(t)$ and $\boldsymbol{v}(t)$ to get

$$p(t) = \boldsymbol{i}^T \boldsymbol{v} \tag{7.26}$$

Now,

$$\boldsymbol{i}(t) = \boldsymbol{C}^T \boldsymbol{I}_m(t) \quad \text{and} \quad \boldsymbol{v}(t) = \boldsymbol{A}^T \boldsymbol{V}_n(t) \tag{7.27}$$

Hence

$$\boldsymbol{i}^T(t) = \boldsymbol{I}_m{}^T \boldsymbol{C} \quad \text{and} \quad \boldsymbol{i}^T(t)\, \boldsymbol{v}(t) = \boldsymbol{I}_m{}^T(t)(\boldsymbol{C}\boldsymbol{A}^T)\boldsymbol{V}_n(t) = 0, \text{ since } \boldsymbol{C}\boldsymbol{A}^T = 0 \tag{7.28}$$

This proves Tellegen's theorem.

Conservation of energy: Since instantaneous power $p_r(t)$ is the rate of energy w_r at time t,

$$p_r(t) = \frac{dw_r(t)}{dt}$$

But by Tellegen's theorem,

$$\Sigma\, p_r = 0$$

and, therefore, $\Sigma\, w_r$ is constant.

The principle of conservation of energy is thus included in Kirchhoff's laws. In other words, Kirchhoff's law for a network are essentially a formulation of the principle of conservation of energy, in terms of currents and voltages.

SUMMARY

Topology provides a solid foundation for an efficient way of writing independent loop or node equations in complicated networks. While the examples in the chapter do not bring out in full the significance of the topological approach, they point to a sound basis for solving network problems.

Solved Problems

P7.1 In the given resistive network shown in Fig. P7.1, write the tie set matrix and obtain the equilibrium equations on the mesh basis, and solve for the mesh currents.

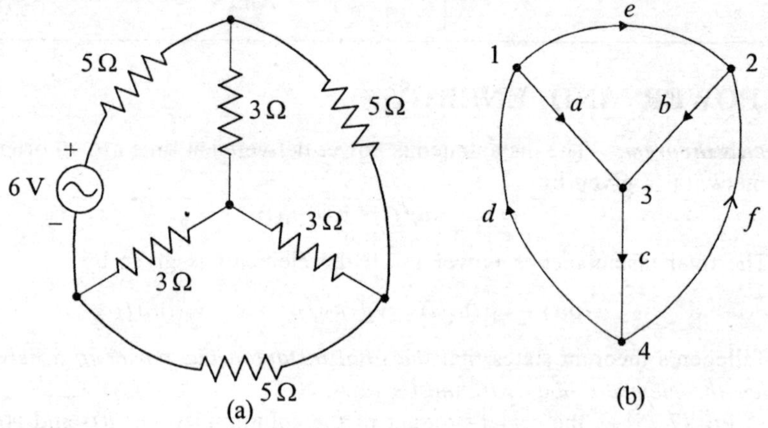

Fig. P7.1

SOLUTION:

An oriented graph for the given circuit is shown in Fig. P7.1(b).
The circuit matrix is:

$$C = \begin{array}{c} \begin{matrix} a & b & c & d & e & f \end{matrix} \\ \begin{bmatrix} 1 & 0 & 1 & 1 & 0 & 0 \\ -1 & 1 & 0 & 0 & 1 & 0 \\ 0 & 1 & 1 & 0 & 0 & 1 \end{bmatrix} \end{array}$$

The v-i relations for the edges are written by inspection,

$$\begin{bmatrix} V_a \\ V_b \\ V_c \\ V_d \\ V_e \\ V_f \end{bmatrix} = \begin{bmatrix} 3 & 0 & 0 & 0 & 0 & 0 \\ 0 & 3 & 0 & 0 & 0 & 0 \\ 0 & 0 & 3 & 0 & 0 & 0 \\ 0 & 0 & 0 & 5 & 0 & 0 \\ 0 & 0 & 0 & 0 & 5 & 0 \\ 0 & 0 & 0 & 0 & 0 & 5 \end{bmatrix} \begin{bmatrix} I_a \\ I_b \\ I_c \\ I_d \\ I_e \\ I_f \end{bmatrix} + \begin{bmatrix} 0 \\ 0 \\ 0 \\ -6 \\ 0 \\ 0 \end{bmatrix}$$

The fundamental loop equations are:

$$\begin{bmatrix} 1 & 0 & 1 & 1 & 0 & 0 \\ -1 & 1 & 0 & 0 & 1 & 0 \\ 0 & 1 & 1 & 0 & 0 & 1 \end{bmatrix} \begin{bmatrix} 3 & 0 & 0 & 0 & 0 & 0 \\ 0 & 3 & 0 & 0 & 0 & 0 \\ 0 & 0 & 3 & 0 & 0 & 0 \\ 0 & 0 & 0 & 5 & 0 & 0 \\ 0 & 0 & 0 & 0 & 5 & 0 \\ 0 & 0 & 0 & 0 & 0 & 5 \end{bmatrix} \begin{bmatrix} 1 & -1 & 0 \\ 0 & 1 & 1 \\ 1 & 0 & 1 \\ 1 & 0 & 0 \\ 0 & 1 & 0 \\ 0 & 0 & 1 \end{bmatrix} I_m$$

$$= -\begin{bmatrix} 1 & 0 & 1 & 1 & 0 & 0 \\ -1 & 1 & 0 & 0 & 1 & 0 \\ 0 & 1 & 1 & 0 & 0 & 1 \end{bmatrix} \begin{bmatrix} 0 \\ 0 \\ 0 \\ -6 \\ 0 \\ 0 \end{bmatrix}$$

or

$$\begin{bmatrix} 11 & -3 & -3 \\ -3 & 11 & -3 \\ -3 & -3 & 11 \end{bmatrix} \begin{bmatrix} I_{m1} \\ I_{m2} \\ I_{m3} \end{bmatrix} = \begin{bmatrix} 6 \\ 0 \\ 0 \end{bmatrix}$$

Solving by Cramer's method,

$$I_{m1} = 0.685 \text{ A}$$

$$I_{m2} = 0.257 \text{ A}$$

$$I_{m3} = 0.257 \text{ A} \qquad \textbf{Ans.}$$

P7.2 Verify Tellegen's theorem for the bridge circuit shown in Fig. P7.2.

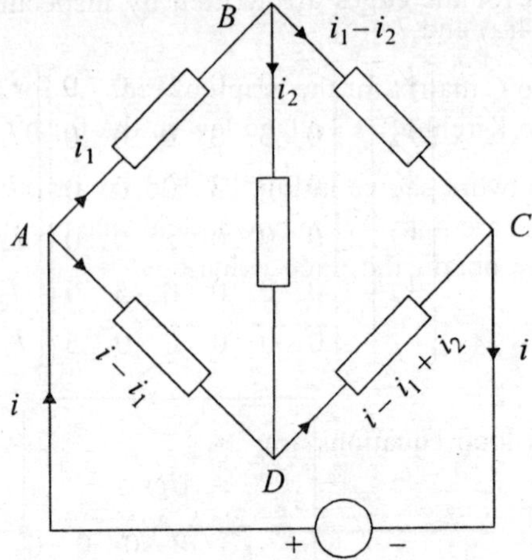

Fig. P7.2

SOLUTION:

With node C as reference, the branch currents are written down using KCL. Let v_A, v_B, v_C be the node voltages.

The branch voltages are: (by KVL)

$$v_{AB} = v_A - v_B \qquad v_{BC} = v_B \qquad v_{AD} = v_A - v_D$$
$$v_{DC} = v_D \qquad v_{BD} = v_B - v_D \qquad v_{CA} = -v_A$$

The sum of the instantaneous power in the branches is expressed as:

$$p = (v_A - v_B)i_1 + v_B(i_1 - i_2) + (v_A - v_D)(i - i_1) + v_D(i - i_1 + i_2) + (-v_A)i + (v_B - v_D)i_2 \equiv 0$$

Thus, Tellegen's theorem is verified. **Ans.**

Note that it is essential to have the current directions marked strictly according to KCL. Power in a branch is taken as the product of the potential difference across the branch and the current flowing from the node at higher potential to that at lower potential.

Exercises

E7.1 Prove that the number of edges $\boldsymbol{e}$ and nodes $\boldsymbol{n}$ in a path are related by the equation:

$$e = n - 1$$

E7.2 Obtain the graph associated with the incidence matrix

$$\begin{bmatrix} 1 & 0 & 0 & 1 & 0 \\ 0 & 1 & 0 & -1 & 1 \\ 0 & 0 & 1 & 0 & -1 \\ -1 & -1 & -1 & 0 & 0 \end{bmatrix}$$

[**Hint:** See Section 7.2.]

E7.3 For the oriented graph in Fig. 7.9, derive the fundamental cut-sets using the trees shown in Fig. 7.4(a) and 7.4(c).

E7.4 Obtain the C matrix for the graph of Fig. 7.9 for the trees shown in Fig. 7.4(a) and Fig. 7.4(c). Write Kirchhoff's voltage law in the form $CV_e = 0$.

E7.5 For the network shown in Fig. E7.1, draw its oriented graph, choose a tree so that all current sources are links. Write down incidence matrix A. Form the *v-i* relations for the edges and finally obtain the node equations.

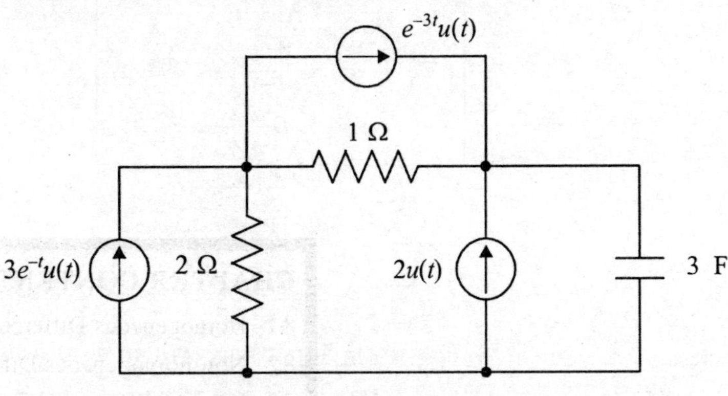

Fig. E7.1

Differential Equations—Solution by Classical Methods

8

CHAPTER CONTENTS

Network analysis is mostly based on the solution of *ordinary differential equations with constant coefficients*, of the form:

$$a_n(d^n x/dt^n) + a_{n-1}(d^{n-1}x/dt^{n-1}) + \cdots + a_1(dx/dt) + a_0 x = f(t) \tag{8.1}$$

The ***order*** of the equation is n, the order of the highest derivative term. The term $f(t)$ is the ***forcing function*** and is independent of $x(t)$. When $f(t)$ is identically zero, the equation is said to be ***homogeneous***.

8.1 HOMOGENEOUS DIFFERENTIAL EQUATIONS

There are several methods available for the solution of homogeneous linear differential equations with constant coefficients. The classical method consists of solving the **characteristic equation** obtained by putting $f(t) = 0$ in Eq. (8.1)

$$a_n m^n + a_{n-1} m^{n-1} + \cdots + a_1 m + a_0 = 0 \tag{8.2}$$

and then introducing the initial conditions.

8.1.1 Single Real Root for the Characteristic Equation

The differential Eq. (8.1) reduces to the form:

$$a_1(dx/dt) + a_0x = 0 \tag{8.3}$$

Solve Eq. (8.3) given the initial conditions $x(0) = x_0$.

The characteristic equation is:

$$a_1m + a_0 = 0$$

which gives $$m = -\frac{a_0}{a_1}$$

and $x(t)$ takes the form:

$$x = Ae^{mt}$$
$$= Ae^{-(a_0/a_1)t}$$

The arbitrary constant A is evaluated by making use of the initial condition. At $t = 0$,

$$x_0 = Ae^0$$

Hence $$A = x_0$$

and the general solution is:

$$x = x_0 e^{-(a_0/a_1)t} \tag{8.4}$$

8.1.2 Distinct Real Roots

The differential equation takes the form:

$$a_2\frac{d^2x}{dt^2} + a_1\frac{dx}{dt} + a_0x = 0 \tag{8.5}$$

EXAMPLE 8.1 Solve the equation

$$\frac{d^2x}{dt^2} + 3\frac{dx}{dt} + 2x = 0$$

with the initial conditions $x(0) = 3$ and $x'(0) = -1$.

SOLUTION:

The characteristic equation is:

$$m^2 + 3m + 2 = 0$$

which gives $$m_1 = -1, \quad m_2 = -2$$

and $x(t)$ takes the form:

$$x = A_1e^{m_1t} + A_2e^{m_2t} \tag{8.6}$$

The arbitrary constants A_1 and A_2 are evaluated by making use of the initial conditions.

At $t = 0$, $$3 = A_1 + A_2 \tag{8.7a}$$

Taking the derivative of x in Eq. (8.6),

$$x'(t) = m_1A_1e^{m_1t} + m_2A_2e^{m_2t}$$

At $t = 0$,

$$-1 = m_1A_1 + m_2A_2$$

$$= -A_1 - 2A_2 \tag{8.7b}$$

From Eqs. (8.7a and 8.7b),

$$A_1 = 5 \qquad A_2 = -2$$

Thus, the general solution is:

$$x = 5e^{-t} - 2e^{-2t} \qquad \textbf{Ans.}$$

8.1.3 Multiple Roots

If there is a root of multiplicity two in the characteristic equation, the solution can be shown to be of the form

$$A_1e^{m_1t} + A_2te^{m_1t} \tag{8.8}$$

EXAMPLE 8.2 Solve the equation

$$x''(t) - 8x'(t) + 16x(t) = 0 \tag{8.9}$$

with the initial conditions $x(0) = 2$ and $x'(0) = 4$.

SOLUTION:
The characteristic equation is:

$$m^2 - 8m + 16 = 0$$

The roots are $m_1 = m_2 = 4$

The solution takes the form:

$$x(t) = A_1e^{4t} + A_2te^{4t} \tag{8.10}$$

The arbitrary constants are evaluated using the initial conditions,

$$x(0) = 2 = A_1 \tag{8.11a}$$

Differentiating Eq. (8.10),

$$x'(t) = 4A_1e^{4t} + A_2t4e^{4t} + A_2e^{4t}$$

and

$$x'(0) = 4A_1 + A_2 = 4 \tag{8.11b}$$

From Eqs. (8.11a and 8.11b),

$$A_1 = 2, \quad A_2 = -4$$

Thus, the general solution is:

$$2e^{4t} - 4te^{4t} \qquad \textbf{Ans.}$$

8.1.4 Complex Conjugate Roots

If there are roots of the form $m_1, m_2 = \sigma \pm j\omega$, the solution is:

$$x(t) = A_1e^{(\sigma+j\omega)t} + A_2e^{(\sigma-j\omega)t}, \tag{8.12}$$

which can be put in a more convenient form:

$$x(t) = e^{\sigma t} \{B_1 \sin \omega t + B_2 \cos \omega t\} \tag{8.13}$$

where B_1 and B_2 are arbitrary constants to be evaluated from given initial conditions.

EXAMPLE 8.3 Solve the equation

$$x''(t) + 2x'(t) + 5x(t) = 0 \tag{8.14}$$

with the initial conditions $x(0) = 1$ and $x'(0) = 0$.

SOLUTION:

The characteristic equation is:

$$m^2 + 2m + 5 = 0$$

which has roots $m_1, m_2 = -1 \pm j2$.

The solution is, therefore, of the form:

$$x(t) = e^{-t}\{B_1 \sin 2t + B_2 \cos 2t\} \tag{8.15}$$

The arbitrary constants are found from Eq. (8.15).

At $t = 0$, $$x(0) = B_2 = 1$$

Taking the derivative of Eq. (8.15),

$$x'(t) = 2B_1e^{-t} \cos 2t - e^{-t} B_1 \sin 2t - 2B_2e^{-t} \sin 2t - B_2e^{-t} \cos 2t$$

At $t = 0$, $$x'(0) = 2B_1 - B_2 = 0$$

It follows $$B_1 = \frac{1}{2} \qquad B_2 = 1$$

The general solution is:

$$x(t) = e^{-t}\{\frac{1}{2} \sin 2t + \cos 2t\} \qquad \textbf{Ans.}$$

8.2 NON-HOMOGENEOUS DIFFERENTIAL EQUATIONS

In a non-homogeneous equation, such as Eq. (8.1), the forcing function $f(t)$ is not identically zero for all t. The classical method of solution consists of finding the **complementary function** and the **particular integral; the general solution** is the sum of the two, provided it satisfies the initial conditions.

Equation (8.1) may be put in the form:

$$F(D)x = f(t)$$

where $$D = d/dt$$

and F is a function of degree n in D. The complementary function (C.F.) is the solution of the homogeneous equation obtained by putting $f(t) = 0$.

The particular integral (P.I.) is given by

$$\frac{f(t)}{F(D)} \tag{8.16}$$

Table 8.1 summarizes useful forcing functions and the corresponding particular integrals.

Table 8.1 Forcing Functions and Particular Integrals

	$f(t)$	$P.I.$	
1.	be^{at}	$\dfrac{be^{at}}{F(a)}$	if $F(a) \neq 0$
2.	e^{at}	$\dfrac{te^{at}}{\phi(a)}$	if $(D - a)\,\phi(D) = F(D)$
3.	$\sin at$	$\dfrac{\sin at}{F(-a^2)}$	if $F(-a^2) \neq 0$
4.	$\cos at$	$\dfrac{\cos at}{F(-a^2)}$	if $F(-a^2) \neq 0$
5.	$\sin at$	$\dfrac{t \cos at}{2a}$	if $F(-a^2) = 0$
6.	$\cos at$	$\dfrac{-t \sin at}{2a}$	if $F(-a^2) = 0$
7.	$e^{at}\, V$	$e^{at}\, \dfrac{1}{F(D+a)}\, V$	
8.	t^m	Expand $1/F(D)$ retaining up to m^{th} power (m is positive integer)	
9.	b	$\dfrac{b}{F(0)}$ since b may be taken as $b \times e^{0t}$ in item 1.	

EXAMPLE 8.4 Find the particular integral for the following differential equations with the corresponding forcing function:

SOLUTION:

(i) $(D^2 - 2D + 2)\, x = e^{2t}$

$$\text{Particular integral} = \frac{1}{D^2 - 2D + 2} e^{2t} = \frac{e^{2t}}{2^2 - 2 \times 2 + 2} = \frac{1}{2} e^{2t} \tag{8.17}$$

See item 1, Table 8.1.

(ii) $(D^2 - 5D + 6)\, x = e^{3t}$

$$\text{Particular integral} = \frac{e^{3t}}{(D-3)(D-2)} = \frac{te^{3t}}{(3-2)} = te^{3t} \tag{8.18}$$

See item 2, Table 8.1.

(iii) $(D^2 + 10)\, x = \sin 3t$

$$\text{Particular integral} = \frac{\sin 3t}{(D^2+10)} = \frac{\sin 3t}{-3^2+10} = \sin 3t \tag{8.19}$$

See item 3, Table 8.1.

(iv) $(D^2 + 10)\, x = \cos 3t$

$$\text{Particular integral} = \frac{\cos 3t}{(D^2+10)} = \frac{\cos 3t}{-3^2+10} = \cos 3t \tag{8.20}$$

See item 4, Table 8.1.

(v) $(D^2 + 9)\, x = \sin 3t$

$$\text{Particular integral} = \frac{\sin 3t}{(D^2+9)} = \frac{-t\cos 3t}{2\times 3} = \frac{-t\cos 3t}{6} \tag{8.21}$$

See item 5, Table 8.1.

(vi) $(D^2 + 9)\, x = \cos 3t$,

$$\text{Particular integral} = \frac{\cos 3t}{(D^2+9)} = \frac{-t\sin 3t}{2\times 3} = \frac{-t\sin 3t}{6} \tag{8.22}$$

See item 6, Table 8.1.

(vii) $(D^2 + 1)\, x = t^2$

$$\text{Particular integral} = \frac{t^2}{(1+D^2)} = (1 + D^2)^{-1}\, t^2 = (1 - D^2 + \cdots)\, t^2$$

$$= t^2 - 2 \tag{8.23}$$

See item 7, Table 8.1.

(viii) $(D^2 - 2D + 2)\, x = e^t\, t^3$

$$\text{Particular integral} = e^t \frac{1}{(D+1)^2 - 2(D+1) + 2}\, t^3$$

$$\text{P.I.} = e^t \frac{1}{(D^2+1)}\, t^3$$

$$= e^t(1 + D^2)^{-1}\, t^3$$

$$= e^t(1 - D^2 + D^4 - \cdots)\, t^3$$

$$= e^t\,(t^3 - 6t) \tag{8.24}$$

See item 8, Table 8.1.

EXAMPLE 8.5 Obtain the general solution for the differential equation:

$$(D^2 + 2D + 5)\, x = e^t$$

with the initial conditions $x(0) = 1$ and $x'(0) = 0$.

SOLUTION:

The characteristic equation is:

$$m^2 + 2m + 5 = 0$$

roots of which are m_1, $m_2 = -1 \pm j2$

The complementary function is:

$$\text{C.F.} = e^{-t} (B_1 \sin 2t + B_2 \cos 2t)$$

Then

$$\text{Particular integral} = \frac{e^t}{(D^2 + 2D + 5)} = \frac{1}{1^2 + 2 \times 1 + 5} = \frac{1}{8} e^t$$

The incomplete solution is:

$$\text{C.F.} + \text{P.I.} = \frac{1}{8} e^t + e^{-t} (B_1 \sin 2t + B_2 \cos 2t) \qquad (8.25)$$

The arbitrary constants B_1 and B_2 are now evaluated using the initial conditions.

At $t = 0$

$$1 = 1/8 + B_2 \qquad (8.26)$$

Differentiating Eq. (8.25),

$$x'(t) = \frac{1}{8} e^t + B_1 e^{-t}\, 2 \cos 2t + B_1 \sin 2t(-e^{-t}) - B_2 e^{-t}\, 2 \sin 2t + B_2 \cos 2t(-e^{-t})$$

At $t = 0$

$$0 = 1/8 + 2B_1 - B_2 \qquad (8.27)$$

From Eqs. (8.26) and (8.27),

$$B_1 = 3/8 \qquad B_2 = 7/8$$

Finally, the general solution is:

$$x(t) = \frac{1}{8} e^t + e^{-t}(3/8 \sin 2t + 7/8 \cos 2t) \qquad \textbf{Ans.}$$

8.3 INTEGRO-DIFFERENTIAL EQUATIONS

There are electrical circuits which are governed by integro-differential equations which involve differential coefficients and integrals.

EXAMPLE 8.6 Solve the integro-differential equation:

$$i + di/dt + \int_0^t i dt = 1 \qquad (8.28)$$

with the initial conditions $i(0) = 0$ and $i'(0) = 1$

SOLUTION:

Method 1

Differentiating Eq. (8.28) with respect to t,

$$\frac{d^2 i}{dt^2} + \frac{di}{dt} + i = 0 \qquad (8.29)$$

The roots of the characteristic equations are m_1, $m_2 = -\frac{1}{2} \pm j\sqrt{3}/2$

The complementary function is:

$$i(t) = e^{-t/2}(B_1 \sin \sqrt{3}t/2 + B_2 \cos \sqrt{3}t/2) \quad (8.30)$$

At $t = 0$,

$$0 = B_2 \quad (8.31)$$

Differentiating Eq. (8.30),

$$i'(t) = B_1 e^{-t/2} \sqrt{3}/2 \cos \sqrt{3}t/2 + B_1 \sin \sqrt{3}t/2 \times (-\frac{1}{2})e^{-t/2} + \text{two terms containing } B_2$$

At $t = 0$,

$$1 = \sqrt{3}/2 B_1 \quad (8.32)$$

$$B_1 = 2/\sqrt{3}\, B_2 = 0$$

The general solution is:

$$i = e^{-t/2} \cdot 2/\sqrt{3} \sin \sqrt{3}t/2 \qquad \textbf{Ans.} \quad (8.33)$$

Method 2

The complementary function is now expressed in terms of exponentials of the roots of the characteristic equation,

$$i = A_1 e^{m_1 t} + A_2 e^{m_2 t} \quad (8.34)$$

where $m_1, m_2 = -\frac{1}{2} \pm j\sqrt{3}/2$

At $t = 0$,

$$0 = A_1 + A_2 \quad (8.35)$$

Differentiating Eq. (8.34),

$$i' = A_1 m_1 e^{m_1 t} + A_2 m_2 e^{m_2 t} \quad (8.36)$$

At $t = 0$,

$$1 = A_1 m_1 + A_2 m_2 \quad (8.37)$$

Solving Eqs. (8.35) and (8.37),

$$A_1 = \frac{1}{m_1 - m_2} = \frac{1}{\sqrt{3}j}$$

and

$$A_2 = \frac{1}{m_2 - m_1} = \frac{1}{-\sqrt{3}j}$$

Substituting the values of m_1 and m_2 in A_1 and A_2.

The general solution is thus

$$i = \frac{1}{\sqrt{3}} e^{(-1/2)t} \left[\frac{1}{j} e^{j(\sqrt{3}/2)t} - \frac{1}{j} e^{-j(\sqrt{3}/2)t} \right]$$

$$\therefore \quad i = e^{-\frac{1}{2}t} \cdot 2/\sqrt{3} \sin (\sqrt{3}t/2) \qquad \textbf{Ans.}$$

8.4 DIFFERENTIAL EQUATIONS GOVERNING ELECTRICAL CIRCUITS

EXAMPLE 8.7 Solve the electrical circuit shown in Fig. 8.1.

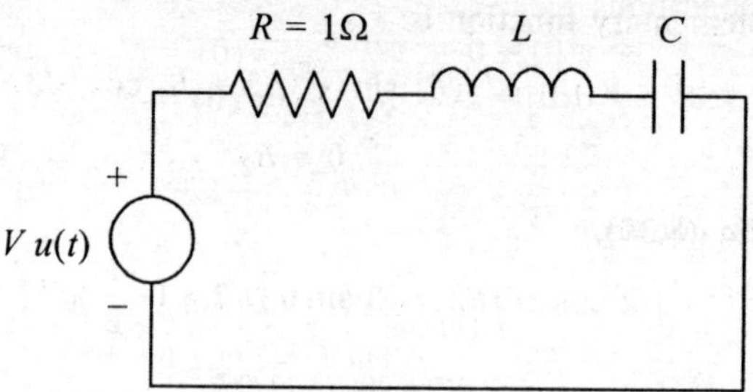

Fig. 8.1

SOLUTION:
The equation governing the circuit is:

$$Ri + L\frac{di}{dt} + \frac{1}{C}\int_0^t i\,dt = Vu(t) \tag{8.38}$$

With $R = 1\ \Omega$, $L = 1$ H, $C = 1$ F and $V = 1$ V, Eq. (8.38) reduces to

$$i + \frac{di}{dt} + \int_0^t i\,dt = 1 \tag{8.39}$$

It is assumed that the circuit has no energy stored in it when the voltage is switched on. The inductance acts as a short-circuit and the capacitance as an open-circuit at time $t = 0$.
Hence

$$i\,(0) = 0$$

$i'(0)$ is found by making use of Eq. (8.38),

$$0 + L\frac{di}{dt} + 0 = V$$

and

$$i'(0) = V/L = 1/1 = 1$$

Equation (8.38) with the initial conditions $i(0) = 0$ and $i'(0) = 1$ is seen to be identical with the integro-differential equation (8.28). Hence the current in the circuit is given by

$$i = e^{-\frac{1}{2}t} \cdot 2/\sqrt{3}\ \sin(\sqrt{3}t/2) \quad \textbf{Ans.}$$

8.5 SIMULTANEOUS EQUATIONS

8.5.1 Homogeneous Simultaneous Equations

The technique adopted for homogeneous simultaneous equations is illustrated in Example 8.8.

EXAMPLE 8.8 Solve the simultaneous equations

$$3dx/dt + 6x + 2dy/dt = 0$$
$$dx/dt + 2x + dy/dt + y = 0$$

with the initial conditions

$$x(0) = 0 \qquad y(0) = 1$$
$$x'(0) = 2 \qquad y'(0) = -3$$

SOLUTION:
The characteristic equation is:

$$F(m) = \begin{vmatrix} 3m+6 & 2m \\ m+2 & m+1 \end{vmatrix} = 0$$

Evaluating the determinant,

$$F(m) = m^2 + 5m + 6 = 0$$

or
$$(m + 2)\,(m + 3\,) = 0$$

Hence the incomplete solution is:

$$x(t) = A_1e^{-2t} + A_2e^{-3t}$$
$$y(t) = A_3e^{-2t} + A_4e^{-3t}$$

Using the initial conditions,
At $t = 0$

$$0 = A_1 + A_2 \tag{8.40}$$
$$2 = -2A_1 - 3A_2 \tag{8.41}$$

Solving Eqs. (8.40) and (8.41),

$$A_1 = 2 \quad A_2 = -2$$

Thus,

$$x(t) = 2e^{-2t} - 2e^{-3t} \qquad \textbf{Ans.}$$

$y(t)$ is also found by using the initial conditions,
At $t = 0$

$$1 = A_3 + A_4 \tag{8.42}$$
$$-3 = -2A_3 - 3A_4 \tag{8.43}$$

Solving Eqs. (8.42) and (8.43),

$$A_3 = 0 \quad A_4 = 1$$

Finally,

$$y(t) = e^{-3t} \qquad \textbf{Ans.}$$

8.5.2 Non-homogeneous Simultaneous Equations

It can be solved as shown in Example 8.9.

EXAMPLE 8.9 Solve the equations

$$3dx/dt + 6x + 2dy/dt = 5e^{5t}$$

$$dx/dt + 2x + dy/dt + y = 0$$

with initial conditions

$$x(0_+) = 0 \qquad y(0_+) = -1$$
$$x'(0_+) = 1 \qquad y'(0_+) = 0$$

SOLUTION:

In order to find the particular integral the given equations are put in the matrix form:

$$\begin{bmatrix} 3D+6 & 2D \\ D+2 & D+1 \end{bmatrix}\begin{bmatrix} x \\ y \end{bmatrix} = \begin{bmatrix} 5e^{5t} \\ 0 \end{bmatrix}$$

By Cramer's Rule,

$$x = \frac{\begin{vmatrix} 5 & 2D \\ 0 & D+1 \end{vmatrix}}{\begin{vmatrix} 3D+6 & 2D \\ D+2 & D+1 \end{vmatrix}} e^{5t}$$

Putting $D = 5$ (See item 1, Table 8.1)

$$x = \frac{\begin{vmatrix} 5 & 2\times 5 \\ 0 & 5+1 \end{vmatrix}}{\begin{vmatrix} 3\times 5+6 & 2\times 5 \\ 5+2 & 5+1 \end{vmatrix}} e^{5t}$$

$$= \frac{5(5+1)}{21\times 6 - 7\times 10} = \frac{30}{56}$$

$$= \frac{15}{28}$$

Similarly,

$$y = \frac{\begin{vmatrix} 3D+6 & 5 \\ D+2 & 0 \end{vmatrix}}{\begin{vmatrix} 3D+6 & 2D \\ D+2 & D+1 \end{vmatrix}} e^{5t}$$

Putting $D = 5$

$$= -\frac{35}{56} = -\frac{5}{8}$$

The complementary function has been found to be (See Example 8.8)

$$x(t) = 2e^{-2t} - 2e^{-3t}$$

$$y(t) = e^{-3t}$$

The general solution is:

$$\frac{15}{28}e^{5t} + 2e^{-2t} - 2e^{-3t} - \frac{5}{8}e^{5t} + e^{-3t}$$

SUMMARY

- The chapter was limited to the solution of ordinary differential equations using classical methods. Transform methods will be introduced in Chapter 9.

Solved Problems

P8.1 Solve the *homogeneous linear differential equation* of the first order:

$$dy/dx + P(x)y = 0 \tag{1}$$

SOLUTION:
From (1),

$$dy/y = -P(x)\, dx$$

Integrating both sides,

$$\ln y = -\int P(x)\, dx + C$$

where C is an arbitrary constant.
Thus,

$$y = e^{-\int P(x)dx + C} = e^{-\int P(x)dx} e^{C}$$

$$= K e^{-\int P(x)dx} \qquad \textbf{Ans.}$$

where K is an arbitrary constant.

P8.2 Solve

$$dy/dx + P(x)y = Q(x) \tag{1}$$

SOLUTION:
Assume

$$y = u(x)\, v(x)$$

where u and v are functions of x to be determined.
Substituting y in Eq. (1)

$$v\, du/dx + u\, dv/dx + P(x)\, uv = Q(x)$$

which may be written in the form:

$$v\, [du/dx + P(x)u] + u\, dv/dx = Q(x)$$

It can be shown that

$$y = C e^{-\int P dx} + e^{-\int P dx} \int e^{\int P(x)dx} Q(x)\, dx \tag{2}$$

Consider the solution of $x\, dy/dx - 2y = 2x$

or $$dy/dx - 2y/x = 2$$

From Eq. (2),

$$y = Ce^{2\ln x} + e^{2\ln x} \int e^{2\ln x}\, 2\, dx$$

$$= Ce^{2\ln x} + e^{2\ln x}[e^{-2\ln x}(-x/2)2]$$

$$= Ce^{2\ln x} - x$$

Check the solution by putting the value of y in Eq. (2).

$$dy/dx = Ce^{2\ln x}\, 2/x - 1$$

and the LHS of Eq. (2) $= xCe^{2\ln x}\, 2/x - 2Ce^{2\ln x} + 2x = 2x$ **Ans.**

P8.3 Find the complementary function of the equation

$$(D^2 - 3D + 2)y = x\,e^x$$

See item 7, Table 8.1.

$$y = e^x \frac{1}{(D-1)(D-2)} x$$

$$= e^x \frac{1}{(D+1-1)(D+1-2)} x$$

$$= e^x \frac{1}{D(D-1)} x$$

$$= e^x \frac{1}{-D(1-D)} x$$

$$= e^x \frac{(1-D)^{-1}}{-D} x$$

$$\text{CF} = e^x \frac{1}{-D}(1 + D + D^2 + \cdots)x$$

$$= e^x \left(-\frac{1}{D} - 1 - D\right)x$$

$$= e^x \left(\frac{-x^2}{2} - x - 1\right) \quad \textbf{Ans.}$$

Exercises

E8.1 Solve the following differential equation:

$$\frac{d^2y}{dt^2} - 2a\frac{dy}{dt} + (a^2 + b^2)y = 0$$

with the initial conditions $y(0) = 0 \quad y'(0) = 1$

[*Ans:* $b^{-1} e^{at} \sin bt$]

E8.2 Find the particular integral of the following equation:

$$(D^2 + 3D + 2)y = x^3 + x$$

Verify the solution by substituting for y in the given equation.

$$\left[\textbf{Hint}: \frac{x^3 + x}{(D+1)(D+2)}\right]$$

[*Ans:* $y = 1/8\,(4x^3 - 18x^2 + 46x - 51)$]

E8.3 Solve $\dfrac{d^3y}{dt^3} - \dfrac{d^2y}{dt^2} + \dfrac{dy}{dt} - y = \cos t + \sin t - 1$

With initial conditions $y(0) = y''(0) = 0$ and $y'(0) = 1$.

[*Ans:* $y = 1 - \cos t + (1 - \tfrac{1}{2}t) \sin t$]

Laplace Transforms in Network Analysis

9

CHAPTER CONTENTS

In Chapter 8, classical methods were used for solving differential equations in the time domain. In this chapter, differential equations will be *transformed* to the *frequency domain* where the independent variable is the *complex frequency* **s**. The resulting algebraic equations, which also take into consideration the initial conditions of the problem are then solved. Thereafter an inverse transformation gives the final time-domain solution. This inverse is obtained by looking up a table of transform pairs or in the alternative by partial fraction techniques. There is also the direct evaluation of the inverse from the defining integral which involves contour integration.

The use of Laplace transforms may be compared to the use of logarithms in arithmetical operations. For example, the product of two numbers a and b is obtained by taking the logarithms of the factors and adding them up. The inverse logarithm of this sum gives the product $a \times b$.

If $$c = a \times b$$

$$\log c = \log a + \log b$$

and $$c = \log^{-1} (\log a + \log b)$$

Thus, the use of logarithms converts the multiplication problem into an addition problem. (which is much simpler). Laplace transforms likewise convert the solution of **differential** equations into that of **algebraic** equations.

Laplace transformation is the most extensively used technique in solving differential equations and integro-differential equations and consequently in the solution of electrical networks. Besides there are several other applications listed in Reference 13. Transform methods specific to ac circuits are considered in detail in Chapter 10.

Two other transforms are in common use: Fourier transform and Z-transform; these are briefly covered in Chapter 13 and Chapter 17, respectively.

9.1 SINGULARITY FUNCTIONS

There are three special functions related to Laplace transformation, namely

- Heaviside unit step function denoted by $u(t)$
- Dirac delta function denoted by $\delta(t)$
- Unit ramp function denoted by $r(t)$

These functions are defined in this section and their Laplace transforms are derived in the following sections.

The **unit step function** is defined as:

$$\begin{aligned} u(t) &= 1 \qquad t > 0 \\ &= 0 \qquad t < 0 \end{aligned} \tag{9.1}$$

and is shown in Fig. 9.1(a).

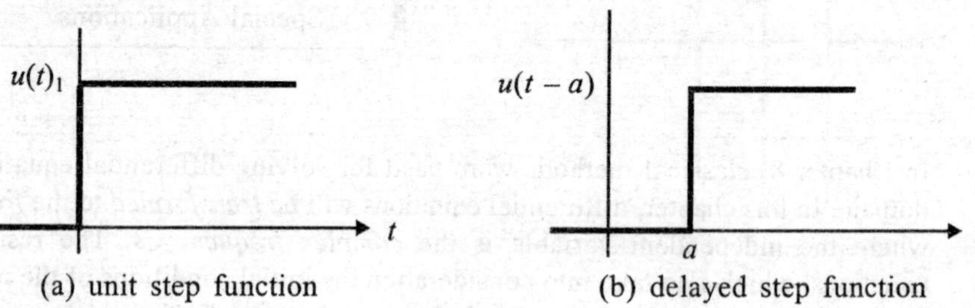

(a) unit step function (b) delayed step function

Fig. 9.1

This function can be used to describe a switch which is closed at $t = 0$ and connects a 1 V source to an electrical circuit.

A **time shifted unit step function** is defined as:

$$u(t-a) = \begin{cases} 1 & t > a \\ 0 & t < a \end{cases}$$

and is shown in Fig. 9.1(b). Unit step functions serve to describe many common waveforms.

EXAMPLE 9.1 The gate function $f(t)$ in Fig. 9.2(a) may be expressed as the sum of two step functions as shown in Fig. 9.2(b).

$$f(t) = 5u(t) - 5u(t-2)$$

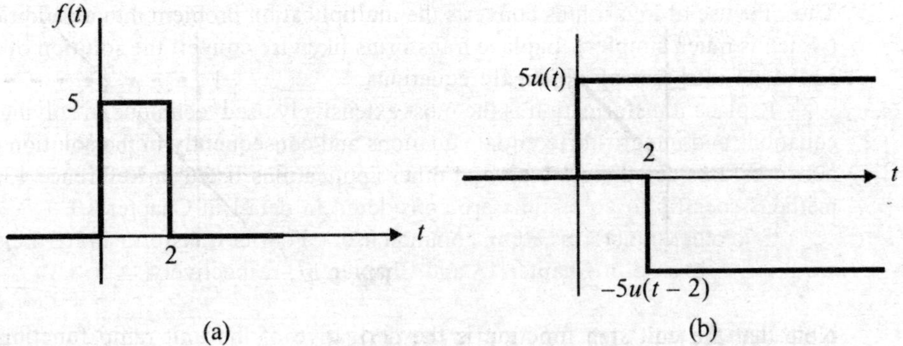

Fig. 9.2 A gate function.

EXAMPLE 9.2 The function $f(t)$ in Fig. 9.3(a) may be expressed as the sum of three step functions shown in Fig. 9.3(b).

$$f(t) = 5u(t) - 10u(t-2) + 5u(t-4)$$

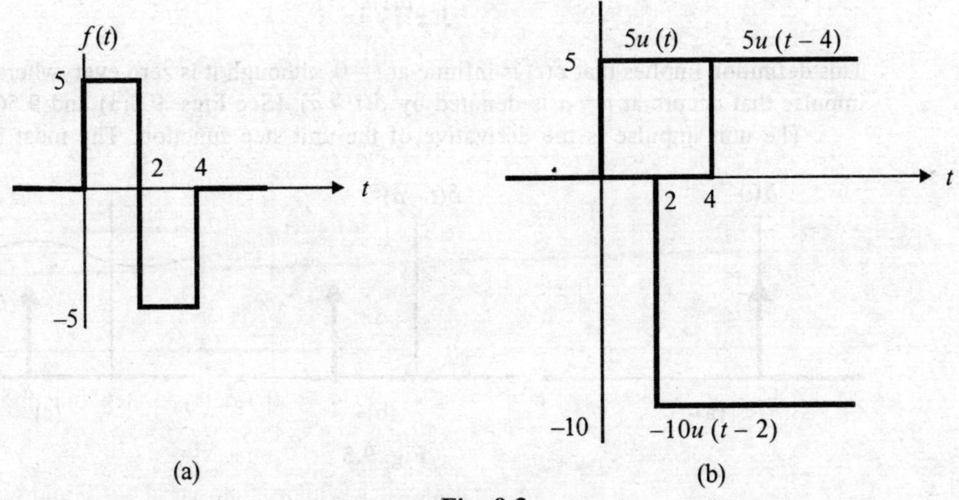

Fig. 9.3

The **unit ramp function** $r(t)$ is defined as:

$$r(t) = t\,u(t) \tag{9.2}$$

and is shown in Fig. 9.4(a).

EXAMPLE 9.3 The function $r(t)$ may be expressed as the product of two functions as shown in Fig. 9.4.

$$r(t) = t[u(t) - u(t-1)]$$

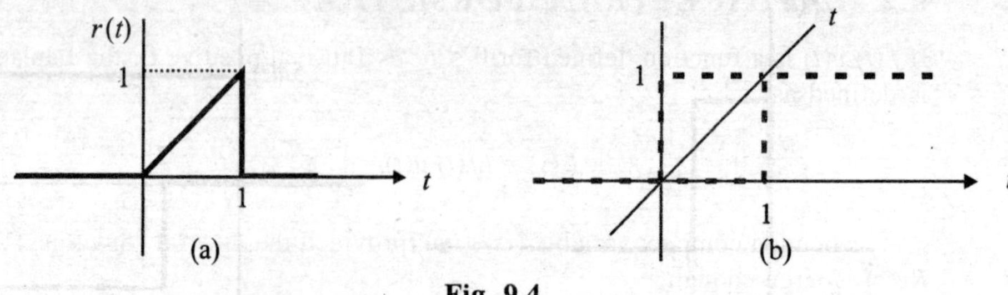

Fig. 9.4

Note that the unit step function is the derivative of the unit ramp function.

The unit impulse function (also known as Dirac delta function) is denoted by $\delta(t)$. In certain physical systems there arises the need for describing an entity which has a very small extend in either space or time but which has a significant total value. Such a distribution is idealized in the **Dirac delta function**; and it is defined by

$$\delta(t) = 0 \quad \text{for} \quad t \neq 0 \tag{9.3}$$

$$\int_{-\infty}^{\infty} \delta(t)dt = 1$$

This definition implies that $\delta(t)$ is infinite at $t = 0$, although it is zero everywhere else. An impulse that occurs at $t = a$ is denoted by $\delta(t - a)$. (See Figs. 9.5(a) and 9.5(b)]

The unit impulse is the derivative of the unit step function. The most important

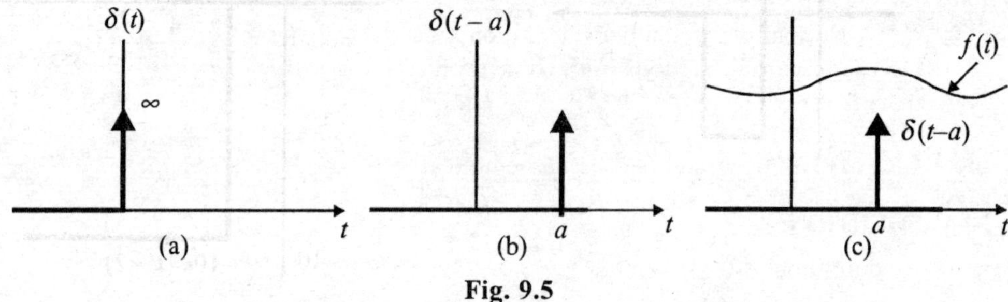

Fig. 9.5

property of the impulse function is the sifting theorem (sampling property) which is expressed as:

$$\int_{-\infty}^{+\infty} f(t)\delta(t-a)dt = f(a) \qquad \text{[See Fig. 9.5(c)]} \tag{9.4}$$

where the function $f(t)$ is continuous at $t = a$. The equation shows that the impulse function sifts out everything except the value of $f(t)$ at $t = a$, the location of the impulse. The validity of Eq. (9.4) may be seen by noting that $\delta(t - a)$ is zero except at $t = a$.

If $a \to 0$, Eq. (9.4) reduces to

$$\int_{-\infty}^{+\infty} f(t)\delta(t)dt = f(0) \tag{9.5}$$

9.2 LAPLACE TRANSFORMATION

If $f(t)\, u(t)$ is a function defined for $0 < t < \infty$ (all real positive t), the Laplace transform is defined as:

$$F(s) = Lf(t)\, u(t) = \int_0^{\infty} e^{st} f(t)\, dt \tag{9.6}$$

s being a complex variable $(\sigma + j\omega)$, provided the integral converges, that is to say $Re(s)$ is large enough.

Sufficient conditions for the existence of the defining integral are:

(a) $f(t)\, u(t)$ must be continuous in the interval $0 < t < \infty$

(b) $f(t)\, u(t)$ must be of exponential order, i.e.,

$$\lim_{t \to \infty} |f(t)| \le Me^{ct} \tag{9.7}$$

where c can be found for $M > 0$ such that Eq. (9.7) holds. Then the integral in Eq. (9.6) converges for all $\sigma > c$. The common functions occurring in network analysis satisfy these conditions.

The time domain $f(t)\, u(t)$ is said to be **transformed** into frequency domain $F(s)$.

Conversely, $$L^{-1} F(s) = f(t)\, u(t) \tag{9.8}$$

The direct evaluation of $f(t)$ from $F(s)$ involves the complex contour integral,

$$f(t)\, u(t) = \frac{1}{2\pi j} \int_{\sigma - j\infty}^{\sigma + j\infty} F(s)\, e^{st}\, ds, \ \sigma \text{ being real and positive.} \tag{9.9}$$

Normally, the contour integration which is rather cumbersome can be avoided by the use of partial fractions in the transform and looking up a Table of Laplace transform pairs.

EXAMPLE 9.4 Find $Le^{at}\, u(t)$, a being real.

SOLUTION:

By definition

$$F(s) = \int_{0-}^{\infty} e^{at}\, dt$$

$$= \left[\frac{e^{-(s-a)i}}{(s-a)} \right]_{0-}^{\infty}$$

$$= \frac{1}{(s-a)} \qquad (\sigma > a) \tag{9.10}$$

Putting $a = 0$ in Eq. (9.10),

$$Lu(t) = \frac{1}{s} \qquad (\sigma > 0) \qquad \textbf{Ans.} \tag{9.11}$$

9.3 PROPERTIES OF LAPLACE TRANSFORM

Several important properties of the Laplace transform are stated hereinafter. See Reference 1 for a formal proof for these properties.

9.3.1 Linearity

$$L(af) = aL(f) \tag{9.12}$$

where f is a function of time and a is any constant.

$$L(f+g) = L(f) + L(g) \tag{9.13}$$

where f and g are functions of time.

EXAMPLE 9.5 Find $L \sin at$

$$L \sin at = L\left(\frac{e^{jat} - e^{-jat}}{2j}\right)$$

$$= \frac{1}{2j}\left[\frac{1}{s-ja} - \frac{1}{s+ja}\right]$$

$$= \frac{a}{s^2+a^2} \tag{9.14}$$

EXAMPLE 9.6 Find $L \cos at$

$$L \cos at = L\left(\frac{e^{jat} + e^{-jat}}{2}\right)$$

$$= \frac{1}{2}\left[\frac{1}{s-ja} + \frac{1}{s+ja}\right]$$

$$= \frac{s}{s^2+a^2} \tag{9.15}$$

EXAMPLE 9.7 Find $L \cosh at$

$$L \cosh at = L\left(\frac{e^{at} + e^{-at}}{2}\right)$$

$$= \frac{1}{2}\left[\frac{1}{s-a} + \frac{1}{s+a}\right]$$

$$= \frac{s}{s^2-a^2} \tag{9.16}$$

EXAMPLE 9.8 Find L sinh at

$$L \sinh at = L\left(\frac{e^{at} - e^{-at}}{2}\right)$$

$$= \frac{1}{2}\left[\frac{1}{s-a} - \frac{1}{s+a}\right]$$

$$= \frac{a}{s^2 - a^2} \tag{9.17}$$

9.3.2 Scale Changing

$$L\, f(kt) = 1/k\; F(s/k) \tag{9.18}$$

By definition of Laplace transform,

$$L\, f(kt) = \int_0^\infty e^{-st} f(kt)\, dt$$

Putting $u = kt$

$$Lf(u) = \int_0^\infty e^{-su/k} f(u)\, du/k$$

$$= 1/k\, F(s/k)$$

or

$$L\, f(kt) = 1/k\; F(s/k)$$

EXAMPLE 9.9 Given $L \sin t = 1/(s^2 + 1)$, prove that

$$L \sin at = \frac{a}{s^2 + a^2}$$

SOLUTION:
Since

$$L \sin at = \frac{1}{a}\left(\frac{1}{(s/a)^2 + 1}\right)$$

$$= \frac{a}{s^2 + a^2}$$

which agrees with Eq. (9.14).

9.3.3 Differentiation with Respect to Parameter

If $f(t, c) = F(s, c)$
Then

$$L\frac{\delta f}{\delta c} = \frac{\delta}{\delta c}(F) \tag{9.19}$$

EXAMPLE 9.10 Derive $L\ t \sin at$ from the transform $L \cos at$.

SOLUTION:

Since $$L \cos at = s/(s^2 + a^2)$$

Equation (9.19) gives,

$$L(-t) \sin at = -\frac{2as}{(s^2 + a^2)^2}$$

or

$$Lt \sin at = \frac{2as}{(s^2 + a^2)^2} \quad \textbf{Ans.} \tag{9.20}$$

EXAMPLE 9.11 Using $L\ e^{at} = 1/(s - a)$. Show that

$$Lt^n e^{at} = \frac{n!}{(s-a)^{n+1}} \tag{9.21}$$

SOLUTION:

Put $a = 0$ in Eq. (9.21),

$$Lt^n = \frac{n!}{s^{n+1}}, \text{ n is positive integer.} \tag{9.22}$$

9.3.4 s-shift

$$L\ e^{at} f(t) = F(s - a) \tag{9.23}$$

By definition of Laplace transform,

$$\int_0^\infty e^{(a-s)t} f(t)\,dt = F(s-a)$$

EXAMPLE 9.12

$$Le^{-kt} \sin at = \frac{a}{(s+k)^2 + a^2} \tag{9.24}$$

and $$Le^{-kt} \cos at = \frac{s+k}{(s+k)^2 + a^2} \tag{9.25}$$

9.3.5 Time Differentiation

$$L f'(t) = s\,F(s) - f(0-) \tag{9.26}$$

For $$L f'(t) = \int_{0-}^\infty e^{-st} f'(t)\,dt$$

$$[e^{-st} f(t)]_{0-}^{\infty} - \int_{0-}^{\infty} f(t) - se^{-st} dt = sF(s) - f(0-)$$

and $$L f''(t) = sf'(t) - f'(0-) = s^2 F(s) - sf(0-) - f'(0) \qquad (9.27)$$

EXAMPLE 9.13 Laplace transform of unit impulse function is given by

$$L\delta = L\frac{\partial u(t)}{\partial t} = s(1/s) - 0 = 1 \qquad (9.28)$$

and $L\delta = 1$

EXAMPLE 9.14 Derive the Laplace transform of cos *at* given the transform of *s* sin *at*.

SOLUTION:

$$L \cos at = L\frac{d}{dt}\sin at$$

$$= sL \sin at - \sin 0$$

$$= \frac{sa}{s^2 + a^2}$$

and $$L \cos at = \frac{s}{s^2 + a^2}$$

which agrees with Eq. (9.15).

9.3.6 Time Integration

Integration in time domain is transformed into division by *s* in the frequency domain.

If $$g(t) = \int_{0-}^{t} f(t)\,dt \qquad (9.29)$$

Then

$$G(s) = \frac{1}{s}F(s)$$

From Eq. (9.29),

$$f(t) = dg/dt$$

and $$F(s) = L f(t) = s\,G(s) - g(0-)$$

Since $g(0-) = 0$

$$G(s) = L\int_{0-}^{t} f(t)\,dt = \frac{1}{s}F(s)$$

which is the required result.

EXAMPLE 9.15 Show that

$$L^{-1}\frac{1}{s(s-a)}=\frac{e^{at}-1}{a}$$

SOLUTION:

Method 1

Using Eq. (9.29),

$$\text{LHS becomes } \int_{0-}^{t} e^{at}\,dt=\frac{e^{at}-1}{a}$$

Method 2

Using partial fractions (see section below)

$$L^{-1}\frac{1}{s(s-a)}=L^{-1}\left[\frac{1/a}{s-a}-\frac{1/a}{s}\right]$$

$$=\frac{1}{a}(e^{at}-1)$$

9.3.7 Indefinite Integral

If
$$h(t)=\int_{..}^{t} f(t)\,dt=h(0)+\int_{0}^{t} f(t)\,\mathrm{dt}$$

Then

$$L\int_{..}^{t} f(t)\,dt=\frac{h(0)}{s}+\frac{F(s)}{s} \tag{9.30}$$

9.3.8 Complex Frequency Differentiation

$$L\{-tf(t)\}=\frac{dF(s)}{ds} \tag{9.31}$$

For
$$\text{RHS}=\int_{0-}^{\infty} f(t)\frac{d}{ds}e^{-st}\,dt=\int_{0-}^{\infty}-tf(t)e^{-st}\,dt$$

$$=L\{-t\,f(t)\}$$

EXAMPLE 9.16 Show that

$$Ltu(t)=1/s^2$$

SOLUTION:

From Eq. (9.11),

$$Lu(t)=1/s$$

Hence
$$L-tu(t)=\frac{d}{ds}\,(1/s)=-1/s^2$$

It follows

$$Ltu(t) = 1/s^2$$

Similarly,

$$Lt^2u(t) = -\frac{d}{ds}\ (1/s^2) = 2/s^3$$

EXAMPLE 9.17 Show that

$$Lte^{-at}dt = \frac{1}{(s+a)^2} \tag{9.32}$$

SOLUTION:
From Eq. (9.10),

$$Le^{-at} = \frac{1}{(s+a)}$$

Using Eq. (9.31),

$$Lte^{-at} = -\frac{d}{ds}\frac{1}{(s+a)}$$

$$= \frac{1}{(s+a)^2}$$

Similarly,

$$Lt^2\ e^{-at} = \frac{2}{(s+a)^3}$$

9.3.9 Complex Frequency Integration

$$L\frac{f(t)}{t} = \int_s^{\infty} F(s)\,ds \tag{9.33}$$

$$\text{RHS} = \int_s^{\infty}\int_{0-}^{\infty} f(t)\,e^{-st}\,dt\,ds$$

$$= \int_{0-}^{\infty} f(t)\,dt \int_s^{\infty} e^{-st}\,ds$$

$$= \int_{0-}^{\infty} \frac{f(t)}{t} e^{-st}\,dt$$

Hence

$$L\frac{f(t)}{t} = \int_s^{\infty} F(s)\,ds$$

EXAMPLE 9.18 Prove that

$$L\frac{\sin t}{t} = \tan^{-1}(1/s) \tag{9.34}$$

SOLUTION:
From Eq. (9.14),

$$L \sin t = \frac{1}{s^2+1}$$

Using Eq. (9.33),

$$L\frac{\sin t}{t} = \int_s^{\infty} 1/(s^2+1)\,ds$$

$$= \left[\tan^{-1} s\right]_s^{\infty}$$

$$= \pi/2 - \tan^{-1} s$$

$$= \tan^{-1}(1/s)$$

See Table 9.1 for important transform pairs and Table 9.2 for important properties of Laplace transforms. Page (148)

9.3.10 Convolution Property or Faltung Theorem

If
$$F(s) = L\,f(t) \text{ and } G(s) = L\,g(t)$$
Then

$$L\{\int_0^t f(x)\ g(t-x)dx\} = F(s)\ G(s) \text{ (Substituting a dummy variable } x \text{ for } t) \quad (9.35)$$

Since the value of a definite integral is independent of the variable of integration, it is convenient to use the letter x instead of t.

The integral in the equation is denoted by $f(t) * g(t)$ and is called the **convolution** of $f(t)$ and $g(t)$. Equation (9.35) is often written as:

$$L\{f(t) * g(t)\} = F(s)\ G(s) \quad (9.36)$$

Corollary to Faltung theorem: By applying the time-differentiation property to Eq. (9.35),

$$L\frac{d}{dt}\int_0^t f(x).\,g(t-x)\,dx = sF(s)\ G(s)$$

By change of variable $x' = t - x$, the convolution becomes

$$f(t) * g(t) = \int_0^t g(x')f(t-x') = g(t) * f(t)$$

which shows that convolution is commutative.

EXAMPLE 9.19

$$L^{-1}1/(s+1)(s+2) = L^{-1}\frac{1}{(s+1)} * L^{-1}\frac{1}{(s+2)}$$

SOLUTION:.

$$\text{RHS} = e^{-t}u(t) * e^{-2t}u(t)$$

$$= \int_{0-}^{t} e^{-x}u(x)e^{-2(t-x)}u(t-x)\,dx$$

$$= e^{-2t}\int_{0-}^{t} e^{x}\,dx$$

$$= e^{-t} - e^{-2t}$$

The convolution property can serve as a powerful tool to find the **inverse** Laplace transform when $F(s)$ is given as the product of two simple functions.

9.3.11 Initial Value and Final Value Theorems

The initial value theorem states that

$$f(0+) = \lim_{s\to\infty} sF(s)$$

where $f(0+)$ is the value of $f(t)$ at $t = 0+$

This theorem is true whether

$$f(0-) = f(0+) \text{ or } f(0-) \neq f(0+)$$

EXAMPLE 9.20 The current in a circuit is given by

$$I(s) = \frac{s^2 + s + 7}{(s^2 + 4s + 3)(s + 8)}$$

Using the initial value theorem find the value at $t = 0+$. Check with calculated value.

SOLUTION:

$$\text{Initial value} = \lim_{s\to\infty} sI(s) = \lim \frac{s^3 + \cdots}{s^3 + \cdots} = 1$$

which may be checked by finding $i(t)$ and determining $i(0+)$ as follows:

$I(s)$ may be written as:

$$\frac{s^2 + s + 7}{(s+1)(s+3)(s+8)} = \frac{1/2}{(s+1)} - \frac{13/10}{(s+3)} + \frac{9/5}{(s+8)},$$

making use of the cover-up rule.

Inverting

$$i(t) = \frac{1}{2}e^{-t} - \frac{13}{10}e^{-3t} + \frac{9}{5}e^{-8t}$$

and

$$i(0+) = \frac{1}{2} - \frac{13}{10} + \frac{9}{5}$$

$$= 1 \qquad \textbf{Ans.}$$

The final value theorem states that

$$\lim_{t\to\infty} f(t) = \lim_{s\to 0} sF(s)$$

provided $sF(s)$ has no poles on the $j\omega$ axis or inside the right half s-plane (see Chapter 12).

Table 9.1 Laplace Transform Pairs

	$f(t)u(t)$	$F(s)$	Type
1.	$u(t)$	$1/s$	Step
2.	$\delta(t)$	1	Impulse
3.	$t\,u(t)$	$1/s^2$	Ramp
4.	$t^n u(t)$ $(n = 0, 1, 2, \ldots,)$	$\dfrac{n!}{s^{n+1}}$	
5.	$e^{-at}u(t)$	$1/(s+a)$	Exponential
6.	$t^n e^{-at}u(t)$	$\dfrac{n!}{(s+a)^{n+1}}$	
7.	$\sin \omega t\, u(t)$	$\omega/(s^2+\omega^2)$	Sine
8.	$\cos \omega t\, u(t)$	$s/(s^2+\omega^2)$	Cosine
9.	$te^{-at}\, u(t)$	$1/(s+a)^2$	Damped ramp
10.	$e^{-at} \sin \omega t\, u(t)$	$\dfrac{\omega}{(s+a)^2+\omega^2}$	Damped sine
11.	$e^{-at} \cos \omega t\, u(t)$	$\dfrac{s+a}{(s+a)^2+\omega^2}$	Damped cosine
12.	$\sinh at\, u(t)$	$a/(s^2-a^2)$	Hyperbolic sine
13.	$\cosh at\, u(t)$	$s/(s^2-a^2)$	Hyperbolic cosine

EXAMPLE 9.21 Apply the final value theorem to the function $I(s)$ of Example 9.20 to find the final value.

SOLUTION:

$$\lim_{s\to 0} sF(s) = \lim \frac{s(s^2+s+7)}{(s+1)(s+3)(s+8)} = \lim \frac{7s}{24} = 0$$

The inverse transform $i(t)$ in Eq. (9.37) shows clearly that $i(t)_{t\to\infty} = 0$.

Table 9.2 Short List of Operational Transforms

Operation	$F(t)$	$F(s)$
1. Linearity	$A_1 f_1(t) + a_2 f_2(t)$	$a_1F_1(s) + a_2F_2(s)$
2. First derivative (time)	$\dfrac{df(t)}{dt} = f'(t)$	$sF(s) - f(0-)$
3. Second derivative (time)	$\dfrac{d^2 f(t)}{dt^2}$	$s^2F(s) - sf(0-) - f'(0-)$
4. Time integral	$\int_0^t f(x)\,dx$	$\dfrac{F(s)}{s}$
5. Translation in time	$f(t-a)\,u(t-a),\ a > 0$	$e^{-as}F(s)$
6. Translation in s domain	$e^{-at}f(t)$	$F(s+a)$

7. Scale changing	$f(at)\ a > 0$	$\frac{1}{a}F(s/a)$
8. Derivative (s)	$tf(t)$	$\frac{-dF(s)}{ds}$
9. s integral	$\frac{f(t)}{t}$	$\int_s^\infty F(u)\,du$
10. Initial value	$\lim_{t\to 0+} f(t)$	$\lim_{s\to\infty} sF(s)$
11. Final value	$\lim_{t\to\infty} f(t)$	$\lim_{s\to 0} s\,F(s)$
12. Convolution in the t domain	$fs(t) * g(t)$	$F(s)\,.\,G(s)$

9.4 INVERSE LAPLACE TRANSFORM

9.4.1 Partial Fraction Expansion

Simple poles only: Given a transform $F(s)$, which is a **proper** rational function, where the degree of the numerator is less than that of the denominator,

$$\frac{N(s)}{D(s)} = \frac{N(s)}{(s-p_1)(s-p_2)\cdots(s-p_m)}$$

$$= \frac{A_1}{(s-p_1)} + \frac{A_2}{(s-p_2)} + \cdots + \frac{A_m}{(s-p_m)} \tag{9.37}$$

One method for calculating the coefficients $A_1, A_2, \ldots, A_m$ is to substitute convenient values for s in Eq. (9.37) and to obtain enough equations to solve for the unknown coefficients.

EXAMPLE 9.22 Let $F(s) = 1/(s+1)(s+2) = \dfrac{A_1}{(s+1)} + \dfrac{A_2}{(s+2)}$

Put $s = 0$ to obtain

$$\frac{1}{1\times 2} = \frac{A_1}{1} + \frac{A_2}{2}$$

and $s = 1$ to obtain

$$\frac{1}{2\times 3} = \frac{A_1}{2} + \frac{A_2}{3}$$

Solving these two equations,

$$A_1 = 1 \qquad A_2 = -1$$

and $$f(t) = e^{-t} - e^{-2t}$$

The solution agrees with that in Example 9.19, which used the convolution property.

Cover-up method

EXAMPLE 9.23 To evaluate A_1 in Example 9.22, cover up the factor $s + 1$ in the denominator and evaluate the rest of $F(s)$ for $s = -1$.

Thus,

$$A_1 = \left.\frac{1}{(s+1)(s+2)}\right|_{s=-1}$$

$$= 1/(-1+2) = 1$$

Similarly,

$$A_2 = \left.\frac{1}{(s+1)(s+2)}\right|_{s=-2}$$

$$= 1/(-2+1)$$

$$= -1$$

Heaviside expansion theorem: If $F(s)$ has the form $P(s)/Q(s)$, where $P(s)$ and $Q(s)$ are polynomials in s, and the degree of $Q(s)$ is at least one higher than the degree of $P(s)$, further $Q(s)$ has m distinct zeros.

$$L^{-1}\frac{P(s)}{Q(s)} = \sum \frac{P(s_r)e^{srt}}{Q'(s_r)}$$

EXAMPLE 9.24 To determine the inverse transform of

$$\frac{P(s)}{Q(s)} = \frac{s}{s^2 + a^2}$$

$$Q'(s) = 2s \quad \text{and} \quad s = \pm ja$$

Hence

$$L^{-1}\frac{s}{s^2+a^2} = \frac{-a^2e^{jat}}{-2a^2} + \frac{-a^2e^{-jat}}{-2a^2}$$

$$= \frac{e^{jat} + e^{-jat}}{2} = \cos at$$

Multiple poles:

EXAMPLE 9.25 Evaluate the coefficients A_1, A_2, A_3 for the given $F(s)$ with multiple poles.

$$F(s) = \frac{s}{(s+1)(s+2)^2} = \frac{A_1}{(s+1)} + \frac{A_2}{(s+2)^2} + \frac{A_3}{(s+2)} \tag{9.38}$$

$$A_1 = \left.\frac{s}{(s+2)^2}\right|_{s=-1} = -1$$

$$A_2 = \left.\frac{s}{(s+1)}\right|_{s=-2} = 2$$

To find A_3, put $s = 0$ in Eq. (9.38), which gives

$$0 = A_1 + \frac{A_2}{4} + \frac{A_3}{2}$$

and $$A_3 = 1$$

Finally, $$F(s) = \frac{-1}{(s+1)} + \frac{2}{(s+2)^2} + \frac{1}{(s+2)}$$

Complex conjugate poles: It is perhaps best to complete the square in the denominator.

EXAMPLE 9.26 Find the inverse transform of $F(s)$ with complex conjugate poles.

$$F(s) = \frac{(s+1)}{(s^2+s+1)} = \frac{(s+1)}{(s+1/2)^2+3/4}$$

$$= \frac{s+1/2}{(s+1/2)^2+3/4} + \frac{1/\sqrt{3}\,.\,\sqrt{3}/2}{(s+1/2)^2+3/4}$$

and the inverse transform $f(t)$ is:

$$f(t) = e^{(-1/2)t}\cos\frac{\sqrt{3}}{2}t + 1/\sqrt{3}\ e^{-1/2t}\sin\frac{\sqrt{3}}{2}t$$

EXAMPLE 9.27 Find the inverse of $F(s) = \dfrac{1}{s^2(s^2+1)}$.

SOLUTION:

Since

$$L^{-1}\frac{1}{(s^2+1)} = \sin t$$

Then

$$L^{-1}\frac{1}{s(s^2+1)} = \int_0^t \sin x\,dx = 1 - \cos t$$

$$L^{-1}\frac{1}{s^2(s^2+1)} = \int_0^t (1-\cos x)\,dx = t - \sin t \qquad \textbf{Ans.}$$

Check

$$L(t - \sin t) = \frac{1}{s} - \frac{1}{s^2+1}$$

Two useful relationships in partial fraction expansion are the following:

1. Sum of all the residues = 0 if $m > n + 1$ (see Example 9.25)
2. Sum of all the residues = K if $m = n + 1$ (sec Example 9.20)

where $F(s) = K\dfrac{P(s)}{Q(s)}$ and $P(s)$ is of degree n and $Q(s)$ is of degree m.

Note that the residue is the multiplier of $1/(s - p_i)$ *only* whether p_i is a simple or a multiple pole.

EXAMPLE 9.28 Express the following transform in partial fractions:

$$Y(s) = \frac{f(s)}{(s-\alpha)^m g(s)}$$

SOLUTION:

$$Y(s) = \frac{A_0}{(s-\alpha)^m} + \frac{A_1}{(s-\alpha)^{m-1}} + \cdots + \frac{A_{m-1}}{(s-\alpha)} + \frac{f_1}{g}$$

Multiplying by $(s - \alpha)^m$

$$Y = A_0 + (s-\alpha)A_1 + \cdots + (s-\alpha)^{m-1} A_{m-1} + (s-\alpha)^m \frac{f_1}{g}$$

Thus,

$$A_0 = \left(\frac{f}{g}\right)_\alpha \quad A_1 = \left(\frac{f}{g}\right)_\alpha^1 \quad r!A_r = \left(\frac{f}{g}\right)_\alpha^r$$ **Ans.**

EXAMPLE 9.29 Show that in the inverse of $Y(s) = \dfrac{f(s)}{[(s+k)^2 + a^2]g(s)}$,

$y(t)$ has the term $\text{Im}\left(\dfrac{f}{ag}\right)_\alpha^{e^{\alpha t}}$, where $\alpha = -k + ja$

SOLUTION:

$$Y(s) = \frac{f}{ag}\frac{a}{(s+k)^2 + a^2}$$

$$= \frac{f}{ag} Le^{-kt} \sin at$$

$$= \text{Im}\frac{f}{ag} Le^{\alpha t} = \text{Im}\frac{f}{ag}\frac{1}{s-\alpha}$$

$y(t)$ has therefore the term

$$\text{Im}\left(\frac{f}{ag}\right)_\alpha^{e^{\alpha t}}$$ **Ans.**

9.5 APPLICATIONS OF LAPLACE TRANSFORM

Chapter 8 covered the classical methods of solving differential equations. In the following sections, the Laplace transform technique will be introduced through a number of examples, some of them being a reworking of the problems of Chapter 8.

The applications of Laplace transforms in network analysis are so important that they are separately considered in Chapter 10 on ac networks.

9.5.1 Solution of Linear Ordinary Constant Coefficient Differential Equations

The feasibility of applying Laplace transforms in network analysis is illustrated in the examples below.

Single root for the characteristic equation

EXAMPLE 9.30 Solve $a_1 dx/dt + a_0 x = 0$

with the initial condition $x(0) = x_0$

SOLUTION:
Taking the Laplace transform term by term,

$$a_1[s\,X(s) - x(0)] + a_0\,X(s) = 0$$

or

$$X(s)\,[a_1 s + a_0] = a_1 x_0$$

or

$$X(s) = \frac{a_1 x_0}{a_1 s + a_0} = \frac{x_0}{s + a_0/a_1}$$

Inverting,

$$x(t) = x_0 e^{-(a_0/a_1)t} \qquad \textbf{Ans.}$$

which agrees with the solution obtained by the classical method in section 8.1.1.

Distinct real roots

EXAMPLE 9.31

$$\text{Solve } \frac{d^2x}{dt^2} + 3\frac{dx}{dt} + 2x = 0$$

with the initial conditions $x(0) = 3$ and $x'(0) = -1$

SOLUTION:
Taking the Laplace transform term by term,

$$s^2X - sx(0) - x'(0) + 3\,[sX - x(0)] + 2X = 0$$

or

$$X\,[s^2 + 3s + 2] = 3s - 1 + 9$$

or

$$X = \frac{3s + 8}{(s+1)(s+2)} = \frac{5}{(s+1)} - \frac{2}{(s+2)}$$

Inverting,

$$x(t) = 5\,e^{-t} - 2e^{-2t} \qquad \textbf{Ans.}$$

as in section 8.1.2.

Multiple roots

EXAMPLE 9.32 Solve $D^2x - 8Dx + 16x = 0$
with the initial conditions $x(0) = 2$ and $x'(0) = 4$

where D denotes the operation $\frac{d}{dx}$.

SOLUTION:

Taking Laplace transforms,

$$[s^2X - s\,x(0) - x'(0)] - 8\,[sX - x(0)] + 16X = 0$$

or
$$X[s^2 - 8s + 16] = 2s + 4 - 16$$

or
$$X = \frac{2s - 12}{s^2 - 8s + 16}$$

$$= \frac{2(s-6)}{(s-4)^2} = \frac{2(s-4-2)}{(s-4)^2}$$

$$= \frac{2}{(s-4)} - \frac{4}{(s-4)^2}$$

Inverting,
$$x(t) = 2e^{4t} - 4te^{4t} \qquad \textbf{Ans.}$$
as in section 8.1.3.

EXAMPLE 9.33 Solve $\frac{md^2y}{dt^2} + Ky = 0$

with initial conditions $y(0) = h$ and $y'(0) = 0$

SOLUTION:
Taking Laplace transform for each term and making use of Eq. (9.27),

$$m\{s^2Y(s) - sy(0) - y'(0)\} + KY(s) = 0$$

or
$$ms^2Y(s) - msh - 0 + KY(s) = 0$$

or
$$Y(s) = \frac{msh}{ms^2 + K} = \frac{hs}{s^2 + (K/m)} = \frac{hs}{s^2 + \omega^2}$$

where $\omega^2 = K/m$.

Taking the inverse transform,

$$y(t) = h\cos\omega t \qquad \textbf{Ans.}$$

Complex conjugate roots

EXAMPLE 9.34 Solve $D^2x + 2Dx + 5x = 0$

with the initial conditions $x(0) = 1$ and $x'(0) = 0$

SOLUTION:
Taking Laplace transform,

$$X\,[s^2 + 2s + 5] = s + 2$$

or

$$X = \frac{s+2}{(s+1)^2+4}$$

$$= \frac{s+1+1}{(s+1)^2+4}$$

Inverting,

$$x(t) = e^{-t}\,[\cos 2t + \tfrac{1}{2}\sin 2t] \qquad \textbf{Ans.}$$

as in section 8.1.4.

Non-homogeneous linear differential equations

EXAMPLE 9.35 Obtain the **general** solution for the non-homogeneous differential equation

$$(D^2 + 2D + 5)x = e^t$$

with initial conditions $x(0) = 1$ and $x'(0) = 0$.

SOLUTION:
Taking Laplace transforms of both sides,

$$(s^2 + 2s + 5)\,X = \frac{1}{(s-1)} + (s + 2)$$

or

$$X = \frac{1}{[(s+1)^2+4](s-1)} + \frac{(s+2)}{(s+1)^2+4}$$

$$X = \frac{1/8}{s-1} - \frac{(1/8)s+3/8}{(s+1)^2+4} + \frac{s+2}{(s+1)^2+4}$$

or

$$= \frac{1/8}{s-1} + \frac{(7/8)(s+1)+3/4}{(s+1)^2+4}$$

Thus,

$$x(t) = 1/8\;e^t + 7/8\;e^{-t}\cos 2t + (3/8)\;e^{-t}\sin 2t \qquad \textbf{Ans.}$$

EXAMPLE 9.36 $(D^2 + 4D + 3)y = te^{-2t}$

with initial conditions $y(0) = 3$ and $y'(0) = -2$

SOLUTION:
Taking Laplace transforms of both sides,

$$s^2Y(s) - 3s + 2 + 4sY(s) - 12 + 3Y(s) = \frac{1}{(s+2)^2}$$

or
$$(s^2 + 4s + 3)Y(s) - (3s + 10) = \frac{1}{(s+2)^2}$$

or
$$Y(s) = \frac{1/(s+2)^2 + (3s+10)}{(s^2 + 4s + 3)}$$

$$= \frac{A}{(s+2)^2} + \frac{B}{(s+2)} + \frac{C}{(s+1)} + \frac{D}{(s+3)}$$

A, C and D are found by the cover up rule as –1, 4, –1, respectively.

To find B, put $s = 0$ on both sides,

$$41/12 = -1/4 + B/2 + 4/1 - 1/3$$

and
$$B = 0$$

Thus,
$$y(t) = -te^{-2t} + 4e^{-t} - e^{-3t} \quad \textbf{Ans.}$$

EXAMPLE 9.37 Solve $(D^2 - 1)y = 1 - \cos t$

with initial conditions $y(0) = y'(0) = 0$.

SOLUTION:

Taking Laplace transforms,

$$(s^2 - 1)\,Y(s) = 1/s - \frac{s}{s^2+1} = \frac{1}{s(s^2+1)}$$

which gives

$$Y(s) = \frac{1}{s(s^2+1)(s^2-1)}$$

Dividing both sides by s

$$\frac{Y(s)}{s} = \frac{1}{s^2(s^2+1)(s^2-1)}$$

Putting $s^2 = S$

$$\frac{Y(s)}{s} = \frac{1}{S(S+1)(S-1)}$$

$$= -\frac{1}{S} + \frac{1}{2}(S+1) + \frac{1}{2}(S-1), \text{by cover up rule.}$$

Thus,

$$Y(s) = -\frac{1}{s} + \frac{1}{2}\frac{s}{s^2+1} + \frac{1}{2}\frac{s}{s^2-1}$$

Inverting,

$$y(t) = -1 + \frac{1}{2}\cos t + \frac{1}{2}\cosh t \quad \textbf{Ans.}$$

Integro-differential equations

EXAMPLE 9.38 Obtain the general solution for the integro-differential equation

$$i + di/dt + \int_0^t i\,dt = 1$$

with initial conditions $i(0) = 0$ and $i'(0) = 1$.

SOLUTION:
Taking Laplace transform for both sides of the equation,

$$I + [sI - i(0)] + (1/s)\,I = 1/s$$

or $$I\,[1 + s + 1/s] = 1/s$$

or $$I = \frac{1}{s^2 + s + 1}$$

$$= \frac{1}{(s + 1/2)^2 + 3/4}$$

Inverting,

$$i(t) = \frac{2}{\sqrt{3}} e^{(-1/2)t} \sin\frac{\sqrt{3}}{2} t$$

as in Section 8.3.

9.5.2 Simple Electrical Circuits

In this chapter, Laplace transform method is used for the solution of first order ac circuits. Higher order circuits are considered in Chapter 10.

Natural response of first order RC circuits: The behaviour of a circuit with one energy storage element is described by a first order differential equation.

$$Ri + \frac{1}{C}\int_{-\infty}^{t} i\,dt = 0 \tag{9.39}$$

In the *RC* circuit of Fig. 9.6, the capacitor is charged to a voltage V_0 before the switch is closed.
Taking the Laplace transform of the time domain Eq. (9.39),

$$RI - \frac{V_0}{s} + \frac{I}{Cs} = 0$$

$$I = \frac{V_0/s}{R + 1/Cs} = \frac{V_0}{R(s + 1/RC)}$$

The time function is obtained by taking the inverse transformation,

$$i(t) = \frac{V_0}{R} e^{-t/RC} \tag{9.40}$$

The current decays exponentially from a value V_0/R as shown in Fig. 9.6(b).

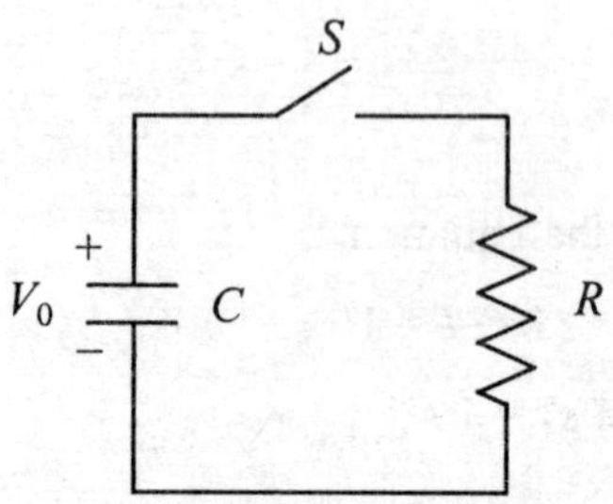

(a) RC circuit with initial charge

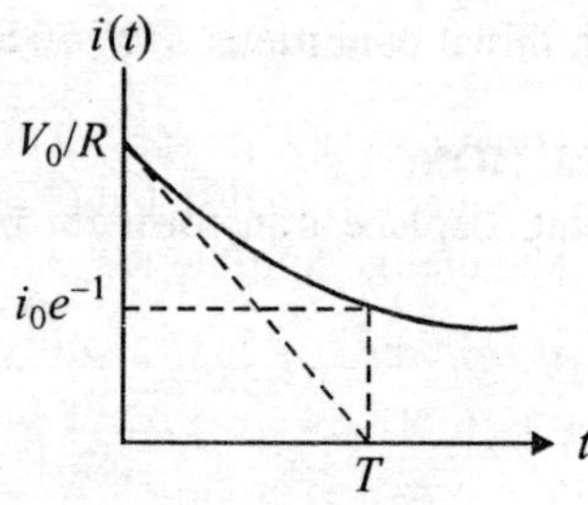

(b) Current-time response of the circuit of (a)

Fig. 9.6

The product RC has the dimension of time and is termed as the **time constant** of the circuit.

At $t = \tau$, $$i = \frac{V_0 e^{-1}}{R} = 0.368$$

Also $$\frac{di}{dt} = -\frac{V_0}{R}\frac{1}{CR} e^{-t/RC}$$

At $t = 0$, $$(di/dt)_{t=0} = -\frac{V_0}{R}\frac{1}{CR}$$

RL circuit: An RL first order circuit is shown in Fig. 9.7. The time domain differential equation is:

$$L\frac{di}{dt} + Ri = 0 \tag{9.41}$$

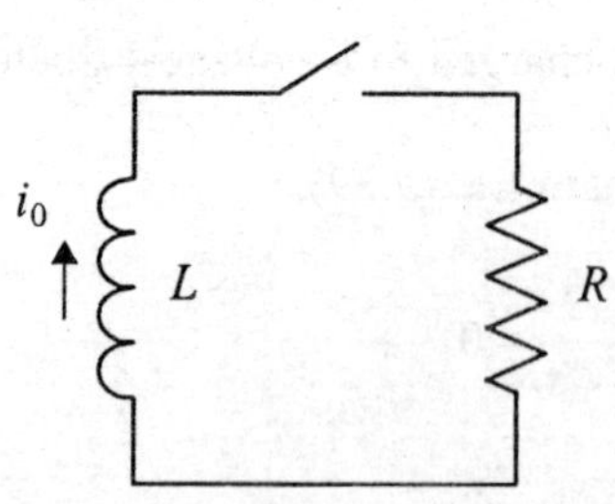

(a) RL circuit with initial current

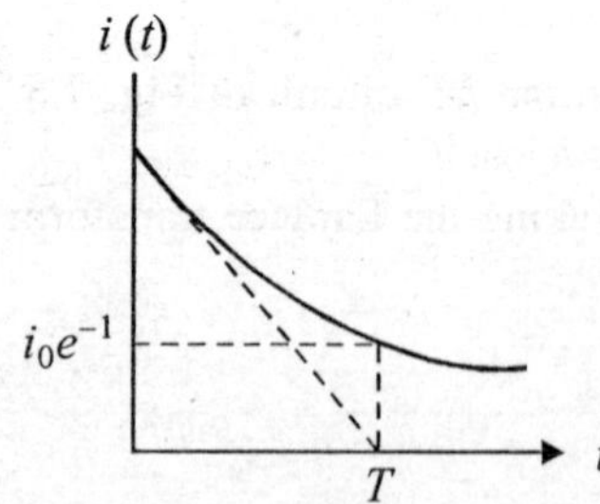

(b) Current-time response of the circuit of (a)

Fig. 9.7

Assuming that the initial current is i_0, the equation transforms to

$$L\,[sI - i_0] + RI = 0$$

which gives

$$I = \frac{L\,i_0}{Ls + R} = \frac{i_0}{s + R/L}$$

and

$$i(t) = i_0\, e^{-R/Lt} \tag{9.42}$$

The time constant $\tau = L/R$ and the natural response of the circuit is shown in Fig. 9.7(b).

Equations (9.40) and (9.42) show that in both *RC* and *RL* circuits the current decays exponentially from the respective initial value.

Total response of first order series RL circuit excited by a step function voltage: The total response of the circuit shown in Fig. 9.8 is obtained directly by using the Laplace

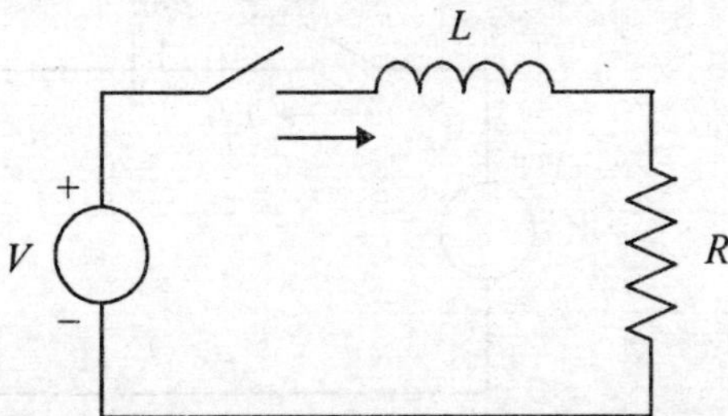

Fig. 9.8 *RL* circuit excited by a step input voltage.

transform. (Compare the classical method of finding complementary function and particular integral separately). The transformed equation is:

$$LsI + RI = V/s$$

which gives

$$I = \frac{V/s}{Ls + R} = \frac{V}{Ls(s + R/L)}$$

$$I = \frac{V}{L}\left[\frac{1}{s} - \frac{1}{s + R/L}\right]$$

$$= \frac{V}{L}\frac{L}{R}\left[\frac{1}{s} - \frac{1}{s + R/L}\right]$$

Inverting,

$$i(t) = V/R\left(1 - e^{-\frac{R}{L}t}\right)$$

The current in the circuit increases from its initial value ($i = 0$) to the final value (steady state value) of V/R when $t = \infty$.

At time $t = \tau = L/R$, the current has increased to $V/R\,[1 - e^{-1}] = 0.632\, V/R$

At time $t = 0$ $(di/dt)_{t=0}$, $i = V/R\,[R/L] = V/R[1/\tau]$

The voltage on the resistance R is:

$$V_R = Ri = V(1 - e^{-(R/L)t})$$

The current time response is plotted in Fig. 9.9.

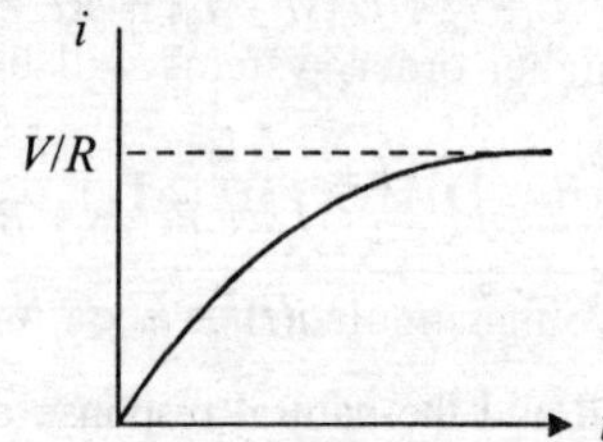

Fig. 9.9 Current-time response of the circuit of Fig. 9.8.

Total response of a series RC circuit:

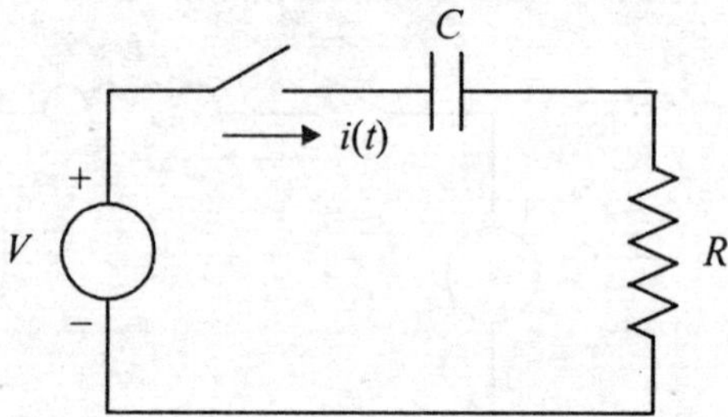

Fig. 9.10 *RC* circuit excited by a step input voltage.

The transform equation is:

$$RI + I/Cs = V/s$$

$$I = \frac{V}{s}\frac{Cs}{RCs+1}$$

$$= \frac{V}{R(s+1/RC)}$$

Inverting,

$$i(t) = (V/R)\, e^{-t/RC}$$

The response is a decaying exponential shown in Fig. 9.11.

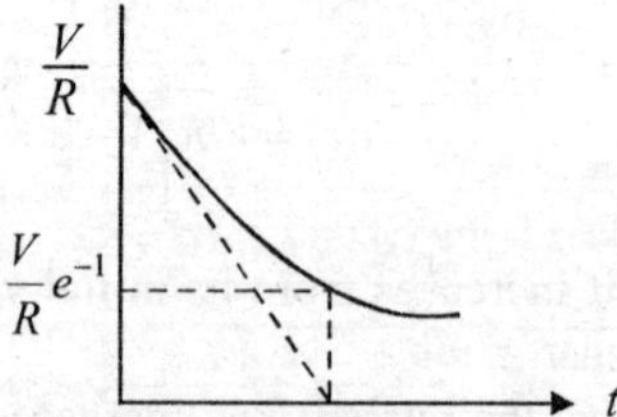

Fig. 9.11 Current-time response of the circuit of Fig. 9.10.

At time $t = \tau = RC$, the current has decreased to 0.368 of its initial value.

Note that the steady state value is zero; the capacitance acts as an open circuit at $t = \infty$.

As stated earlier higher order systems will be discussed in Chapter 10.

9.5.3 Simultaneous Differential Equations

EXAMPLE 9.39 Solve the simultaneous equations

$$3Dx + 6x + 2Dy = 0$$

$$Dx + 2x + Dy + y = 0$$

with initial conditions $x(0) = 0, \quad x'(0) = 2$

$$y(0) = 1, \quad y'(0) = -3$$

SOLUTION:
Taking Laplace transforms,

$$\begin{bmatrix} 3s+6 & 2s \\ s+2 & s+1 \end{bmatrix} \begin{bmatrix} X \\ Y \end{bmatrix} = \begin{bmatrix} 2 \\ 1 \end{bmatrix}$$

Solving for X and Y,

$$X = \frac{\begin{vmatrix} 2 & 2s \\ 1 & s+1 \end{vmatrix}}{\begin{vmatrix} 3s+6 & 2s \\ s+2 & s+1 \end{vmatrix}}$$

$$= \frac{2}{s^2 + 5s + 6} = \frac{2}{(s+2)(s+3)}$$

$$= \frac{2}{(s+2)} - \frac{2}{(s+3)}$$

Inverting,

$$x(t) = 2e^{-2t} - 2e^{-3t}$$

Similarly,

$$Y = \frac{s+2}{(s+2)(s+3)}$$

and

$$y(t) = e^{-3t} \qquad \textbf{Ans.}$$

9.5.4 Ordinary Differential Equations with Variable Coefficients

EXAMPLE 9.40 Solve $tx'' + 2x' + tx = 0$

SOLUTION:

$$x(0) = 1$$

Taking the Laplace transform of each term,

$$-\frac{d}{ds}\{s^2X - sx(0) - x'(0)\} + 2\{sX - x(0)\} - \frac{d}{ds}X = 0$$

or $$-s^2X' - 2sX + 1 + 2sX - 2 - X' = 0$$

which gives $$X' = -1/(s^2 + 1)$$

Integrating,

$$X = -\tan^{-1}s + C$$

Since $X = 0$ as $s \to \infty$

$$C = \pi/2$$

$$X = \pi/2 - \tan^{-1}s$$

$$= \tan^{-1}(1/s)$$

By Example 9.18,

$$x = L^{-1}\tan^{-1}(1/s)$$

$$= \frac{\sin t}{t}$$ **Ans.**

which is the required solution.

9.5.5 Evaluation of Certain Integrals

EXAMPLE 9.41

Prove that $\displaystyle\int_0^\infty \frac{\sin t}{t}dt = \pi/2$

SOLUTION:

From Example 9.18,

$$L\frac{\sin t}{t} = \tan^{-1}1/s$$

which means

$$\int_0^\infty e^{-st}\frac{\sin t}{t}dt = \tan^{-1}1/s$$

Let $s \to 0$

$$\int_0^\infty \frac{\sin t}{t}dt = \tan^{-1}\infty = \pi/2$$

Hence the result.

9.6 SPECIAL APPLICATIONS

9.6.1 Boundary Value Problems

Mathematical models of problems in Science and Engineering which lead to the partial differential equations (satisfying boundary conditions) constitute the boundary value problem. Laplace transformation is of great use in dealing with these problems. A simple example is given as follows:

EXAMPLE 9.42 Solution of Laplace's equation in one dimension: Consider the parallel conductors of Fig. 9.12 where $v = 0$ at $z = 0$ and $v = 100$ V at $z = d$.

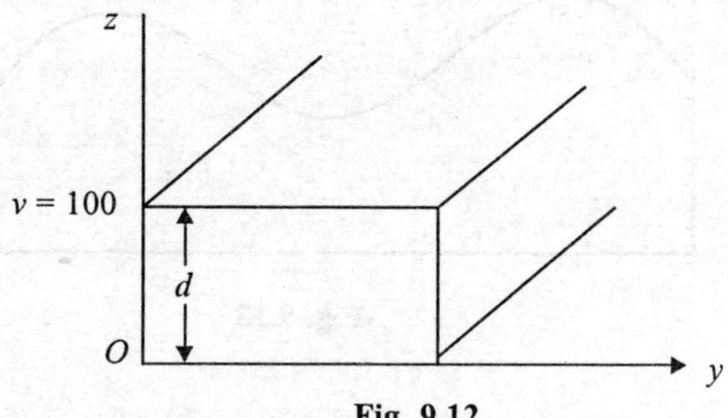

Fig. 9.12

The potential in the region between the plates is governed by **Laplace's equation**.

$$\frac{\partial^2 v}{\partial x^2} + \frac{\partial^2 v}{\partial y^2} + \frac{\partial^2 v}{\partial z^2} = 0$$

In this case the potential can vary only with z.
Hence

$$\frac{d^2 v}{dz^2} = 0$$

Taking Laplace transform,

$$s^2 V - s\, v(0) - v'(0) = 0$$

From the initial conditions, $v(0) = 0$, $v'(0) = $ constant a.

$$\text{since} \quad \frac{d^2 v}{dz^2} = 0$$

and, therefore, dv/dz is a constant.

$$V = \frac{a}{s^2}$$

Inverting,

$$v = za$$

Also $$100 = da$$

or $$a = 100/d$$

and $$v = \frac{100\,z}{d}$$

9.6.2 Transforms of Periodic Functions

If $y = \phi(t) + \phi(t - T) + \phi(t - 2T) + \cdots \infty$ as shown in Fig. 9.13.

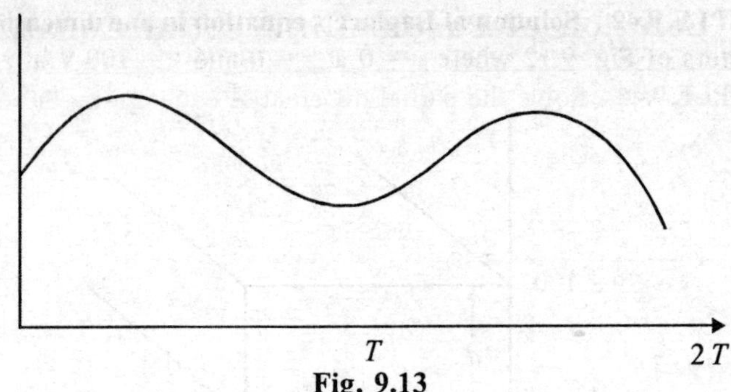

Fig. 9.13

Then

$$Y(s) = \phi(s) + e^{-sT}\,\phi(s) + e^{-2sT}\,\phi(s) + \cdots$$

which is an infinite geometric series with common ratio e^{-sT}
Hence

$$Y(s) = \frac{\phi(s)}{1 - e^{-sT}}$$

where $$\phi(s) = \int_0^T e^{-st}\,y\,dt$$

EXAMPLE 9.43 Find $Y(s)$ for the rectangular wave shown in Fig. 9.14.

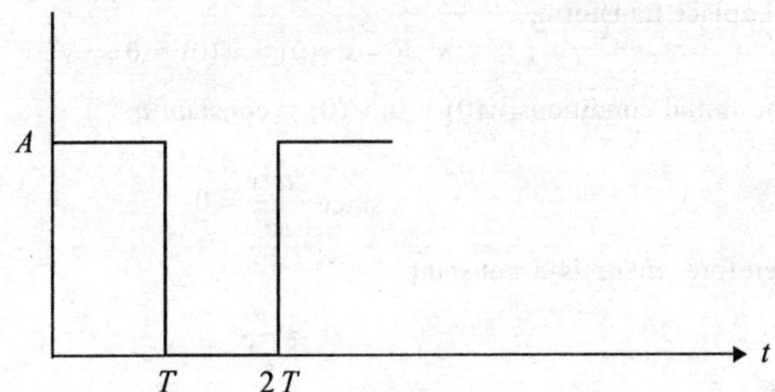

Fig. 9.14

SOLUTION:

$$\phi = \frac{A}{s}(1 - e^{-sT})$$

$$Y(s) = \frac{A(1 - e^{-sT})}{s(1 - e^{-2sT})}$$

$$= \frac{A}{s(1 + e^{-sT})}$$

9.6.3 Solution of Partial Differential Equations

EXAMPLE 9.44 Solve the partial differential equation by the method of separation of variables.

$$\frac{\partial u}{\partial x} + \frac{\partial u}{\partial y} - 2(x+y)u = 0$$

$$\frac{\partial u}{\partial x} - 2xu + \frac{\partial u}{\partial y} - 2yu = 0$$

SOLUTION:

The variables may be separated in this equation giving

$$\frac{\partial u}{\partial x} - 2xu = k$$

$$\frac{\partial u}{\partial y} - 2yu = -k$$

In the first equation, $P = -2x$ and $Q = k$

and in the second equation $P = -2y$ and $Q = -k$

The solution is, therefore,

$$u = ce^{-\int 2x\,dx} + e^{-\int 2x\,dx} \int k\, e^{\int 2x\,dx}$$

$$= ce^{-x} + e^{-x} k \int e^{x}$$

Similarly,

$$\frac{\partial u}{\partial y} = ce^{-y} + e^{y+2y}$$

Hence the general solution is:

$$u(x, y) = ce^{x+y+c(x-y)} \qquad \textbf{Ans.}$$

Transmission lines: To illustrate the use of Laplace transform method of solving partial differential equations, consider the problem of determining the distribution of current and potential difference along a transmission line of infinite length represented in Fig. 9.15.

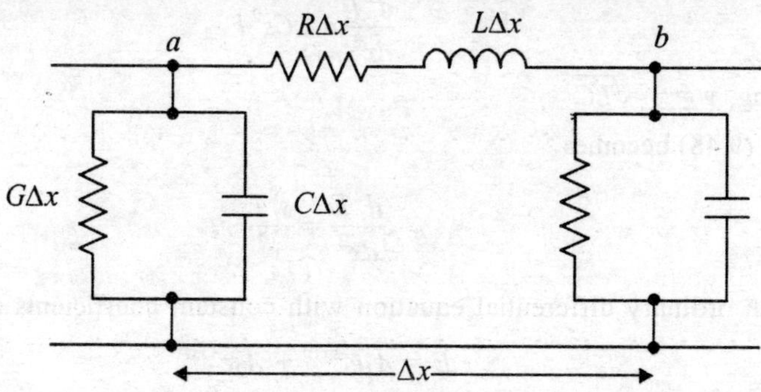

Fig. 9.15

Take a length Δx of the transmission line at a distance x from the sending end. The primary constants of the line are R, G, L, C. If the line is dissipation less $R = 0$, $G = 0$.

The element Δx is governed by the following partial differential equations:

$$\frac{\partial i}{\partial x} = \frac{C\partial e}{\partial t} \tag{9.43}$$

$$\frac{\partial e}{\partial x} = L\frac{\partial e}{\partial t} \tag{9.44}$$

Taking Laplace transforms,

$$Le(x, t) = E(x, s) \tag{9.45}$$

$$Li(x, t) = I(x, s) \tag{9.46}$$

E and I are Laplace transforms.

Assume at $t = 0$, the current and potential distribution of the line are zero. Hence

$$L\frac{\partial e}{\partial t} = sE, \qquad L\frac{\partial i}{\partial t} = sI$$

Since in Eqs. (9.45) and (9.46), x is a parameter.

$$L\frac{\partial e}{\partial x} = \frac{dE}{dx}, \qquad L\frac{\partial i}{\partial x} = \frac{di}{dx}$$

Equations (9.43) and (9.44) transform into

$$-\frac{dE}{dx} = LsI$$

$$-\frac{dI}{dx} = CsE$$

Eliminating I from these two equations,

$$\frac{d^2E}{dx^2} = LCs^2E \tag{9.47}$$

Eliminating E,

$$\frac{d^2I}{dx^2} = LCs^2I \tag{9.48}$$

On putting, $v = 1/\sqrt{LC}$

Equation (9.48) becomes

$$\frac{d^2E}{dx^2} = \frac{s^2E}{v^2}$$

This is an ordinary differential equation with constant coefficients and its solution is:

$$E = A_1e^{-sx/v} + A_2e^{+sx/v}$$

where A_1 and A_2 are arbitrary constants.

If the line is of infinite length, it can be shown that

$$E(x, s) = E_0\, e^{-sx/v}$$

$$L^{-1} E_0\, e^{-sx/v} = E_0 \qquad t > x/v \text{ (using the } t \text{ shift theorem)}$$

This represents a potential wave of amplitude E_0 travelling with velocity v to the receiving end.

SUMMARY

In this chapter, Laplace transform and its more important properties are discussed. A set of transform pairs has been given in Table 9.2. Methods of extending the table by making use of the properties of Laplace transform and the partial fraction technique are indicated. Methods for obtaining inverse transforms have also been included. Applications of the transform cover a wide range, including solution of differential equations governing first order electrical circuits; higher order systems have been reserved to Chapter 10.

Solved Problems

P9.1 Obtain the general solution of the equation:

$$2x(t) + 4 + 10\int_0^t x(t) = 4e^{-t}u(t)$$

SOLUTION:

Taking the Laplace transform,

$$2X(s) + 4/s + \frac{10X(s)}{s} = \frac{4}{s+1}$$

Rearranging

$$(2 + 10/s)\, X(s) = \frac{4}{s+1} - \frac{4}{s}$$

$$X(s) = \frac{2s}{(s+1)(s+5)} - \frac{2}{(s+5)}$$

$$= \frac{A}{(s+1)} + \frac{B}{(s+5)} - \frac{2}{(s+5)}$$

$$= -\frac{1/2}{(s+1)} + \frac{1/2}{(s+5)}$$

Inverting,

$$x(t) = -\frac{1}{2}e^{-t} + \frac{1}{2}e^{-5t} \qquad \textbf{Ans.}$$

P9.2 Find the inverse of $F(s) = \dfrac{s+10}{s^4 + s^2}$.

SOLUTION:

Method 1

$$F(s) = \frac{s+10}{s^2(s^2+1)}$$

$$= \frac{1}{s(s^2+1)} + \frac{10}{s^2(s^2+1)}$$

Since

$$L^{-1}\frac{1}{s^2+1} = \sin t$$

$$L^{-1}\frac{1}{s(s^2+1)} = \int_0^t -\sin t\, dt = -\cos t + 1$$

(See item 9, Table 9.2)

and $$L^{-1}\frac{1}{s^2(s^2+1)} = \int_0^t (-\cos t + 1) = [-\sin t + t]_0^t = -\sin t + t$$

Now, time function $f(t) = 1 - \cos t + 10(-\sin t + t)$
$= 1 + 10t - \cos t - 10 \sin t$ **Ans.**

Method 2

$$F(s) = \frac{s+10}{s^2(s^2+1)}$$

Partial fraction expansion $$F(s) = \frac{10}{s^2} + \frac{1}{s} - \frac{s+10}{s^2+1}$$

Inverting,

$$f(t) = 10t + 1 - 10 \sin t - \cos t$$ **Ans.**

P9.3 Find $$L^{-1}\frac{1}{(s+a)(s+b)}$$

SOLUTION:

Since $$L^{-1}\frac{1}{(s+a)} = e^{-at} \quad L^{-1}\frac{1}{s+b} = e^{-bt}$$

The required inverse

$$= \int_0^t e^{-ax} e^{-b(t-x)} dx$$

$$= e^{-bt}\left[\frac{e^{x(b-a)}}{b-a}\right]_0^t$$

$$= \frac{1}{(b-a)}\left[e^{-at} - e^{-bt}\right]$$ **Ans.**

Check by partial fraction method.

P9.4 Find the current in the circuit of Fig. P9.1(a) if the input is the square pulse shown in Fig. P9.1(b).

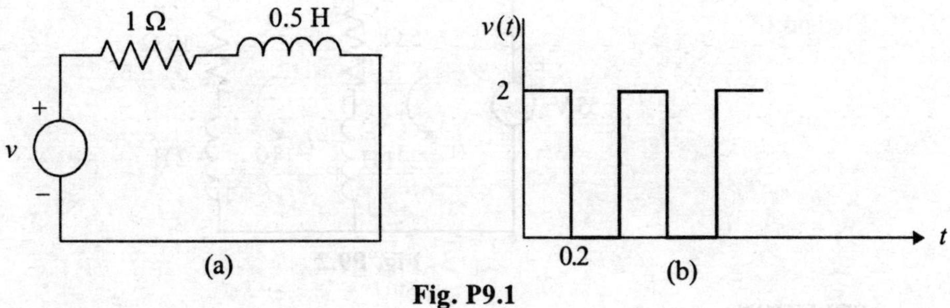

Fig. P9.1

SOLUTION:
The differential equation governing the circuit is:

$$Ri + L\frac{di}{dt} = v(t)$$

Taking Laplace transforms of both sides,

$$I(s)\,[R + sL] = V(s) = \frac{2}{s} - \frac{2e^{-0.2s}}{s}$$

or

$$I(s) = \frac{2/s - 2/s\, e^{-0.2s}}{(0.5s + 1)}$$

$$= \frac{2[1/s - 1/s\, e^{-0.2s}]}{0.5(s + 2)}$$

$$= \frac{4[1/s(1 - e^{-0.2s})]}{s + 2}$$

$$= \frac{4(1 - e^{-0.2s})}{s(s + 2)}$$

$$= [2/s - 2/(s + 2)](1 - e^{-0.2s})$$

Inverting,

$$i(t) = [2 - 2e^{-2t}]\, u(t) - [2 - 2e^{-2(t-0.2)}]\, u(t - 0.2) \qquad \textbf{Ans.}$$

P9.5 In the circuit of Fig. P9.2, find the current from the source if the initial currents are zero.

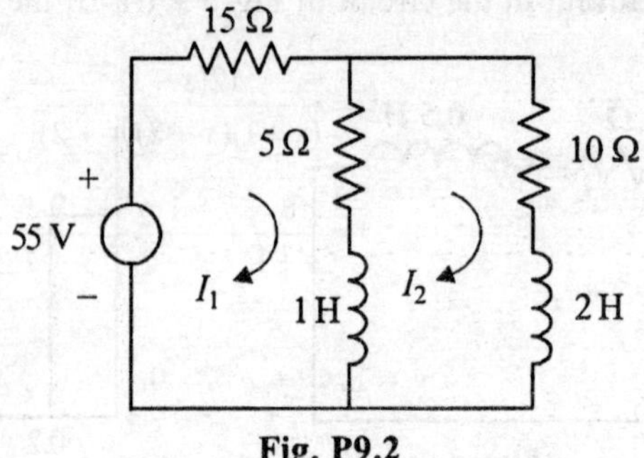

Fig. P9.2

SOLUTION:

Let I_1 I_2 be the currents in the two meshes. They are governed by the equations,

$$\begin{bmatrix} s+20 & -(s+5) \\ -(s+5) & 3s+15 \end{bmatrix}\begin{bmatrix} I_1 \\ I_2 \end{bmatrix} = \begin{bmatrix} 55/s \\ 0 \end{bmatrix}$$

Solving by Cramer's rule,

$$I_1 = \frac{55(3s+15)}{s(2s^2+65s+275)} = 3\left[\frac{1}{s} - \frac{2}{2s+55}\right]$$

Inverting,

$$i_1(t) = 3 - 3e^{(-55/2)t}$$

Evidently the source current equals

$$i_1(t) = 3 - 3e^{(-55/2)t}$$

with a steady state value of 3 and transient part $-3e^{(-55/2)t}$ **Ans.**

P9.6 Solve the simultaneous differential equations

$$(5D - 1)y - (2D - 1)z = 4e^{-2x}$$

$$(D + 7)y - 3z = 12e^{-2x}$$

where D denoted d/dx.

SOLUTION:

Taking Laplace transforms

$$(5s - 1)Y - (2s - 1)Z = 4/(s + 2)$$

$$(s + 7)Y - 3sZ = 12/(s + 2)$$

Assuming zero initial conditions for y and z.
Solving the equations by Cramers rule,

$$Y = \frac{\begin{vmatrix} 4/s+2 & -(2s-1) \\ 12/s+2 & -3 \end{vmatrix}}{2s^2 - 2s - 4}$$

which simplifies to

$$= \frac{12(s-1)}{(s+1)(s-2)(s+2)}$$

$$= \frac{8}{s+1} + \frac{1}{s-2} - \frac{9}{s+2}$$

Inverting,

$$y = 8e^{-x} + e^{2x} - 9e^{-2x} \quad \textbf{Ans.}$$

Similarly,

$$Z = \frac{\begin{vmatrix} 5s-1 & 4/(s+2) \\ s+7 & 12/(s+2) \end{vmatrix}}{2s^2 - 2s - 4}$$

which simplifies to

$$= \frac{4(7s-5)}{(s-2)(s+1)(s+2)}$$

$$= \frac{16}{s+1} + \frac{3}{s-2} - \frac{19}{s+2}$$

Inverting,

$$z = 16e^{-x} + 3e^{2x} - 19e^{-2x} \quad \textbf{Ans.}$$

The solution for y and z can be checked by substitution in the first of the original equations.

P9.7 Solve the simultaneous differential equations

$$(3D + 7)y + (3D + 5)z = 0 \qquad (1)$$

$$(4D + 11)y - (D + 5)z = 0 \qquad (2)$$

SOLUTION:

For the homogeneous equations, the determinant of the coefficient matrix must be zero. Thus,

$$(15D^2 + 75D + 90)y = 0$$

or

$$(D + 2)(D + 3)y = 0$$

which has the solution

$$y = Ae^{-2x} + Be^{-3x}$$

and

$$z = A_1e^{-2x} + B_1e^{-3x}$$

Substitute the values of y and z back into Eq. (1) above, which gives

$$(A - A_1)e^{-2x} + (-2B - 4B_1)e^{-3x} = 0$$

Hence Eq. (1) is satisfied only if

$$A_1 = A \quad \text{and} \quad B_1 = -\frac{1}{2}B$$

The general solution is:

$$y = Ae^{-2x} + Be^{-3x}$$

$$z = Ae^{-2x} - \frac{1}{2}Be^{-3x} \qquad \textbf{Ans.}$$

P9.8 Solve the simultaneous equations

$$(D + 2)\,x - Dy = 0 \tag{1}$$

$$3Dx + (3D + 2)y = 0 \tag{2}$$

with the initial conditions $x(0) = 1$, $y(0) = -3$.

SOLUTION:

Taking Laplace transforms,

$$\begin{bmatrix} s+2 & -s \\ 3s & 3s+2 \end{bmatrix}\begin{bmatrix} x \\ y \end{bmatrix} = \begin{bmatrix} 4 \\ -6 \end{bmatrix}$$

or

$$X = \frac{\begin{vmatrix} 4 & -s \\ -6 & 3s+2 \end{vmatrix}}{6s^2 + 8s + 4}$$

$$X = \frac{6s + 8}{6s^2 + 8s + 4}$$

or

$$= \frac{s + 2/3}{(s + 2/3)^2 + 2/9} + 2/3 \times 3/\sqrt{2}\,\frac{\sqrt{2}/3}{(s + 2/3)^2 + 2/9}$$

Inverting,

$$x = e^{(-2/3)t}\cos(\sqrt{2}/3)t + \sqrt{2}e^{(-2/3)t}\sin(\sqrt{2}/3)t \qquad \textbf{Ans.}$$

and

$$Y = \frac{\begin{vmatrix} s+2 & 4 \\ 3s & -6 \end{vmatrix}}{6s^2 + 8s + 4}$$

$$= \frac{-18s - 12}{6s^2 + 8s + 4}$$

$$= \frac{-3(s + 2/3)}{(s + 2/3)^2 + 2/9}$$

Inverting,

$$y = -3e^{(-2/3)t}\cos(\sqrt{2}/3)t \qquad \textbf{Ans.}$$

The arbitrary constants found above may be verified from Eq. (1).

P9.9 Solve

$$x^2 \frac{d^2 y}{dx^2} + 7x \frac{dy}{dx} + 8y = 0$$

SOLUTION:

The Laplace transform can be used in solving some ordinary differential equations in which the coefficients are variable. However, it is feasible to seek trial solutions of the form $y = A\,x^q$, which gives

$$x\frac{dy}{dx} = Aqx^q, \quad x^2 \frac{d^2 y}{dx^2} = Aq(q-1)x^q$$

On substitution of these expressions, the given equation reduces to an algebraic equation for q.

The trial solution $y = Ax^q$ gives

$$A\{q(q-1) + 7q + 8\}x^q = 0$$

That is,

$$q^2 + 6q + 8 = 0$$

Hence

$$q = -2 \text{ or } q = -4$$

The general solution is:

$$y = Ax^{-2} + Bx^{-4} \quad \textbf{Ans.}$$

P9.10 Solve the simultaneous equations

$$(D^2 + 2)x - Dy = 0$$

$$3Dx + (3D + 2)y = 0$$

with the initial conditions $x(0) = 1$, $x'(0) = -1$ and $y(0) = -3$.

SOLUTION:

Taking Laplace transforms,

$$(s^2 + 2)X - sY = sx(0) + x'(0) - y(0) = s - 1 + 3 = s + 2$$

or

$$3sX + (3s + 2)Y = 3x(0) + 3y(0) = 3 - 9 = -6$$

Solving,

$$X = \frac{\begin{vmatrix} s+2 & -s \\ -6 & 3s+2 \end{vmatrix}}{\begin{vmatrix} s^2+2 & -s \\ 3s & 3s+2 \end{vmatrix}} = \frac{3s^2 + 2s + 4}{3s^3 + 5s^2 + 6s + 4} = \frac{1}{s+1}$$

and

$$Y = \frac{\begin{vmatrix} s^2+2 & s+2 \\ 3s & -6 \end{vmatrix}}{\begin{vmatrix} s^2+2 & -s \\ 3s & 3s+2 \end{vmatrix}} = \frac{-3}{s+1}$$

Taking inverse transform,

$$x = e^{-t}$$
$$y = -3e^{-t} \qquad \textbf{Ans.}$$

Exercises

E9.1 Use the relation $\frac{d}{dt}te^t = (t+1)e^t$ to find $L(te^t)$.

[**Hint:** Use the derivative formula $L[f'(t)] = sF(s) - f(0)$.]

E9.2 Find $L(t^2e^t)$.

[**Hint:** See item 3, Table 9.2.]

E9.3 Draw the graph of $(t - t_0)\, u(t - t_0)$.

E9.4 Represent the periodic function shown in Fig. E9.1 by a series of step functions.

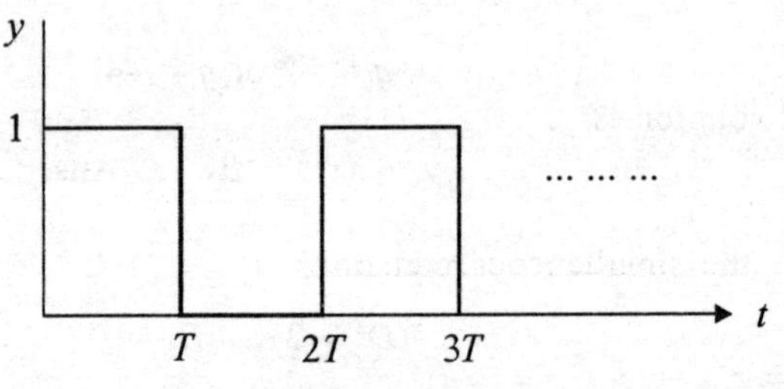

Fig. E9.1

[*Ans:* $y = u(t) - u(t - T) + u(t - 2T) - u(t - 3T) + \cdots$]

E9.5 In Example 9.43, show that $Y(s) = \frac{A}{s}(1 - e^{-sT} + e^{-2sT} - \cdots)$

and hence

$$y = A\,[u(t) - u(t - T) + u(t - 2T) - \cdots]$$

E9.6 Show that $L^{-1}\frac{1}{(s^2 + a^2)^2} = \frac{\sin at}{2a^3} - t\frac{\cos at}{2a^2}$.

E9.7 Solve

$$Dy - Dz - 2y + 2z = \sin t$$
$$D^2y + 2Dz + y = 0$$

with the conditions $y(0) = y'(0) = z(0) = 0$.

[*Ans:* $y = (1/9)e^{-t} + (4/45)e^{2t} - 1/5 \cos t - 2/5 \sin t + (1/3)te^{-t}$

$z = (1/9)e^{-t} - (1/9)e^{2t} + (1/3)te^{-t}$]

E9.8 Solve the simultaneous differential equations

$$t\frac{dx}{dt} + y = 0, \qquad t\frac{dy}{dt} + x = 0$$

such that $x = 1$ and $y = -1$ when $t = 1$.

[*Ans:* $x = t, \quad y = -t$]

E9.9 Solve the simultaneous differential equations

$$\frac{dx}{dt} = y, \quad \frac{dy}{dt} = z, \quad \frac{dz}{dt} = x$$

Given that $x = 1$, $y = 1$ and $z = 1$ when $t = 0$.

[*Ans:* $x = y = z = e^t$]

E9.10 Solve the equation $\dfrac{d^2y}{dt^2} - y = 1$
with the initial conditions $y(0) = 0$, $y'(0) = 1$.

[*Ans:* $y(t) = e^t - 1$]

E9.11 Use the convolution property to find

$$L^{-1}\frac{s}{(s^2+1)(s^2+4)}.$$

[*Ans:* 1/3 $(\cos t - \cos 2t)$]

Steady State Sinusoidal Response of Higher Order Networks

10

CHAPTER CONTENTS

In Chapter 4, a number of techniques were introduced for analyzing dc circuits. These techniques are now extended to ac circuits through the *complex notation*. It will be seen that the theorems for ac networks are practically in line with those for dc networks.

10.1 COMPLEX NOTATION

Voltage, current and impedance: In Chapter 5, simple ac circuits were solved with the aid of phasor diagrams. With complicated networks, considerable simplification is possible if complex algebra is used. The method is based on the principle that a phasor such as voltage or current may be described by two coordinates—one along the x-axis and the other along the y-axis. Making use of the j operator (which when applied to a phasor alters its direction by 90° anticlockwise). Thus the voltage phasor in Fig. 10.1 may be represented by

$$E = E_p + jE_q = E(\cos\delta + j\sin\delta) \qquad \text{(In rectangular form)}$$

or

$$E = E\angle\delta \qquad \text{(In the polar form)}$$

$E\angle\delta$ is really an abbreviation of the exponential form $Ee^{j\delta}$

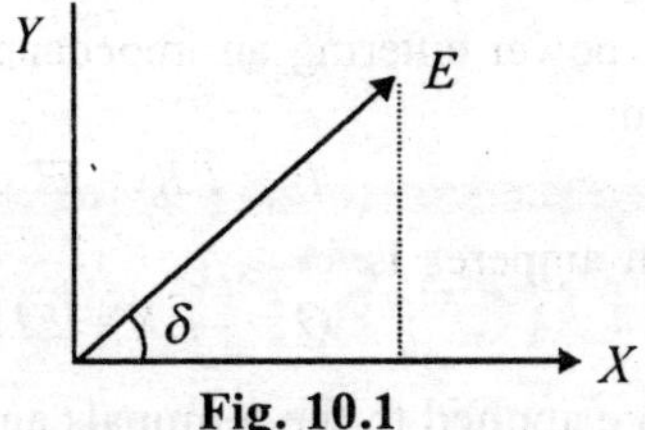

Fig. 10.1

Current phasors are expressed in similar forms.

Inorder to obtain the proper expression for **impedance**, consider a circuit with a resistance R in series with an inductive reactance X_L. The phasor diagram in Fig. 10.2 shows that

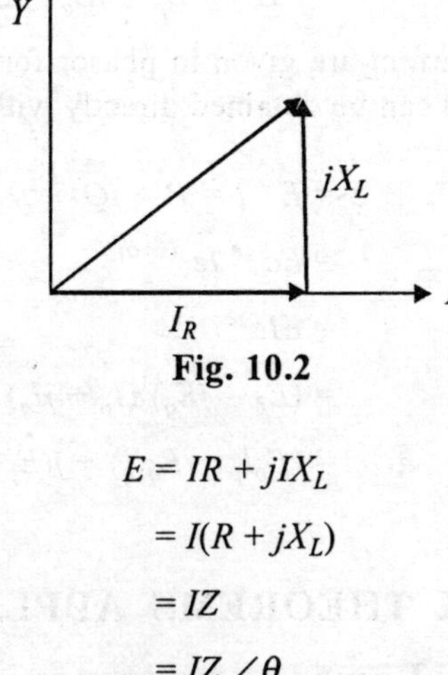

Fig. 10.2

$$E = IR + jIX_L$$
$$= I(R + jX_L)$$
$$= IZ$$
$$= IZ\angle\theta$$

Thus, the impedance is expressed as:

$$Z = R + jX_L = Z\angle\,\theta = Z(\cos\,\theta + j\,\sin\,\theta)$$

and
$$\tan\,\theta = X_L/R$$

Admittance, which is the reciprocal of impedance, may be written as:

$$Y = \frac{1}{Z} = \frac{1}{Z}e^{-j\theta}$$
$$= Y(\cos\,\theta - j\sin\,\theta)$$
$$= G + jB$$

Just like impedance, admittance may be resolved into a real component (conductance) and an imaginary component (susceptance). Note that the quantity represented by the susceptance has the opposite sign to the reactance X and that the angle of Y is equal to the magnitude but opposite in sign to the angle of Z.

Volt amperes: The power entering an impedance Z that carries a current I is given by the scalar expression:

$$P = I^2R = EI \cos \phi$$

and the reactive volt amperes is:

$$Q = -I^2X = EI \sin \phi$$

where E is the voltage applied to the terminals and ϕ is the angle by which current **leads** the voltage. These equations make the reactive volt amperes positive when current is leading and negative when it is lagging.

A driving voltage $E = E_p + jE_q = Ee^{j\delta}$ has for its conjugate

$$E^* = E_p - jE_q = Ee^{-j\delta}$$

If voltage and current are given in phasor form, the real and reactive components of the volt amperes (S) can be obtained directly with the correct sign, by multiplying E^* by I. Thus,

$$\begin{aligned} S &= E^* I = P + jQ \\ &= Ee^{-j\delta} Ie^{j(\delta+\phi)} \\ &= EIe^{j\phi} \\ &= (E_p - jE_q)(I_p + jI_q) \\ &= (E_pI_p + E_qI_q) + j(E_pI_q - E_qI_p) \end{aligned}$$

10.2 NETWORK THEOREMS APPLIED TO AC NETWORKS

10.2.1 Network Laws

1. Branch law: The phasor voltage drop V caused by a phasor current I in an impedance Z is given by the phasor equation $V = ZI$.

2. Junction law: The phasor sum of all the currents entering any junction (node) in a network is zero.

$$\Sigma_{p=1}^{m} I_{pn} = 0$$

3. Mesh law: The phasor sum of all the driving voltages in a closed path (or mesh) in a network is equal to the phasor sum of the products of the impedance and the current in the component branches.

$$\Sigma E = \Sigma ZI$$

10.2.2 General Network Theorems

1. Superposition theorem: The current that flows in any branch due to the simultaneous action of several driving voltages at various points in a network is equal to the phasor sum of the component currents which would be caused by each voltage acting alone with all the others short circuited.

2. Reciprocity theorem: If a driving voltage in one branch in a network produces a current in any other branch, the same driving voltage in the second branch would produce the same current in the first.

3. Substitution theorem: An impedance Z carrying a current I flowing in a given direction can be replaced by a driving voltage of value $|Z|\ |I|$ acting in the opposite direction, and conversely.

4. Compensation theorem: If a network is modified by making a change ΔZ in the impedance in one of its branches, the resultant current in any other branch is equal to the current that would be produced by a driving voltage of value $-\Delta ZI$ acting in series with the modified branch, where I is the previous current in the modified branch.

10.2.3 Network Reduction Theorems

Passive networks:

Three-terminal star-delta theorem. The modified theorems are:

$$Z_{12} = \frac{Z_{10} + Z_{20} + Z_{10}Z_{20}}{Z_{30}} \quad \text{(Star-delta)}$$

with similar results for Z_{23} and Z_{31}.

and
$$Z_{10} = \frac{Z_{12}Z_{31}}{Z_{12} + Z_{23} + Z_{31}} \quad \text{(Delta-star)}$$

See Fig. 10.3 and compare with the corresponding expressions for resistive networks (See Section 4.2).

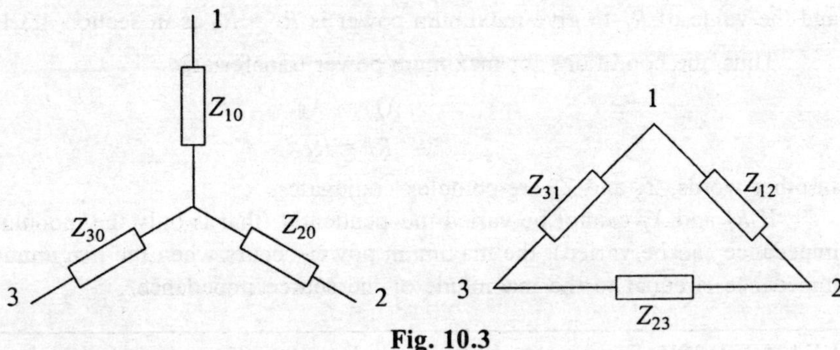

Fig. 10.3

Active networks:

1. Nodal voltage theorem (Millman's theorem). If the voltages between any node zero, and a number of points 1, 2, 3 in a network are known, the voltage between node zero and a second node n to which these points are connected is given by

$$V_{n0} = \frac{V_{10}Y_1 + V_{20}Y_2 + V_{30}Y_3}{Y_1 + Y_2 + Y_3}$$

This theorem is very useful in finding the neutral potential in an unbalanced three-phase circuit.

2. Thevenin's theorem and Norton's theorem. Procedure identical with that in section 4.2.2. As an illustration of the method, refer Examples 10.5 and 10.6.

Most of the theorems applicable to ac networks are simple extensions to the corresponding theorems for resistive networks—resistance being replaced by impedance; the maximum power transfer theorem is however an exception.

3. Maximum power transfer theorem. In section 4.3.1, it was found that the maximum power transfer in a dc circuit occurs when the load resistance is equal to the source resistance. However, in the ac circuit of Fig. 10.4, the maximum current and hence maximum

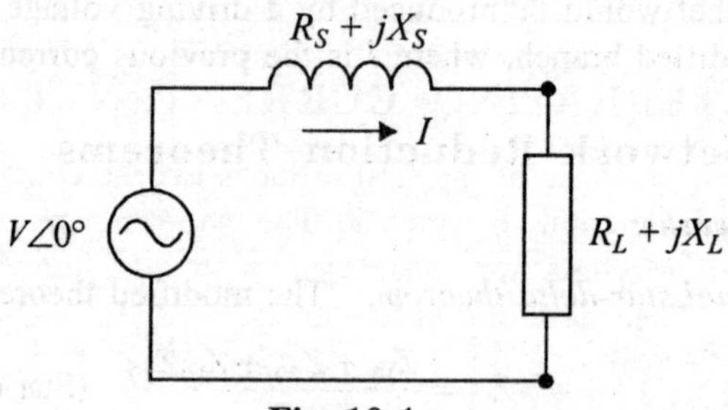

Fig. 10.4

power is developed when current is in phase with the supply voltage (resonant condition) that is when $X_L = -X_S$.
Thus,

$$P_L = \frac{V^2 R_L}{(R_S + R_L)^2}$$

and the value of R_L to give maximum power is $R_L = R_S$ as in section 4.3.1.

Thus, the conditions for maximum power transfer are:

$$X_L = -X_S$$
$$R_L = R_S$$

In other words, Z_S and Z_L are complex conjugates.

If R_L and X_L cannot be varied independently (that is only the modulus of the load impedance can be varied), the maximum power occurs when the magnitude of the load impedance is equal to the magnitude of the source impedance.

EXAMPLE 10.1 For the network shown in Fig. 10.5, determine the impedance of the load which will dissipate maximum power and determine the maximum power.

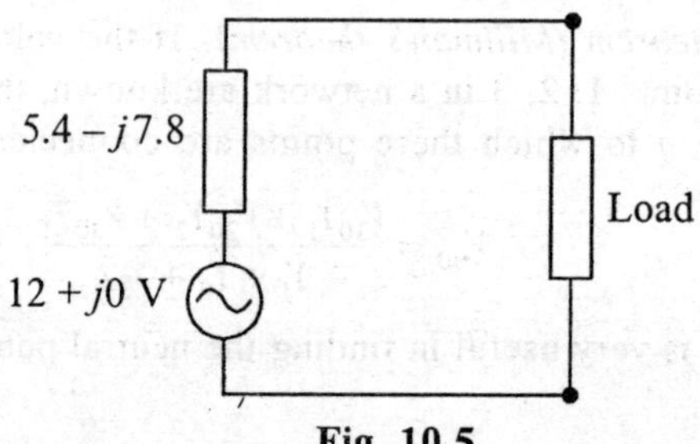

Fig. 10.5

SOLUTION:

For the power transfer to be maximum, the load impedance should be complex conjugate of the source impedance. Hence

$$Z_L = 5.4 + j\,7.8$$

Current $$I = \frac{V}{Z_{\text{total}}} = \frac{12 + j0}{5.4 - j7.8 + 5.4 + j7.8} = \frac{12 + j0}{10.8 + j0}$$

$$= 1.11 \text{ A}$$

$$P_L = I^2 R = (1.11)^2\, 5.4 = 6.65 \text{ W} \qquad \textbf{Ans.}$$

10.3 ALTERNATING CURRENT NETWORKS

Direct current response of the first order networks containing only one storage element has been considered in Chapter 9. The natural response and step input response of second order networks are considered in this section. The response under alternating current source will be found in Section 10.4.

10.3.1 Second Order Networks

Networks containing two storage elements (a capacitance and an inductance) can be described by a differential equation with two arbitrary constants, which may be determined from the initial conditions existing in the network.

Natural response of second order networks: The initial energy stored is represented by an initial current through the inductance and an initial voltage across the capacitance. The application of Kirchhoff's voltage law to the circuit gives

$$L\frac{di}{dt} + Ri + \frac{1}{C}\int_{-\infty}^{t} i\,dt = 0 \qquad (10.1)$$

In the series circuit of Fig. 10.6(a), the natural response corresponds to the energy stored within the circuit. The initial energy stored in the network is represented by an initial current flowing through the inductance and an initial voltage existing in the capacitance.

or $$i_{t=0} = I_0 = \frac{1}{L}\int_{-\infty}^{0} v_L\,dt$$

and $$v_{C\,t=0} = V_0 = \frac{1}{C}\int_{-\infty}^{0} i\,dt$$

Initially, Eq. (10.1) is:

$$L\left(\frac{di}{dt}\right)_{t=0} + Ri_{t=0} + \frac{1}{C}\int_{-\infty}^{0} i\,dt$$

Hence

$$\left(\frac{di}{dt}\right)_{t=0} = -\frac{1}{L}\,(RI_0 + V_0)$$

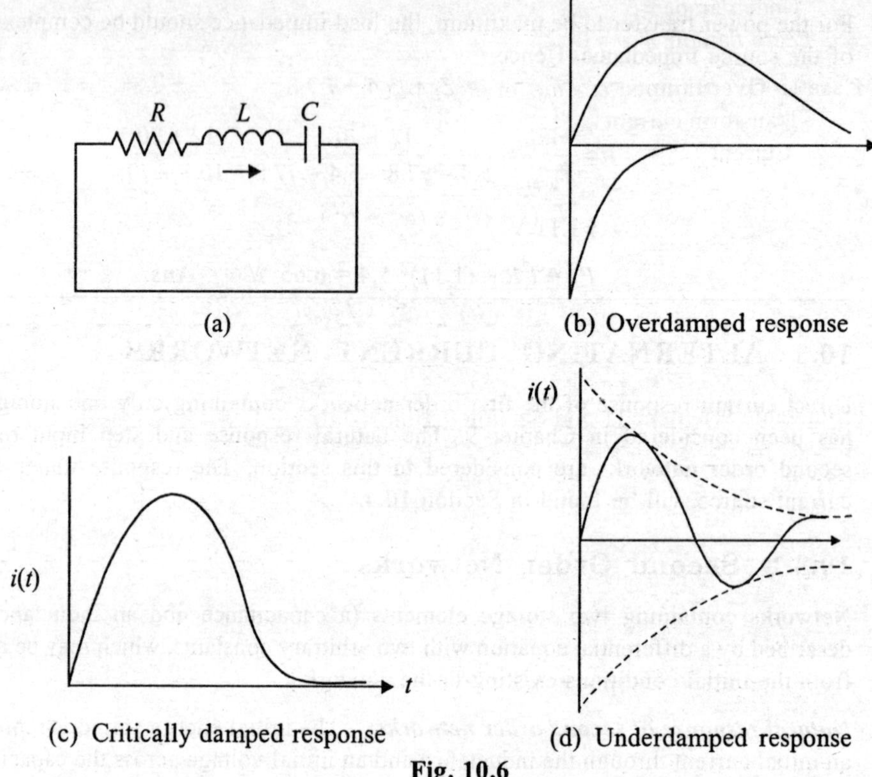

(a) (b) Overdamped response

(c) Critically damped response (d) Underdamped response

Fig. 10.6

The initial value of the current and its first derivative are, therefore, known in terms of initial stored energy.

The transform equation is:

$$L(sI - 0) + RI + \frac{V_0}{s} + \frac{I}{C_s} = 0$$

or

$$I = \frac{V_0/s}{Ls + R + 1/Cs}$$

$$= \frac{V_0}{Ls^2 + Rs + 1/C} \tag{10.1a}$$

$$= \frac{V_0}{L(s^2 + 2\alpha s + \omega_0^2)}$$

where $\alpha = R/2L$, and $\omega_0^2 = 1/LC$.

There are four different kinds of responses depending upon the relative values of α and ω_0.

Overdamped $\alpha > \omega_0$

Critically damped $\alpha = \omega_0$

Underdamped $\alpha < \omega_0$
Zero damping $\alpha = 0$

Case 1: Overdamped $\alpha > \omega_0$, or $R^2 > 4\,LC$
Transform current

$$I = \frac{V_0}{L}\frac{1}{(s+\alpha)^2 + (\omega_0^2 - \alpha^2)}$$

$$= \frac{V_0}{L}.\frac{1}{(s+\alpha)^2 - \omega_d^2} = \frac{V_0}{L\omega_d}\frac{\omega_d}{(s+\alpha)^2 - \omega_d^2}$$

where $\omega_d^2 = \alpha^2 - \omega_0^2$

From Table 9.1, item 11 and Section 9.3 on s-shift,

$$i(t) = \frac{V_0}{L\omega_d} e^{-\alpha t} \sinh \omega_d t$$

The response of the circuit is shown in Fig. 10.6(b).

The response may also be put as:

$$i(t) = \frac{V_0}{V\omega_d} e^{-\alpha t}\left(e^{\omega dt} - e^{-\omega dt}\right)$$

$$= \frac{V_0}{2L\omega_d}\left(e^{\alpha + \omega dt} - e^{-\alpha - \omega dt}\right)$$

Case 2: Critically damped $\alpha = \omega_0$ or $R^2 = 4LC$

$$I = \frac{V_0}{L}\frac{1}{(s+\alpha)^2}$$

Using Section 9.3,

$$i(t) = \frac{V_0}{L} te^{-\alpha t}$$

The response is shown in Fig. 10.6(c).

Case 3: Underdamped $\alpha < \omega_0$ or $R^2 < 4LC$

$$I = \frac{V_0}{L}\frac{1}{(s+\alpha)^2 + (\omega_0^2 - \alpha^2)}$$

$$= \frac{V_0}{L\omega_d}\frac{\omega_d}{(s+\alpha)^2 + \omega_d^2}$$

Inverting,

$$i(t) = \frac{V_0 e^{-\alpha t}}{L\omega_d} \sin \omega_d t$$

Note: ω_d^2 is now taken as $\omega_0^2 - \alpha^2$.

The response is shown in Fig. 10.6(d).

Case 4: Zero damping $\alpha = 0$ or $R = 0$

$$I = \frac{V_0}{L}\frac{1}{s^2 + \omega_0^2}$$

$$= \frac{V_0}{L\omega_0}\frac{\omega_0}{s^2 + \omega_0^2}$$

Then

$$i(t) = \frac{V_0}{L\omega_0}\sin \omega_0 t$$

The natural response of the circuit is seen to be a continuous oscillation at angular frequency ω_0.

Important definitions in this section are:

Damping constant $\alpha = R/2L$

Undamped natural frequency $\omega_0 = 1/\sqrt{(LC)}$

Damped natural frequency $\omega_d^2 = \omega_0^2 - \alpha^2$ for underdamped system.

Step response of second order circuit:

The differential equation governing the circuit is:

$$Ri + L\frac{di}{dt} + \frac{1}{C}\int_{-\infty}^{t} i\,dt = V\,u(t)$$

The initial conditions are:

$$i(0) = 0$$

$$\frac{di}{dt} = \frac{V}{L} \text{ as in Eq. (10.1)}$$

Taking Laplace transforms,

$$RI + L(sI - 0) + \frac{I}{Cs} - \frac{V_0}{s} = \frac{V}{s}$$

$$I = \frac{V/s + V_0/s}{Ls + R + 1/Cs} = \frac{V + V_0}{L(s^2 + R/s + 1/LC)}$$

The solution is completed as in Section 10.3.

EXAMPLE 10.2 Solve the electrical circuit shown in Fig. 10.7 with

$$R = 1\,\Omega \quad L = 1\text{ H} \quad C = 1\text{ F} \quad \text{and} \quad V = 1\text{ V}$$

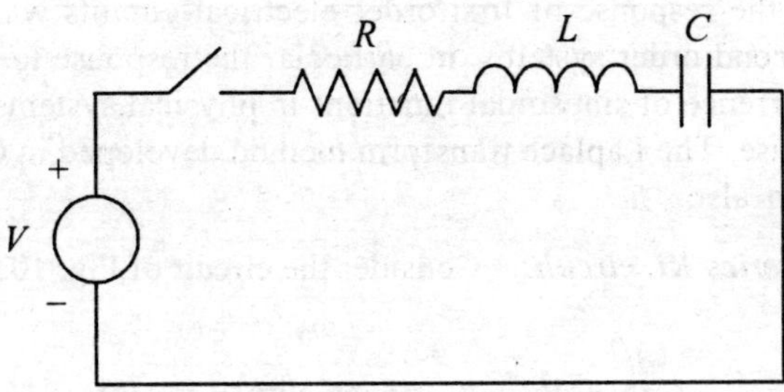

Fig. 10.7

SOLUTION:
Inductance and capacitance initially contain no energy and the source a step input voltage. The equation governing the circuit is:

$$Ri + L\frac{di}{dt} + \int_0^t i\,dt = V$$

which reduces to

$$i + \frac{di}{dt} + \int_0^t i\,dt = 1 \qquad (10.2)$$

$$\alpha = \frac{R}{2L} = \frac{1}{2}$$

$$\omega_0 = \frac{1}{\sqrt{(LC)}} = 1$$

Since $\omega_0 > \alpha$, the circuit is underdamped.

$$\omega_d^2 = \omega_0^2 - \alpha^2 = \frac{3}{4}$$

and

$$\omega_d = \sqrt{3}/2$$

Initial conditions are:

$$i_{t=0} = 0$$

$$\left(\frac{di}{dt}\right)_{t=0} = \frac{V}{L} = 1$$

Equation (10.2) is identical with the integro-differential equation in Example 9.38 and with the same initial conditions.
Hence

$$i = \frac{2}{\sqrt{3}}\,e^{-\frac{1}{2}t}\sin\frac{\sqrt{3}\,t}{2} \qquad \textbf{Ans.}$$

10.4 STEADY STATE SINUSOIDAL RESPONSE OF ELECTRICAL CIRCUITS

In Chapter 9 the response of first order electrical circuits was considered. This section deals with second order systems, in particular the response to sinusoidal excitation—the frequent occurrence of sinusoidal functions in physical systems calls for special attention to such response. The Laplace transform method developed in Chapter 9 will be followed in this section also.

Response of series RL circuit: Consider the circuit of Fig. 10.8. The equation governing the circuit is:

$$L\frac{di}{dt} + Ri = V_m \sin \omega t \text{ with } R = 1\ \Omega \qquad L = 1\ \text{H}$$

Take the initial conditions as:

$$i(0) = 0$$

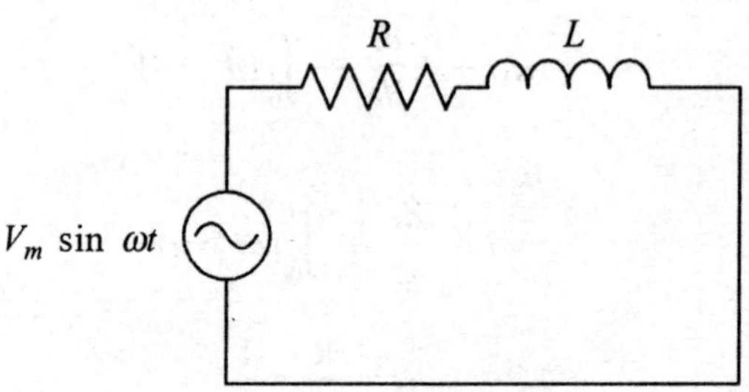

Fig. 10.8

The response has been obtained by three different methods so as to enable the reader to have a better insight into the problem.

1. Taking Laplace transforms of both sides of the given equation:

$$I(s) = V_m \frac{\omega}{(s+1)(s^2+\omega^2)} \tag{10.3}$$

Splitting into partial fractions,

$$I(s) = V_m \frac{\omega}{\omega^2+1}\left[\frac{1}{s+1} + \frac{1}{s^2+\omega^2} - \frac{s}{s^2+\omega^2}\right]$$

and

$$i(t) = V_m \frac{\omega}{\omega^2+1} e^{-t} + \frac{V_m}{\sqrt{\omega^2+1}} \sin(\omega t - \theta)$$

where $\theta = \tan^{-1}\omega/1$.

2. From Example 9.29, if $I(s) = \dfrac{f}{[(s+k)^2 + a^2]g(s)}$, the inverse $i(t)$ has the term:

$$\text{Im}\,[flag]_{\alpha}e^{\alpha t}$$

where $\alpha = -k + ja$

Let $f = V_m\omega$, $g = s + 1$, $a = \omega$ and $\alpha = 0 + j\omega$

The inverse $i(t)$ then has the term.

$$\text{Im}\left[\frac{V_m\omega}{\omega(s+1)}\right]_{j\omega}e^{j\omega t}$$

which works out to $\dfrac{1}{\sqrt{1+\omega^2}}\sin t\,(\omega t - \theta)$.

This is the steady state solution corresponding to the factor $s^2 + \omega^2$.

The transient solution which corresponds to the factor $s + 1$ in the transform may be found by the cover up rule as:

$$\frac{V_m\omega}{1+\omega^2}e^{-t}$$

3. Convolution method

$$L^{-1}\frac{V_m}{(s+1)}\frac{\omega}{(s^2+\omega^2)} = V_m\int_0^t e^{-t}\sin\omega t\,dt$$

$$= V_m\int_0^t e^{-(t-x)}\sin\omega x\,dx$$

$$= V_m e^{-t}\int_0^t e^{+x}\sin\omega x\,dx$$

$$= V_m\frac{\omega}{\omega^2+1}e^{-t} + \frac{V_m}{\sqrt{\omega^2+1}}\sin(\omega t - \theta)$$

Response of series RLC circuit:

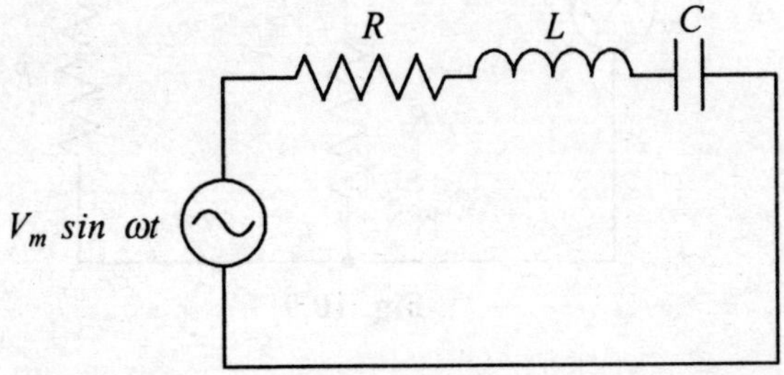

R L C

V_m sin ωt

The equation which governs the series circuit is:

$$Ri + L\frac{di}{dt} + \frac{1}{C}\int_{-\infty}^{t} i\,dt = V_m\sin\omega t$$

with $i(0) = 0$.

Taking Laplace transforms,

$$\left[R + sL + \frac{1}{Cs}\right] I(s) = V_m \frac{\omega}{s^2 + \omega^2}$$

$$I(s) = \frac{V_m \omega}{(s^2 + \omega^2)} \frac{Cs}{(RCs + LCs^2 + 1)}$$

$$= \frac{V_m \omega}{(s^2 + \omega^2)} \frac{s}{L\left(s^2 + \left(\frac{R}{L}\right)s + \frac{1}{LC}\right)}$$

The steady state response (which corresponds to the $s^2 + \omega^2$ in the transform $i(t)$) is given by

$$\text{Im}\left[\frac{f}{ag}\right]_{\alpha}^{e^{\alpha t}}$$

where $f = \frac{V_m \omega s}{L}$, $a = \omega$, $g = s^2 + \frac{R}{L}s + \frac{1}{LC}$ and $\alpha = j\omega$.

On simplification steady state response becomes

$$\frac{V_m}{\sqrt{R^2 + (\omega L - 1/\omega C)^2}} \sin\left[\omega t + \tan^{-1} - \frac{(\omega L - 1/\omega C)}{R}\right]$$

EXAMPLE 10.3 Find the condition under which $V_L = V_{R2}$ in Fig. 10.9.
SOLUTION:

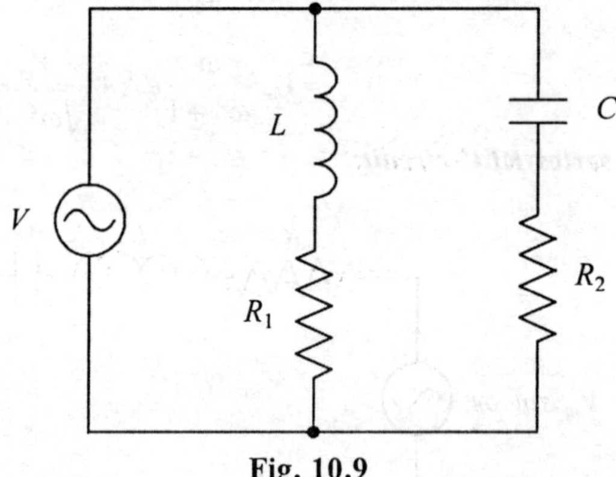

Fig. 10.9

$$V_L = \frac{Vj\omega L}{R_1 + j\omega L}$$

and

$$V_{R2} = \frac{VR_2}{R_2 + 1/j\omega C}$$

If V_L is to be equal to V_{R2} then

$$\frac{j\omega L}{R_1 + j\omega L} = \frac{R_2}{R_2 + 1/j\omega C}$$

On simplification,

$$R_1 R_2 = L/C$$

which is the required condition. **Ans.**

EXAMPLE 10.4 Determine the voltage V_1 in Fig. 10.10, using superposition theorem.

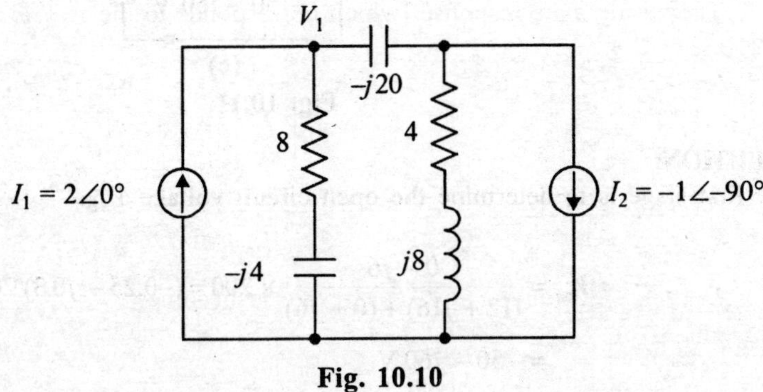

Fig. 10.10

SOLUTION:
Voltage due to current I_1 alone,

$$V_1' = \frac{2\angle 0°(8 - j4)(-j20 + 4 + j8)}{8 - j4 - j20 + 4 + j8}$$

$$= 8 - j8 \text{ V}$$

Voltage due to current I_2,

$$V_1'' = \frac{-1\angle -90°(4 + j8)}{8 - j4 - j20 + 4 + j8}(8 - j4 - j20)$$

$$= -4 \text{ V}$$

Hence

$$V = V_1' + V_1''$$

$$= 4 - j8 \text{ V} \quad \textbf{Ans.}$$

Thevenin equivalent circuits

EXAMPLE 10.5 Determine the Thevenin equivalent circuit for the network supplying the load shown in Fig. 10.11(a).

R $X_L = 16\ \Omega$ 200 V 12 Ω $X_c = 6\ \Omega$ Load

(a)

$12 + j16\ \Omega$ $0 + j6\ \Omega$ $200 + j0$ V_{oc}

(b)

$(1.77 - 7.48j)\Omega$ Load $-50 - j60$ V

(c)

Fig. 10.11

SOLUTION:
The first stage is to determine the open circuit voltage V_{AB}.

$$V_{oc} = \frac{0 - j6.}{(12 + j16) + (0 - j6)} \times 200 = (-0.25 - j0.3)\ 200$$
$$= -50 - j60 \text{ V}$$

To determine the internal impedance, short circuit the source voltage and consider the circuit shown in Fig. 10.11(b).

$$Z_{\text{in}} = \frac{(0 - j6)(12 + j16)}{(0 - j6) + (12 + j16)} = 1.77 - j7.48 \qquad \textbf{Ans.}$$

The Thevenin equivalent circuit is shown in Fig. 10.11(c).

Norton equivalent circuits

EXAMPLE 10.6 Find the Norton equivalent at the terminals *ab* of the circuit shown in Fig. 10.12.

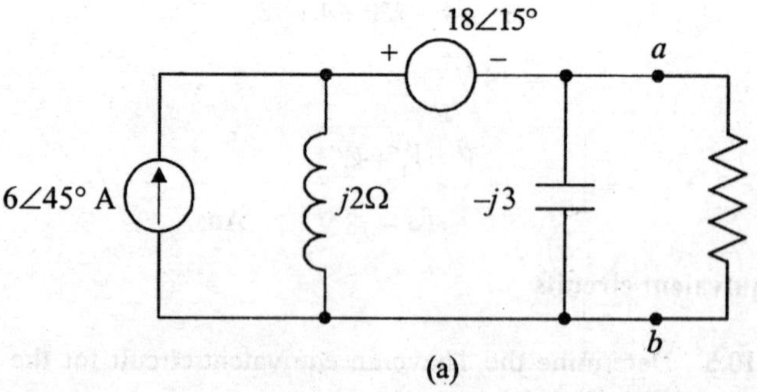

(a)

SOLUTION:
From Fig. 10.12(b),

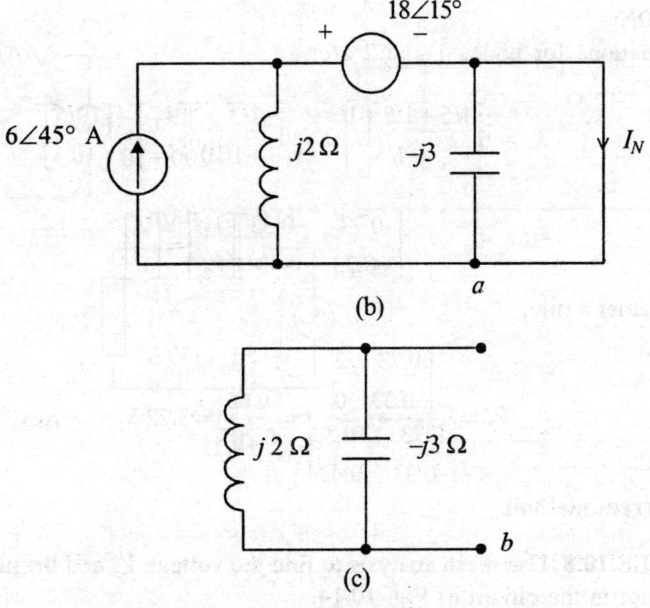

Fig. 10.12

$$I_N = 6\angle 45° - I_1$$

$$= 6\angle 45° - \frac{18\angle 15°}{2\angle 90°} = 13.1\angle 81.6° \text{ A}$$

From Fig. 10.12(c)

$$Z_{Th} = \frac{j2(-j3)}{j2 - j3} = j6\ \Omega$$

$$Y_N = \frac{1}{Z_{Th}} = \frac{1}{j6} = 0.17\angle -90° \text{ siemens} \qquad \textbf{Ans.}$$

Node voltage method

EXAMPLE 10.7 Find the voltage across the 10 Ω resistance in the circuit of Fig. 10.13 using nodal analysis.

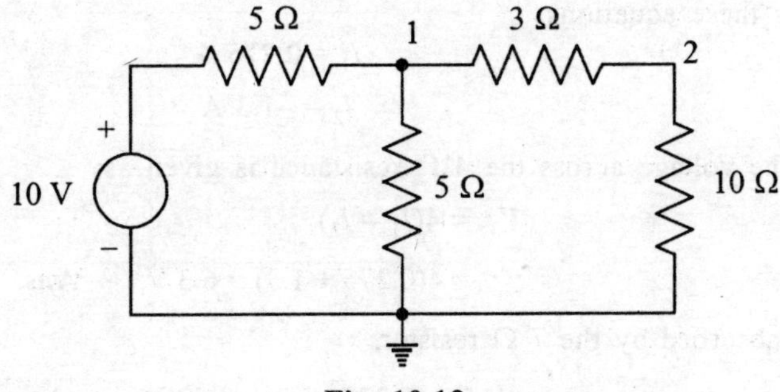

Fig. 10.13

SOLUTION:

Node equations for nodes 1 and 2 are:

$$\begin{bmatrix} 1/5+1/5+1/3 & -1/3 \\ -1/3 & 1/3+1/10 \end{bmatrix} \begin{bmatrix} V_1 \\ V_2 \end{bmatrix} = \begin{bmatrix} 10/5 \\ 0 \end{bmatrix}$$

or

$$\begin{bmatrix} 0.73 & -0.33 \\ -0.33 & 0.43 \end{bmatrix} \begin{bmatrix} V_1 \\ V_2 \end{bmatrix} = \begin{bmatrix} 2 \\ 0 \end{bmatrix}$$

Using Cramer's rule,

$$V_2 = \frac{\begin{vmatrix} 0.73 & 2 \\ -0.33 & 0 \end{vmatrix}}{\begin{vmatrix} 0.73 & -0.33 \\ -0.33 & 0.43 \end{vmatrix}} = \frac{0.66}{0.205} = 3.22 \text{ V} \qquad \textbf{Ans.}$$

Mesh current method

EXAMPLE 10.8 Use mesh analysis to find the voltage V_3 and the power absorbed in the 7 Ω resistor in the circuit of Fig. 10.14.

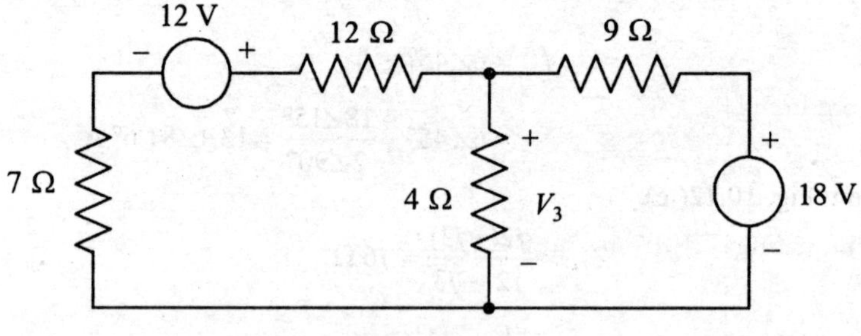

Fig. 10.14

Writing KVL for the two meshes,

$$12I_1 + 4\,(I_1 - I_2) + 7I_1 = 12$$

$$9I_2 + 4(I_2 - I_1) = -18$$

Solving these equations,

$$I_1 = 0.275 \text{ A}$$

$$I_2 = -1.3 \text{ A}$$

The voltage across the 4 Ω resistance is given as:

$$V_3 = 4(I_1 - I_2)$$

$$= 4(0.275 + 1.3) = 6.3 \text{ V} \qquad \textbf{Ans.}$$

Power absorbed by the 7 Ω resistor,

$$= I_1^2\, 7 = 0.275^2 \times 7 = 0.53 \text{ W} \qquad \textbf{Ans.}$$

Magnetically coupled circuits

EXAMPLE 10.9 Determine the currents I_1 and I_2 at time t after closing the switch in the circuit of Fig. 10.15. Specify the initial and final values of the currents.

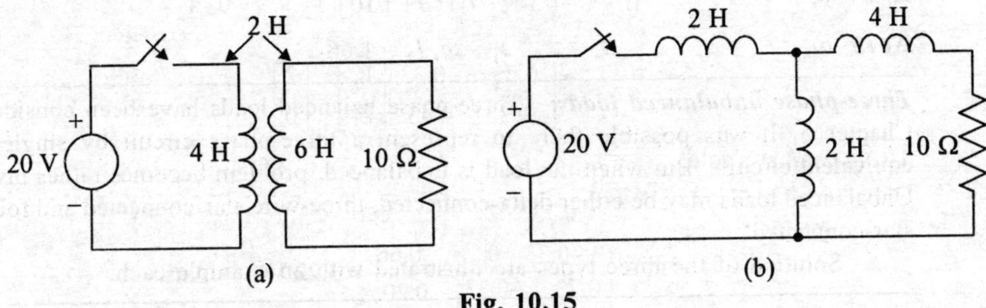

Fig. 10.15

Figure 10.15(b) is the conductively coupled equivalent circuit with mesh currents I_1 and I_2. The corresponding equations are:

$$4\frac{dI_1}{dt} - 2\frac{dI_2}{dt} = 20 \tag{1}$$

$$6\frac{dI_2}{dt} + 6I_2 - \frac{2dI_1}{dt} = 0 \tag{2}$$

From Eq. (1),

$$\frac{dI_1}{dt} = 5 + 0.5\frac{dI_2}{dt} \tag{3}$$

Eliminating $\frac{dI_1}{dt}$ from Eq. (2),

$$5\frac{dI_2}{dt} + 6I_2 = 10$$

This is a standard first order equation with initial condition $I_2 = 0$.
Thus,

$$I_2 = \frac{10}{6} - \frac{10}{6}e^{-1.2t}$$

$$= 1.67 - 1.67\,e^{-1.2t}$$

Substituting $\frac{dI_2}{dt}$ in Eq. (3),

$\frac{dI_1}{dt}$ is obtained as:

$$\frac{dI_1}{dt} = 5 + e^{-1.2t}$$

On integration,

$$I_1 = 5t + \frac{1}{1.2}e^{-1.2t} + C$$

where C is an arbitrary constant.

Using the condition, $I_1 = 0$ at $t = 0$,

$$C = 0$$

$$\therefore \quad I_1 = 5t + 0.83e^{-1.2t}$$

At $t = 0$, $\quad I_1 = I_2 = 0$

At $t = \infty$, $\quad I_1 = \infty,\ I_2 = 1.66$

Three-phase unbalanced loads: Three-phase balanced loads have been considered in Chapter 5. It was possible there to represent a three-phase circuit by single phase equivalent circuits. But when the load is unbalanced, problem becomes rather involved. Unbalanced loads may be either delta-connected, three-wire star-connected and four-wire star-connected.

Solution of the three types are illustrated with an example each.

EXAMPLE 10.10 A delta-connected load is shown in Fig. 10.16. Supply voltage is 400 V at 50 Hz. Calculate: (a) the phase currents and (b) the line currents.

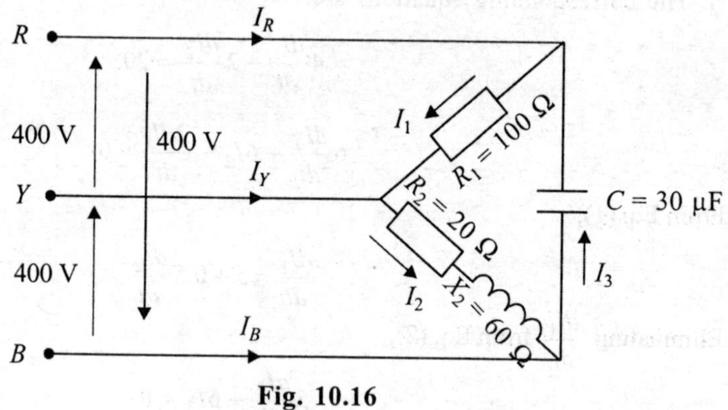

Fig. 10.16

SOLUTION:

Assume the phase sequence $R\ Y\ B$. The phase voltages are V_{RY}, V_{YB}, V_{BR}. If I_1, I_2, I_3 are the phase currents in the loads then

$$I_1 = \frac{400\angle 0}{100 + j0} = 4\angle 0 \text{ A}$$

$$I_2 = \frac{400\angle -120}{20 + j60} = 6.32\angle -191.56 \text{ A}$$

$$I_3 = 400\ \angle -240 \times j2\pi \times 50 \times 30 \times 10^{-6} = 3.77\angle -150 \text{ A} \qquad \textbf{Ans.}$$

The line currents are:

$$I_R = I_1 - I_3$$

$$= 4\angle 0 - 3.77\ \angle -150$$

$$= 7.5\ \angle 14.59 \text{ A}$$

$$I_Y = I_2 - I_1$$
$$= 6.32 \angle -191.56 - 4\angle 0$$
$$= 10.26\angle 172.95 \text{ A}$$
$$I_B = I_3 - I_2$$
$$= 3.77\angle -150 - 6.32 \angle -191.56$$
$$= 4.3\angle -47.1 \text{ A} \qquad \textbf{Ans.}$$

Verify $I_R + I_Y + I_B = 0$.

EXAMPLE 10.11 A symmetrical three-phase 400 V, three-wire supply feeds an unbalanced star-connected load with load impedances $Z_R = 5\angle 0\ \Omega$, $Z_Y = 2\angle 90\ \Omega$ and $Z_B = 4\angle -90\ \Omega$. Find the line currents and the displacement neutral voltage.

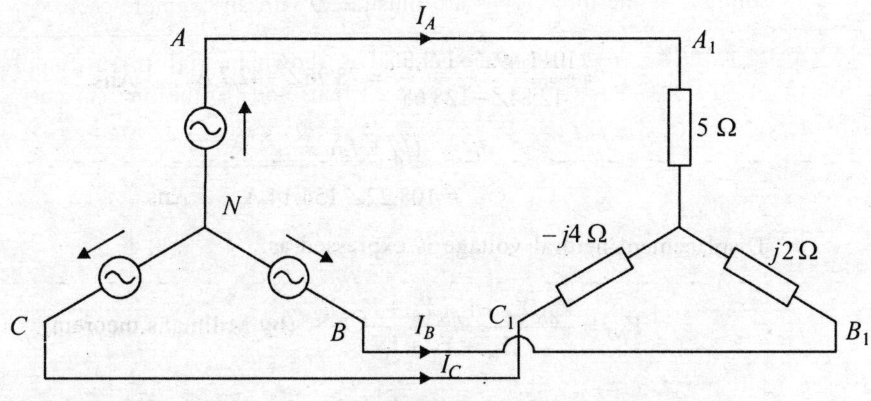

Fig. 10.17

SOLUTION:

V_{AB} is taken as the reference.

$$V_{AB} = 400\angle 0$$
$$V_{BC} = 400\angle -120$$
$$V_{CA} = 400\angle -240$$

For the circuit $A\ A_1\ B_1\ B$,

$$400\angle 0 = 5I_A - j2\ I_B$$

For the circuit $B\ B_1\ C_1\ C$,

$$400\angle -120 = j2\ I_B - (-j4)\ I_C$$

A similar equation for circuit $C\ C_1\ A_1\ A$ will not be an independent equation. But use may be made of the equation:

$$I_A + I_B + I_C = 0$$

The above equations can be solved by Cramer's rule.
The solution is:

$$I_A = \frac{\begin{vmatrix} 400 & -2\angle 90 & 0 \\ 400\angle -120 & 2\angle 90 & 4\angle 90 \\ 0 & 1 & 1 \end{vmatrix}}{D}$$

where

$$D = \begin{vmatrix} 5 & -2\angle 90 & 0 \\ 0 & 2\angle 90 & 4\angle 90 \\ 1 & 1 & 1 \end{vmatrix}$$

$$I_A = \frac{1385.64\angle -60}{12.81\angle -128.65} = 108.16\angle 68.65 \text{ A} \qquad \textbf{Ans.}$$

$$I_B = \frac{\begin{vmatrix} 5 & 400 & 0 \\ 0 & 400\angle -120 & 4\angle 90 \\ 1 & 0 & 1 \end{vmatrix}}{D}$$

$$= \frac{1011.49\angle -171.35}{12.81\angle -128.65} = 78.96\angle -42.7 \text{ A} \qquad \textbf{Ans.}$$

$$I_C = -(I_A + I_B)$$

$$= 108.22\angle 154.14 \text{ A} \qquad \textbf{Ans.}$$

Displacement neutral voltage is expressed as:

$$V_{N'} = \frac{V_{AN}Y_A + V_{BN}Y_B + V_{CN}Y_C}{Y_A + Y_B + Y_C} \quad \text{(by Millmans theorem)}$$

$$= \frac{\frac{400}{\sqrt{3}} \times \frac{1}{5} + \frac{400\angle -120}{\sqrt{3}} \times \frac{1}{j2} + \frac{400\angle -240}{\sqrt{3}} \times \frac{-1}{j4}}{\frac{1}{5} + \frac{1}{j2} - \frac{1}{j4}}$$

$$= \frac{107.75\angle 164.45}{0.32\angle -51.34}$$

$$= 336.72\angle 215.79 \text{ V} \qquad \textbf{Ans.}$$

EXAMPLE 10.12 A three-phase 400 V, four-wire supply feeds an unbalanced star connected load with impedances as $Z_A = 5\angle 0\ \Omega$, $Z_B = 2\angle 90\ \Omega$ and $Z_C = 4\angle -90\ \Omega$. Find the line currents.

SOLUTION:

$$I_A = \frac{400\angle 0}{\sqrt{3} \times 5} = 46.18\angle 0 \text{ A} \qquad \textbf{Ans.}$$

$$I_B = \frac{400\angle -120}{\sqrt{3} \times j2} = 115.47\angle -210 \text{ A} \qquad \textbf{Ans.}$$

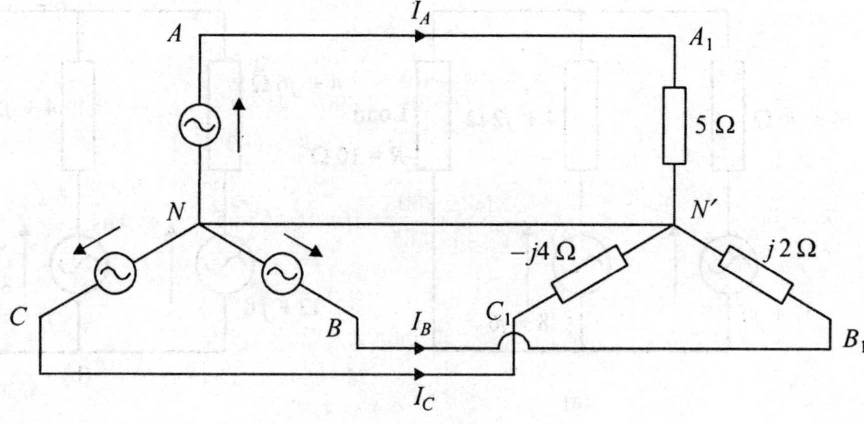

Fig. 10.18

$$I_C = \frac{400\angle -240}{\sqrt{3} \times (-j4)} = 57.74\angle -150 \text{ A} \qquad \textbf{Ans.}$$

$$I_N = -(I_A + I_B + I_C)$$

$$= -\ 107.75\angle 164.45 \text{ A}$$

SUMMARY

- The complex notation introduced in this chapter expedites ac network analysis as it is evidenced by the illustrative examples and problems.
- Network theorems applied to ac networks have been stated. The complex notation enables dc network techniques to be used as well with ac networks.
- The transient response, steady state response of first order and second order networks are included in last sections of the chapter.

Solved Problems

P10.1 For the circuit shown in Fig. P10.1(a) calculate the load current using Thevenin's equivalent circuits.

SOLUTION:

In Fig. P10.1(b), the circulating current is given by

$$\frac{12 + j0 - 8 + j6}{(4 + j6) + (4 + j2)} = 0.625 + j0.125$$

$$V_{oc} = (12 + j0) - (0.625 + j0.125)(4 + 6j)$$

$$= 10.25 - j4.25 \text{ V}$$

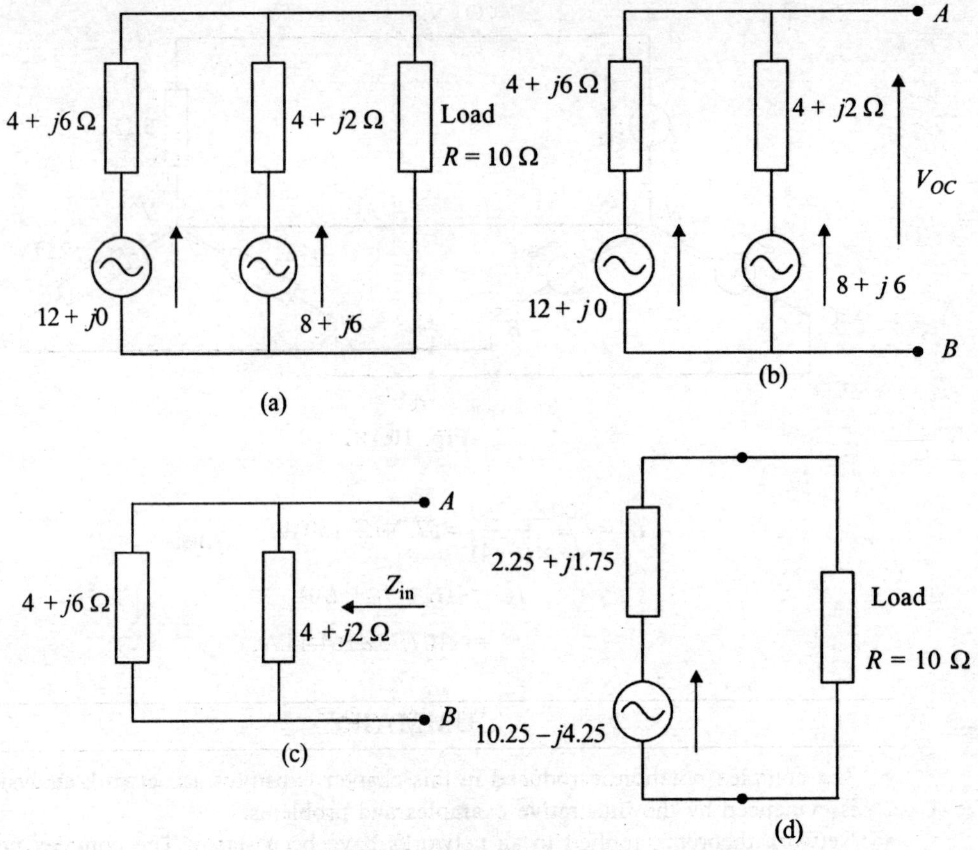

(a)

(b)

(c)

(d)

Fig. P10.1

The Thevenin impedance is obtained by short-circuiting the voltages and using the circuit shown in Fig. P10.1(c).

$$Z_{Th} = \frac{(4+j6)(4+j2)}{(4+j6)+(4+j2)} = 2.25 + j1.75$$

The Thevenin equivalent circuit is shown in Fig. P10.1(d).

The load current is given by

$$I = \frac{10.25 - j4.25}{(2.25 + j1.75) + (10 + j0)} = \frac{10.25 - j4.25}{12.25 + j1.75}$$

Therefore, magnitude $I = 0.9$ A **Ans.**

P10.2 Obtain the Norton equivalent of the circuit of Fig. P10.2 at the terminals *ab*.

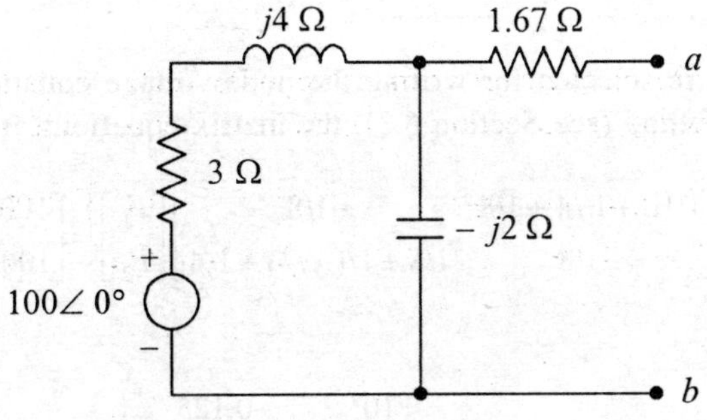

Fig. P10.2

SOLUTION:

$$I_N = \frac{100\angle 0^\circ}{(3+j4)+\dfrac{1.67(-j2)}{1.67-j2}} \times \frac{-j2}{1.67-j2}$$

$$= -15.05\angle 101.52^\circ \text{ A}$$

$$Z_N = 1.67 + \frac{(-j2)(3+j4)}{-j2+3+j4}$$

$$= 3.68\angle -45.19^\circ \ \Omega$$

and $$Y_N = \frac{1}{Z_N} = 0.27\angle 45.19^\circ \text{ siemens}$$

0.27∠45.19° S

−15.05∠101.52° A

Ans.

P10.3 Use the node voltage method to obtain the current I in the network of Fig. P10.3.

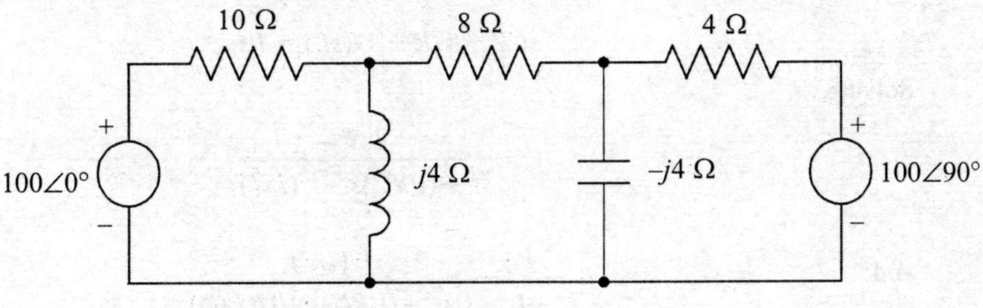

Fig. P10.3

SOLUTION:

Nodes 1 and 2 are selected for writing the node voltage equations. By the rules of the nodal voltage method (see Section 6.3), the matrix equations are:

$$\begin{bmatrix} 1/10 + 1/j4 + 1/8 & -1/8 \\ -1/8 & 1/8 + 1/(-j4) + 1/4 \end{bmatrix} \begin{bmatrix} V_1 \\ V_2 \end{bmatrix} = \begin{bmatrix} 100\angle 0°/10 \\ 100\angle 90°/4 \end{bmatrix}$$

which gives

$$V_1 = \frac{\begin{vmatrix} 10 & -0.125 \\ 25j & 0.375 + 0.25j \end{vmatrix}}{\begin{vmatrix} 0.225 - 0.25j & -0.125 \\ -0.125 & 0.375 + 0.25j \end{vmatrix}}$$

$$= \frac{6.76\angle 56.31}{0.136\angle -15.95}$$

$$= 49.7\angle 72.26 \text{ V}$$

and

$$I = V_1/j2$$
$$= 49.7\angle 72.26/\ 2\angle 90$$
$$= 24.85\angle -17.74 \text{ A} \qquad \textbf{Ans.}$$

P10.4 Determine the step response of a parallel *RCL* circuit of Fig. P10.4(a) by the node voltage method. Find I_L.

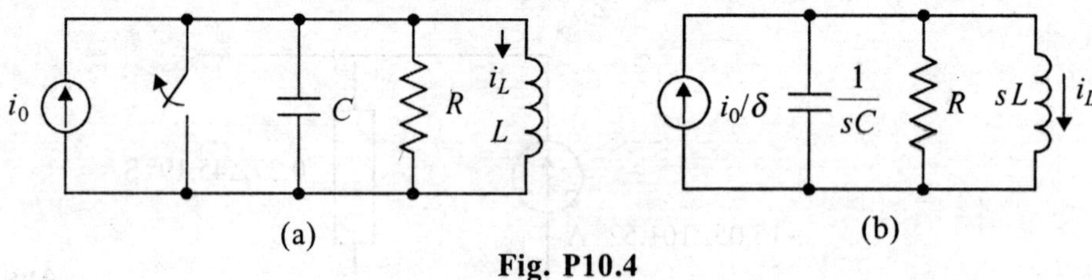

Fig. P10.4

SOLUTION:

The *s* domain equivalent circuit is shown in Fig. P10.4(b). Summing the currents at the top node gives

$$sCV + V/R + V/(sL) = I/s$$

Solving,

$$V = \frac{I/C}{s^2 + (1/RC)s + 1/(LC)}$$

and

$$I_L = \frac{V}{sL} = \frac{I/(LC)}{s\{s^2 + (1/RC)s + 1/(LC)\}}$$

Using the numerical values of the parameters,

$$I_L = \frac{192 \times 10^5}{s(s^2 + 64000s + 16 \times 10^8)}$$

$$= \frac{A}{s} + \frac{Bs + C}{(s + 32000)^2 + 576 \times 10^6}$$

Partial fraction coefficients are:

$$A = 12 \times 10^{-3}$$
$$B = -12 \times 10^{-3}$$
$$C = -768$$

Taking the inverse transform,

$$I_L = 12 \times 10^{-3} - e^{-32000t} (12 \times 10^{-3} \cos 24000t + 16 \times 10^{-3} \sin 24000t) \text{ A}$$

$$= 12 - e^{-32000t} (12 \cos 24000t + 16 \sin 24000t) \text{ mA} \qquad \textbf{Ans.}$$

P10.5 In the circuit of Fig. P10.5, if $V_1 = 10\angle 0$, find all the other currents and voltages using mesh currents.

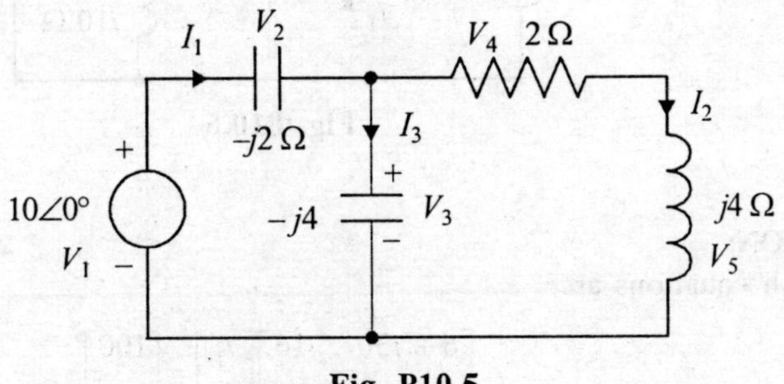

Fig. P10.5

SOLUTION:

In matrix form, the mesh current equations are:

$$\begin{bmatrix} -j6 & j4 \\ j4 & 2 + j4 - j4 \end{bmatrix} \begin{bmatrix} I_1 \\ I_2 \end{bmatrix} = \begin{bmatrix} 10 \\ 0 \end{bmatrix}$$

$$I_1 = \frac{\begin{vmatrix} 10 & j4 \\ 0 & 2 \end{vmatrix}}{\begin{vmatrix} -j6 & j4 \\ j4 & 2 \end{vmatrix}} = \frac{20}{20\angle -36.87} = 1\angle 36.87$$

Similarly,

$$I_2 = \frac{-40\angle 90}{20\angle -36.87} = -2\angle 126.87$$

Now, the branch currents are:

$$i_1 = I_1 = 1\angle 36.87 \text{ A}$$
$$i_2 = I_2 = -2\angle 126.87 \text{ A}$$
$$i_3 = I_1 - I_2 = 2.24\angle 100.3 \text{ A} \qquad \textbf{Ans.}$$

and the voltages are:

$$V_2 = I_1(-j2) = -2\angle 126.87 \text{ V}$$
$$V_3 = I_3(-j4) = -8.96\angle 190.3 \text{ V}$$
$$V_4 = I_2(2) = -4\angle 126.87 \text{ V}$$
$$V_5 = I_2(j4) = -8\angle 216.87 \text{ V} \qquad \textbf{Ans.}$$

P10.6 In the circuit of Fig. P10.6, find the voltage across the 10 Ω reactance.

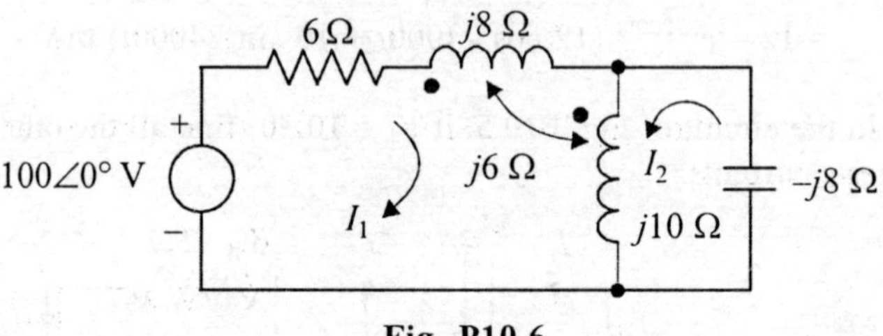

Fig. P10.6

SOLUTION:

The mesh equations are:

$$\begin{bmatrix} 6 + j30 & j16 \\ j16 & -j6 \end{bmatrix} \begin{bmatrix} I_1 \\ I_2 \end{bmatrix} = \begin{bmatrix} 100 \\ 0 \end{bmatrix}$$

Solving these equations by Cramer's rule,

$$I_1 = \frac{-600\angle 90}{437.48\angle -4.72} = -1.37\angle 94.7 \text{ A}$$

and

$$I_2 = \frac{-1600\angle 90}{437.48\angle -4.72} = -3.66\angle 94.7 \text{ A}$$

The voltage across 10 Ω reactance is:

$$V = -I_2(-j16) = -58.56\angle 184.7 \text{ V} \qquad \textbf{Ans.}$$

Exercises

E10.1 Find the Thevenin equivalent for the circuit to the left of terminals A and B in Fig. E10.1 and calculate the current in and the voltage across the 1.5 Ω load resistance. Determine the value of R_L for which power transferred is maximum.

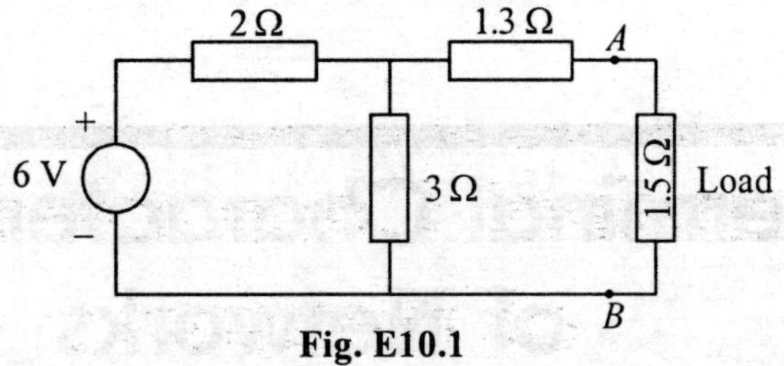

Fig. E10.1

[*Ans:* $V_{Th} = 6$ V, $R_{Th} = 2.5\ \Omega$, $I = \dfrac{6}{2.5 + 1.5} = 1.5$ A, $V = IR_L = 2.25$ V

The load resistance for which the power transfer is maximum is = 2.5 Ω]

E10.2 The Thevenin equivalent values of a network are $V_{Th} = 400\angle 30°$ V and $Z_{Th} = 10\angle 10°\ \Omega$. Determine the parameters of the Norton equivalent circuit.

[**Ans:** $I_N = 40\angle 20°$ A, $Y_N = 0.1\angle -10°$ siemens]

Terminal Characteristics of Networks

11

CHAPTER CONTENTS

This chapter deals with the mathematical representation of input output relationships (the external characterization) of networks.

A port in a network is a pair of terminals for which the current entering one terminal is equal to the current leaving the other terminal.

For example, Fig. 11.1(a) shows a one-port network whereas Fig. 11.1(b) shows a two-port network.

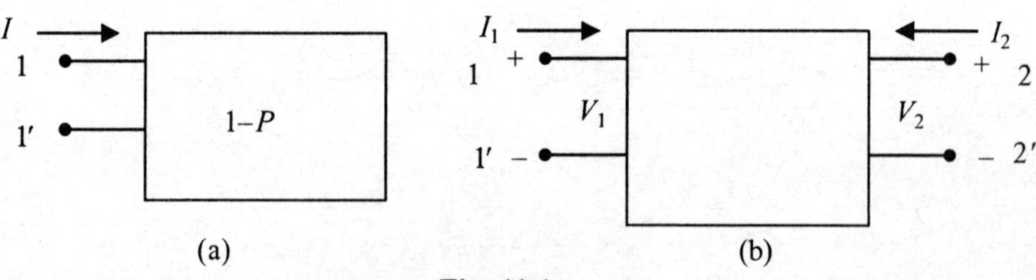

Fig. 11.1

11.1 ONE-PORT NETWORK

A one-port network is shown in Fig. 11.1(a). The port variables are $V(s)$ and $I(s)$.

Driving point terminals are 1–1′. With a current source $i(t)$ as the input, the zero-state response is $v(t)$, and in the s domain,

$$V(s) = Z_{dp}(s)\, I(s) \tag{11.1}$$

where $Z_{dp}(s)$ is by definition the *driving point impedance* seen by the source at 1–1′. Examples of *dp* impedances for $1-P^s$ are:

$$R, \quad Ls \quad \text{and} \quad \frac{1}{Cs}.$$

If the input is a voltage source $v(t)$, the zero-state response is $i(t)$ and

$$I(s) = Y_{dp}(s)\, V(s) \tag{11.2}$$

where $Y_{dp}(s)$ is by definition the *dp admittance* of the $1-P$. For the single elements, the *dp* admittances are:

$$G\,(=1/R), \quad \frac{1}{Ls} \quad \text{and} \quad Cs$$

In general, for $1-P^s$

$$Y_{dp}(s) = Z_{dp}^{-1}\,(s)$$

Equations (11.1) and (11.2) are special cases of the form:

$$R(s) = H(s)\, E(s)$$

where $R(s)$ is the zero-state response, $E(s)$ is the excitation (input) and $H(s)$ is the network function.

Methods of calculating dp functions:

Series connection

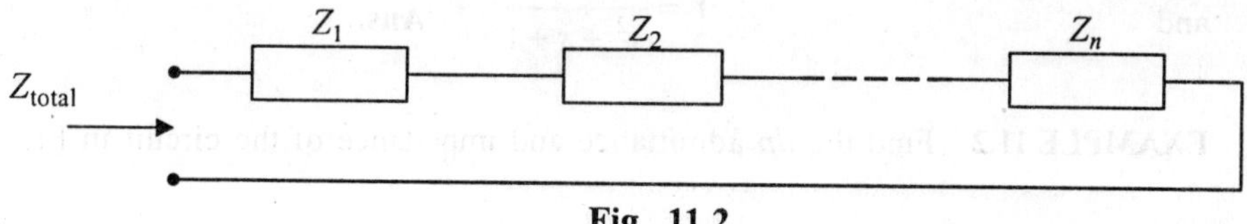

Fig. 11.2

The total impedance is:

$$Z(s) = \sum_{k=1}^{n} Z_k(s)$$

Parallel connection

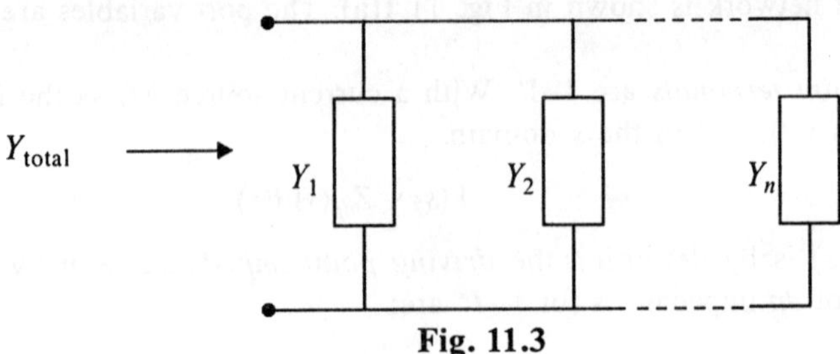

Fig. 11.3

The total admittance is:

$$Y(s) = \sum_{k=1}^{n} Y_k(s)$$

EXAMPLE 11.1 Find the *dp* impedance and admittance of the network in Fig. 11.4.

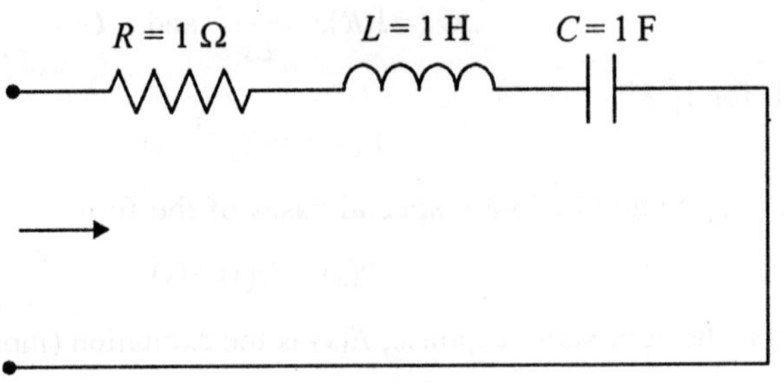

Fig. 11.4

SOLUTION:

$$Z = 1 + s + 1/s = \frac{s^2 + s + 1}{s} \qquad \textbf{Ans.}$$

and

$$Y = \frac{s}{s^2 + s + 1} \qquad \textbf{Ans.}$$

EXAMPLE 11.2 Find the *dp* admittance and impedance of the circuit in Fig. 11.5.

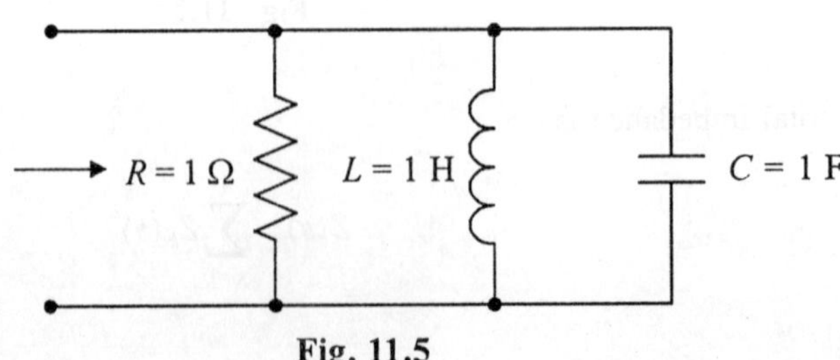

Fig. 11.5

SOLUTION:

$$Y = 1 + 1/s + s = \frac{s^2 + s + 1}{s}$$

and
$$Z = \frac{s}{s^2 + s + 1}$$

11.2 TWO-PORT NETWORK

Figure 11.1(b) is redrawn here which shows a two-port network.

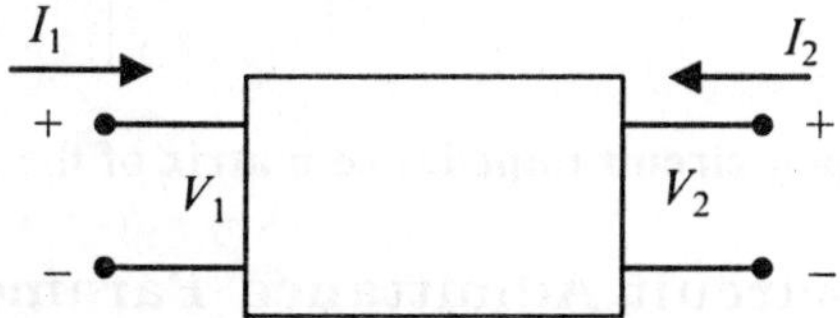

The four variables at the ports are represented by the phasors V_1, I_1, V_2 and I_2. There are six possible ways in which the four variables are related depending upon which two of the four are regarded as independent and which are dependent.

11.2.1 Open-circuit Impedance Parameters

If the port currents I_1 and I_2 are regarded as the independent variables, the port voltages are by superposition:

$$V_1 = z_{11}I_1 + z_{12}I_2 \tag{11.3}$$

$$V_2 = z_{21}I_1 + z_{22}I_2 \tag{11.4}$$

The open-circuit impedance of port 1 is:

$$z_{11} = \frac{V_1}{I_1} \quad \text{with } I_2 = 0 \quad \text{(Port 2 on open circuit)}$$

The open-circuit forward transfer impedance is:

$$z_{21} = \frac{V_2}{I_1} \quad \text{with } I_2 = 0 \quad \text{(Port 2 on open circuit)}$$

The open-circuit driving point impedance of port 2 is:

$$z_{22} = \frac{V_2}{I_2} \quad \text{with } I_1 = 0 \quad \text{(Port 1 on open circuit)}$$

The open-circuit reverse transfer impedance is:

$$z_{12} = \frac{V_1}{I_2} \quad \text{with } I_1 = 0 \quad \text{(Port 1 on open circuit)}$$

For passive circuits, by the reciprocity theorem,

$$z_{12} = z_{21} \tag{11.5}$$

Thus, any linear passive two-port network with the exception of the gyrator* (see Reference 5) may be characterized by three z-parameters. These are called *open-circuit impedance* parameters. Since they are measured at one port while the other port is *open*.

In matrix form,

$$\begin{bmatrix} V_1 \\ V_2 \end{bmatrix} = \begin{bmatrix} z_{11} & z_{12} \\ z_{21} & z_{22} \end{bmatrix} \begin{bmatrix} I_1 \\ I_2 \end{bmatrix} \tag{11.6}$$

where

$$z_{oc} = \begin{bmatrix} z_{11} & z_{12} \\ z_{21} & z_{22} \end{bmatrix} \tag{11.7}$$

is defined as the **open-circuit impedance matrix** of the network.

11.2.2 Short-circuit Admittance Parameters

If the port voltages are regarded as the independent variables, the port currents are:

$$I_1 = y_{11}V_1 + y_{12}V_2 \tag{11.8}$$

$$I_2 = y_{21}V_1 + y_{22}V_2 \tag{11.9}$$

The *short-circuit admittance* parameters are defined by Eqs. (11.8) and (11.9). The short-circuit driving point admittance of port 1 is:

$$y_{11} = \frac{I_1}{V_1} \quad \text{with } V_2 = 0 \quad \text{(Port 2 short circuited)}$$

The short-circuit forward transfer admittance is:

$$y_{21} = \frac{I_2}{V_1} \quad \text{with } V_2 = 0 \quad \text{(Port 2 short circuited)}$$

The short-circuit driving point admittance of port 2 is:

$$y_{22} = \frac{I_2}{V_2} \quad \text{with } V_1 = 0 \quad \text{(Port 1 short circuited)}$$

The short-circuit reverse transfer admittance is:

$$y_{12} = \frac{I_1}{V_2} \quad \text{with } V_1 = 0 \quad \text{(Port 1 short circuited)}$$

In matrix form, Eqs. (11.8) and (11.9) may be put as:

$$\begin{bmatrix} I_1 \\ I_2 \end{bmatrix} = \begin{bmatrix} y_{11} & y_{12} \\ y_{21} & y_{22} \end{bmatrix} \begin{bmatrix} V_1 \\ V_2 \end{bmatrix} \tag{11.10}$$

where

$$y_{sc} = \begin{bmatrix} y_{11} & y_{12} \\ y_{21} & y_{22} \end{bmatrix} \tag{11.11}$$

is defined as **short-circuit admittance matrix**.

For passive circuits, by the reciprocity theorem,

$$y_{12} = y_{21} \tag{11.12}$$

and only three y-parameters are required for the passive networks.

11.2.3 Hybrid Parameters

The current I_1 and voltage V_2 are taken as the independent variables. The characteristic equations are:

$$V_1 = h_{11}I_1 + h_{12}V_2 \tag{11.13}$$

$$I_2 = h_{21}I_1 + h_{22}V_2 \tag{11.14}$$

The hybrid or h parameters are:

$$\text{Short circuit input impedance } h_{11} = \frac{V_1}{I_1} \quad (\text{with } V_2 = 0)$$

$$\text{Short circuit forward current gain } h_{21} = \frac{I_2}{I_1} \quad (\text{with } V_2 = 0)$$

$$\text{Open circuit output admittance } h_{22} = \frac{I_2}{V_2} \quad (\text{with } I_1 = 0)$$

$$\text{Open circuit reverse voltage gain } h_{12} = \frac{V_1}{V_2} \quad (\text{with } I_1 = 0)$$

The h parameters have mixed (hybrid) units—h_{11} is an impedance, h_{12} is dimensionless, h_{21} is dimensionless and h_{22} is an admittance.

The characteristic equations (11.13) and (11.14) are particularly useful in the analysis of transistor circuits.

11.2.4 Inverse Hybrid Parameters

These parameters result from V_1 and I_2 being taken as the independent variables. Hence

$$I_1 = g_{11}V_1 + g_{12}I_2 \tag{11.15}$$

$$V_2 = g_{21}V_1 + g_{22}I_2 \tag{11.16}$$

The g parameters are:

$$\text{Open circuit input admittance } g_{11} = \frac{I_1}{V_1} \quad (\text{with } I_2 = 0)$$

$$\text{Open circuit forward voltage gain } g_{21} = \frac{V_2}{V_1} \quad (\text{with } I_2 = 0)$$

$$\text{Short circuit output impedance } g_{22} = \frac{V_2}{I_2} \quad (\text{with } V_1 = 0)$$

$$\text{Short circuit reverse current gain } g_{12} = \frac{I_1}{I_2} \quad (\text{with } V_1 = 0)$$

Note that g_{11} is an admittance, g_{12} is dimensionless, g_{21} is dimensionless and g_{22} is an impedance.

11.2.5 Transmission Parameters (Chain Parameters or *T* Parameters)

Characteristic equations which are more useful for the transmission line analysis are derived by taking V_2 and I_2 as the independent variables. Thus

$$V_1 = AV_2 + B(-I_2) \tag{11.17}$$

$$I_1 = CV_2 + D(-I_2) \tag{11.18}$$

The negative sign is used in these equations because classical transmission line analysis considers the current at port 2 in the reverse direction to that shown in Fig. 11.1(b).

In matrix form,

$$\begin{bmatrix} V_1 \\ I_1 \end{bmatrix} = \begin{bmatrix} A & B \\ C & D \end{bmatrix} \begin{bmatrix} V_2 \\ -I_2 \end{bmatrix}$$

where A is dimensionless, B is an impedance, C is an admittance and D is dimensionless.

It can be shown that $ABCD$ parameters satisfy the condition $AD - BC = 1$ clearly this is the determinant of the T matrix.

11.2.6 Reverse Transmission Parameters

The reverse transmission parameters describe V_2 and $-I_2$ in terms of V_1 and I_1.

$$\begin{bmatrix} V_2 \\ -I_2 \end{bmatrix} = \begin{bmatrix} E & F \\ G & H \end{bmatrix} \begin{bmatrix} V_1 \\ I_1 \end{bmatrix} \tag{11.19}$$

The reverse transmission matrix is the inverse of the chain parameter matrix.

11.2.7 Relationships among the Two-port Parameters

The six sets of parameters are related to one another (see Table 11.1). Any one set may be derived in terms of any other set. For example, from Eqs. (11.8) and (11.13),

$$h_{11} = 1/y_{11}$$

and

$$h_{12} = y_{12}/y_{11}$$

Table 11.1 Summary of Terminal Parameters

z parameters $V_1 = z_{11}I_1 + z_{12}I_2$ $V_2 = z_{21}I_1 + z_{22}I_2$	y parameters $I_1 = y_{11}V_1 + y_{12}V_2$ $I_2 = y_{21}V_1 + y_{22}V_2$
h parameters $V_1 = h_{11}I_1 + h_{12}V_2$ $I_2 = h_{21}I_1 + h_{22}V_2$	g parameters $I_1 = g_{11}V_1 + g_{12}I_2$ $V_2 = g_{21}V_1 + g_{22}I_2$
T parameters $V_1 = AV_2 + B(-I_2)$ $I_1 = CV_2 + D(-I_2)$	

EXAMPLE 11.3 Calculate the open-circuit impedance parameters for the resistive T network shown in Fig. 11.6.

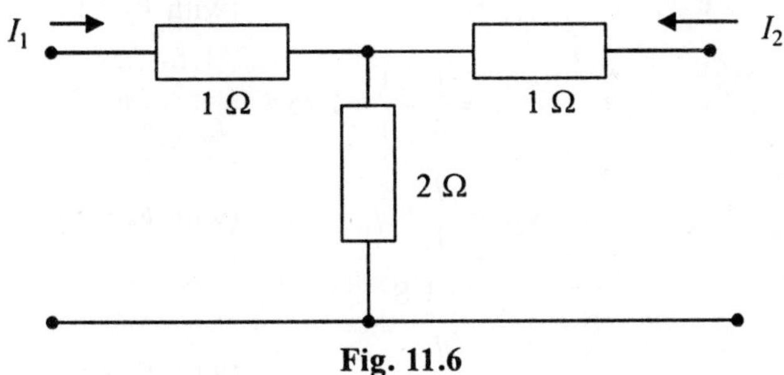

Fig. 11.6

SOLUTION:
By definition,

$$z_{11} = \frac{V_1}{I_1} = 1 + 2 = 3\ \Omega \qquad \text{(with } I_2 = 0\text{)}$$

$$z_{21} = \frac{V_2}{I_1} = 2\ \Omega \qquad \text{(with } I_2 = 0\text{)}$$

Similarly

$$z_{22} = 2 + 1 = 3\ \Omega$$

$$z_{12} = 2\ \Omega$$

The impedance matrix is given by

$$z_{0c} = \begin{bmatrix} 3 & 2 \\ 2 & 3 \end{bmatrix} \qquad \textbf{Ans.}$$

Note that the circuit is reciprocal.

EXAMPLE 11.4 Calculate the y parameters for the network shown in Fig. 11.7.

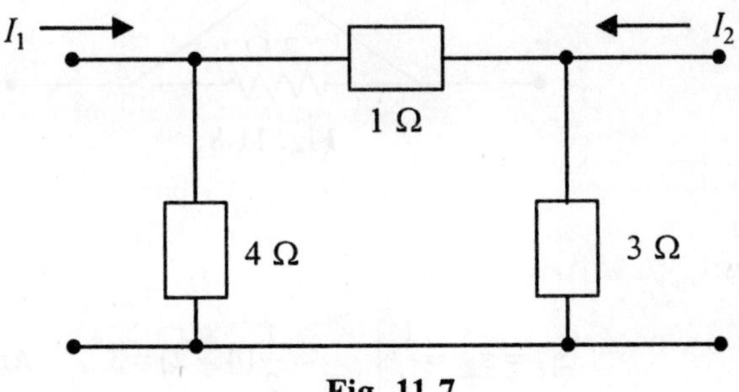

Fig. 11.7

From the equations for the y parameters, y_{11} and y_{21} are found when port 2 is short circuited and y_{22} and y_{12} are found when port 1 is short circuited.

From section 11.2.2,

$$y_{11} = \frac{I_1}{V_1} \qquad \text{(with } V_2 = 0\text{)}$$

$$= \frac{1}{4} + \frac{1}{1} = 1.25 \text{ S}$$

$$y_{21} = \frac{I_2}{V_1} \qquad \text{(with } V_2 = 0\text{)}$$

$$= -1 \text{ S}$$

$$y_{22} = \frac{I_2}{V_2} \qquad \text{(with } V_1 = 0\text{)}$$

$$= \frac{1}{3} + \frac{1}{1} = 1.333 \text{ S}$$

$$y_{12} = \frac{I_1}{V_2} \qquad \text{(with } V_1 = 0\text{)}$$

$$= -1 \text{ S}$$

The admittance matrix is:

$$\begin{bmatrix} 1.25 & -1 \\ -1 & 1.33 \end{bmatrix}$$

Note that the circuit is non-symmetric.

EXAMPLE 11.5 Find the *z, y, h, ABCD* and *g* parameters of the circuit shown in Fig. 11.8.

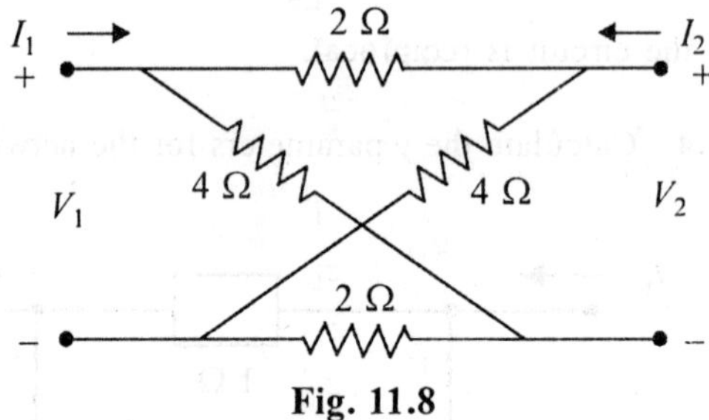

Fig. 11.8

SOLUTION:

z parameters:

$$z_{11} = z_{22} = \left.\frac{V_1}{I_1}\right|_{I_2=0} = \frac{1}{2}(4+2) = 3 \qquad \textbf{Ans.}$$

$$z_{21} = z_{12} = \left.\frac{V_2}{I_1}\right|_{I_2=0} = \left(\frac{4I_1}{2} - \frac{2I_1}{2}\right) / I_1 = 1 \qquad \textbf{Ans.}$$

y parameters:

$$y_{11} = \frac{z_{22}}{\Delta_z} = 3/8$$

$$y_{22} = 3/8$$

$$y_{12} = \frac{-z_{21}}{\Delta_z} = -1/8 \qquad \textbf{Ans.}$$

h parameters:

$$h_{11} = \frac{\Delta_z}{z_{22}} = 8/3$$

$$h_{12} = \frac{z_{12}}{z_{22}} = 1/3$$

$$h_{21} = \frac{-z_{21}}{z_{22}} = -1/3$$

$$h_{22} = \frac{1}{z_{22}} = 1/3 \qquad \textbf{Ans.}$$

ABCD parameters:

$$A = \frac{z_{22}}{z_{12}} = 3/1 = 3$$

$$B = \frac{\Delta_z}{z_{21}} = 8/1 = 8$$

$$C = \frac{1}{z_{21}} = 1$$

$$D = \frac{z_{22}}{z_{21}} = 3/1 = 3 \qquad \textbf{Ans.}$$

g parameters:

$$g_{11} = \frac{1}{z_{11}} = 1/3$$

$$g_{12} = \frac{z_{12}}{z_{11}} = -1/3$$

$$g_{21} = \frac{z_{21}}{z_{11}} = 1/3$$

$$g_{22} = \frac{\Delta_z}{z_{22}} = 8/-3 \qquad \textbf{Ans.}$$

EXAMPLE 11.6 Calculate the T parameters for the two-port network as shown in Fig. 11.9.

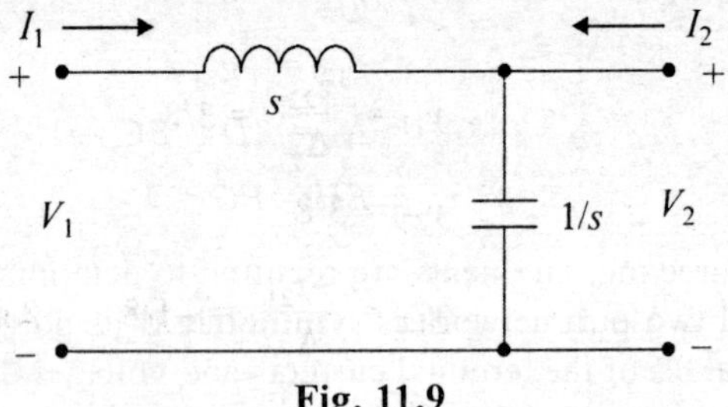

Fig. 11.9

SOLUTION:

$$A = \frac{V_1}{V_2} = s^2 + 1 \qquad (\text{with } I_2 = 0)$$

$$C = \frac{I_1}{V_2} = s \qquad (\text{with } I_2 = 0)$$

$$D = \frac{-I_1}{I_2} = 1 \qquad (\text{with } V_2 = 0)$$

$$B = \frac{-V_1}{I_2} = s \qquad (\text{with } V_2 = 0) \qquad \textbf{Ans.}$$

An easier method is to obtain the T parameters directly from the Kirchhoff's law equations by suitable rearrangement of the variables.

EXAMPLE 11.7 Show that the inverse transmission matrix reduces to $\begin{bmatrix} D & -B \\ -C & A \end{bmatrix}$.

SOLUTION:

The inverse of the T matrix is equal to $\begin{bmatrix} A & B \\ C & D \end{bmatrix}^{-1}$

$$= \frac{Adj\,T}{Det\,T}$$

$$= \frac{\begin{bmatrix} D & -B \\ -C & A \end{bmatrix}}{1}$$

since $AD - BC = 1$ for reciprocal networks.

11.2.8 Reciprocal Two-port Networks

A two-port network is **reciprocal** if the interchange of an ideal voltage source at one port with an ideal ammeter at the other port produces the same ammeter readings. See Chapter 4.

If a two-port network is reciprocal, the following relations exist among the port parameters:

$$z_{12} = z_{21}$$
$$y_{12} = y_{21}$$

$$h_{12} = -h_{21}$$

$$g_{12} = -g_{21}$$

$$\Delta_T = AD - BC = 1$$

$$EH - FG = 1$$

Note that only three measurements are required to determine all the two-port parameters.

Reciprocal two-port network is **symmetric** if its ports can be interchanged without disturbing the values of the terminal currents and voltages. Examples of symmetrical two-port networks are shown in Fig. 11.10.

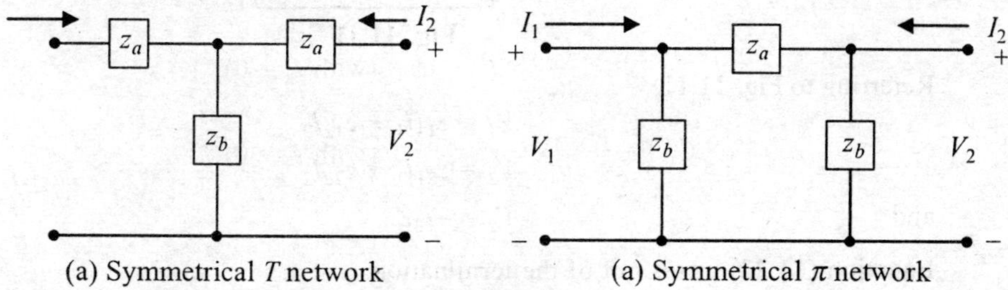

(a) Symmetrical *T* network (a) Symmetrical π network

Fig. 11.10

In such cases, the following additional relationships exist:

$$z_{11} = z_{22}$$

$$y_{11} = y_{22}$$

$$\Delta_h = 1$$

$$\Delta_g = 1$$

$$A = D$$

$$E = H$$

For a symmetric network, only two measurements are required to determine all the two-port parameters.

11.2.9 Terminated Two-port Networks

There are certain characteristics of a terminated two-port network which define its terminal behaviour. They include

1. Input impedance $Z_{in} = \dfrac{V_1}{I_1}$

2. Output impedance $Z_o = \dfrac{V_2}{I_2}$

Input impedance in terms of z parameters and the load impedance:

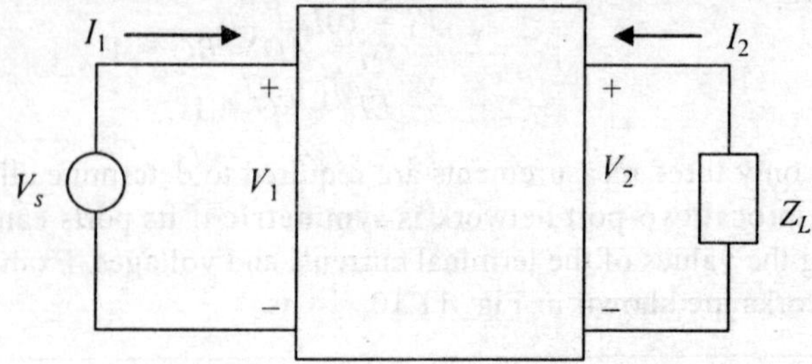

Fig. 11.11

Referring to Fig. 11.11,

$$V_1 = z_{11}I_1 + z_{12}I_2 \tag{11.20}$$

$$V_2 = z_{21}I_1 + z_{22}I_2 \tag{11.21}$$

and

$$V_2 = -I_2 Z_L \tag{11.22}$$

Equation (11.22) arises out of the termination.

Replacing V_2 with $-I_2 Z_L$ in Eq. (11.21) and solving for I_2,

$$I_2 = \frac{-z_{21}I_1}{z_{22} + Z_L}$$

From Eq. (11.20),

$$V_1 = z_{11}I_1 - \frac{z_{12}z_{21}}{z_{22} + Z_L} I_1$$

Input impedance

$$Z_{in} = \frac{V_1}{I_1} = z_{11} - \frac{z_{12}z_{21}}{z_{22} + Z_L} = \frac{\Delta_z + z_{11}Z_L}{z_{22} + Z_L}$$

Note that if $Z_L = \infty$ (Port 2 on open circuit)

$$Z_{in} = z_{11}$$

EXAMPLE 11.8 The voltages V_1 and V_2 at the two ports of a network, shown in Fig. 11.12, are:

$$V_1 = 10I_1 + 4I_2$$

$$V_2 = 4I_1 + 2I_2$$

A load resistor of 6 Ω is connected across port 2. Find the input impedance.

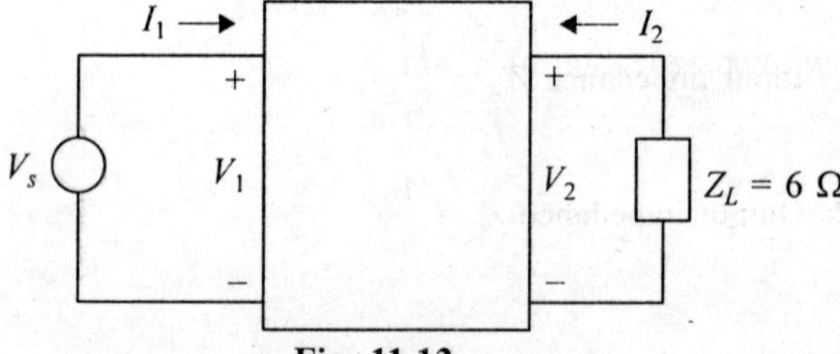

Fig. 11.12

SOLUTION:

Solving the equations,

$$V_1 = 10I_1 + 4I_2$$

$$V_2 = 4I_1 + 2I_2$$

and

$$V_2 = -I_2 Z_L = -6I_2$$

$$V_1 = 8I_1$$

Hence input impedance $= \dfrac{V_1}{I_1} = 8\ \Omega$ **Ans.**

11.2.10 Interconnection of Two-port Networks

Many problems in circuit theory (such as filters, delay lines and transducers) involve the determination of the current and potential along a chain of identical symmetrical two-port networks, which may be interconnected in series, shunt or cascade.

Series connection: z parameters are most appropriate for getting the overall parameters of the series connection shown in Fig. 11.13.

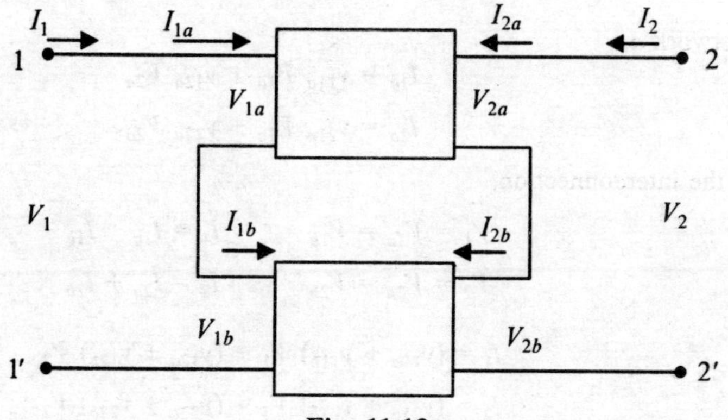

Fig. 11.13

For network a,

$$V_{1a} = z_{11a} I_{1a} + z_{12a} I_{2a}$$
$$V_{2a} = z_{21a} I_{1a} + z_{22a} I_{2a}$$

For network b,

$$V_{1b} = z_{11b} I_{1b} + z_{12b} I_{2b}$$
$$V_{2b} = z_{21b} I_{1b} + z_{22b} I_{2b}$$

From the interconnection of a and b

$$I_1 = I_{1a} = I_{1b} \qquad V_1 = V_{1a} + V_{1b}$$
$$I_2 = I_{2a} = I_{2b} \qquad V_2 = V_{2a} + V_{2b}$$

Hence

$$V_1 = (z_{11a} + z_{11b})I_1 + (z_{12a} + z_{12b})I_2$$
$$V_2 = (z_{21a} + z_{21b})I_1 + (z_{22a} + z_{22b})I_2$$

which results in matrix:

$$z = z_a + z_b$$

Parallel connection: Figure 11.14 shows a parallel connection of two-port networks. a and b with short circuit admittance parameters.

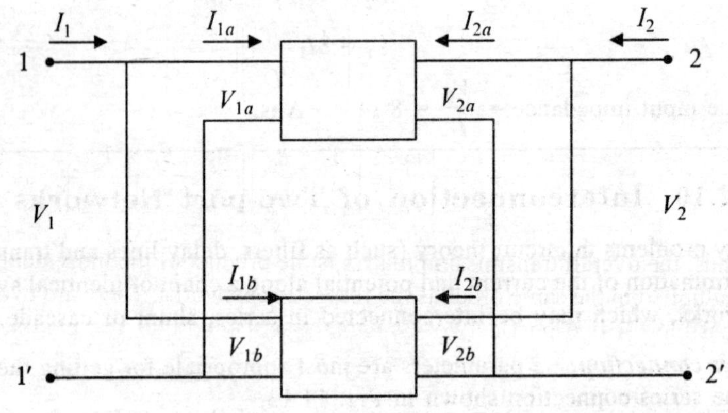

Fig. 11.14

For network a,

$$I_{1a} = y_{11a}\, V_{1a} + y_{12a}\, V_{2a}$$

$$I_{2a} = y_{21a}\, V_{1a} + y_{22a}\, V_{2a}$$

From the interconnection,

$$V_1 = V_{1a} = V_{1b} \qquad I_1 = I_{1a} + I_{1b}$$

$$V_2 = V_{2a} = V_{2b} \qquad I_2 = I_{2a} + I_{2b}$$

Hence

$$I_1 = (y_{11a} + y_{11b})\, V_1 + (y_{12a} + y_{12b})\, V_2$$

$$I_2 = (y_{21a} + y_{21b})\, V_1 + (y_{22a} + y_{22b})\, V_2$$

which gives

$$y = y_a + y_b$$

Cascade connection: The T parameters are most appropriate for certain applications. The diagram for cascade interconnection is shown in Fig. 11.15.

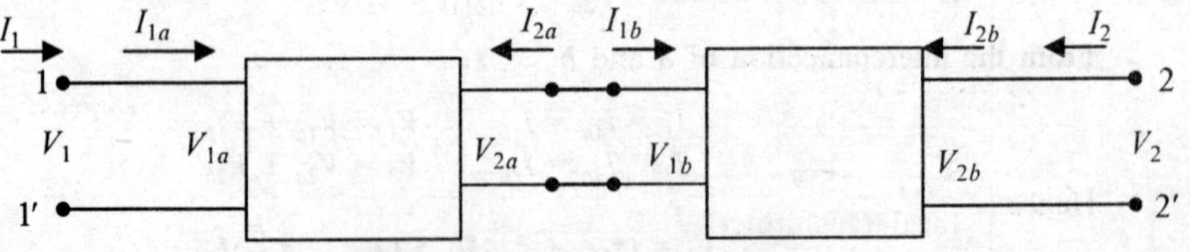

Fig. 11.15

$$\begin{bmatrix} V_1 \\ I_1 \end{bmatrix} = \begin{bmatrix} V_{1a} \\ I_{1a} \end{bmatrix} = \begin{bmatrix} A_a & B_a \\ C_a & D_a \end{bmatrix} \begin{bmatrix} V_{2a} \\ -I_{2a} \end{bmatrix}$$

$$= \begin{bmatrix} A_a & B_a \\ C_a & D_a \end{bmatrix} \begin{bmatrix} V_{1b} \\ I_{1b} \end{bmatrix}$$

$$= \begin{bmatrix} A_b & B_b \\ C_b & D_b \end{bmatrix} \begin{bmatrix} V_{2b} \\ -I_{2b} \end{bmatrix}$$

or

$$\begin{bmatrix} V_1 \\ I_1 \end{bmatrix} = \begin{bmatrix} A_a & B_a \\ C_a & D_a \end{bmatrix} \begin{bmatrix} A_b & B_b \\ C_b & D_b \end{bmatrix} \begin{bmatrix} V_2 \\ -I_2 \end{bmatrix}$$

Thus, the overall transmission matrix is the product of the individual transmission matrices, multiplied in the same order as the respective networks. There are two other interconnections possible for two-port networks, namely series-parallel and parallel-series.

Table 11.2 Transmission Matrices of Fundamental Two-port Structures

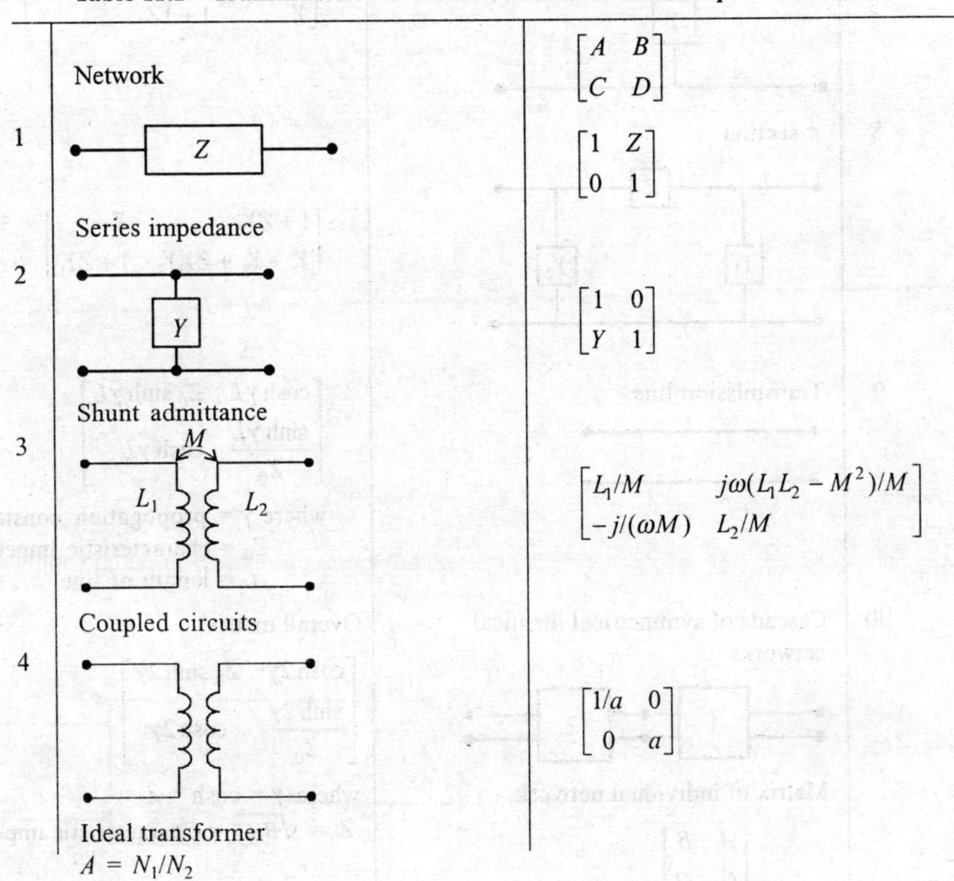

	Network	$\begin{bmatrix} A & B \\ C & D \end{bmatrix}$
1	Series impedance	$\begin{bmatrix} 1 & Z \\ 0 & 1 \end{bmatrix}$
2	Shunt admittance	$\begin{bmatrix} 1 & 0 \\ Y & 1 \end{bmatrix}$
3	Coupled circuits	$\begin{bmatrix} L_1/M & j\omega(L_1L_2 - M^2)/M \\ -j/(\omega M) & L_2/M \end{bmatrix}$
4	Ideal transformer $A = N_1/N_2$	$\begin{bmatrix} 1/a & 0 \\ 0 & a \end{bmatrix}$

Table 11.2 Transmission Matrices of Fundamental Two-port Structures (*cont.*)

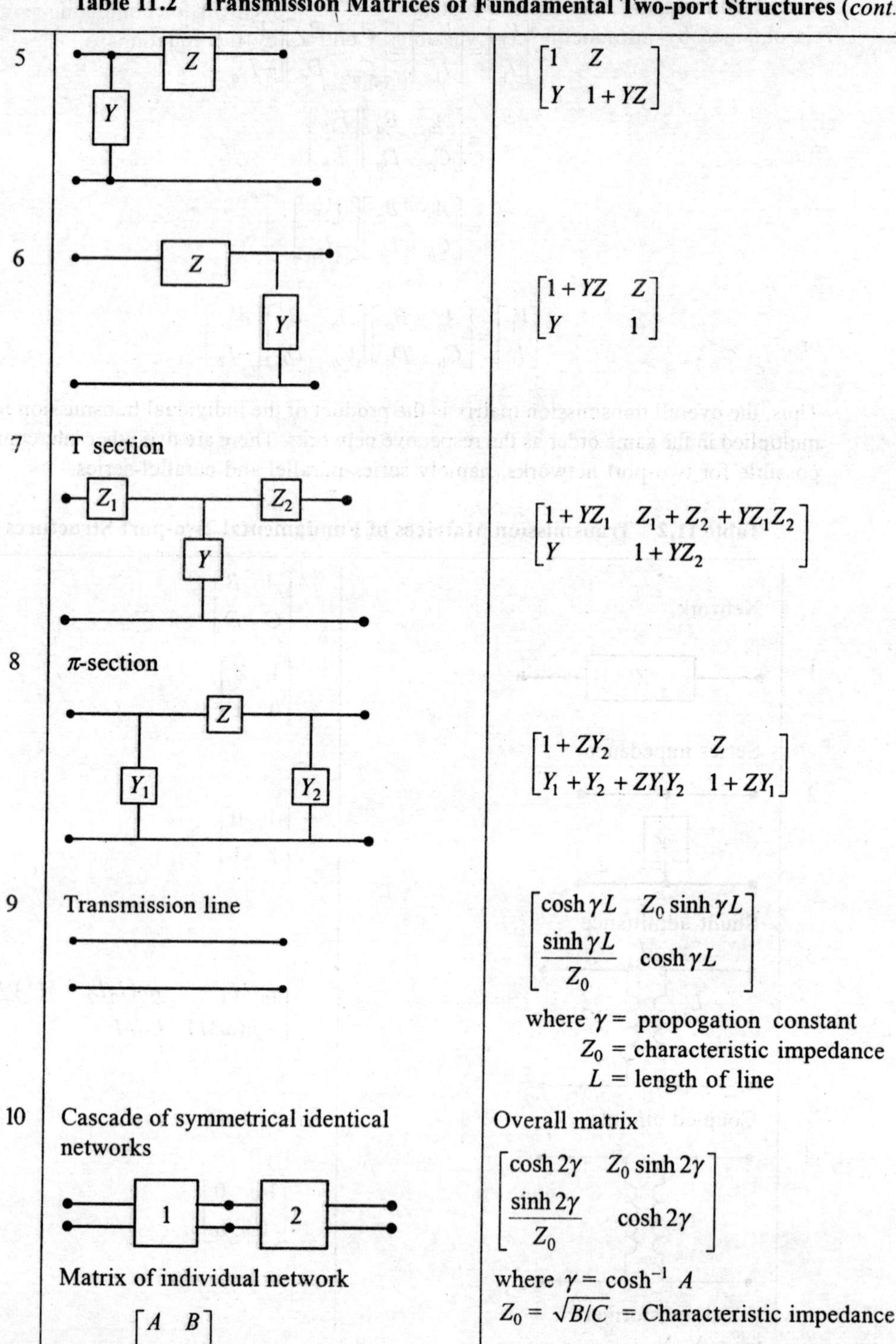

	Network	Transmission matrix
5		$\begin{bmatrix} 1 & Z \\ Y & 1+YZ \end{bmatrix}$
6		$\begin{bmatrix} 1+YZ & Z \\ Y & 1 \end{bmatrix}$
7	T section	$\begin{bmatrix} 1+YZ_1 & Z_1+Z_2+YZ_1Z_2 \\ Y & 1+YZ_2 \end{bmatrix}$
8	π-section	$\begin{bmatrix} 1+ZY_2 & Z \\ Y_1+Y_2+ZY_1Y_2 & 1+ZY_1 \end{bmatrix}$
9	Transmission line	$\begin{bmatrix} \cosh \gamma L & Z_0 \sinh \gamma L \\ \dfrac{\sinh \gamma L}{Z_0} & \cosh \gamma L \end{bmatrix}$ where γ = propogation constant Z_0 = characteristic impedance L = length of line
10	Cascade of symmetrical identical networks Matrix of individual network $\begin{bmatrix} A & B \\ C & D \end{bmatrix}$	Overall matrix $\begin{bmatrix} \cosh 2\gamma & Z_0 \sinh 2\gamma \\ \dfrac{\sinh 2\gamma}{Z_0} & \cosh 2\gamma \end{bmatrix}$ where $\gamma = \cosh^{-1} A$ $Z_0 = \sqrt{B/C}$ = Characteristic impedance

Note: A form of the transmission matrix T that is convenient for computing powers of T is obtained by introducing new variables γ and Z_0 by the equations:

$$\gamma = \cosh^{-1} A$$

and

$$Z_0 = \sqrt{B/C}$$

Thus,

$$T = \begin{bmatrix} \cosh\gamma & Z_0 \sinh\gamma \\ \dfrac{\sinh\gamma}{Z_0} & \cosh\gamma \end{bmatrix}$$

Now,

$$T^2 = T(T) = \begin{bmatrix} \cosh 2\gamma & Z_0 \sinh 2\gamma \\ \dfrac{\sinh 2\gamma}{Z_0} & \cosh 2\gamma \end{bmatrix} \tag{11.23}$$

This can be extended to obtain $T^r = \begin{bmatrix} \cosh r\gamma & Z_0 \sinh r\gamma \\ \dfrac{\sinh r\gamma}{Z_0} & \cosh r\gamma \end{bmatrix}$

where $r = 0, \pm 1, \pm 2, \ldots,$

Equation (11.23) is very useful in the study of four terminal networks and associated structures such as transmission lines.

Table 11.3 Two-port Parameters in Terms of

	Z	y	h	g	T
z	z_{11} z_{12} z_{21} z_{22}	y_{22}/Δ_y $-y_{12}/\Delta_y$ $-y_{21}/\Delta_y$ y_{11}/Δ_y	Δ_h/h_{22} h_{12}/h_{22} $-h_{21}/h_{22}$ $1/h_{22}$	$1/g_{11}$ $-g_{12}/g_{11}$ g_{12}/g_{11} Δ_g/g_{11}	A/C Δ_T/C $1/C$ D/C
y	z_{22}/Δ_z $-z_{12}/\Delta_z$ $-z_{21}/\Delta_z$ z_{11}/Δ_z	y_{11} y_{12} y_{21} y_{22}	$1/h_{11}$ $-h_{12}/h_{11}$ h_{12}/h_{11} Δ_h/h_{11}	Δ_g/g_{22} g_{12}/g_{22} $-g_{21}/g_{22}$ $1/g_{22}$	D/B $-\Delta_T/B$ $-1/B$ A/B
h	Δ_z/z_{22} z_{12}/z_{22} $-z_{21}/z_{22}$ $1/z_{22}$	$1/y_{11}$ $-y_{12}/y_{11}$ y_{12}/y_{11} Δ_y/y_{11}	h_{11} h_{12} h_{21} h_{22}	g_{22}/Δ_g $-g_{12}/\Delta_g$ $-g_{21}/\Delta_g$ g_{11}/Δ_g	B/D Δ_T/D $-1/D$ C/D
g	$1/z_{11}$ $-z_{12}/z_{11}$ z_{12}/z_{11} Δ_z/z_{11}	Δ_y/y_{22} y_{12}/y_{22} $-y_{21}/y_{22}$ $1/y_{22}$	h_{22}/Δ_h $-h_{12}/\Delta_h$ $-h_{21}/\Delta_h$ h_{11}/Δ_h	g_{11} g_{12} g_{21} g_{22}	C/A $-\Delta_T/A$ $1/A$ B/A
T	z_{11}/z_{21} Δ_z/z_{21} $1/z_{21}$ z_{22}/z_{21}	$-y_{22}/y_{21}$ $-1/y_{21}$ $-\Delta_y/y_{21}$ $-y_{11}/y_{21}$	$-\Delta_h/h_{21}$ $-h_{11}/h_{21}$ $-h_{22}/h_{21}$ $-1/h_{21}$	$1/g_{21}$ g_{22}/g_{21} g_{11}/g_{21} Δ_g/g_{21}	A B C D

EXAMPLE 11.9 Find the input impedance of a network in terms of its T parameters.

SOLUTION:
From the defining equations for the transmission parameters,

$$Z_{in} = \frac{AV_2 - BI_2}{CV_2 - DI_2}$$

Since $V_2 = -Z_L I_2$, where Z_L is the load impedance.

$$\therefore \quad Z_{\text{in}} = \frac{A(-Z_L I_2) - BI_2}{C(-Z_L I_2) - DI_2}$$

$$= \frac{AZ_L + B}{CZ_L + D} \quad \textbf{Ans.} \qquad (11.24)$$

EXAMPLE 11.10 The z parameters for the resistive two ports in Fig. 11.16 are:

$$z_{oca} = \begin{bmatrix} 10 & -50 \\ 100 & 1000 \end{bmatrix}$$

$$z_{ocb} = \begin{bmatrix} 8 & -40 \\ 200 & 1600 \end{bmatrix}$$

Find the input impedance Z_{io} if (a) the load impedance Z_L is infinity (b) Z_L = zero

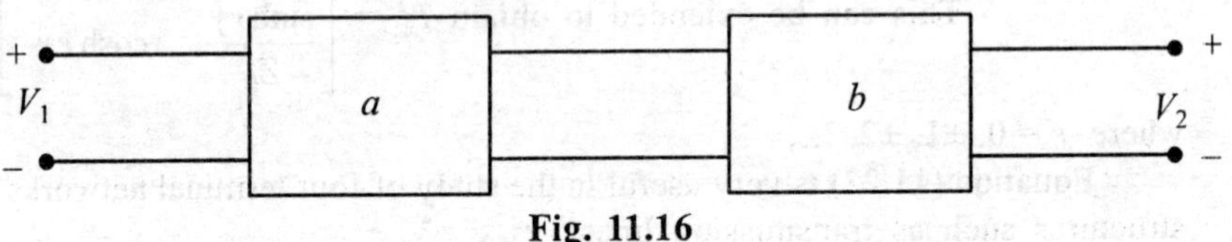

Fig. 11.16

SOLUTION:
The transmission parameters are:

$$A_a = 0.1 \qquad A_b = 0.04$$
$$B_a = 150\ \Omega \qquad B_b = 104\ \Omega$$
$$C_a = 1 \times 10^{-2}\ \text{S} \qquad C_b = 5\ \text{mS}$$
$$D_a = 10 \qquad D_b = 8$$

The T parameters for the cascade connection are:

$$A = 0.754 \qquad C = 50.4\ \text{mS}$$
$$B = 1210.4\ \Omega \qquad D = 81.04$$

(a) If $Z_L = \infty$, from Eq. (11.24), we get $Z_{io} = \dfrac{A}{C}$

Hence

$$Z_{io} = \frac{0.754}{50.4 \times 10^{-3}} = 14.96\ \Omega \quad \textbf{Ans.}$$

(b) If $Z_L = 0$

$$Z_{io} = \frac{B}{D} = \frac{1210.4}{81.04} = 14.93\ \Omega \quad \textbf{Ans.}$$

EXAMPLE 11.11 Find the transmission parameters for the network in Fig. 11.17.

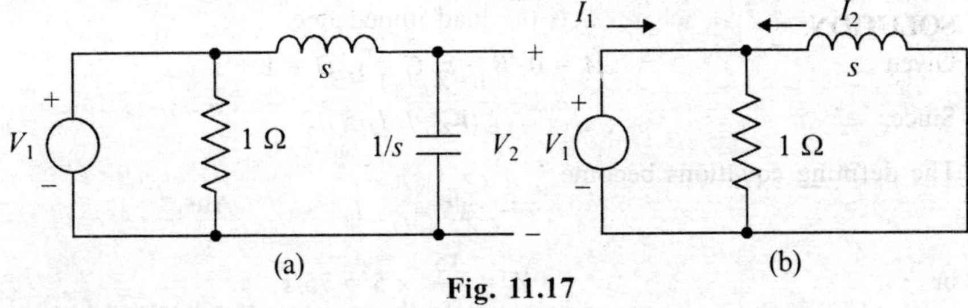

Fig. 11.17

SOLUTION:
From the defining equation,

$$A = \left.\frac{V_1}{V_2}\right|_{I_2=0} = s^2 + 1 \quad \textbf{Ans.}$$

$$\text{Current } I_1 = \frac{V_1}{1} + \frac{V_1}{s + 1/s}$$

$$= \frac{V_1(1 + s + s^2)}{1 + s^2}$$

On solving for V_1 and substituting in the defining equation for C,

$$C = \frac{I_1}{V_2} = 1 + s + s^2 \qquad \textbf{Ans.}$$

The parameters B and D may be calculated using Fig. 11.17(b).

$$B = -\frac{V_1}{I_2} = s \quad \textbf{Ans.}$$

The capacitor has been removed from the circuit since no current will flow through it. The ratio of the currents is then obtained as:

$$D = -\frac{I_1}{I_2} = 1 + s \qquad \textbf{Ans.}$$

EXAMPLE 11.12 Find V_2 if the input voltage is 10 V at a frequency of $\omega = 1$.

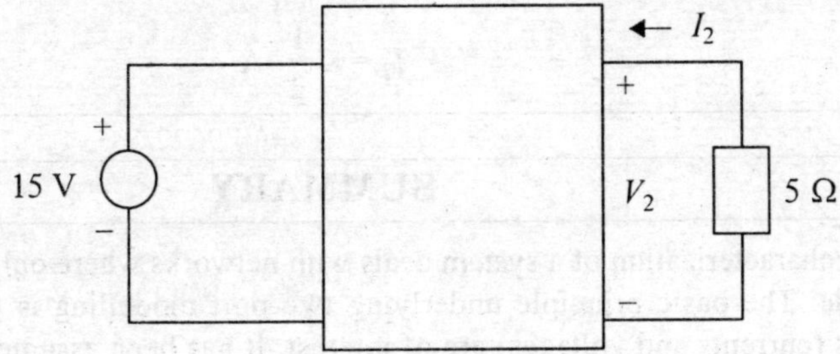

Fig. 11.18

SOLUTION:

Given $A = 0, B = s, C = s, D = 1 + s$

Since, $V_2 = -I_2 5$

The defining equations become

$$15 = -s\, I_2$$

or
$$V_2 = \frac{15}{s} \times 5 = 75/s$$

At frequency $\omega = 1$,

$$V_2 = \frac{75}{j} = -j75 \text{ V} \quad \textbf{Ans.}$$

EXAMPLE 11.13 Find the current I_2 in the circuit of Fig. 11.19, if the z parameters are:

$$z_{11} = 10 \qquad z_{12} = 5$$
$$z_{21} = 5 \qquad z_{22} = 10$$

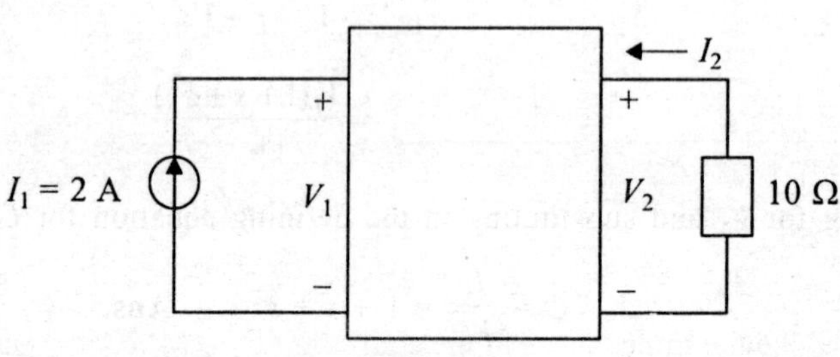

Fig. 11.19

SOLUTION:

The characteristic equations for impedance parameters are:

$$V_1 = z_{11}I_1 + z_{12}I_2$$
$$V_2 = z_{21}I_1 + z_{22}I_2$$

By the terminal constraint at port 2,

$$V_2 = -10I_2$$

From the second equation,

$$-10I_2 = 5 \times 2 + 10I_2$$

which gives

$$I_2 = -\frac{1}{2} \text{ A}$$

SUMMARY

External characterization of a system deals with networks where only several terminals are accessible. The basic principle underlying two-port modelling is that only the terminal variables (currents and voltages) are of interest. It has been assumed that no independent sources are inside the network between the ports. The current into the port is equal to the current out of the port and no external connections exist between the input and the output ports.

Out of the four terminal variables (V_1, I_1, V_2, I_2), two are independent and, therefore, there are six possible sets of simultaneous equations involving the four terminal variables. These are called **parameter equations**. They are written down in the s domain, dc values are obtained by setting $s = 0$ and the sinusoidal steady state values are obtained by setting $s = j\omega$. Any set of parameters may be calculated or measured by making open circuit and short circuit conditions at the input and the output ports. The effect of reciprocity and symmetry on the two-port parameters are considered in some detail. Large networks can be divided into subnetworks by means of interconnected two-port models.

Solved Problems

P11.1 Find the equivalent T network for a network with given z parameters.

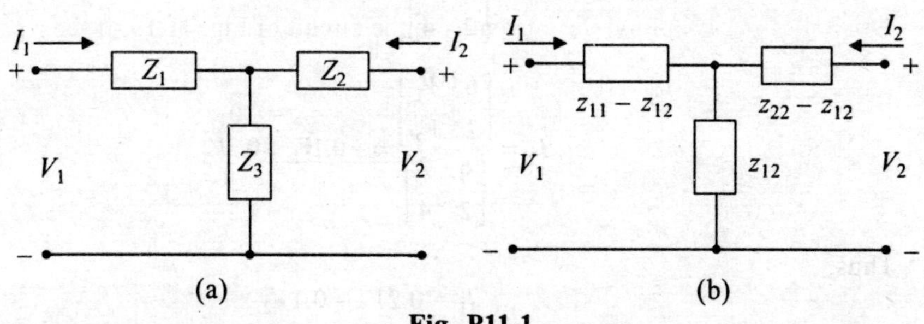

Fig. P11.1

SOLUTION:

Let Z_1, Z_2, Z_3 be the impedances in the equivalent T network. To express these impedances in terms of the open circuit parameters, let us consider Fig. P11.1(a).

$$z_{11} = \frac{V_1}{I_1} \quad \text{with } I_2 = 0 \quad \text{(Port 2 on open circuit)}$$
$$= Z_1 + Z_3$$

and

$$z_{21} = \frac{V_2}{I_1} \quad \text{with } I_2 = 0 \quad \text{(Port 2 on open circuit)}$$
$$= Z_3$$

Similarly,

$$z_{22} = Z_2 + Z_3$$

and

$$z_{12} = Z_3$$

Solving the above four equations,

$$Z_3 = z_{12} = z_{21} = 3$$

$$Z_1 = z_{11} - z_{12} = s + 3 - 3 = s$$

and

$$Z_2 = z_{22} - z_{12} = 3 + \frac{1}{2s} - 3 = \frac{1}{2s}$$

These are shown in Fig. P11.1(b). **Ans.**

P11.2 Find the *y* parameters if the voltages at the 2 ports, V_1 and V_2, of a two-port network are given by the equations:

$$V_1 = 6I_1 + 2I_2$$

$$V_2 = 2I_1 + 4I_2$$

SOLUTION:

The *y* parameter equations are obtained by solving the given equations for I_1 and I_2. By Cramer's rule,

$$I_1 = \frac{\begin{bmatrix} V_1 & 2 \\ V_2 & 4 \end{bmatrix}}{\begin{bmatrix} 6 & 2 \\ 2 & 4 \end{bmatrix}} = 0.2V_1 - 0.1V_2$$

$$I_2 = \frac{\begin{bmatrix} 6 & V_1 \\ 2 & V_2 \end{bmatrix}}{\begin{bmatrix} 6 & 2 \\ 2 & 4 \end{bmatrix}} = -0.1V_1 + 0.3V_2$$

Thus,

$$I_1 = 0.2V_1 - 0.1V_2$$

$$I_2 = -0.1V_1 + 0.3V_2$$

The *y* parameters are, therefore,

$$y_{11} = 0.2$$

$$y_{22} = 0.3$$

$$y_{12} = y_{21} = -0.1 \qquad \textbf{Ans.}$$

P11.3 Find the *z* parameters of the bridged *T* network as shown in Fig. P.11.2.

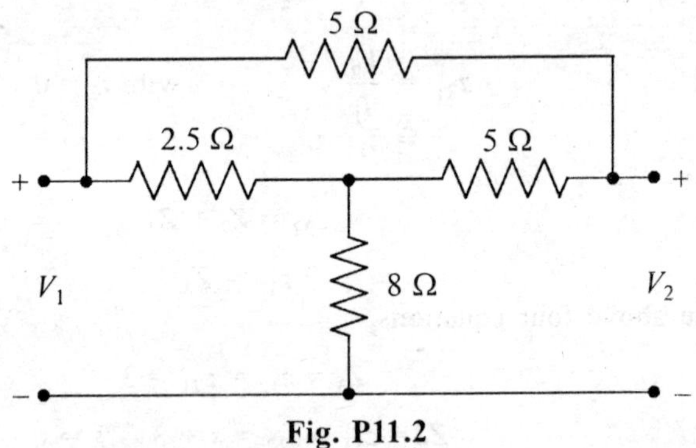

Fig. P11.2

SOLUTION:

$$z_{11} = \frac{V_1}{I_1} \qquad \text{with } I_2 = 0$$

$$= 10 \parallel 2.5 + 8 = 10\ \Omega$$

$$z_{21} = \frac{V_2}{I_1} \quad \text{with } I_2 = 0$$

$$= \frac{8I_1 + \dfrac{2.5 \times 5}{12.5} I_1}{I_1} = 9\ \Omega$$

$$z_{22} = \frac{V_2}{I_2} \quad \text{with } I_1 = 0$$

$$= 5 \parallel 7.5 + 8 = 11\ \Omega$$

$$z_{12} = \frac{V_1}{I_2} \quad \text{with } I_1 = 0$$

$$= \frac{8I_2 + \dfrac{5}{12.5} \times 2.5 I_2}{I_2} = 9\ \Omega$$

In matrix form,

$$z_{oc} = \begin{bmatrix} 10 & 9 \\ 9 & 11 \end{bmatrix} \quad \textbf{Ans.}$$

P11.4 Find the *h* parameters of a two-port resistance network from the following test data: With port 2 short circuited and 25 mV applied to port 1, the current into port 1 is 25 μA and the current into port 2 is 1 mA. With 5 mV applied to port 2 and port 1 open, the current in the port 2 is 0.125 μA and the voltage across port 1 is 2.5 μV.

SOLUTION:

$$h_{11} = \left.\frac{V_1}{I_1}\right|_{V_2=0} = \frac{25 \times 10^{-3}}{25 \times 10^{-6}} = 1000\ \Omega$$

$$h_{21} = \left.\frac{I_2}{I_1}\right|_{V_2=0} = \frac{1 \times 10^{-3}}{25 \times 10^{-6}} = 40$$

$$h_{22} = \left.\frac{I_2}{V_2}\right|_{I_1=0} = \frac{0.125 \times 10^{-6}}{5 \times 10^{-3}} = 25\ \mu\text{S}$$

$$h_{12} = \left.\frac{V_1}{V_2}\right|_{I_1=0} = \frac{2.5 \times 10^{-6}}{5 \times 10^{-3}} = 5 \times 10^{-4} \quad \textbf{Ans.}$$

P11.5 Find the *T* parameters for the network shown in Fig. P11.3.

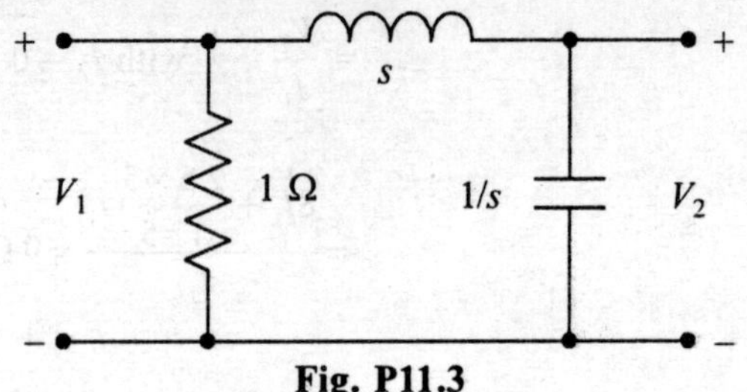

Fig. P11.3

SOLUTION:

With port 2 open (by voltage division)

$$\frac{V_2}{V_1} = \frac{1/s}{s + 1/s} = \frac{1}{1 + s^2} \tag{1}$$

Hence

$$A = 1 + s^2 \quad \textbf{Ans.}$$

Now,

$$I_1 = \frac{V_1}{1} + \frac{V_1}{s + 1/s} = \frac{V_1(1 + s^2 + s)}{1 + s^2}$$

Substituting for V_1 into Eq. (1),

$$C = 1 + s^2 + s \quad \textbf{Ans.}$$

$$B = -\frac{V_1}{I_2} = s \quad \textbf{Ans.}$$

$$D = -\frac{I_1}{I_2} = 1 + s \quad \textbf{Ans.}$$

P11.6 Find a T realization (equivalent T) for the z parameters:

$$\begin{bmatrix} z_{11} & z_{12} \\ z_{21} & z_{22} \end{bmatrix} = \begin{bmatrix} 2 + s & s \\ s & 3 + s \end{bmatrix}$$

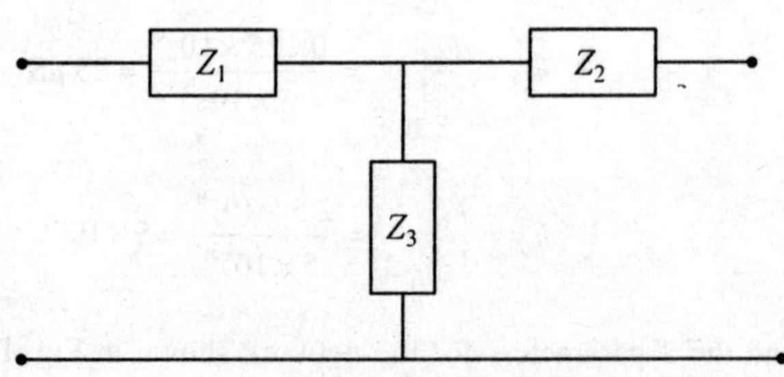

Fig. P11.4

Let the T network required be as shown in Fig. P11.4.

SOLUTION:

See Problem P11.1

$$Z_1 = z_{11} - z_{12} = 2 + s - s = 2\ \Omega$$
$$Z_2 = z_{22} - z_{12} = 3 + s - s = 3\ \Omega$$
$$Z_3 = z_{12} = s \qquad \textbf{Ans.}$$

P11.7 Determine the impedance parameters in the circuit shown in Fig. P11.5.

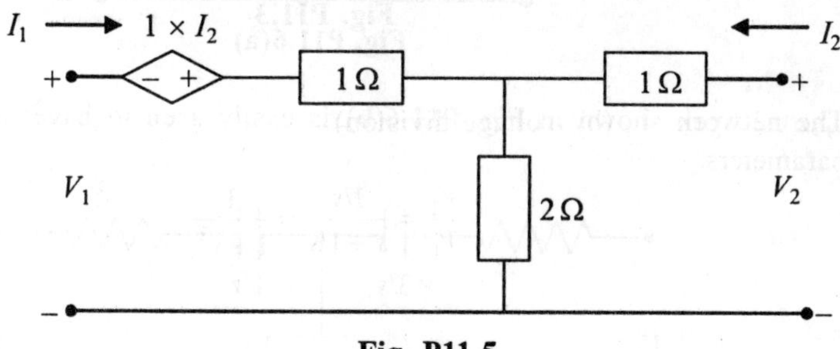

Fig. P11.5

SOLUTION:

Apply KVL around the two loops to obtain,

$$V_1 = 1I_1 - I_2 + (I_1 + I_2)2$$
$$= 3I_1 + I_2$$
$$V_2 = 1I_2 + (I_1 + I_2)2$$
$$= 2I_1 + 3I_2$$

It follows

$$z_{11} = V_1/I_1 \qquad (\text{with } I_2 = 0)$$
$$= 3$$
$$z_{21} = V_2/I_1 \qquad (\text{with } I_2 = 0)$$
$$= 2$$
$$z_{22} = V_2/I_2 \qquad (\text{with } I_1 = 0)$$
$$= 3$$
$$z_{12} = V_1/I_2 \qquad (\text{with } I_1 = 0) \qquad \textbf{Ans.}$$
$$= 1$$

P11.8 Find a network that realizes each of the following set of z parameters:

(a) $\begin{bmatrix} 4 & 3 \\ 3 & 5 \end{bmatrix}$

(b) $\begin{bmatrix} 2+4/s & 1/s \\ 1/s & 3+2/s \end{bmatrix}$

(c) $\begin{bmatrix} 4 & 2 \\ 3 & 5 \end{bmatrix}$

(a) By inspection, the z parameters of the reciprocal network shown in Fig. P11.6(a) can be seen to have the same values as given in the problem.

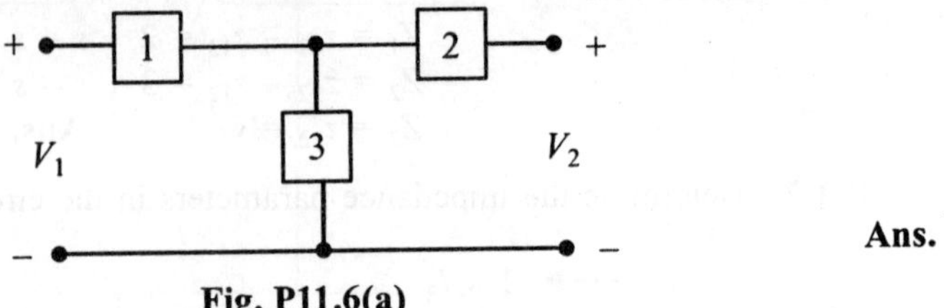

Ans.

Fig. P11.6(a)

(b) The network shown in Fig. P11.6(b) is easily seen to have the given impedance parameters.

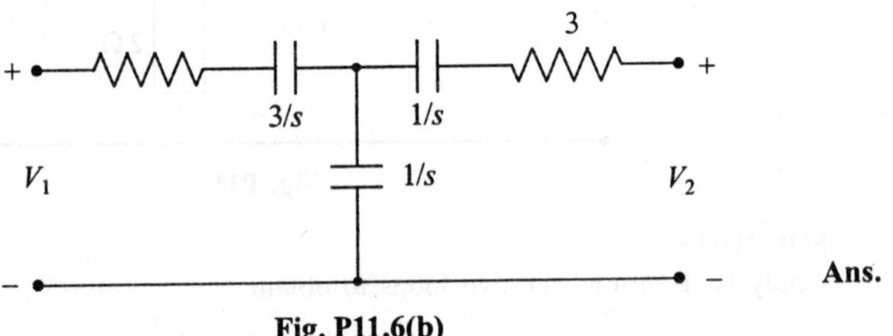

Ans.

Fig. P11.6(b)

(c) Since $z_{12} \neq z_{21}$, the circuit is non-reciprocal. The z parameters of the circuit with a current dependent voltage source introduced, as shown in Fig. P11.6(c), are obtained by applying KVL around the two loops.

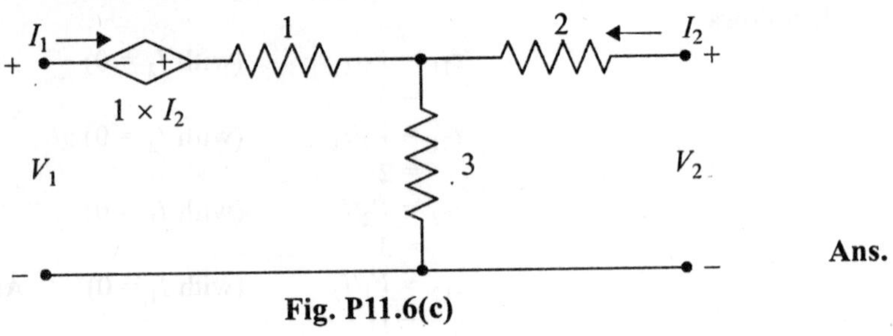

Ans.

Fig. P11.6(c)

$$V_1 = 1(I_1 - I_2) + 3(I_1 + I_2)$$
$$V_2 = 2I_2 + 3(I_1 + I_2)$$

Rearranging,

$$V_1 = 4I_1 + 2I_2$$
$$V_2 = 3I_1 + 5I_2$$

The z matrix, therefore, is:

$$\begin{bmatrix} 4 & 2 \\ 3 & 5 \end{bmatrix}$$

P11.9 Find the *ABCD* parameters of the circuit shown in Fig. P11.7(a).

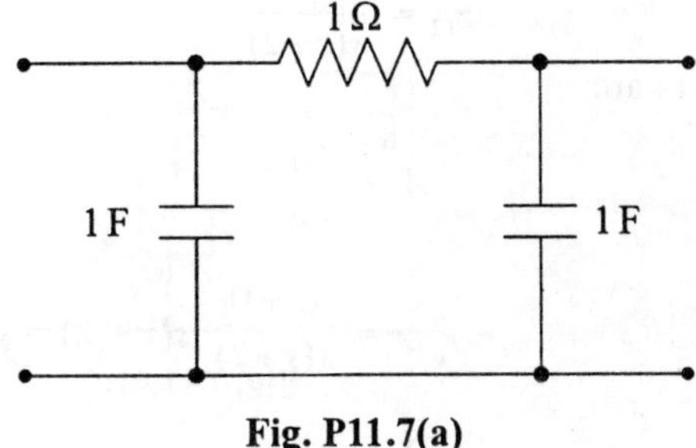

Fig. P11.7(a)

SOLUTION:

Figure P11.7(b) gives the *s* domain network.

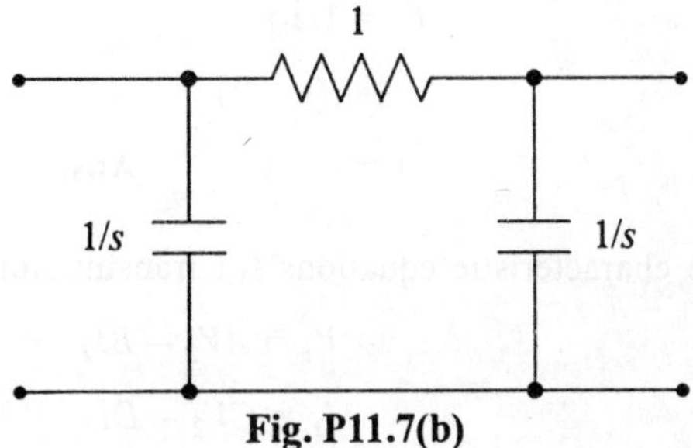

Fig. P11.7(b)

Figure P11.7(c) shows the equivalent *T* network.

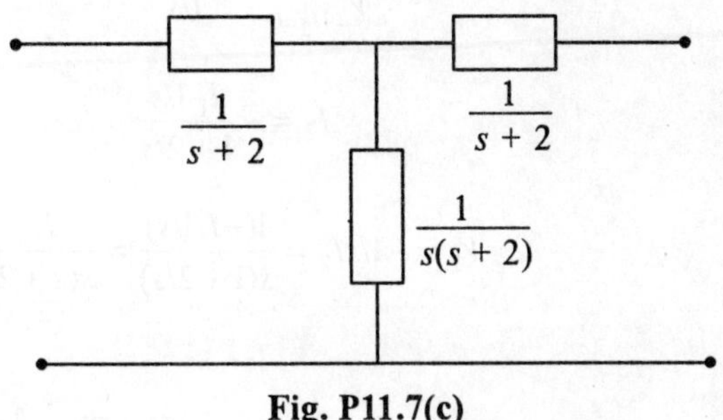

Fig. P11.7(c)

$$z_{11} = \frac{1}{s+2}\ (1 + 1/s) = \frac{s+1}{s(s+2)}$$

$$z_{22} = \frac{s+1}{s(s+2)}$$

$$z_{21} = \frac{1}{s(s+2)}$$

$$z_{12} = \frac{1}{s(s+2)}$$

The T parameters are:

$$A = \frac{z_{11}}{z_{21}}$$

$$= \frac{(s+1)}{s(s+2)} s(s+2) = s+1$$

$$B = \Delta_z/z_{21}$$

$$= \frac{1}{s(s+2).1/s(s+2)} = 1$$

$$C = 1/z_{21}$$

$$= s(s+2)$$

$$D = s+1 \qquad \textbf{Ans.}$$

Aliter

Write down the characteristic equations for transmission parameters.

$$V_1 = AV_2 - BI_2$$

$$I_1 = CV_2 - DI_2$$

Now,

$$A = \left.\frac{V_1}{V_2}\right|_{I_2=0} = \frac{1+1/s}{1/s} = s+1$$

$$I_2 = -\frac{I_1\ 1/s}{1+2/s}$$

and

$$V_2 = -1/sI_2 = \frac{1(-I_1 1/s)}{s(1+2/s)} = \frac{I_1}{s(s+2}$$

Hence

$$C = \left.\frac{I_1}{V_2}\right|_{I_2=0} = s(s+2)$$

$$D = A$$

and

$$B = -\left.\frac{V_1}{I_2}\right|_{V_2=0} = \frac{-V_1}{-V_1/1} = 1$$

P11.10 For the symmetrical two-port network (refer Fig. P11.8), find z parameters, $ABCD$ parameters and iterative impedance and propagation constant.

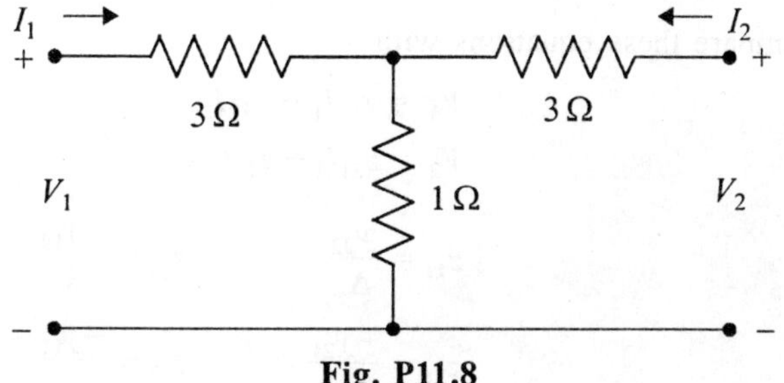

Fig. P11.8

SOLUTION:
By inspection,

$$z_{11} = V_1/I_1 \qquad \text{(with } I_2 = 0\text{)}$$
$$= 3 + 1 = 4$$
$$z_{21} = V_2/I_1 \qquad \text{(with } I_2 = 0\text{)}$$
$$= 1$$
$$z_{22} = V_2/I_2 \qquad \text{(with } I_1 = 0\text{)}$$
$$= 3 + 1 = 4$$
$$z_{12} = V_1/I_2 \qquad \text{(with } I_1 = 0\text{)} \qquad \textbf{Ans.}$$
$$= 1$$

The *ABCD* parameters are:

$$A = z_{11}/z_{21}$$
$$= 4/1 = 4$$
$$B = \Delta_z/z_{21}$$
$$= \frac{16-1}{1} = 15\ \Omega$$
$$C = 1/z_{21}$$
$$= 1$$
$$D = z_{22}/z_{21}$$
$$= 4/1 = 4 \qquad \textbf{Ans.}$$

Characteristic impedance $\sqrt{B/C} = \sqrt{15}/1 = 3.873$ **Ans.**
Propagation constant $\gamma = \cosh^{-1} A = \cosh^{-1} 4 = 2.065$ **Ans.**

P11.11 Express the z parameters in terms of y parameters.

$$I_1 = y_{11}V_1 + y_{12}V_2$$
$$I_2 = y_{21}V_1 + y_{22}V_2$$

SOLUTION:
Solving these for the voltages by using Cramer's rule,

$$V_1 = \frac{y_{22}}{\Delta_y}I_1 - \frac{y_{12}}{\Delta_y}I_2$$

and

$$V_2 = \frac{-y_{21}}{\Delta_y}I_1 + \frac{y_{11}}{\Delta_y}I_2$$

where Δ_y is the determinant of the admittance matrix.

Compare these equations with

$$V_1 = z_{11}I_1 + z_{12}I_2$$
$$V_2 = z_{21}I_1 + z_{22}I_2$$

Hence $\quad z_{11} = \dfrac{y_{22}}{\Delta_y} \qquad z_{12} = \dfrac{-y_{12}}{\Delta_y}$

$$z_{21} = \frac{-y_{21}}{\Delta_y} \qquad z_{22} = \frac{-y_{11}}{\Delta_y}$$

Conversely,

$$y_{11} = \frac{z_{22}}{\Delta_z} \qquad y_{12} = \frac{-z_{21}}{\Delta_z}$$

$$y_{21} = \frac{-z_{12}}{\Delta_z} \qquad y_{22} = \frac{z_{11}}{\Delta_z}$$

P11.12 A two-port network has the open circuit impedance matrix:

$$\begin{bmatrix} 20 & 10 \\ 10 & 24 \end{bmatrix}$$

Calculate the y parameters for the same circuit.

SOLUTION:

Determinant of the impedance matrix is:

$$\Delta_z = z_{11}z_{22} - z_{12}^2 = 380$$

From interrelations between z and y parameters,

$$y_{11} = \frac{z_{22}}{\Delta_z} = 0.063 \text{ S}$$

$$y_{22} = \frac{z_{11}}{\Delta_z} = 0.052 \text{ S}$$

$$y_{12} = y_{21} = \frac{-z_{12}}{\Delta_z} = 0.026 \text{ S} \qquad \textbf{Ans.}$$

Exercises

E11.1 Find the z parameters of the bridged T network shown in Fig. E11.1.

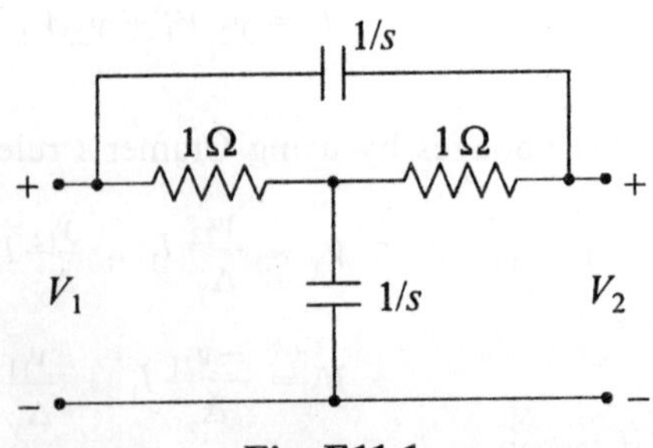

Fig. E11.1

$$\left[\textit{Ans:}\ z_{11} = \frac{1+s}{2s+1} + \frac{1}{s},\ z_{22} = \frac{s^2+3s+1}{s(2s+1)},\ z_{21} = \frac{s}{2s+1} + \frac{1}{s},\ z_{12} = \frac{s^2+2s+1}{s(2s+1)}\right]$$

E11.2 Show that the T parameters may be obtained from the z parameters.

$$\left[\textit{Ans:}\quad A = \frac{z_{11}}{z_{21}},\qquad B = \frac{\Delta_z}{z_{21}},\quad C = \frac{1}{z_{21}},\quad D = \frac{z_{22}}{z_{21}}\right]$$

E11.3 Find a π realization for the y parameters given by

$$\begin{bmatrix} y_{11} & y_{12} \\ y_{21} & y_{22} \end{bmatrix} = \begin{bmatrix} 2+3s & -1 \\ -1 & 3+4s \end{bmatrix}$$

[**Hint:** $Y_1 = y_{11} + y_{12} = 1 + 3s$

$Y_2 = y_{22} + y_{12} = 2 + 4s$

$Y_3 = -y_{12} = 1$]

Ans: The s domain values of the parameters are marked in Fig. E11.2(a) and the frequency domain values are given in Fig. E11.2(b).

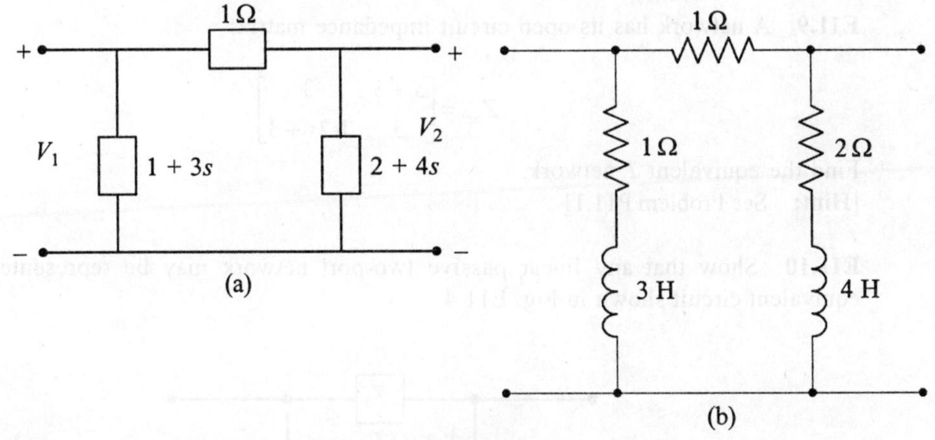

Fig. E11.2

E11.4 For a reciprocal two-port network, show that $AD - BC = 1$.

[**Hint:** Convert the LHS to z parameters and use the relation, $z_{12} = z_{21}$ for a reciprocal network.]

E11.5 Prove Eq. (11.23) governing cascade connection of two two-port networks.

E11.6 Draw a circuit model for the hybrid parameters defined by Eqs. (11.13) and (11.14).

E11.7 Find the T parameters as a function of ω and specify their values when $\omega = 1$.

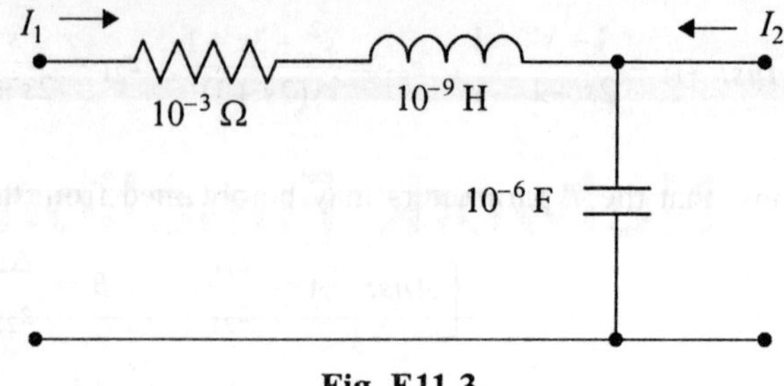

Fig. E11.3

Ans: $A = 1 - 10^{-9}\omega^2 + j10^{-9}\omega$, at $\omega = 1$ $A = 1$
$B = 10^{-3}(1 + j\omega)$, at $\omega = 1$ $B = 10^{-3}(1 + j)$
$C = 10^{-6} j\omega$, at $\omega = 1$ $C = 10^{-6} j$
$D = 1$

E11.8 Find the T parameters of a two-port device whose z parameters are:

$$z_{11} = s \qquad z_{12} = z_{21} = 10s \quad \text{and} \quad z_{22} = 100\ s$$

[*Ans:* $A = 0.1$, $B = 0$, $C = 10^{-1}/s$ and $D = 10$]

E11.9 A network has its open circuit impedance matrix:

$$Z_{oc} = \begin{bmatrix} s+3 & 3 \\ 3 & 1/2s+3 \end{bmatrix}$$

Find the equivalent T network.
[**Hint:** See Problem P11.1]

E11.10 Show that any linear passive two-port network may be represented by a π equivalent circuit shown in Fig. E11.4.

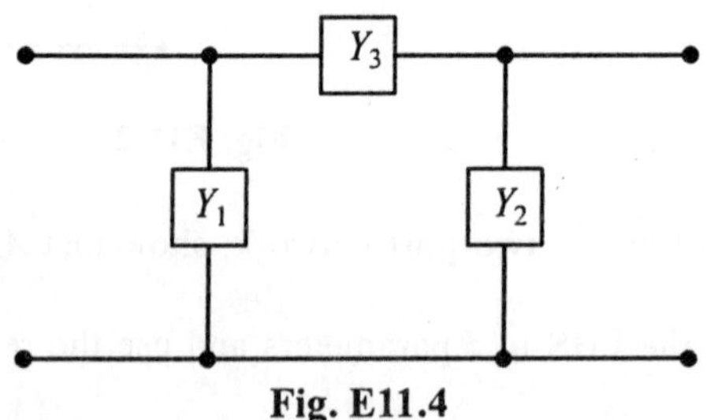

Fig. E11.4

[**Hint:** Use the y parameters of the network to find the equivalent π network.]

[*Ans:* $Y_1 = y_{11} + y_{12}$, $Y_2 = y_{22} + y_{12}$, $Y_3 = -y_{12}$]

Network Functions

12

CHAPTER CONTENTS

Some of the mathematical functions associated with Laplace transform methods in the analysis of LTI networks are dealt with in this chapter.

A ***network function*** $H(s)$ is defined as:

$$R(s) = H(s)\,E(s) \tag{12.1}$$

where $R(s)$ is the zero state transform response and $E(s)$ is the transform excitation.

Depending upon the type of the excitation and the response, network functions are classified as follows:

Network function	Response/Excitation	Symbol
Impedance	Voltage/Current	$Z(s)$
Admittance	Current/Voltage	$Y(s)$
Current transfer	Current/Current	$\alpha(s)$
Voltage transfer	Voltage/Voltage	$G(s)$

The network function $H(s)$ is used as a common symbol for specific network functions —impedance, admittance, current transfer and voltage transfer. When the excitation and the response are at the same port, the impedance $Z(s)$ is referred to as ***driving point*** (*dp*) impedance $Z_{dp}(s)$ or the driving point (*dp*) admittance $Y_{dp}(s)$. When the excitation is at one port and the response is at a different port, the impedance is the transfer impedance $Z_{tr}(s)$ or a transfer admittance $Y_{tr}(s) \cdot \alpha(s)$ and G(s) are ***transfer functions***.

12.1 DRIVING POINT FUNCTIONS

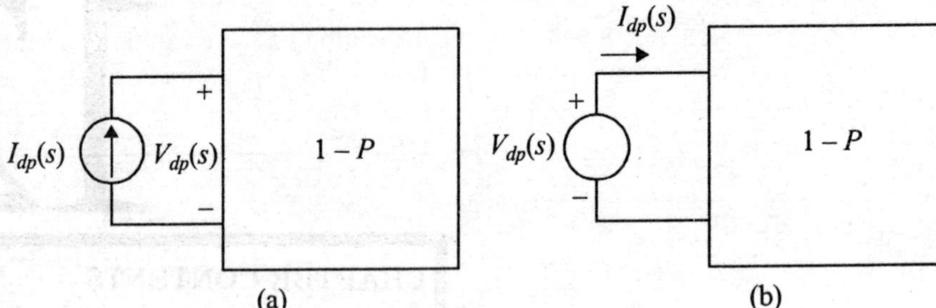

Fig. 12.1 Driving point functions.

The networks in Fig. 12.1 contain passive elements and possibly dependent energy sources. If the *dp* terminals are excited by a current source i_{dp}, the response is v_{dp}. The network function is:

$$Z_{dp}(s) = V_{dp}(s)/I_{dp}(s) \tag{12.2}$$

as shown in Fig. 12.1(a).
If the excitation is by a voltage source with a response i_{dp},

$$Y_{dp}(s) = I_{dp}(s)/V_{dp}(s) \tag{12.3}$$

Thus, the reciprocal of one network function (*dp*) is also a network function. In general, the reciprocal of $H(s)$ is *not* a network function for the same network.

EXAMPLE 12.1 Find $Z_{dp}(s)$ for the network shown in Fig. 12.2.

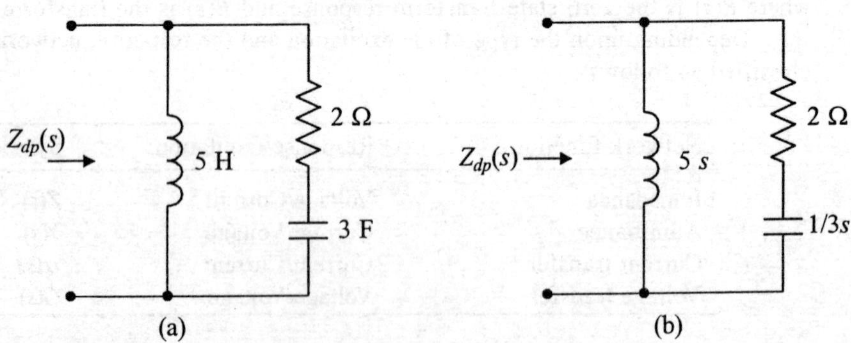

Fig 12.2

SOLUTION:
The network is a simple series and parallel combination of passive elements. The corresponding transformed network is given in Fig. 12.2(b).

$$Z_{dp}(s) = \frac{5s(2 + 1/(3s))}{5s + 2 + 1/3s} = \frac{30s^2 + 5s}{15s^2 + 6s + 1} \quad \textbf{Ans.}$$

12.2 TRANSFER FUNCTIONS

In the case of transfer functions proper identification of the excitation and the response is critical (unlike *dp* functions where excitation with i_{dp} or v_{dp} make no difference in the computations).

EXAMPLE 12.2 In the ladder network of Fig. 12.3, find the voltage transfer function:

$$G_{21}(s) = \frac{V_2(s)}{V_1(s)}$$

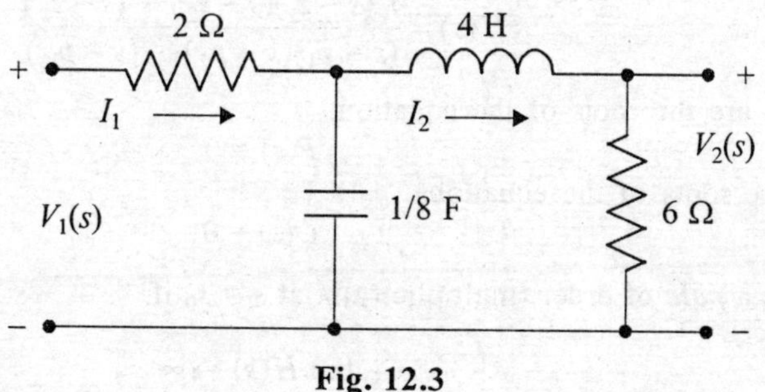

Fig. 12.3

SOLUTION:
The mesh equations are:

$$\begin{bmatrix} 2 + 8/s & -8/s \\ -8/s & 6 + 4s + 8/s \end{bmatrix} \begin{bmatrix} I_1 \\ I_2 \end{bmatrix} = \begin{bmatrix} V_1(s) \\ 0 \end{bmatrix}$$

and
$$V_2(s) = 6I_2$$

Solving the above equations,

$$I_2 = \frac{V_1(s)8/s}{(8s^3 + 44s^2 + 64s)/s^2}$$

$$= \frac{2V_1(s)}{2s^2 + 11s + 16}$$

Therefore,

$$\frac{V_2(s)}{V_1(s)} = \frac{12}{2s^2 + 11s + 16} \quad \textbf{Ans.}$$

12.3 PROPERTIES OF NETWORK FUNCTIONS—POLES AND ZEROS

For LTI networks, $H(s)$ is a ***rational function*** (ratio of two polynomials).

$$H(s) = K\frac{(s^n + a_1 s^{n-1} + \cdots + a_{n-1}s + a_n)}{s^m + b_1 s^{m-1} + \cdots + b_{m-1}s + b_m}$$

$$= KP(s)/Q(s) \tag{12.4}$$

Note that all the coefficients in $H(s)$ are real because they are related to the values of the network elements. This leads to an important property of $H(s)$, namely

Property 1: $H(s)$ is real when s is real. Equation (12.4) may be put in the form:

$$H(s) = \frac{K(s - z_1)(s - z_2)\cdots(s - z_n)}{(s - p_1)(s - p_2)\cdots(s - p_m)} \tag{12.5}$$

where z_i are the roots of the equation,

$$P(s) = 0 \tag{12.6}$$

p_j are the roots of the equation,

$$Q(s) = 0 \tag{12.7}$$

$H(s)$ has a ***pole*** of order (multiplicity) k at $s = s_0$ if

$$\lim_{s \to s_0} H(s) \to \infty \tag{12.8}$$

if $k = 1$, the pole is ***simple***. From Eq. (12.5), $H(s)$ is seen to have finite poles at $s = p_1$, $s = p_2$, ..., $s = p_m$, these are simple poles if $p_1 \neq p_2 \neq \cdots \neq p_m$. If k of these are repeated then this is a pole of order k. If n is greater than m,

$$\lim_{s \to \infty} H(s) = \lim_{s \to \infty} H(s)\, s^{n-m} \tag{12.9}$$

$H(s)$ has therefore a pole of order $(n - m)$ at $s = \infty$.

Again $H(s)$ has a ***zero*** of order (multiplicity) k at $s = s_0$ if $1/H(s)$ has a pole of order k. If $k = 1$, the zero is called simple. Since $H(s)$ has zeros at $s = z_1$, $s = z_2$, ..., $s = z_n$. These zeros are simple if they are all distinct. If k of them are repeated, the zero is of order k. Further if $n < m$,

$$\lim_{s \to \infty} 1/H(s) = \lim_{s \to \infty} Ks^{m-n} \tag{12.10}$$

and hence $H(s)$ has a zero of $m - n$ at $s = \infty$.

In the complex s plane, a pole is denoted by a small cross and a zero by a small circle.

EXAMPLE 12.3 Find the poles and zeros of $H(s) = \dfrac{s^2 + 2s + 5}{s^2 + 4s + 3}$.

SOLUTION:
Factorizing,

$$H(s) = \frac{(s+1+j2)(s+1-j2)}{(s+1)(s+3)}$$

There are two simple poles at –1 and –3; zeros at $-1 \pm j2$. **Ans.**

Property 2: Poles and zeros are either real or complex conjugate. This is because the coefficients in the rational function are real.

Property 3: The number of poles equals the number of zeros.
This number is n or m whichever is greater and is given by the highest power of s in $H(s)$.

EXAMPLE 12.4 Draw the pole zero diagram for the function $H(s) = \dfrac{s^3 - 2s^2 + 2s}{(s+1)^2(s^2+4)}$.

SOLUTION:
In factored form,

$$H(s) = \frac{s(s-1+j1)(s-1-j1)}{(s+1)^2(s+j2)(s-j2)}$$

The poles are at

$$s = -1(\text{double}),\ s = -j2,\ s = j2$$

and the zeros are at

$$s = 0,\ s = -1 + j1,\ s = -1 - j1,\ s = \infty$$

The poles and zeros of $H(s)$ are shown in Fig. 12.4.

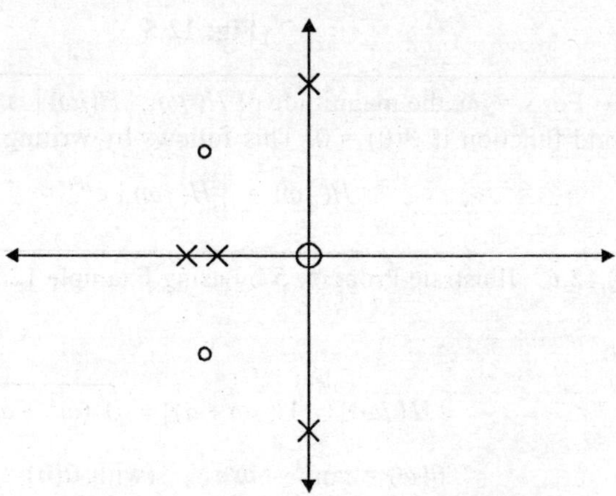

Ans.

Fig. 12.4 Pole-zero diagram of $H(s)$.

Property 4: Relates to the evaluation of $H(s)$ for purely imaginary $s = j\omega$. $H(j\omega)$ will then be a complex number. $H(j\omega) = R(\omega) + jX(\omega)$, with the real part $R(\omega)$ and the imaginary part $X(\omega)$ both being functions of ω. It can be shown that the real part $R(\omega)$ is an even function and the imaginary part is odd function.

The following example will illustrate Property 4.

EXAMPLE 12.5 Illustrate Property 4 for $H(s) = 1/(s + a)$.

SOLUTION:
For $s = j\omega$,

$$H(j\omega) = \frac{1}{j\omega + a} = \frac{-j\omega + a}{a^2 + \omega^2}$$

$$= \frac{1}{a + \omega^2} + j - \frac{\omega}{1 + \omega^2} = R(\omega) + jX(\omega)$$

$R(\omega)$ is evidently an even function and $X(\omega)$ is an odd function as required in Property 4. Plots of $R(\omega)$ and $X(\omega)$ are shown in Fig. 12.5(a). **Ans.**

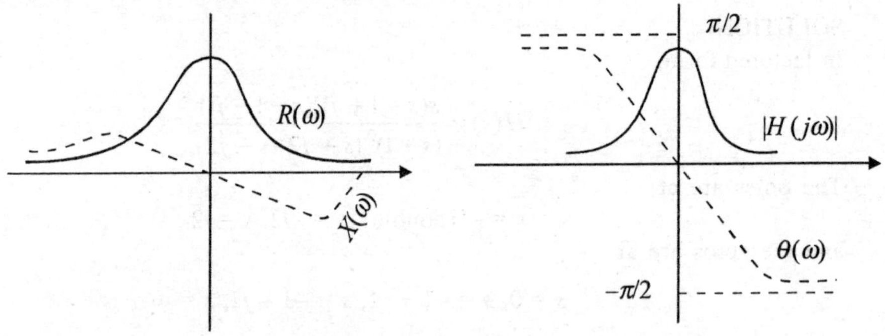

(a) Poles of $R(\omega)$ and $X(\omega)$ (b) Poles of $|H(j\omega)|$ and $\theta(\omega)$

Fig. 12.5

Property 5: For $s = j\omega$, the magnitude of $H(j\omega)$, $|H(j\omega)|$ is an even function. The angle $\theta(\omega)$ is an odd function if $\theta(0) = 0$. This follows by writing $H(j\omega)$ in its polar form as:

$$H(j\omega) = |H(j\omega)| e^{j\theta\omega}$$

EXAMPLE 12.6 Illustrate Property 5 by using Example 12.5.

SOLUTION:

$$|H(j\omega)| = |1/(j\omega + a)| = \sqrt{1/(\omega^2 + a^2)}$$

and

$$\theta(\omega) = \tan^{-1} - \omega/a \qquad (\text{with } \theta(0) = 0)$$

Plots of $|H(j\omega)|$ and $\theta(\omega)$ are shown in Fig. 12.5(b). **Ans.**

12.4 NATURAL FREQUENCIES

The zero-input response is the response of the network with no inputs. This response will be dependent on the initial state of the network and on its topology and element values. It may be noted that this corresponds to the complementary function in the solution of homogeneous differential equations, with the inputs set to zero.

EXAMPLE 12.7 In the network shown in Fig. 12.6, the initial state is given by $I_{L1(0-)}$. Find the natural frequencies of the current through the 1 Ω common resistor.

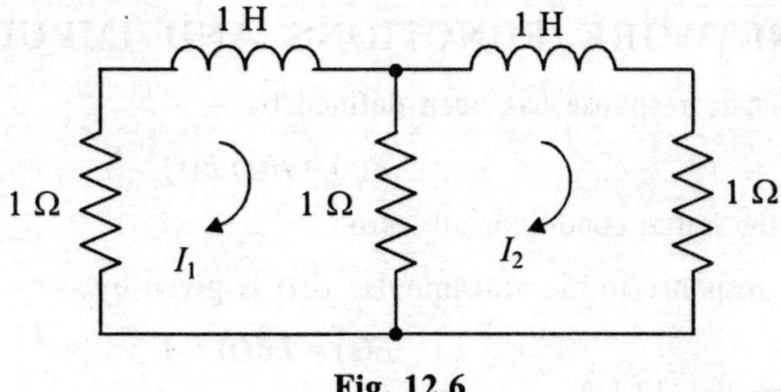

Fig. 12.6

SOLUTION:

The transform equations are:

$$\begin{vmatrix} s+2 & -1 \\ -1 & s+2 \end{vmatrix} \begin{vmatrix} I_1 \\ I_2 \end{vmatrix} = \begin{vmatrix} I_{L1(0-)} \\ 0 \end{vmatrix}$$

On solving,

$$I_1 = \frac{\begin{vmatrix} I_{L1(0-)} & -1 \\ 0 & s+2 \end{vmatrix}}{\begin{vmatrix} s+2 & -1 \\ -1 & s+2 \end{vmatrix}}$$

$$= \frac{I_{L1(0-)}(s+2)}{s^2+4s+3}$$

$$I_2 = \frac{I_{L1(0-)}}{s^2+4s+3}$$

The natural frequencies of $i_1(t)$ and of $i_2(t)$ are the roots of

$$\det Z = s^2 + 4s + 3 = 0$$

that is

$$s_1 = -1, \qquad s_2 = -3$$

The current through the common resistor is given by

$$I(s) = I_1(s) - I_2(s) = \frac{2I_{L1(0-)}}{s^2 + 4s + 3}$$

The natural frequencies of the current in the common resistor are:

$$s_1 = -1 \text{ Np/s} \quad \text{and} \quad s_2 = -3 \text{ Np/s} \qquad \textbf{Ans.}$$

12.5 NETWORK FUNCTIONS AND IMPULSE RESPONSE

The zero state response has been defined by

$$R(s) = H(s)\, E(s) \tag{12.11}$$

provided the initial conditions are zero.

The response to the unit impulse $\delta(t)$ is given by

$$E(s) = L\delta(t) = 1$$

Hence, from Eq. (12.11),

$$R(s) = H(s)$$

This means $H(s)$ is the transform of the impulse response of the network, denoted by $h(t)$.

or
$$L^{-1}\, R(s)]_{E(s)=1} = L^{-1}H(s) = h(t)$$

EXAMPLE 12.8 Find the unit impulse response of the one-port network shown in Fig. 12.7.

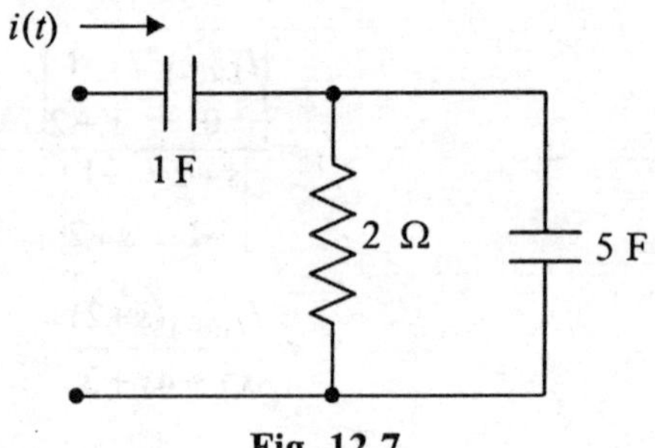

Fig. 12.7

SOLUTION:

Driving point impedance is given by

$$Z_{dp}(s) = \frac{1}{s} + \frac{2(1/5s)}{2 + 1/(5s)}$$

$$= \frac{1}{s} + \frac{0.4}{s + 0.2}$$

$$\therefore \quad h(t) = L^{-1}\, Z_{dp}(s) = (1 + 0.4e^{-0.2t})u(t) \qquad \textbf{Ans.}$$

12.6 COMPLETE RESPONSE

For all linear systems, the complete response is the zero state response plus the zero input response. See Example 10.2 where the complete response of a series *RL* circuit with a sinusoidal input has been found as:

$$\text{Complete response } i(t) = \frac{\omega}{1+\omega^2} e^{-t} + \frac{1}{\sqrt{1+\omega^2}} \sin(\omega t - \phi)$$

Note:

(1) The steady state response is due to the zero-state response alone.
(2) The zero-state response will resemble the input—a constant if the input is constant, a sinusoid if the input is a sinusoid (of the same frequency) with a possible phase shift, an exponential if the input is exponential.
(3) The zero-input response resembles the impulse response.
(4) The complete response may consist entirely of a transient part.

SUMMARY

- The chapter has defined network functions, their classification and properties. The concept of natural frequencies and impulse response have been introduced.

Solved Problems

P12.1 Find the driving point admittance of the one-port network shown in Fig. P12.1.

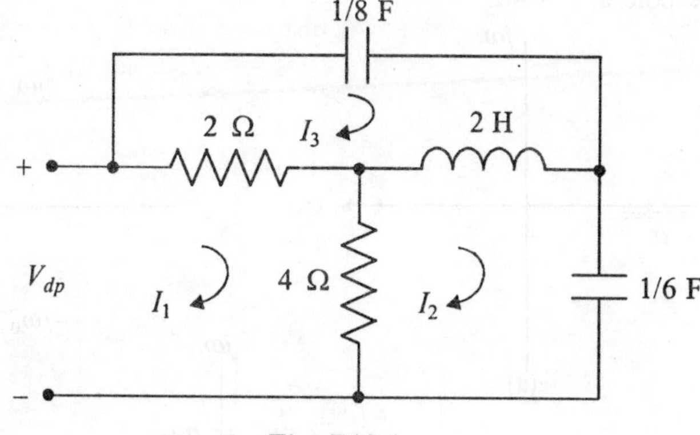

Fig. P12.1

SOLUTION:

Since there are no series or parallel combinations in the network, the mesh current method is applied. The mesh equations are written by inspection, in the form:

$$ZI = E$$

$$\begin{bmatrix} 6 & -4 & -2 \\ -4 & 2s+4+6/s & -2s \\ -2 & -2s & 2s+2+8/s \end{bmatrix} \begin{bmatrix} I_1 \\ I_2 \\ I_3 \end{bmatrix} = \begin{bmatrix} V_{dp} \\ 0 \\ 0 \end{bmatrix}$$

By Cramer's rule,

$$I_1(s) = \frac{\det Z_{11}}{\det Z} = \frac{\begin{vmatrix} 2s+4+6/s & -2s & V_{dp} \\ 2s & 2s+2+8/s & \\ 6 & -4 & -2 \end{vmatrix}}{\begin{vmatrix} -4 & 2s+4+6/s & -2s \\ -2 & -2s & 2s+2+8/s \end{vmatrix}} \quad (1)$$

$$= \frac{3s^3+9s^2+11s+12}{42s^2+28s+72} V_{dp}$$

Since,

$$I_{dp}(s) = I_1(s)$$

$$\therefore \; Y_{dp}(s) = \frac{3s^3+9s^2+11s+12}{42s^2+28s+72} \quad \textbf{Ans.}$$

Equation (1) is valid in general for one-port networks (with no energy sources). In such a case, Z can be written down by inspection and Y_{dp} is obtained from Eq.(1).

P12.2 Construct the pole-zero plot for the following standard signals:

(a) The exponential signal $e^{-\alpha t}$ where $\alpha > 0$.

(b) $\cos \omega_0 t$

(c) $e^{-\alpha t} \cos \omega_0 t$.

SOLUTION:

(a) The signal $e^{-\alpha t}$ has the transform $1/(s+\alpha)$ as shown in Fig. P12.2(a), which has a single pole at $s = -\alpha$.

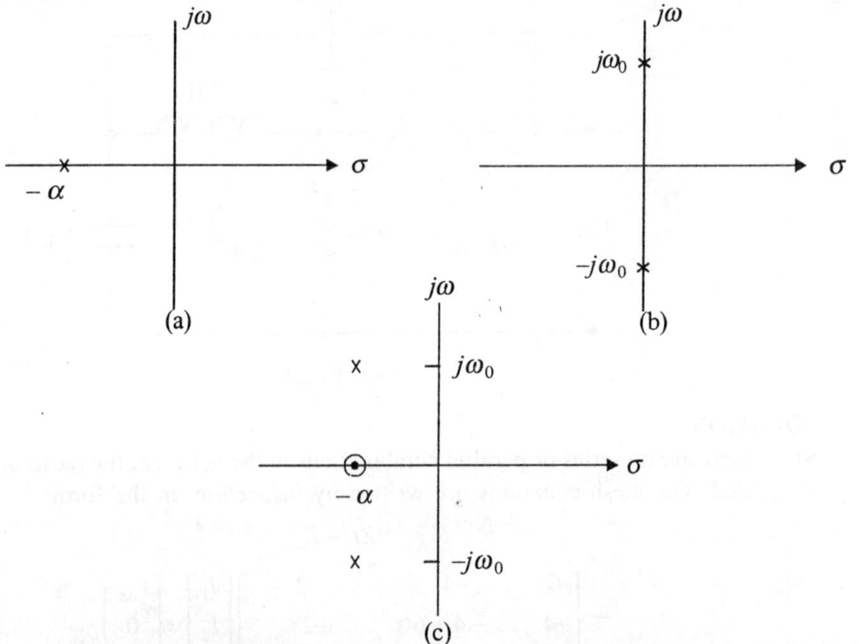

Fig. P12.2 Poles and zeros of functions.

(b) The cosine function has a transform $s/(s^2 + \omega_0^2)$ which has a zero at the origin and a pair of conjugate poles at $s = \pm j\omega_0$ as shown in Fig. P12.2(b).

(c) The transform of $e^{-\alpha t} \cos \omega_0 t = \dfrac{s+\alpha}{(s+\alpha)^2 + \omega_0^2}$

The damped cosine wave is seen to have a zero at $s = -\alpha$ and a pair of conjugate poles at $s = \pm j\omega_0$ as shown in Fig. P12.2(c).

It can be seen that the poles corresponding to decaying exponential waves are on the negative real axis and have zero imaginary parts. The poles and zeros corresponding to undamped sinusoids are on the $j\omega$ axis and have zero real parts. The poles and zeros for damped sinusoids have real and imaginary parts (both non-zero).

P12.3 Find the natural frequencies of the network shown in Fig. P12.3.

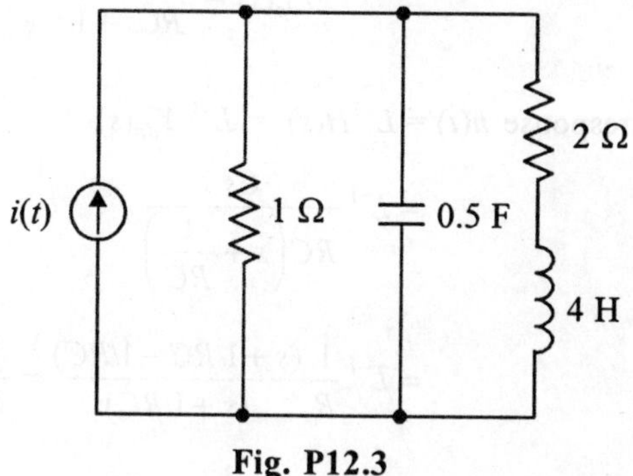

Fig. P12.3

SOLUTION:
The response to the current source is the voltage across it. Hence, the network function:

$$H(s) = \frac{V(s)}{I(s)} = Z(s)$$

Now,

$$\frac{1}{Z(s)} = \frac{1}{1} + \frac{1}{2/s} + \frac{1}{2+4s} = \frac{2s^2 + 5s + 3}{2 + 4s}$$

or

$$Z(s) = \frac{2 + 4s}{2s^2 + 5s + 3}$$

and the natural frequencies are the poles of the network function,

$$s = -1 \text{ Np/s} \quad \text{and} \quad s = -1.5 \text{ Np/s} \qquad \textbf{Ans.}$$

P12.4 Find the impulse response of the *RC* network shown in Fig. P12.4.

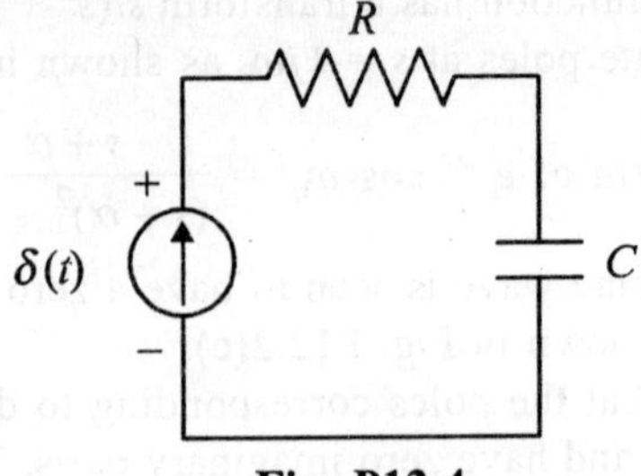

Fig. P12.4

SOLUTION:

$$Z_{dp}(s) = R + 1/Cs$$

$$Y_{dp}(s) = \frac{Cs}{RCs+1}$$

Now,

Impulse response $h(t) = L^{-1}H(s) = L^{-1}\,Y_{dp}(s)$

$$= L^{-1}\frac{Cs}{RC\left(s+\dfrac{1}{RC}\right)}$$

$$= L^{-1}\frac{1}{R}\frac{(s+1/RC-1/RC)}{(s+1/RC)} = \frac{1}{R}\left[\delta(t) - \frac{1}{RC}e^{-t/RC}u(t)\right]$$

$$= \frac{1}{R}\delta(t) - \frac{1}{R^2C}e^{-t/RC}u(t)$$

Alternatively, the impulse response may be found by taking the derivative of its unit step response. The unit step response is:

$$i(t) = \frac{1}{R}e^{-t/RC}u(t)$$

It follows:

$$h(t) = \frac{-1}{R^2C}e^{-t/RC}u(t) + \frac{1}{R}\delta(t) \qquad \textbf{Ans.}$$

P12.5 Find the current $i(t)$ for the network in Fig. P12.5, when the voltage source is $e(t) = 2e^{-0.5t}\,u(t)$ and $v_c(0-) = 0$.

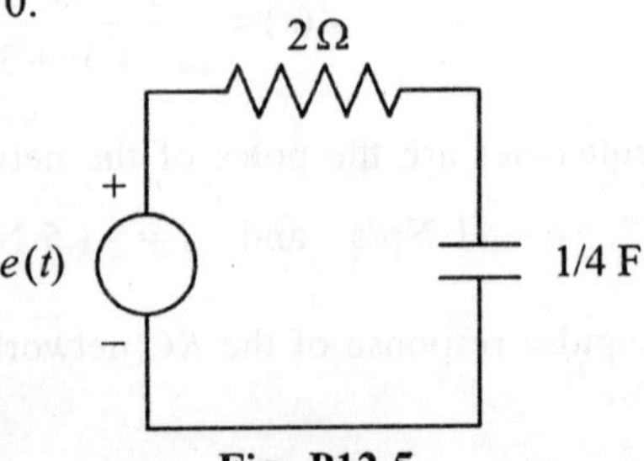

Fig. P12.5

SOLUTION:

The equation governing the circuit is:

$$e(t) = Ri(t) + \frac{1}{C}\int_{0-}^{t} i(t)\,dt + v_c(0-)$$

or

$$2i(t) + 4\int_{0-}^{t} i(t)\,dt = 2e^{-0.5t}u(t)$$

Taking Laplace transforms of both sides,

$$2I(s) + \frac{4I(s)}{s} = \frac{2}{s+0.5}$$

$$I(s) = \frac{s}{(s+0.5)(s+2)}$$

Taking inverse Laplace transform,

$$i(t) = 1.33e^{-2t} - 0.33\,e^{-0.5t}$$ **Ans.**

Exercises

E12.1 In the circuit shown in Fig. E12.1, $L = 1$ H, $R = 20\ \Omega$ and $C = 0.5$ mF. Find the frequency (a) at which the magnitude of the impedance of the circuit is zero (b) at which the admittance of the circuit is wholly conductive.

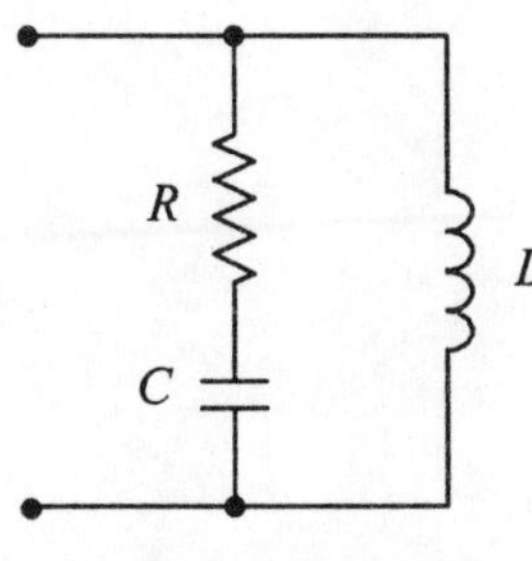

Fig. E12.1

[*Ans:* (a) ω = 0, (b)

E12.2 A series *RLC* circuit has $R = 5\ \Omega$, $L = 1$ H and $C = 1/6$ F. As
conditions, determine the unit impulse voltage response.

[**Hint:** The circuit voltage equation for an uncharged capacitor is:

$$\frac{1}{C}\int_{0}^{t} i(t)\,dt + Ri + L\,di/dt = v(t)$$

Take Laplace transform.]

E12.3 Find the impulse response of the *RL* circuit shown in Fig. E12.2.

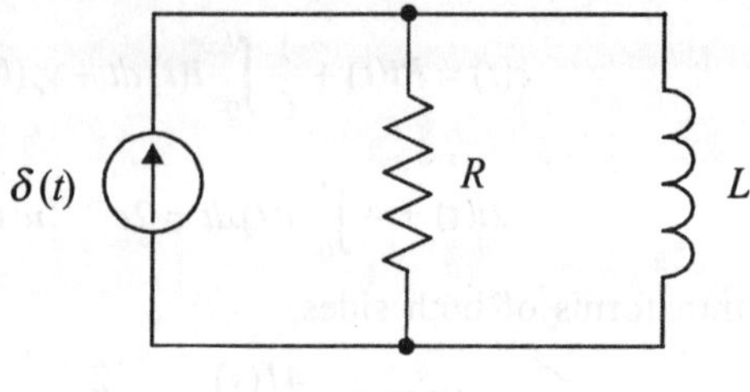

Fig. E12.2

[*Ans:* $(-R^2/L)\, e^{-Rt/L}\, u(t) + R\delta(t)$]

Wave Filters—Frequency Selective Circuits

13

CHAPTER CONTENTS

A filter is a network used to shape the frequency spectrum of an electrical signal. These networks form an essential part of the communication and control systems. Frequency selective circuits are called **filters** because of their ability to filter out certain input signals on the basis of frequency. It may be noted that no practical frequency selective circuit can completely filter out selected frequencies.

Filters *attenuate*, that is to say, weaken the effect of any input signals with frequencies outside a particular band. The major categories of filters are:

- Low pass filter
- High pass filter
- Band pass filter
- Band reject filter
- Amplitude equalizers
- Delay equalizers

Prior to 1960's *passive RLC* components were used in the design of filters for voice and data communication systems. In recent years, hybrid IC technology has opened up the large field of *active RC* filters.

PASSIVE FILTERS

The signals passed from the input to the output fall within a band of frequencies called the **pass band**. Input signals outside this band have their magnitudes attenuated by the circuit and hence prevented from reaching the output terminals. Frequencies not in the pass band are in the stop band. Frequency selective circuits are categorized by the location of the pass band. A frequency response plot shows how the circuits transfer function (amplitude and phase) changes as the source frequency changes. The ideal frequency response plot for the four major categories of filters are shown in Fig. 13.1.

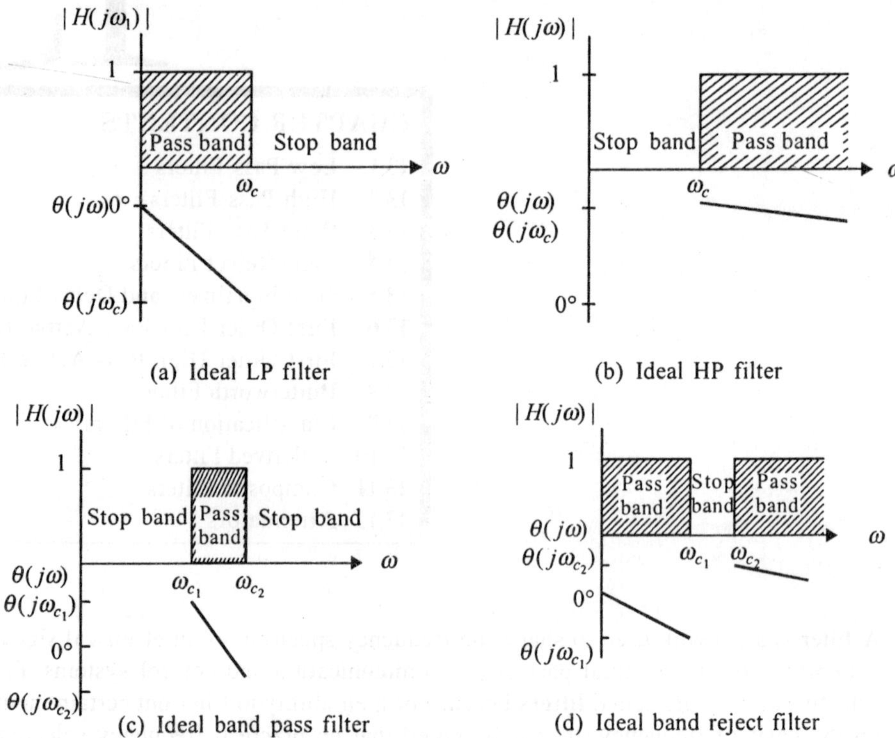

Fig. 13.1 Ideal frequency response of four types of filter circuits.

Parts (a) and (b) illustrate the ideal plots for a low pass and a high pass filter, respectively. Both filters have one pass band and one stop band, which are defined by the *cut off* frequency that separates them. A *low pass filter* passes signals at frequencies lower than the cut off frequency, and a high pass filter passes signals at frequencies higher than the cut off frequency. The two remaining categories of filters have two cut off frequencies. Figure 13.1(c) illustrates the ideal frequency response of a *band pass filter,* which passes signals when the source frequency is within the band defined by the two cut off frequencies. Figure 13.1(d) shows the ideal *band reject filter*, which passes signals only when the source frequency is outside the band defined by the two cut off frequencies.

Part I of this chapter describes passive filters, so called because their capabilities depend only on passive elements—resistors, capacitors and inductors.

13.1 LOW PASS FILTERS

The basic function of a low pass filter is to pass low frequencies with very little loss and to attenuate high frequencies. A familiar application of low pass filters is in the tone control of hi-fi amplifiers. The treble control varies the cut off frequency of a LP filter and is used to attenuate the high frequency noise.

Two circuits that behave as low pass filters are the series *RL* circuit and the series *RC* circuit.

13.1.1 The Series *RL* Circuit

A series *RL* circuit is shown in Fig. 13.2.

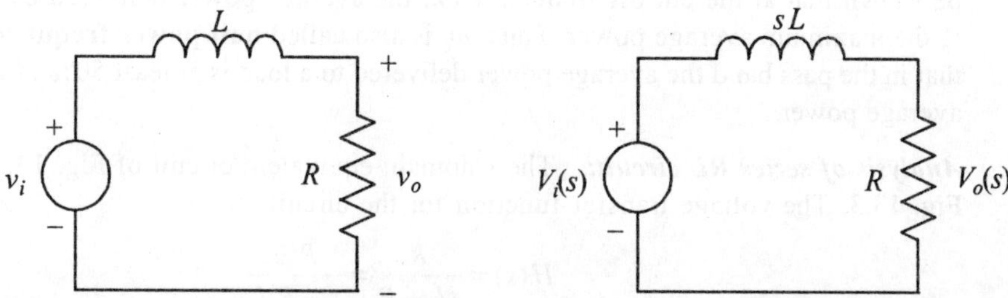

Fig 13.2 A series *RL* low pass filter. **Fig. 13.3** The *s* domain equivalent of the circuit in Fig. 13.2.

The circuit input is a sinusoidal voltage with variable frequency. The output is defined as *the voltage across the resistor*. At low frequencies the impedance of the inductor is very small compared with that of the resistor and the inductor effectively function as a short circuit. As the frequency increases, the impedance of the inductor increases resulting in a corresponding increase in the voltage drop across the inductor and a corresponding decrease in the output voltage. There is also a difference in phase angle between the input and the output voltage.

At high frequencies the inductance impedance is very large, effectively blocking the flow of current in the circuit. Thus, the circuit selectively passes low frequency inputs to the output and it blocks high frequency inputs from reaching the output. The ideal magnitude plot for a low pass filter in Fig 13.1(a) may be compared with the magnitude plot of the *RL* filter in Fig 13.4(a).

It is readily seen that the ideal filter exhibits a discontinuity at the cut off frequency, which creates an abrupt transition into and out of the pass band. Practical low pass filters have a response that changes gradually from pass band to stop band.

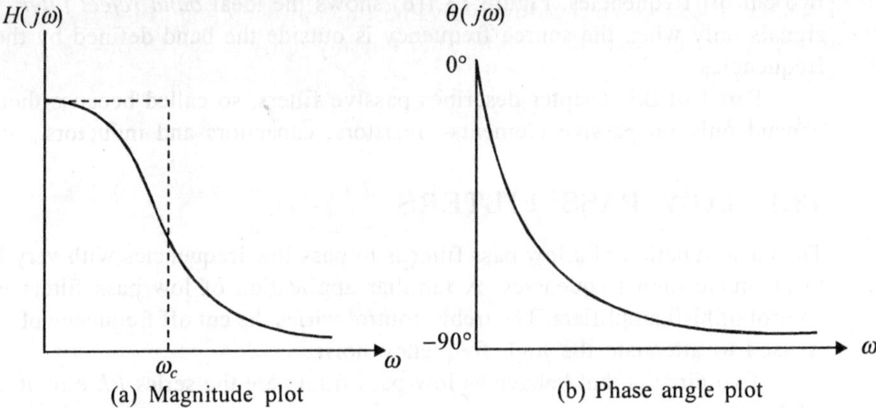

Fig. 13.4 Frequency response for the series *RL* low pass filter in Fig 13.2.

The definition for cut off frequency widely used is the frequency for which the transfer function magnitude is decreased by the factor 1/√2 from its maximum value. It can be shown that at the cut off frequency ω_c, the average power delivered by the circuit is ½ the maximum average power. Thus, ω_c is also called **half power frequency**. It follows that in the pass band the average power delivered to a load is at least 50% of the maximum average power.

Analysis of series RL circuit: The *s* domain equivalent circuit of Fig. 13.2 is shown in Fig. 13.3. The voltage transfer function for the circuit is:

$$H(s) = \frac{R}{sL + R} = \frac{R/L}{s + R/L}$$

The frequency response is obtained by putting $s = j\omega$.

$$H(j\omega) = \frac{R/L}{j\omega + R/L}$$

Hence

$$|H(j\omega)| = \frac{R/L}{\sqrt{\omega^2 + (R/L)^2}} \tag{13.1}$$

$$\theta(j\omega) = -\tan^{-1}\omega L/R \tag{13.2}$$

When $\omega = 0$, $|H(j0)| = 1$. This means at $\omega = 0$ the input voltage is passed to the output terminals without any loss in gain. ω_c may be found by equating $|H(j\omega)|$ to $(1/\sqrt{2})\,H_{max}$, that is to say,

$$(1/\sqrt{2})\,.1 = \frac{R/L}{\sqrt{\omega_c^2 + (R/L)^2}}$$

Solving for ω_c,

$$\omega_c = R/L \tag{13.3}$$

This equation for a low pass filter facilitates the determination of cut off frequency, by proper selection of *R* and *L*. *L* = 100 mH is a preferred choice for the inductance.

EXAMPLE 13.1 Select a suitable value for R in the circuit of Fig. 13.2 such that the resulting circuit could be used to filter out any noise above 10 Hz and pass the electric signals at about 1 Hz.

SOLUTION:
From Eq. (13.3),

$$R = L\omega_c = 100 \times 10^{-3} \times 2\pi \times 10, \text{ choosing } L = 100 \text{ mH}$$
$$= 6.28\ \Omega$$

The magnitude of v_o can be found from the equation:

$$|v_o| = |H(j\omega)||V_i|$$

Thus,

$$v_o(\omega) = \frac{R/L}{\sqrt{\omega^2 + (R/L)^2}}$$

$$= \frac{20\pi}{\sqrt{\omega^2 + 400\pi^2}}|v_i|$$

For an input voltage of 1 V at a frequency of 1 Hz, v_o is 0.995. For an input voltage of 1 V at 10 Hz, v_o is 0.707. **Ans.**

13.1.2 The Series *RC* Circuit

A series *RC* circuit is shown in Fig. 13.5.

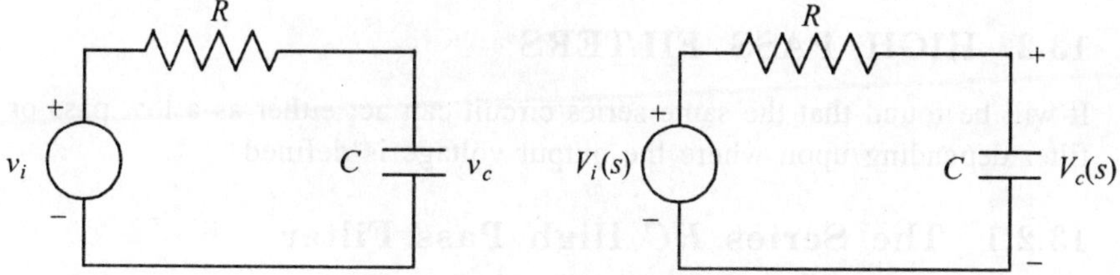

(a) *RC* low pass filter. (b) *s*-domain equivalent circuit of Fig. 13.5(a)

Fig. 13.5 A series *RC* low pass filter.

The transfer function is:

$$H(s) = \frac{1/RC}{s + 1/(RC)}$$

The frequency response is:

$$H(j\omega) = \frac{1/RC}{j\omega + 1/(RC)}$$

and

$$|H(j\omega)| = \frac{1/RC}{\sqrt{\omega^2 + (1/RC)^2}}$$

At cut off frequency, $H(\omega)$ is $1/\sqrt{2}$ times H_{max}.

or
$$(1/\sqrt{2})\cdot 1 = \frac{1/RC}{\sqrt{\omega_c^2 + (1/RC)^2}}$$

Solving for ω_c,
$$\omega_c = 1/RC \tag{13.4}$$

C is chosen as 1 μF.

EXAMPLE 13.2 For the circuit of Fig. 13.5, find the transfer function between the source voltage and the output voltage. Choosing $C = 1$ μF, find the value of R such that the low pass filter will have a cut off frequency of 3 kHz.

SOLUTION:

$$|H(j\omega)| = \frac{1/RC}{\sqrt{\omega^2 + (1/RC)^2}}$$

The value of ω_c is given by Eq. (13.4) as:
$$\omega_c = 1/RC = \frac{1}{R \times 1 \times 10^{-6}}$$

or
$$R = \frac{1}{2\pi(3 \times 10^3)(1 \times 10^{-6})} = 53.05\ \Omega \qquad \textbf{Ans.}$$

13.2 HIGH PASS FILTERS

It will be found that the same series circuit can act either as a low pass or a high pass filter depending upon where the output voltage is defined.

13.2.1 The Series *RC* High Pass Filter

A series *RC* circuit is shown in Fig. 13.6. Note that the output voltage is defined across the resistor. At $\omega = 0$, the capacitor behaves like an open circuit and so there is no current flowing in the resistance. There is no voltage across the resistor and the circuit filters out

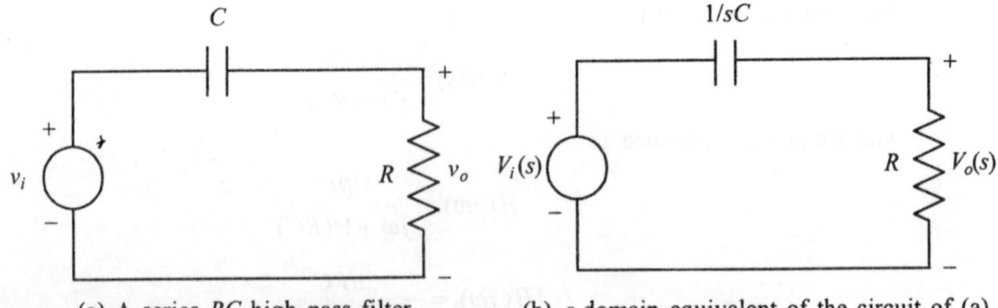

(a) A series *RC* high pass filter (b) *s*-domain equivalent of the circuit of (a)

Fig. 13.6

the low frequency source voltage before it reaches the output. When the frequency of the source is infinite the capacitor behaves like a short circuit and thus there is no voltage across the capacitor; in this condition the input voltage and output voltage are the same.

Analysis of the series RC circuit: From the *s*-domain circuit in Fig. 13.6(b), the transfer function is:

$$H(s) = \frac{s}{s + 1/RC}$$

and the frequency response is:

$$H(j\omega) = \frac{j\omega}{j\omega + 1/RC}$$

$$|H(j\omega)| = \frac{\omega}{\sqrt{\omega^2 + (1/RC)^2}}$$

and
$$\theta(j\omega) = 90 - \tan^{-1} \omega RC$$

For a high pass filter,

$$H_{\max} = |H(j\omega)|_{\omega=\infty} = 1$$

And at the cut off frequency, the magnitude of the transfer function is $(1/\sqrt{2})\, H_{\max}$. Thus,

$$(1/\sqrt{2}).1 = \frac{\omega_c}{\sqrt{\omega_c^2 + (1/RC)^2}}$$

Solving for ω_c,

$$\omega_c = 1/RC$$

Thus, the cut off frequency for the series *RC* circuit has the value 1/*RC*, whether it functions as à low pass filter or as high pass filter. Note the connection between the cut off frequency ω_c and the time constant τ of a circuit.

13.3 BAND PASS FILTERS

The ideal band pass filters having two cut off frequencies ω_{c1} and ω_{c2}, are shown in Fig. 13.1. The cut off frequencies are defined as the *frequencies for which the magnitude of the transfer function equals* $(1/\sqrt{2})\, H_{\max}$.

Centre frequency, bandwidth, quality factor: The centre frequency ω_0 is defined as *the frequency for which the circuit's transfer function is purely real*. The centre frequency is also known as **resonant frequency** and is given as:

$$\omega_0 = \sqrt{\omega_{c1}\, \omega_{c2}}$$

For band pass filters, the magnitude of the transfer function is a maximum at the centre frequency.

The bandwidth β is the width of the pass band and quality factor Q is the ratio of centre frequency to the bandwidth.

It will be shown that out of the five parameters mentioned— ω_{c1}, ω_{c2}, ω_0, β and Q, only two can be specified independently.

13.3.1 The Series *RLC* Circuit

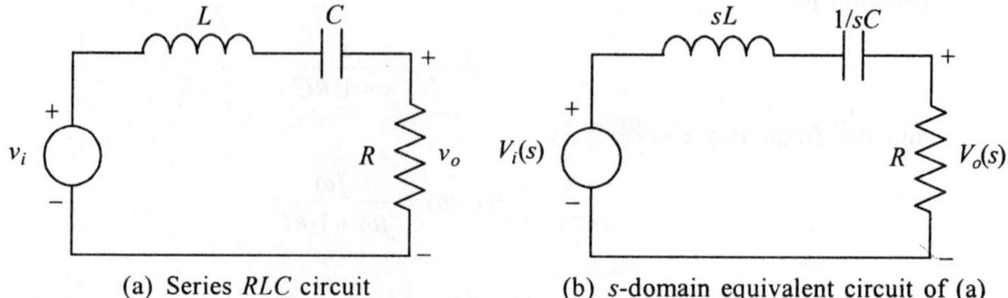

(a) Series *RLC* circuit (b) *s*-domain equivalent circuit of (a)

Fig. 13.7

The transfer function is given by

$$H(s) = \frac{R}{sL + 1/sC + R}$$

$$= \frac{(R/L)s}{s^2 + (R/L)s + 1/LC}$$

Putting $s = j\omega$,

$$|H(j\omega)| = \frac{\omega(R/L)}{\sqrt{(1/LC - \omega^2)^2 + (\omega R/L)^2}} \tag{13.5}$$

and

$$\theta(j\omega) = 90^\circ - \tan^{-1}\frac{\omega(R/L)}{(1/LC) - \omega^2}$$

Calculation of the five parameters that characterize the RLC band pass filter: The transfer function will be real when the frequency of the voltage source makes the sum of the capacitor and inductor impedances zero.

$$j\omega_0 L + \frac{1}{j\omega_0 C} = 0$$

On solving,

$$\omega_0 = \sqrt{1/LC} \tag{13.6}$$

At cut off frequency,

$$H_{\max} = |H(j\omega_0)|$$

$$= \frac{\omega_0(R/L)}{\sqrt{(1/LC - \omega_0^2)^2 + (\omega_0 R/L)^2}}$$

$$= 1 \quad \text{(using Eq. (13.6))}$$

Setting the transfer function magnitude at cut off frequencies equal to $1/\sqrt{2}\ H_{max}$ $(= 1/\sqrt{2} \times 1)$ in Eq. (13.5),

$$1/\sqrt{2} = \frac{\omega_c R/L}{\sqrt{(1/LC - \omega_c^2)^2 + (\omega_c R/L)^2}} \tag{13.7}$$

which yields

$$\omega_c^2 L \pm \omega_c R - 1/C = 0 \tag{13.8}$$

Solving for ω_c and considering only the positive values (which have physical significance):

$$\omega_{c1} = -\frac{R}{2L} + \sqrt{(R/2L)^2 + (1/LC)} \tag{13.9}$$

$$\omega_{c2} = +\frac{R}{2L} + \sqrt{(R/2L)^2 + (1/LC)} \tag{13.10}$$

The centre frequency ω_0 is the geometric mean of the two cut off frequencies:

$$\omega_0 = \sqrt{\omega_{c1} \cdot \omega_{c2}}$$
$$= \sqrt{1/LC} \tag{13.11}$$

Bandwidth is defined as:

$$\beta = \omega_{c2} - \omega_{c1}$$
$$= R/L \qquad \text{(from Eqs. (13.9) and (13.10))} \tag{13.12}$$

The quality factor defined as *the ratio of centre frequency to bandwidth* is:

$$Q = \frac{\omega_0}{\beta} = \frac{\sqrt{1/LC}}{R/L} = \sqrt{\frac{L}{CR^2}} \tag{13.13}$$

In terms of β and ω_0,

$$\omega_{c1} = -\beta/2 + \sqrt{(\beta/2)^2 + \omega_0^2} \tag{13.14}$$

$$\omega_{c2} = \beta/2 + \sqrt{(\beta/2)^2 + \omega_0^2} \tag{13.15}$$

These equations may also be written in terms of the centre frequency and the quality factor.

$$\omega_{c1} = \omega_0 \left[\frac{-1}{2Q} + \sqrt{1 + \left(\frac{1}{2Q}\right)^2} \right] \tag{13.16}$$

$$\omega_{c2} = \omega_0 \left[\frac{1}{2Q} + \sqrt{1 + \left(\frac{1}{2Q}\right)^2} \right] \tag{13.17}$$

EXAMPLE 13.3 Show that the *RLC* circuit in Fig. 13.8 is also a band pass filter by deriving the transfer function $H(s)$. Calculate the centre frequency ω_0, the cut off frequencies ω_{c1} and ω_{c2}, the bandwidth β and the quality factor Q.

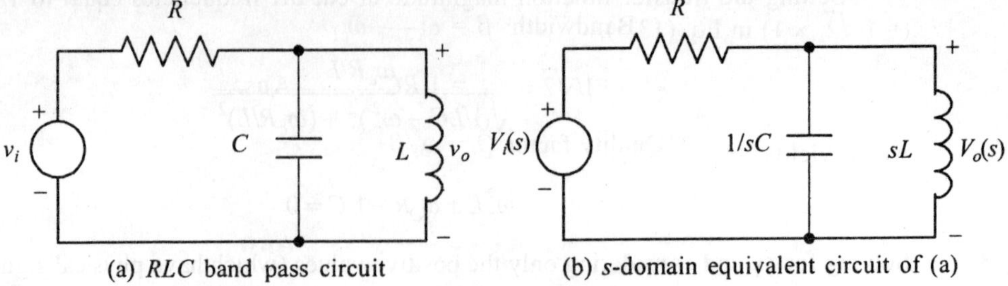

(a) *RLC* band pass circuit (b) *s*-domain equivalent circuit of (a)

Fig. 13.8

SOLUTION:

The transfer function is easily seen to be

$$H(s) = \frac{sL \| 1/(sC)}{R + sL \| 1/(sC)}$$

$$= \frac{s/(RC)}{s^2 + s/(RC) + 1/LC}$$

Substituting $s = j\omega$,

$$|H(j\omega)| = \frac{\omega/RC}{\sqrt{(1/LC - \omega^2)^2 + (\omega/RC)^2}}$$

$$= \frac{1}{\sqrt{1 + \left(\omega RC - \dfrac{1}{\omega L/R}\right)^2}}$$

The magnitude of this is maximum when $\left(\omega RC - \dfrac{1}{\omega L/R}\right) = 0$

Thus,

$$\omega_0 = \sqrt{1/(LC)} \qquad \textbf{Ans.}$$

At the cut off frequencies,

$$(1/\sqrt{2})H_{max} = 1/\sqrt{2}$$

Since

$$H_{max} = |H(j\omega_0)| = 1$$

Thus,

$$\omega_c RC - \frac{1}{\omega_c L/R} = \pm 1$$

$$\omega_{c1} = \frac{-1}{2RC} + \sqrt{\left(\frac{1}{2RC}\right)^2 + \frac{1}{LC}} \qquad \textbf{Ans.}$$

$$\omega_{c2} = \frac{1}{2RC} + \sqrt{\left(\frac{1}{2RC}\right)^2 + \frac{1}{LC}} \qquad \textbf{Ans.}$$

$$\text{Bandwidth } \beta = \omega_{c2} - \omega_{c1}$$

$$= 1/RC \qquad \textbf{Ans.}$$

$$\text{Quality factor } Q = \omega_0/\beta$$

$$= \frac{R^2C}{L} \qquad \textbf{Ans.}$$

Dual-tone-multiple frequency system: An example that illustrates the application of low pass, high pass and band pass filters is the dual-tone-multiple frequency (DTMF) system used in push button telephones. As the telephone number is dialed by the push buttons, a set of signals is transmitted to the telephone exchange.

Usually sixteen buttons (ten decimal digits plus six for special purposes) provide for sixteen distinct signals by using eight signal frequencies in the audio range. The eight frequencies are divided into two groups, four low band and four high band. Pressing a button generates a pair of tones, one from the high band and one from the low band. Each button is, therefore, identified by a unique pair of signal frequencies.

At the telephone exchange, these tones are identified and converted to a suitable set of dc signals for the switching system to connect the caller to the called party. After amplification the two tones are separated into their respective groups. The HP filter passes the high group tones while the LP filter passes the low group tones. The separated tones are then converted to square waves of fixed amplitude using limiters. The next step is to identify the individual tones in the respective groups by using eight band pass filters. Each of the band pass filters passes one tone, rejecting all the neighbouring tones. The band pass filters are followed by the detectors that are energized when their input voltage exceeds a certain threshold voltage. The output of each detector provides the required dc signal.

13.4 BAND REJECT FILTERS

Band reject filters reject a band of frequencies from a signal. The frequency band to be rejected is the stop band [see Fig. 13.1(d)] from ω_{c1} to ω_{c2}. The pass bands extend below ω_{c1} and above ω_{c2}. Band pass filters and band reject filters thus perform complementary functions in the frequency domain. Band reject filters are characterized by the same parameters as band pass filters—the two cut off frequencies, the centre frequency, the band width and the quality factor.

Most common use of band reject filter is in the removal of undesired tones from the frequency spectrum of a signal such as the billing of long distance telephone calls. Calls that are to be toll free (like calls to the operator for information, complaints) are to be prevented from being billed. This is achieved by removing these tones from the voice signal before being transmitted from the exchange to the listener.

13.4.1 The Series *RLC* Circuit

Figure 13.9 shows a series *RLC* circuit.

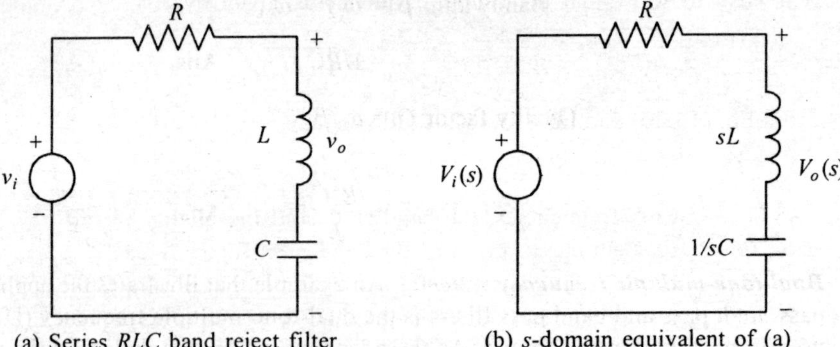

(a) Series *RLC* band reject filter (b) *s*-domain equivalent of (a)

Fig. 13.9

Note that the output voltage is now defined across the inductance capacitance pair. The transfer function is given by

$$H(s) = \frac{sL + 1/(sC)}{R + sL + 1/(sC)}$$

$$= \frac{s^2 + 1/(LC)}{s^2 + \frac{R}{L}s + \frac{1}{LC}}$$

The transfer function magnitude is given by

$$|H(j\omega)| = \left| \frac{\frac{1}{LC} - \omega^2}{\sqrt{\left(\frac{1}{LC} - \omega^2\right)^2 + \left(\frac{\omega R}{L}\right)^2}} \right| \tag{13.18}$$

$$\theta(j\omega) = -\tan^{-1} \frac{\omega R/L}{\frac{1}{LC} - \omega^2} \tag{13.19}$$

For the band reject filter, the centre frequency is still defined as *the frequency for which the sum of the impedances of the capacitance and inductance is zero*. In the band pass filter, the magnitude of the centre frequency was a maximum but in the band reject filter this magnitude is a minimum. (See Fig. 13.10)

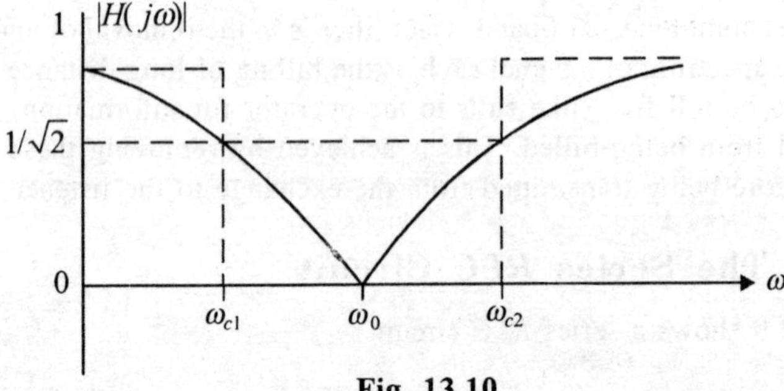

Fig. 13.10

It is easy to show that the centre frequency is given by

$$\omega_0 = \sqrt{1/(LC)} \tag{13.20}$$

From Eq. (13.20) and Eq. (13.18), it follows that

$$|H(j\omega_0)| = 0$$

The cut off frequencies, the bandwidth and the quality factor are defined for the band reject filter similar to the way for band pass filters. Thus,

$$\omega_{c1} = -\frac{R}{2L} + \sqrt{\left(\frac{R}{2L}\right)^2 + \frac{1}{LC}} \tag{13.21}$$

$$\omega_{c2} = \frac{R}{2L} + \sqrt{\left(\frac{R}{2L}\right)^2 + \frac{1}{LC}} \tag{13.22}$$

$$\beta = R/L$$

and

$$Q = \sqrt{\frac{L}{R^2 C}} \tag{13.23}$$

13.5 GAIN EQUALIZERS AND DELAY EQUALIZERS

Gain equalizers are used to shape the gain frequency spectrum of a given signal. The shaping can take the form of an emphasis or a de-emphasis of a band of frequencies. The shapes, however, which they provide are not characterized by a pass band and a stop band. A common application is in the recording and reproduction of music on phonograph records.

Delay equalizers compensate for delay distortions introduced by filters and other parts of the transmission system. The equalizer introduces the necessary delay shape to make the total filter plus equalizer delay as flat as possible. It must be ensured that the delay equalizer will not affect the loss characteristic of the filter.

An example of the application of delay equalizers is in the transmission of data on cables. A second order delay equalizer can be used to compensate for the distortion produced in the cable. With proper choice of the parameters the sum of the delays of the cable and equalizer will be flat.

ACTIVE FILTERS

The use of inductors in passive filters presents problems, thanks to the high Q inductors, which tend to become bulky and expensive below 1 kHz. Active filters make use of resistance, capacitance and active devices (operational amplifiers). These devices can all be integrated so as to achieve

- reduction in size and weight
- increased circuit reliability

- cheaper than passive counter parts
- improvement in performance
- reduction in parasitics owing to the smaller size
- the design process simpler
- realizing of a wider class of functions
- ensuring voltage gains

The chief drawback is the finite bandwidth of the active devices.

Electromechanical filters: Use is made of *mechanical resonance* for the filtering of the electrical signals as shown in the block diagram in Fig. 13.11.

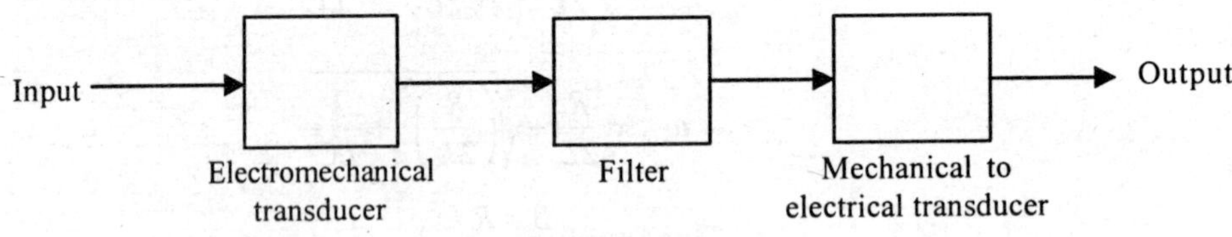

Fig. 13.11

The electrical signal is converted into a mechanical vibration by a transducer and after the filtering, the mechanical vibration is converted back into an electrical signal.

Digital filters: The input analog signal is sampled at uniformly spaced intervals and the sampled values are converted to binary words using an *A/D* converter (Reference 22). The binary representation of the input is then filtered by the digital filter. The output word is finally converted back into an analog signal.

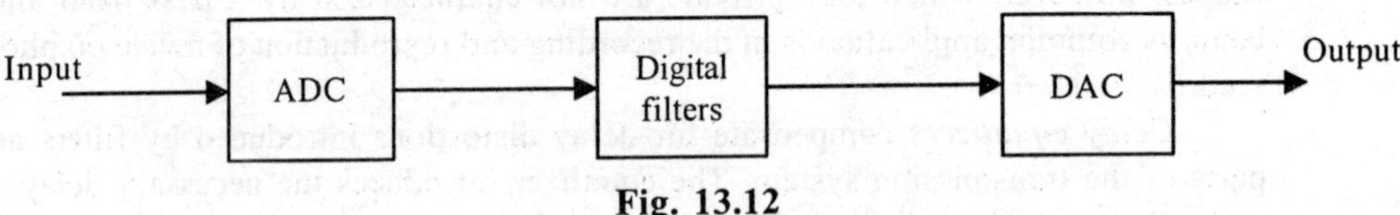

Fig. 13.12

Microwave filters: These are used for transmission at frequencies from 200 MHz to 100 GHz. These consist of transmission lines and wave-guides. Areas of application include radar, satellite communication and telephone carrier systems.

13.6 FIRST ORDER LOW PASS ACTIVE FILTERS

A general op-amp circuit is shown in Appendix C, with an input impedance Z_i and feedback impedance Z_f. A first order low pass filter is shown in Fig. 13.13.

The transfer function for the circuit in Fig. 13.13(a) is expressed as:

$$H(s) = \frac{-Z_f}{Z_1}$$

$$= \frac{-R_2 \| 1/(sC)}{R_1}$$

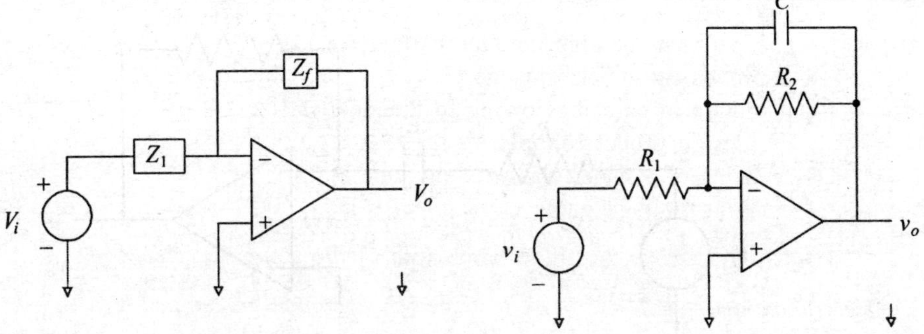

(a) A general op-amp circuit (b) A first order low pass active circuit

Fig. 13.13

$$= \frac{-R_2}{R_1} \frac{\omega_c}{s + \omega_c}$$

$$= \frac{-K\omega_c}{s + \omega_c} \tag{13.24}$$

where $$K = \frac{R_2}{R_1} \text{ and } \omega_c = \frac{1}{R_2 C} \tag{13.25}$$

Note that the gain is set by the ratio R_2/R_1.

EXAMPLE 13.4 In the circuit of Fig. 13.13(b), calculate the values of C and R_2 for given $R_1 = 1\ \Omega$ for the circuit to serve as a low pass filter with gain of 1 in the pass band and cut off frequency of 1 rad/s. Construct the transfer function for this filter.

SOLUTION:

$$R_2 = KR_1 = 1 \times 1 = 1\ \Omega$$

C is calculated from Eq. (13.25),

$$C = \frac{1}{R_2 \omega_c} = \frac{1}{1 \times 1} = 1\ \text{F}$$

From Eq. (13.24),

$$H(s) = -1 \times \frac{1}{s+1} = \frac{-1}{s+1}$$

Such a filter which uses a resistor of 1 Ω, capacitor of 1 F giving a cut off frequency of 1 rad/s, is called **prototype low pass op amp filter**.

13.7 FIRST ORDER HIGH PASS ACTIVE FILTERS

Circuit of Fig. 13.14 is a first order high pass filter.

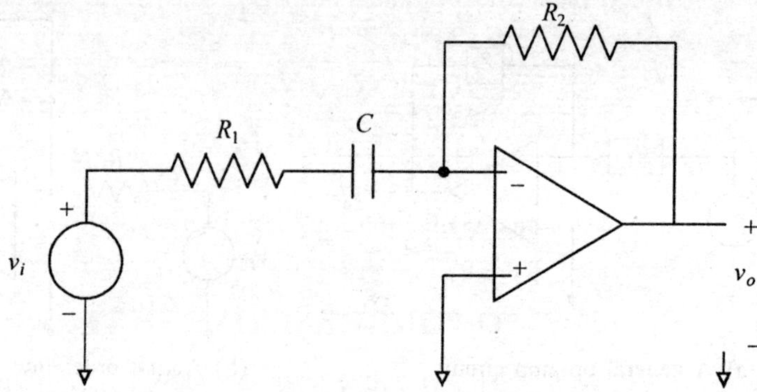

Fig. 13.14 A first–order high–pass active filter.

The transfer function for the circuit is:

$$H(s) = \frac{-Z_f}{Z_1} = \frac{-R_2}{R_1 + 1/sC} = -\frac{Ks}{s + \omega_c} \tag{13.26}$$

where
$$K = \frac{R_2}{R_1} \text{ and } \omega_c = \frac{1}{R_1 C} \tag{13.27}$$

13.8 BUTTERWORTH FILTERS

A unity gain Butterworth low pass filter has a transfer function whose magnitude is given by

$$|H(j\omega)| = \frac{1}{\sqrt{1 + (\omega/\omega_c)^{2n}}}$$

where n is an integer, which denotes the order of the filter.

Note:

- The cut off frequency is ω_c rad/s for all values of n.
- If n is large, the denominator is always close to unity when $\omega < \omega_c$.
- The exponent of ω/ω_c is always even.

It can be shown that

$$H(s)H(-s) = \frac{1}{1 + (-1)^n s^{2n}} \tag{13.27a}$$

EXAMPLE 13.5 Find the Butterworth transfer function for $n = 2$ in Eq. (13.27a).

The roots of the polynomials in Eq. (13.27a) are given by $s^4 = -1$

Hence the four roots are $s_1 = 1\angle 45° = 1/\sqrt{2} + j/\sqrt{2}$

$$s_2 = 1\angle 135° = -1/\sqrt{2} + j/\sqrt{2}$$
$$s_3 = 1\angle 225° = -1/\sqrt{2} - j/\sqrt{2}$$
$$s_4 = 1\angle 315° = 1/\sqrt{2} - j/\sqrt{2}$$

Assigning left-half plane roots s_2 and s_3 to $H(s)$,

$$H(s) = \frac{1}{(s + 1/\sqrt{2} - j/\sqrt{2})(s + 1/\sqrt{2} + j/\sqrt{2})}$$

$$= \frac{1}{(s^2 + \sqrt{2}s + 1)}$$

which is the required transfer function.

TWO-PORT REACTIVE FILTERS

Basic relations in a filter: Filter networks in this class consist of one or more sections of symmetrical T or π networks or their half sections. See Fig. 13.15.

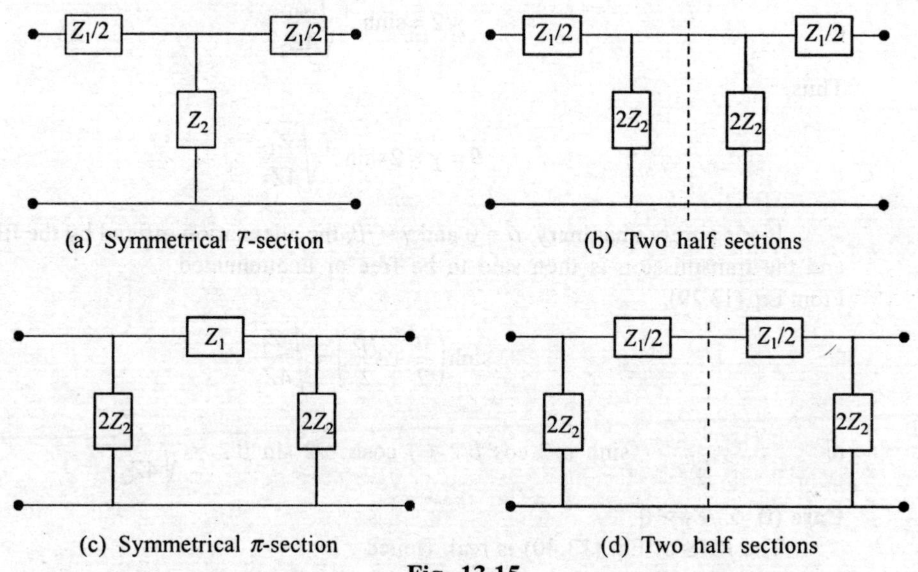

Fig. 13.15

The transfer constant θ of a two-port network in terms of $ABCD$ parameters is:

$$\theta = \cosh^{-1}\sqrt{AD} \quad \text{(see Chapter 11)}$$

For a symmetrical network such as the T-section, the transfer constant θ becomes the propagation constant γ and is given by

$$\theta = \gamma = \alpha + j\beta = \cosh^{-1}\sqrt{AD}$$

But in the T-section

$$A = \frac{z_{11}}{z_{21}} = \frac{Z_2 + Z_1/2}{Z_2} = 1 + \frac{Z_1}{2Z_2}$$

$$D = \frac{z_{22}}{z_{21}} = 1 + \frac{Z_1}{2Z_2}$$

or $$A = D$$

Hence

$$\gamma = \cosh^{-1} A$$

$$= \cosh^{-1}\left(1+\frac{Z_1}{2Z_2}\right) \tag{13.28}$$

$\therefore$ $$\cosh\gamma = 1+\frac{Z_1}{2Z_2}$$

and $$\sinh\gamma/2 = \sqrt{\frac{Z_1}{4Z_2}}$$

or $$\gamma/2 = \sinh^{-1}\sqrt{\frac{Z_1}{4Z_2}}$$

Thus,

$$\theta = \gamma = 2\sinh^{-1}\sqrt{\frac{Z_1}{4Z_2}} \tag{13.29}$$

If γ is purely imaginary, $\alpha = 0$ and $\gamma = j\beta$, the attenuation offered by the filter is zero and the transmission is then said to be free or unattenuated.
From Eq. (13.29),

$$\sinh\left(\frac{\alpha}{2}+\frac{j\beta}{2}\right) = \sqrt{\frac{Z_1}{4Z_2}}$$

or $$\sinh\alpha/2\cos\beta/2 + j\cosh\alpha/2\sin\beta/2 = \sqrt{\frac{Z_1}{4Z_2}} \tag{13.30}$$

Case (i) $Z_1/Z_2 > 0$

The RHS of Eq. (13.30) is real. Hence

$$\sinh\alpha/2.\cos\beta/2 = \sqrt{\frac{Z_1}{4Z_2}} \tag{13.31}$$

and $$\cosh\alpha/2.\sin\beta/2 = 0$$

Since $\cosh\alpha/2$ cannot be zero, $\sin\beta/2 = 0$. So that

$$\beta/2 = \pm n\pi \quad \text{where } n = 0, 1, 2, \ldots,$$

Also $$\cos\beta/2 = 1$$

Hence, from Eq. (13.31),

$$\sinh\alpha/2 = \sqrt{\frac{Z_1}{4Z_2}}$$

$$\alpha = 2\sinh^{-1}\sqrt{\frac{Z_1}{4Z_2}}$$

Thus, when $Z_1/(4Z_2)$ is positive, α is real, that is, attenuation takes place. The frequency range over which $Z_1/(4Z_2) > 0$, constitutes the *attenuation band* or the stop band.

Case (ii)

$$\frac{Z_1}{4Z_2} < 0$$

$\sqrt{\dfrac{Z_1}{4Z_2}}$ is now imaginary.

From Eq. (13.30),

$$\sinh \alpha/2 \cos \beta/2 = 0$$

Either

$$\sinh \alpha/2 = 0 \text{ or } \cos \beta/2 = 0 \tag{13.32}$$

Taking,

$$\sinh \alpha/2 = 0,\ \alpha = 0.$$

The frequency over which $Z_1/(4Z_2) < 0$ and $\alpha = 0$, constitutes the *pass band* or the transmission band. Further since

$$\sinh \alpha/2 = 0,\ \cosh \alpha/2 = 1$$

Hence, from Eq. (13.30),

$$j \sin \beta/2 = \sqrt{Z_1/(4Z_2)}$$

$$\beta = 2 \sin^{-1} \sqrt{-Z_1/(4Z_2)}$$

Therefore, within the pass band, β is a function of Z_1 and Z_2 that is a function of frequency.

When $Z_1/(4Z_2)$ is negative and $|Z_1/(4Z_2)| > 1$, $\sin \beta/2 > 1$ which is absurd. This means that for free transmission,

$$-1 < Z_1/(4Z_2)$$

Thus, for free transmission through a filter $Z_1/(4Z_2)$ must be negative and numerically less than unity. That is

$$-1 < Z_1/(4Z_2) < 0 \tag{13.33}$$

When $\quad Z_1/(4Z_2) < -1, \quad \cos \beta/2 = 0$

Therefore $\quad \sin \beta/2 = \pm 1$

Hence $\quad \beta/2 = (2n - 1)\pi/2, \quad n = 1, 2, 3, \ldots,$

By putting $\cos \beta/2 = 0$ in Eq. (13.30),

$$j \cosh \alpha/2 = \sqrt{Z_1/(4Z_2)}$$

or

$$\cosh \alpha/2 = \sqrt{|Z_1/(4Z_2)|}$$

Hence

$$\alpha = 2 \cosh^{-1} \sqrt{|Z_1/(4Z_2)|}$$

Note that in these derivations the elements are assumed to be pure reactances.

Cut off frequencies: The cut off frequencies are given by $Z_1/(4Z_2) = 0$ and $Z_1/(4Z_2) = -1$. [see Eq. (13.33)].

The characteristic impedance of a filter in pass band can be shown to be

$$Z_{0T} = \sqrt{Z_1 Z_2 (1 + Z_1/(4Z_2)} \qquad \text{(For a } T\text{-section)} \qquad (13.34)$$

and

$$Z_{0\pi} = \sqrt{\frac{Z_1 Z_2}{1 + Z_1 Z_2}} \qquad \text{(For } \pi\text{-section)} \qquad (13.35)$$

It follows

$$Z_{0\pi} = \frac{Z_1 Z_2}{Z_{0T}} \qquad (13.36)$$

Note with Z_1 and Z_2 as pure reactances the characteristic impedance is a pure resistance in the pass band; and in the stop band the characteristic impedances is purely reactive.

13.9 CLASSIFICATION OF FILTERS

Filters may be classified according to their frequency characteristics as in the preliminary concepts in this chapter. Filters may also be classified on the relation between Z_1 and Z_2. On the latter basis, filters are classified as:

- Constant k filters
- m-derived filters

In a constant k filter the series impedance Z_1 and the shunt impedance Z_2 are related by the equation $Z_1 Z_2 = k^2$, where k is independent of frequency. Invariably the product $Z_1 Z_2$ is real and has the dimensions of resistance so that $Z_1 Z_2 = R_0^2$. R_0 is called the **design impedance** of the filter section. The constant k filters are also referred to as prototype filters.

In the m-derived filters $Z_1 Z_2 \neq k^2$ but have the same characteristic impedance as the corresponding constant k sections, however, with much sharper attenuation characteristics.

13.9.1 Constant k Low Pass Filter

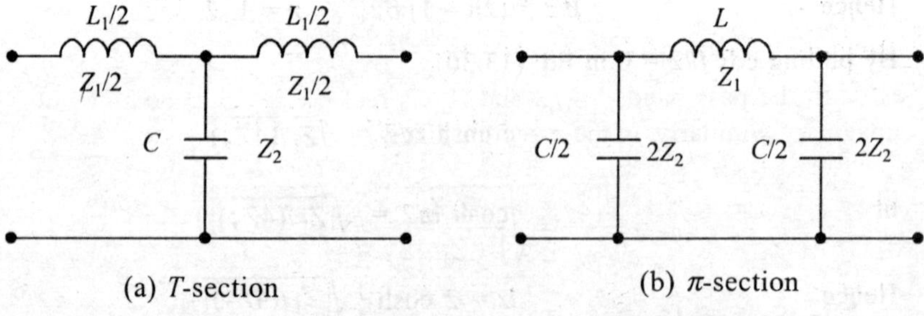

Fig. 13.16 Constant k low pass filters.

In both T- and π-sections,

Total series impedance $Z_1 = j\omega L$

Shunt impedance $Z_2 = 1/(j\omega C)$

Hence $Z_1Z_2 = L/C$

The design impedance is, therefore,

$$R_0 = \sqrt{Z_1Z_2} = \sqrt{L/C}$$

Further

$$\frac{Z_1}{4Z_2} = \frac{-\omega^2 LC}{4}$$

Pass band is given by

$$-1 < Z_1/(4Z_2) < 0$$

The cut off frequencies are, therefore,

$$Z_1/(4Z_2) = 0$$

and

$$Z_1/(4Z_2) = -1$$

The first cut off frequency is given by $Z_1 = 0$, i.e.,

$$j\omega L = 0 \quad \text{or} \quad f_{c1} = 0$$

Second cut off frequency is given by $Z_1/(4Z_2) = -1$, i.e.,

$$(\omega^2 LC)/4 = -1 \quad \text{or} \quad \omega_{c2} = 2/\sqrt{LC} \quad \text{or} \quad f_c = 1/(\pi\sqrt{LC})$$

Thus, the pass band is 0 to $1/(\pi\sqrt{LC})$.

Characteristic impedance: In the T-section filter

$$Z_{0T} = \sqrt{Z_1Z_2(1 + Z_1/(4Z_2))}$$

$$= \sqrt{\frac{L}{C}\left(1 - \frac{\omega^2 LC}{4}\right)}$$

$$= \sqrt{\frac{L}{C}}\sqrt{1 - (f/f_c)^2}$$

$$= R_0\sqrt{1 - (f/f_c)^2}$$

In the pass band, $f < f_{c2}$ and Z_{0T} is real. In the attenuation band, $f > f_c$ and Z_{0T} is imaginary. Similarly, in the π-section filter

$$Z_{0\pi} = \frac{Z_1Z_2}{Z_{0T}} = \sqrt{\frac{Z_1Z_2}{1 + Z_1/4Z_2}} = \frac{\sqrt{L/C}}{\sqrt{1 - (f/f_c)^2}} = \frac{R_0}{\sqrt{1 - (f/f_c)^2}}$$

In the pass band, $f < f_c$ so that $Z_{0\pi}$ is real and positive. In the attenuation band, $f > f_c$ and $Z_{0\pi}$ becomes imaginary.

EXAMPLE 13.6 For the given *T*-section, low pass filter determine the cut off frequency and the nominal characteristic impedance R_0.

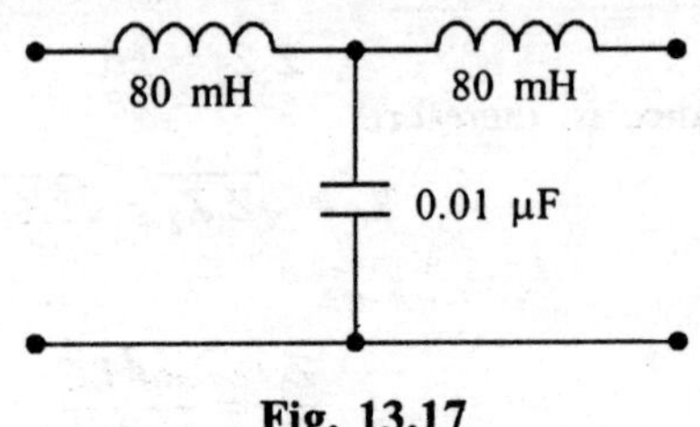

Fig. 13.17

SOLUTION:

Given $L = 160$ mH, $C = 0.01$ μF.

Then $$f_c = \frac{1}{\pi\sqrt{(LC)}} = \frac{1}{\pi\sqrt{(160\times10^{-3}/0.01\times10^{-6})}} = 7.96 \text{ kHz}$$ **Ans.**

and $$R_0 = \sqrt{(L/C)} = \sqrt{160\times10^{-3}/0.01\times10^{-6}} = 4 \text{ k}\Omega$$ **Ans.**

EXAMPLE 13.7 For the given π-section low pass filter, find the cut off frequency and the design resistance R_0.

SOLUTION:

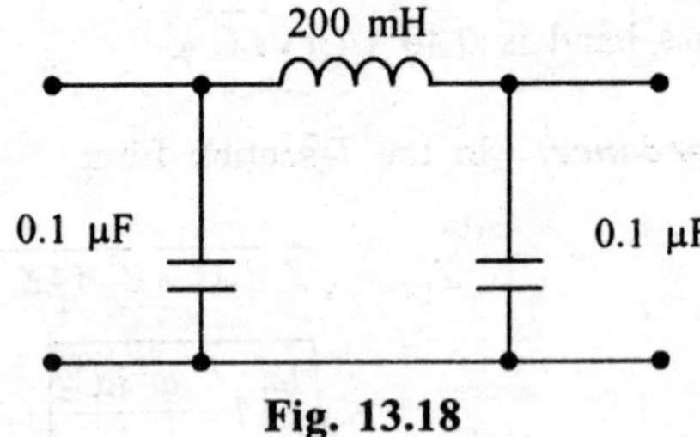

Fig. 13.18

Given $L = 200$ mH, $C = 0.2$ μF

Then

Cut off frequency $$f_c = \frac{1}{(\pi\sqrt{LC})}$$

$$= \frac{1}{\pi\sqrt{(200\times10^{-3}\times0.2\times10^{-6})}} = 1592 \text{ Hz}$$ **Ans.**

and $$R_0 = \sqrt{L/C}$$

$$= \sqrt{(200\times10^{-3}/0.2\times10^{-6}} = 1000\ \Omega$$ **Ans.**

13.9.2 Constant k High Pass Filter

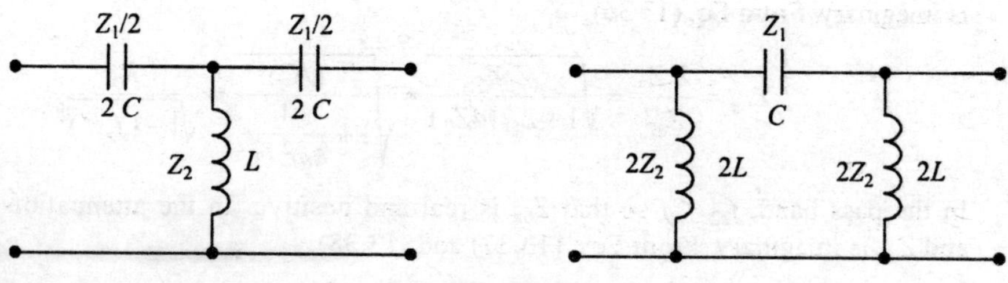

Fig 13.19

Nominal impedance or design impedance: In both *T*- and *π*-sections total series impedance is:

$$Z_1 = \frac{1}{j\omega C}$$

and shunt impedance is expressed as:

$$Z_2 = j\omega L$$

Therefore, $\quad R_0 = \sqrt{Z_1 Z_2} = \sqrt{L/C}$, as in low pass filter

Cut off frequency: Pass band is $-1 < Z_1/(4Z_2) < 0$. Hence cut off frequencies are given by

$$Z_1/(4Z_2) = 0$$

which gives $Z_1 = 1/(j\omega_c C) = 0$ and ω_c is infinity

and

which gives $\quad Z_1/(4Z_2) = -1$

$$\frac{-1}{(\omega_c)^2 4LC} = -1 \quad \text{and} \quad \omega_c = \frac{1}{2\sqrt{(LC)}}$$

or

$$f_c = \frac{1}{4\pi\sqrt{(LC)}}$$

Thus, the pass band of the constant **k** high pass filter extends from $\dfrac{1}{4\pi\sqrt{(LC)}}$ to ∞.

Characteristic impedance Z_0: The characteristic impedance of a *T*-section filter is given by

$$Z_{0T} = \sqrt{Z_1 Z_2 (1 + Z_1/(4Z_2))}$$

$$= \sqrt{\frac{L}{C}\left[1 - \frac{1}{4\omega_c^2 LC}\right]} \qquad (13.37)$$

$$= R_0\sqrt{1 - (f_c/f)^2}$$

In the pass band, $f_{c2} < f$ so that Z_{0T} is real. In the attenuation band, $f < f_{c2}$ and Z_{0T} is imaginary. From Eq. (13.36),

$$Z_{0\pi} = \frac{Z_1 Z_2}{Z_{0T}} = \sqrt{\frac{Z_1 Z_2}{1 + Z_1/(4Z_2)}} = \sqrt{\frac{L/C}{1 + \dfrac{1}{4\omega^2 LC}}} = \frac{R_0}{\sqrt{1-(f_c/f)^2}} \tag{13.38}$$

In the pass band, $f_{c2} < f$ so that $Z_{0\pi}$ is real and positive. In the attenuation band, $f < f_c$ and $Z_{0\pi}$ is imaginary. From Eqs. (13.37) and (13.38),

$$Z_{0T}\, Z_{0\pi} = R_0^2 \tag{13.39}$$

EXAMPLE 13.8 For the given T-section high pass filter shown in Fig. 13.20, find the cut off frequency and nominal characteristic resistance.

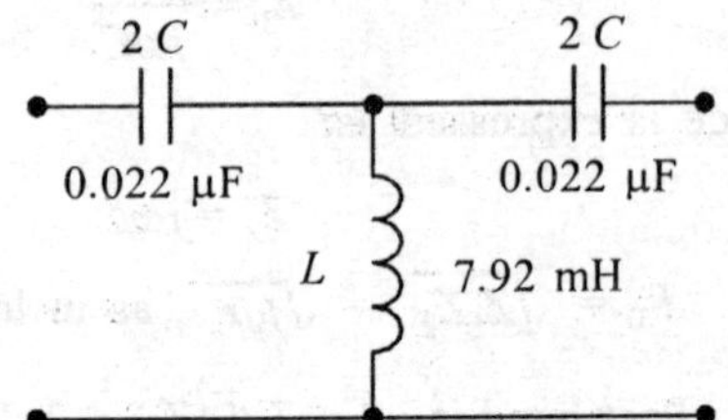

Fig. 13.20 T-section constant k high pass filter.

SOLUTION:

Given $L = 7.92$ mH, $C = 0.022$ μF

Then

$$R_0 = \sqrt{L/C}$$

$$= \sqrt{7.92 \times 10^{-3} / 0.022 \times 10^{-6}}$$

$$= 600\ \Omega \quad \textbf{Ans.}$$

and

$$C = \frac{1}{4\pi R_0 f_c}$$

$$0.022 \times 10^{-6} = \frac{1}{4\pi \times 600 \times f_c}$$

$\therefore$

$$f_c = 6 \text{ kHz} \quad \textbf{Ans.}$$

13.9.3 Design of Constant k Low Pass and High Pass Filters

Low pass filter:

$$R_0 = \sqrt{L/C}$$

and $$f_c = \frac{1}{\pi\sqrt{(LC)}}$$

From the above equations,

$$L = \frac{R_0}{\pi f_c} \tag{13.40}$$

and $$C = \frac{1}{\pi R_0 f_c} \tag{13.41}$$

Equations (13.40) and (13.41) give the values of L and C in terms of f_c and design resistance.

High pass filter:

$$R_0 = \sqrt{(L/C)}$$

and $$f_c = \frac{1}{4\pi\sqrt{(LC)}}$$

Combining these equations,

$$L = \frac{R_0}{4\pi f_c} \tag{13.42}$$

and $$C = \frac{1}{4\pi f_c R_0} \tag{13.43}$$

For constant k high pass filters, Eqs. (13.42) and (13.43) give the values of L and C in terms of f_c and design resistance.

EXAMPLE 13.9 Design a T-section constant k low pass filter having cut off frequency of 10 kHz and nominal characteristic impedance R_0 of 600 Ω.

SOLUTION:

$$L = \frac{R_0}{\pi f_c} = \frac{600}{\pi \times 10 \times 10^3} = 19.09 \text{ mH} \quad \textbf{Ans.}$$

and $$C = \frac{1}{\pi R_0 f_c} = \frac{1}{\pi \times 600 \times 10 \times 10^3} = 0.05\ \mu\text{F} \quad \textbf{Ans.}$$

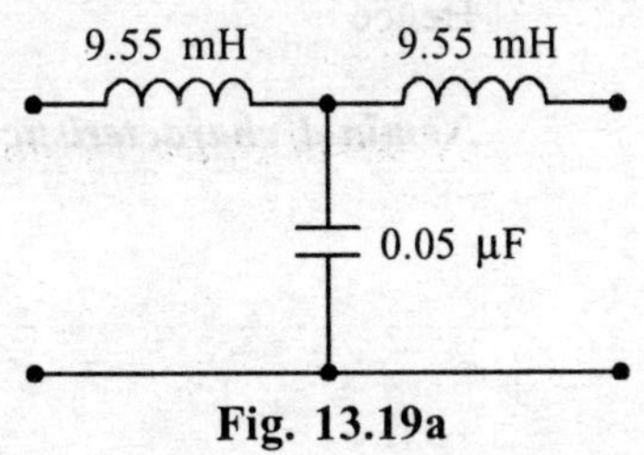

Fig. 13.19a

The parameters are shown in Fig. 13.19a

EXAMPLE 13.10 Design a π-section constant k high pass filter having cut off frequency f_c = 10 kHz and nominal characteristic impedance R_0 = 600 Ω.

SOLUTION:

$$L = \frac{R_0}{4\pi f_c} = \frac{600}{4 \times \pi \times 10 \times 10^3} = 4.77 \text{ mH} \quad \textbf{Ans.}$$

and $$C = \frac{1}{4\pi R_0 f_c} = \frac{1}{4 \times \pi \times 600 \times 10 \times 10^3} = 0.013 \ \mu\text{F} \quad \textbf{Ans.}$$

4.77 mH

6.5 nF

6.5 nF

Fig. 13.19b

The parameters are shown in Fig. 13.19b.

13.9.4 Constant k Band Pass Filters

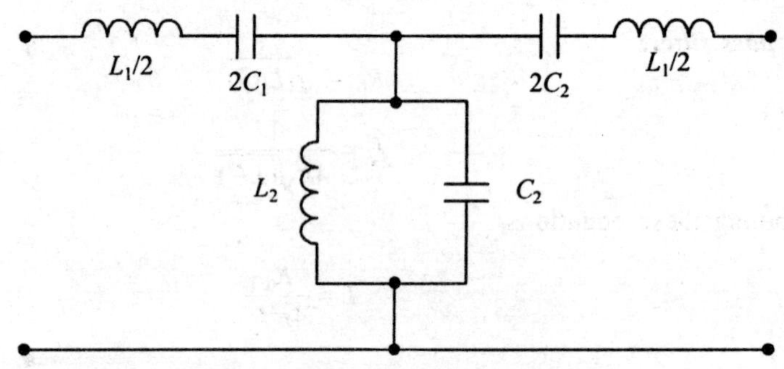

Fig. 13.21 Band pass filter.

The band pass filter may be viewed as a *T*-section filter in which the series impedance Z_1 is a resonance circuit while the shunt impedance Z_2 is a parallel resonance circuit. L_1, C_1, L_2 and C_2 are so chosen that the two resonant frequencies are the same. If this frequency is taken as ω_0 rad/s,

For the series arm $\qquad \omega_0^2 L_1 C_1 = 1$

and For the shunt arm $\qquad \omega_0^2 L_2 C_2 = 1$

Hence $$L_1 C_1 = L_2 C_2 = \frac{1}{\omega_0^2} \tag{13.44}$$

Nominal characteristic resistance R_0:

$$Z_1 = j\omega L_1 + \frac{1}{j\omega C_1} \tag{13.45}$$

$$Z_2 = \frac{j\omega L_2[1/(j\omega C_2)]}{j\omega L_2 + [1/(j\omega C)]} \tag{13.46}$$

Combining Eqs. (13.45), (13.46) and (13.44),

$$Z_1 Z_2 = \frac{L_2}{C_1} = \frac{L_1}{C_2} = R_0^2 \tag{13.46a}$$

Cut off frequencies: Since Z_1Z_2 is independent of frequency, the filter is a constant k filter. The pass band is given by $-1 < Z_1/(4Z_2) < 0$ so the cut off frequency is given by

$$-1 = Z_1/(4Z_2)$$

or

$$Z_1 = -4Z_2$$

$$Z_1^2 = -4Z_1Z_2 = -4R_0^2$$

and

$$Z_1 = \pm j2R_0$$

which defines two cut off frequencies.
At cut off frequency f_{c1},

$$Z_1 = -j2R_0$$

and at frequency f_{c2},

$$Z_1 = +j2R_0$$

Hence

$$\left(\omega_{c1}L_1 - \frac{1}{\omega_{c1}C_1}\right) = -\left(\omega_{c2}L_1 - \frac{1}{\omega_{c2}C_1}\right)$$

or

$$L_1C_1 = \frac{1}{\omega_0^2}$$

It follows

$$f_0^2 = f_{c1}f_{c2}$$

Thus, the resonant frequency of the shunt arm or series arm is the geometric mean of the cut off frequencies.

13.9.5 Design of Constant k Band Pass Filter

The design consists in finding the values of circuit element for given values of nominal characteristic impedance R_0 and cut off frequencies f_{c1} and f_{c2}. It can be shown that

$$f_{c2} - f_{c1} = 4\pi R_0 f_{c1} f_{c2} C_1$$

$$C_1 = \frac{f_{c2} - f_{c1}}{4\pi R_0 f_{c1} f_{c2}}$$

It can also be shown that

$$L_1 = \frac{1}{\omega_0^2 C_1} = \frac{1}{4\pi^2 f_0^2 C_1} = \frac{1}{4\pi^2 f_{c1} f_{c2} C_1} = \frac{R_0}{\pi(f_{c2} - f_{c1})} \qquad (13.46b)$$

and

$$L_2 = C_1 R_0^2$$

Finally,

$$L_2 = \frac{R_0^2 (f_{c2} - f_{c1})}{4\pi R_0 f_{c1} f_{c2}}$$

or
$$L_2 = \frac{R_0(f_{c2} - f_{c1})}{4\pi f_{c1} f_{c2}}$$

and
$$C_2 = \frac{L_1}{R_0^2} = \frac{1}{R_0^2} \frac{R_0}{\pi(f_{c2} - f_{c1})}$$

or
$$C_2 = \frac{1}{\pi R_0 (f_{c2} - f_{c1})}$$

EXAMPLE 13.11 In the band pass circuit of Fig. 13.22 shown below, determine characteristic resistance R_0 and bandwidth.

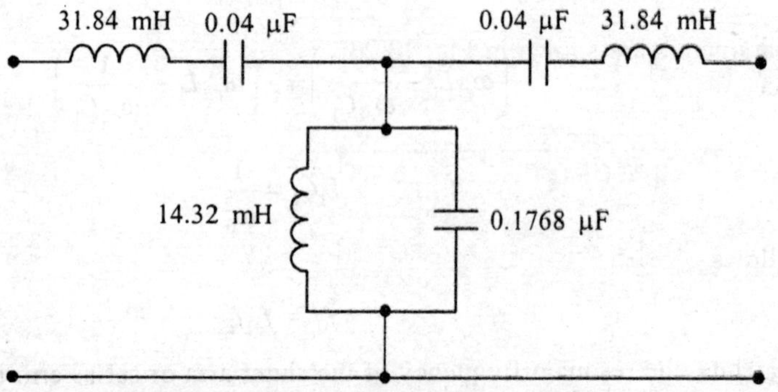

Fig. 13.22 A band pass circuit.

SOLUTION:

Given $L_1 = 63.68$ mH, $C_1 = 0.04$ μF, $L_2 = 14.32$ mH, $C_2 = 0.1768$ μF

From Eq. (13.46a),
$$L_1 = C_2 R_0^2$$

or
$$63.68 \times 10^{-3} = 0.1768 \times 10^{-6} R_0^2$$

Hence
$$R_0 = 600\ \Omega$$

From Eq. (13.46b),
$$L_1 = \frac{R_0}{\pi(f_{c2} - f_{c1})}$$

or
$$63.668 \times 10^{-3} = \frac{600}{\pi(f_{c2} - f_{c1})}$$

or
$$(f_{c2} - f_{c1}) = 3000 \text{ Hz} \qquad \textbf{Ans.}$$

EXAMPLE 13.12 Design a constant k band pass filter with cut off frequencies of 3000 Hz and 5000 Hz and nominal characteristic impedance of 600 Ω.

SOLUTION:

$$L_1 = \frac{R_0}{\pi(f_{c2} - f_{c1})} = \frac{600}{\pi(5000 - 3000)} = 95.49 \text{ mH}$$

$$C_1 = \frac{f_{c2} - f_{c1}}{4\pi R_0 f_{c1} f_{c2}} = \frac{5000 - 3000}{4\pi \times 600 \times 5000 \times 3000} = 0.017\ \mu\text{F}$$

$$L_2 = \frac{R_0(f_{c2} - f_{c1})}{4\pi f_{c1} f_{c2}} = \frac{600(5000 - 3000)}{4\pi \times 5000 \times 3000} = 0.36 \text{ mH}$$

$$C_2 = \frac{1}{\pi R_0(f_{c2} - f_{c1})} = \frac{1}{\pi \times 600 \times (5000 - 3000)} = 0.265\ \mu\text{F}$$

All the parameters are shown in Fig. 13.23.

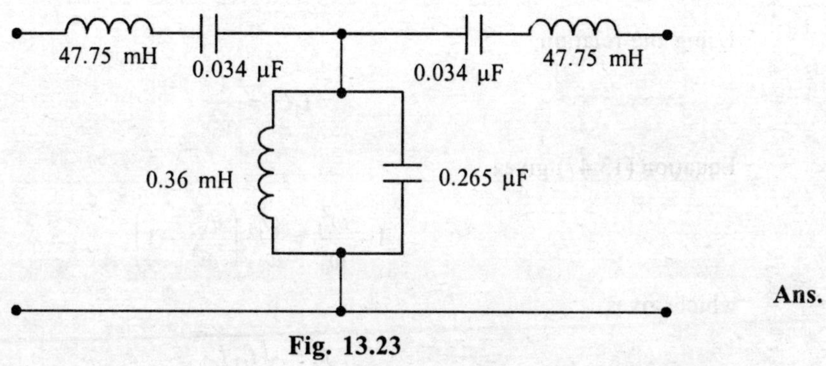

Ans.

Fig. 13.23

13.9.6 Constant k Band Elimination Filter

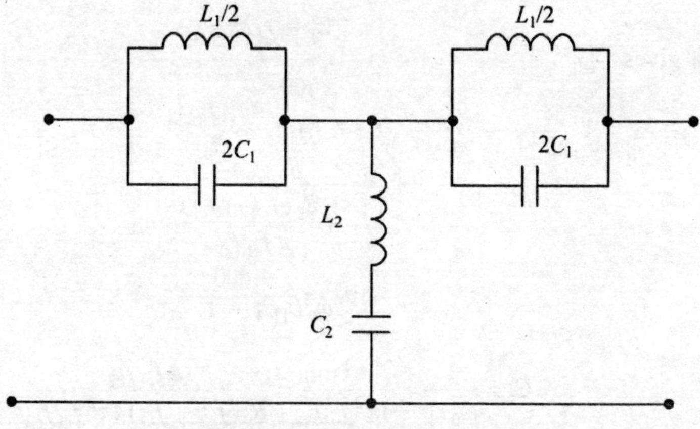

Fig. 13.24 Band stop filter.

The components $L_1C_1L_2$ and C_2 are so chosen that the shunt resonant frequency of the series are equals the series resonance frequency of the shunt arm. If this frequency is denoted by ω_0 rad/s, for the series arm $\omega_0^2L_1C_1 = 1$ and for the shunt arm $\omega_0^2L_2C_2 = 1$. Hence

$$L_1C_1 = L_2C_2 = \frac{1}{\omega_0^2}$$

Characteristic impedance R_0:

$$Z_1Z_2 = \frac{L_1}{C_2} = \frac{L_2}{C_1} = R_0^2$$

Cut off frequencies: Since Z_1Z_2 is independent of frequency, the filter is a constant **k** filter. The pass band of the filter is, therefore, given by $-1 < Z_1/(4Z_2) < 0$. It can be shown (on lines similar to the proof for cut off frequencies for the band pass filter) that

$$1 - \omega_{c1}^2L_1C_1 = \frac{\omega_{c1}}{\omega_{c2}}(\omega_{c2}^2L_1C_1 - 1) \tag{13.47}$$

Using the relation,

$$L_1C_1 = \frac{1}{\omega_0^2}$$

Equation (13.47) gives

$$1 - \frac{\omega_{c1}^2}{\omega_0^2} = \frac{\omega_{c1}}{\omega_{c2}}\left(\frac{\omega_{c2}^2}{\omega_0^2} - 1\right)$$

which gives

$$f_0 = \sqrt{f_{c1}f_{c2}}$$

13.9.7 Design of Constant k Band Elimination Filters

At lower cut off frequency f_{c1},

$$Z_1 = +j2R_0$$

which gives

$$f_{c2} - f_{c1} = \frac{\pi f_{c1}f_{c2}L_1}{R_0}$$

or

$$L_1 = \frac{R_0(f_{c2} - f_{c1})}{\pi f_{c1}f_{c2}}$$

But

$$\omega_0^2L_1C_1 = 1$$

Hence

$$C_1 = \frac{1}{\omega_0^2L_1} = \frac{1}{4\pi^2f_0^2L_1} = \frac{\pi f_{c1}f_{c2}}{R_0(f_{c2} - f_{c1})4\pi^2f_{c1}f_{c2}}$$

or
$$C_1 = \frac{1}{4\pi R_0(f_{c2} - f_{c1})}$$

$$\therefore \quad L_2 = C_1 R_0^2 = \frac{R_0}{4\pi(f_{c2} - f_{c1})}$$

and
$$C_2 = \frac{L_1}{R_0^2} = \frac{(f_{c2} - f_{c1})}{\pi R_0 f_{c1} f_{c2}}$$

13.10 m-DERIVED FILTERS

Two drawbacks of constant k filters are:

- In the attenuation band, the attenuation does not increase rapidly with frequency beyond the cut off frequency.
- In the pass band, the characteristic impedance Z_0 does not remain constant.

There are methods for improving the attenuation characteristics beyond cut off such as connecting two or more similar constant k sections in tandem but these methods do not improve the variation in characteristic impedance Z_0. The use of m-derived filter sections has resulted in very rapid attenuation rise in the attenuation band just beyond cut off. But in order to ensure constant Z_0 over the pass band terminating half sections are used in conjunction with the m-derived filter sections.

13.10.1 m-derived *T*-section

Figure 13.25(a) shows a constant k *T* type filter. While Fig. 13.25(b) shows the corresponding m-derived filter having the same characteristic impedance.

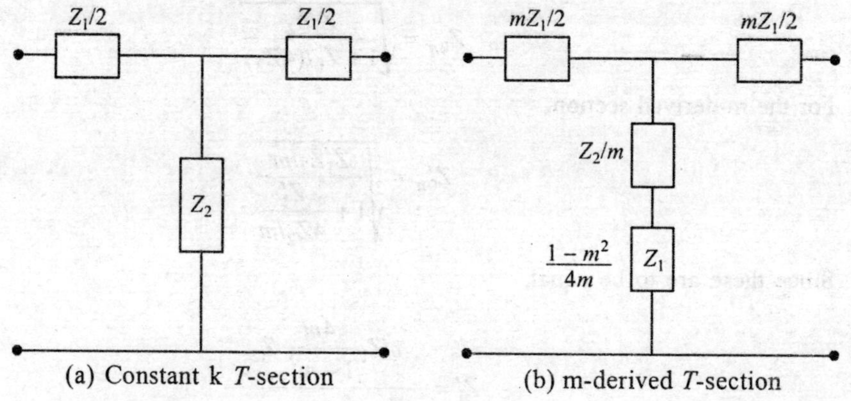

(a) Constant k *T*-section (b) m-derived *T*-section

Fig. 13.25

For the constant k section,

$$Z_{0T} = \sqrt{\frac{Z_1^2}{4} + Z_1 Z_2}$$

For the m-derived filter,

$$Z'_{0T} = \sqrt{\frac{m^2 Z_1^2}{4} + mZ_1 Z'_2}$$

Since these must be equal,

$$\frac{Z_1^2}{4} + Z_1 Z_2 = mZ_1 Z'_2$$

where

$$Z'_2 = \frac{Z_2}{m} + \frac{(1-m^2)Z_1}{4m} \tag{13.48}$$

It follows that an m-derived section may be obtained from a constant k section by multiplying the series arm impedance by m and altering the shunt impedance as given in Eq. (13.48), m having a value between 0 and 1.

13.10.2 m-derived π-section

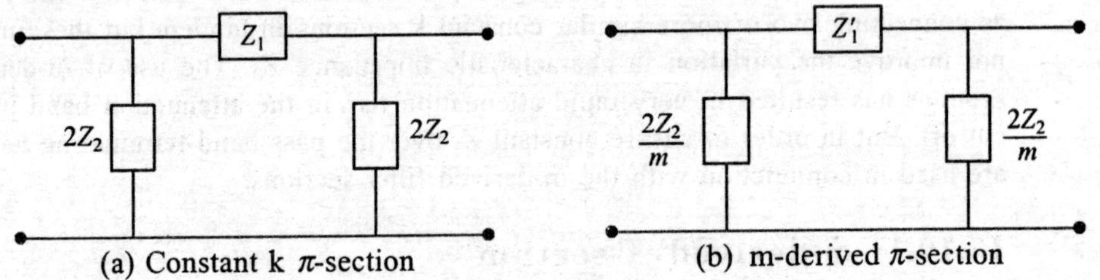

Fig. 13.26

For constant k π-section, the characteristic impedance is given by

$$Z_{0\pi} = \sqrt{\frac{Z_1 Z_2}{1 + Z_1/(4Z_2)}}$$

For the m-derived section,

$$Z'_{0\pi} = \sqrt{\frac{Z'_1 Z_2/m}{1 + \dfrac{Z'_1}{4Z_2/m}}}$$

Since these are to be equal,

$$Z'_1 = \frac{mZ_1 \dfrac{4m}{1-m^2} Z_2}{mZ_1 + \left(\dfrac{4m}{1-m^2}\right) Z_2} \tag{13.49}$$

Equation (13.49) shows that Z'_1 is a parallel combination of mZ_1 and $\dfrac{4mZ_2}{1-m^2}$.

13.10.3 m-derived Low Pass Filters

Figure 13.27 shows an m-derived low pass *T*-section.

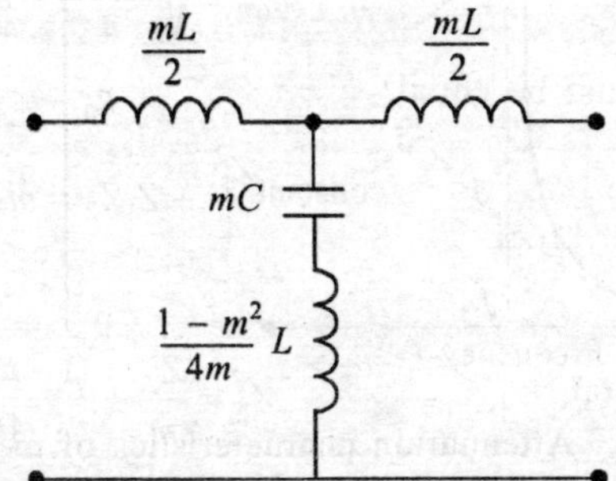

Fig. 13.27 m-derived low pass *T*-section filter.

The values of the reactances are obtained by putting $Z_1 = j\omega L$ and $Z_2 = 1/(j\omega C)$. The cut off frequency can be shown to be

$$f_c = \frac{1}{\pi\sqrt{(LC)}} \tag{13.50}$$

The resonant frequency of the shunt arm can be shown to be

$$f_\infty = \frac{1}{\pi\sqrt{(LC)}\sqrt{(1-m^2)}} \tag{13.51}$$

At this resonant frequency the transmission through the filter is zero that is the attenuation is infinity and hence denoted by f_∞.

From Eqs. (13.50) and (13.51),

$$1 - m^2 = (f_c/f_\infty)^2$$

or

$$m = \sqrt{1-(f_c/f_\infty)^2}$$

Knowing f_c and R_0, the components of the m-derived sections may be calculated.

Attenuation and Z_0 characteristics of m-derived low pass filter: In Fig. 13.28(a), the solid curve gives attenuation verses frequency curve of a m-derived low pass filter. The dotted curve shows the attenuation for the corresponding constant k filter. It is seen that the attenuation remains zero in the frequency range 0 to f_c and it becomes infinite at a frequency f_∞ slightly higher than f_c. By comparison in constant k filter, the attenuation becomes infinity only at infinite frequency. The variation of Z_0 with frequency is shown in Fig. 13.28(b).

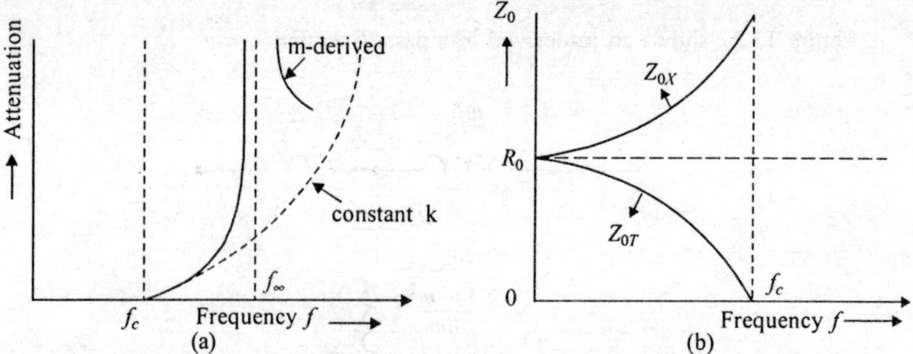

Fig. 13.28 Attenuation characteristics of m-derived low pass filters.

***π-Section*:** Figure 13.29 shows the m-derived low pass π-section.

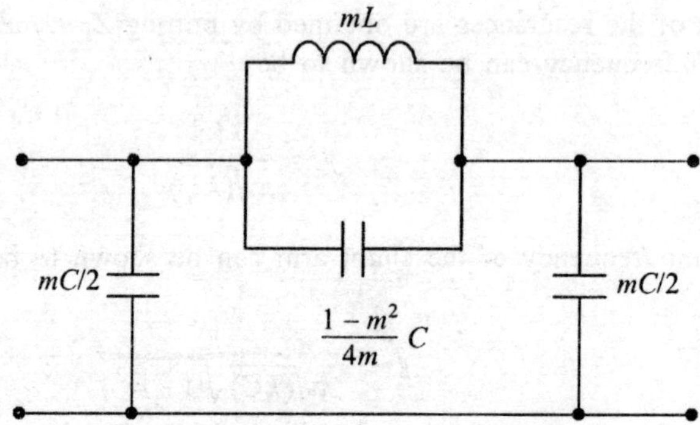

Fig. 13.29 m-derived low pass π-section filter.

Note that the series arm is a parallel LC circuit which offers infinite impedance and hence zero transmission through it at its resonant frequency. It can be shown that

$$f_\infty = \frac{1}{\pi\sqrt{(LC)}} \frac{1}{\sqrt{(1-m^2)}}$$

But cut off frequency is:

$$f_c = \frac{1}{\pi\sqrt{(LC)}}$$

and

$$f_\infty = \frac{f_c}{\sqrt{(1-m^2)}}$$

Note that in both T- and π-section m-derived filters, same equations relate f_∞ and f_c.

EXAMPLE 13.13 Design an m-derived T-section low pass filter having characteristic impedance 600 Ω cut off frequency 4800 Hz and infinite attenuation at 5000 Hz.

SOLUTION:

$$m = \sqrt{1-(f_c/f_\infty)^2}$$

$$= \sqrt{1-(4800/5000)^2}$$

$$= 0.28$$

$$R_0 = 600\ \Omega$$

Series arm inductance is:

$$L = \frac{R_0}{\pi f_c} = \frac{600}{\pi \times 4800} = 39.78 \text{ mH}$$

Shunt capacitance is

$$C = \frac{1}{\pi R_0 f_c} = \frac{1}{\pi \times 600 \times 4800} = 0.11\ \mu\text{F}$$

For the m-derived filter, the elements are:

$$mL/2 = \frac{0.28 \times 39.78 \times 10^{-3}}{2} = 5.56 \text{ mH}$$

$$mC = 0.28 \times 0.11 \times 10^{-6} = 0.03\ \mu\text{F}$$

$$\frac{1-m^2}{4m} L = \frac{1-(0.28)^2}{4 \times 0.28} \times 39.78 \times 10^{-3} = 32.73 \text{ mH}$$ **Ans.**

13.10.4 m-derived High Pass Filters

***T*-section:** Figure 13.30(a) shows a constant k high pass *T*-section filter and Fig. 13.30(b) shows the corresponding m-derived filter.

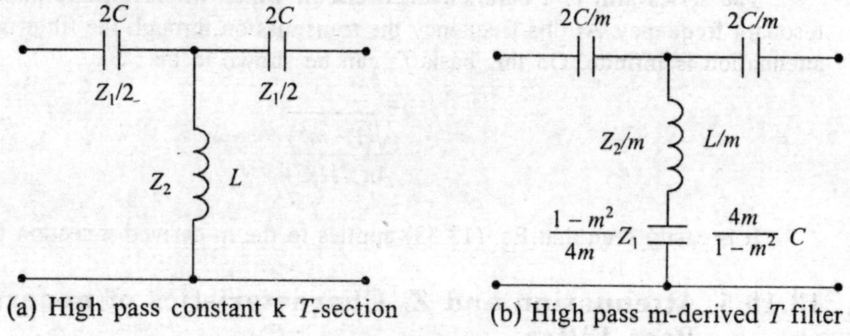

(a) High pass constant k *T*-section

(b) High pass m-derived *T* filter

Fig. 13.30

These values of the reactances are obtained from Fig. 13.25 by putting $Z_1 = 1/(j\omega C)$ and $Z_2 = j\omega L$. Note that the shunt arm is a series *LC* circuit which offers almost zero impedance at resonance. At this resonant frequency f_∞ the transmission through the filter is zero and the attenuation is infinite. It can be shown that

$$f_\infty = \frac{\sqrt{1-m^2}}{4\pi\sqrt{(LC)}} \tag{13.52}$$

But
$$f_c = \frac{1}{4\pi\sqrt{(LC)}}$$

Hence
$$f_\infty = \sqrt{1-m^2}\ f_c$$

and
$$m = \sqrt{1-(f_\infty/f_c)^2} \tag{13.53}$$

Knowing f_c, R_0 and m, the component values of the filter may be calculated.

π-section: Figure 13.31(a) shows a constant k high pass filter while Fig. 13.31(b) shows the corresponding m-derived section.

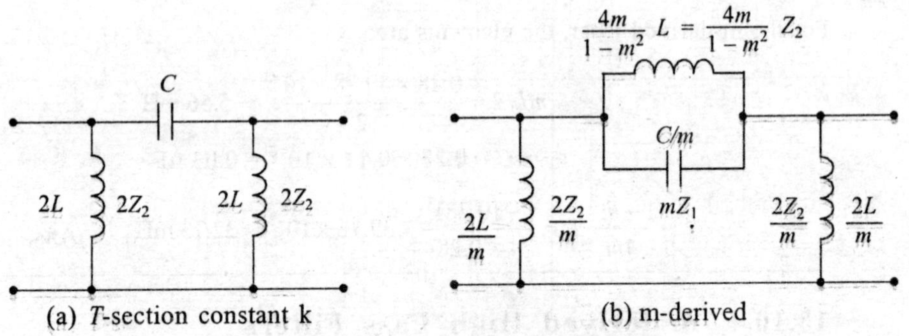

(a) *T*-section constant k (b) m-derived

Fig. 13.31 High pass m-derived filters.

The series arm is a parallel tuned circuit which offers infinite impedance at the resonant frequency. At this frequency the transmission through the filter is zero and the attenuation is infinite. On this basis f_∞ can be shown to be

$$\frac{\sqrt{(1-m^2)}}{4\pi\sqrt{(LC)}}$$

It is easily seen that Eq. (13.53) applies to the m-derived π-section filter as well.

13.10.5 Attenuation and Z_0 Characteristics of m-derived High Pass Filter

In Fig. 13.32, the solid curve gives attenuation versus frequency curve of a m-derived high pass filter. Compare the dotted curve which gives the attenuation for the corresponding constant k filter. It is seen that the attenuation remains zero in the frequency range from f_c to infinity and it becomes infinite at a frequency f_∞ slightly lower than f_c. By comparison in constant k filter the attenuation becomes infinity only at zero frequency.

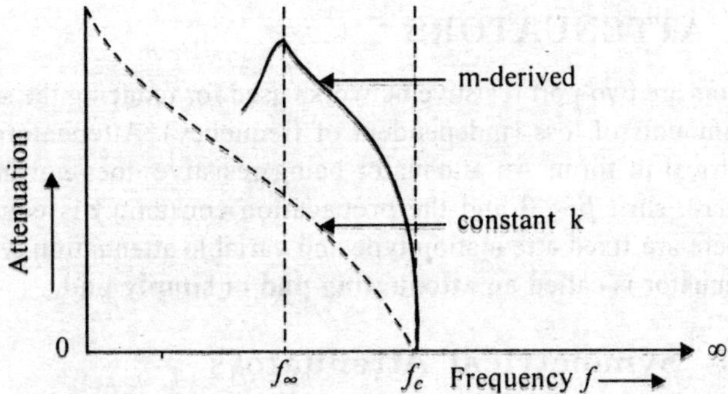

Fig 13.32 Attenuation-frequency curves of m-derived high pass filters.

The characteristics in Fig. 13.33 are the same as those for constant k section filter.

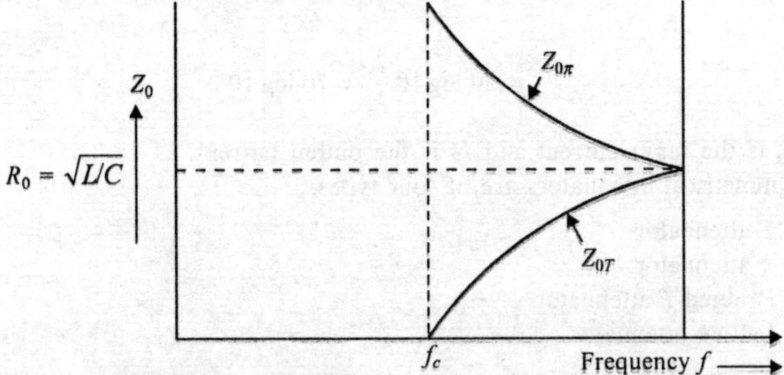

Fig. 13.33 Z_0 versus f of m-derived high pass filter.

13.11 COMPOSITE FILTERS

In order to get the best results from the use of constant k filter and the m derived filter, composite filters are used. Figure 13.34 shows the block schematic diagram of a typical composite filter.

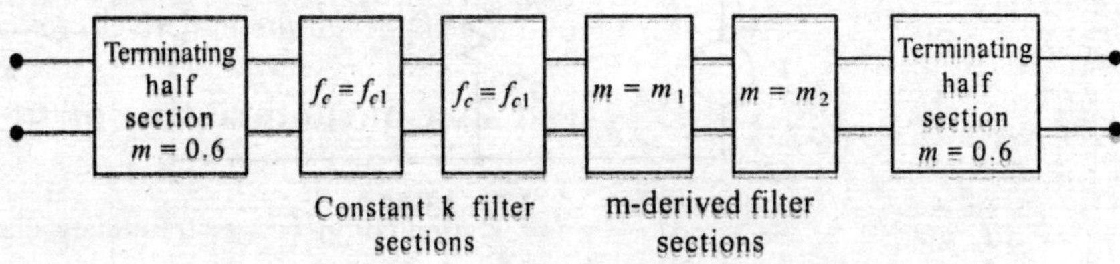

Fig 13.34 Block schematic diagram of a composite filter.

Two terminating half sections provide almost constant input and output impedances.

13.12 ATTENUATORS

Attenuators are two-port resistive networks used for reducing the signal level by introducing known amounts of loss (independent of frequency). Attenuators may be symmetrical or asymmetrical in form. An attenuator being resistive does not introduce any phase shift. Thus, phase shift $\beta = 0$ and the propagation constant γ is equal to α.

There are fixed attenuation type and variable attenuation type attenuators. The fixed type attenuator is called an **attenuating pad** or **simply pad**.

13.12.1 Symmetrical Attenuators

Symmetrical attenuators are placed between two equal impedances of value Z_0, the characteristic impedance of the attenuators. If P_1 is the input power and P_2 is the output power, Attenuation in dB is equal to

$$10 \log 10 \frac{P_1}{P_2}$$

$$= 20 \log 10 \frac{I_s}{I_R} = 20 \log 10\, N$$

where I_s is the input current and I_R is the output current.

Symmetrical attenuators are of four types:

- T-attenuator
- π-attenuator
- Bridged T-attenuator
- Lattice attenuator

Symmetrical T-attenuators: Figure 13.35 shows the symmetrical T-attenuator. An attenuator is designed for values of R_0 and attenuation. The elements R_1 and R_2 are then chosen such that these values of R_0 and attenuation are obtained. Since the T-attenuator is a symmetrical T-network using resistances only $Z_0 = R_0$ and $\gamma = \alpha$. Hence

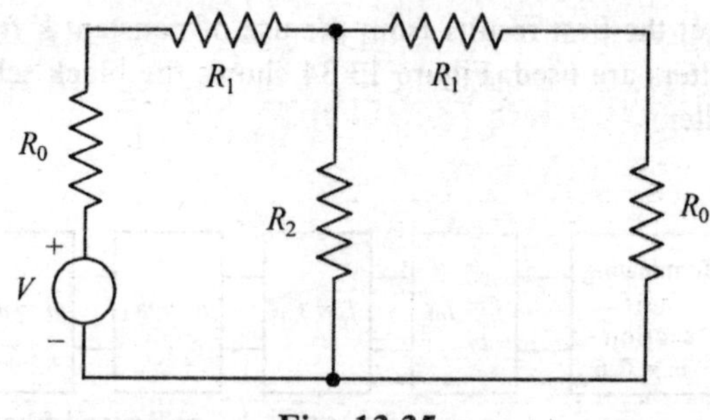

Fig. 13.35

$$R_1 = R_0 \tanh \alpha/2$$

and

$$R_2 = \frac{R_0}{\sinh \alpha/2}$$

It can be shown that

$$R_1 = R_0 \frac{N-1}{N+1} \tag{13.54}$$

$$R_2 = R_0 \frac{2N}{N^2-1} \tag{13.55}$$

These two equations form the basis for design of symmetrical *T*-attenuators.

Symmetrical π-attenuator: Figure 13.36 shows a symmetrical π-attenuator. The source voltage *V* has an internal resistance R_0 and the attenuator has a load resistance R_0 at the output port. R_0 is the characteristic resistance of the attenuator.

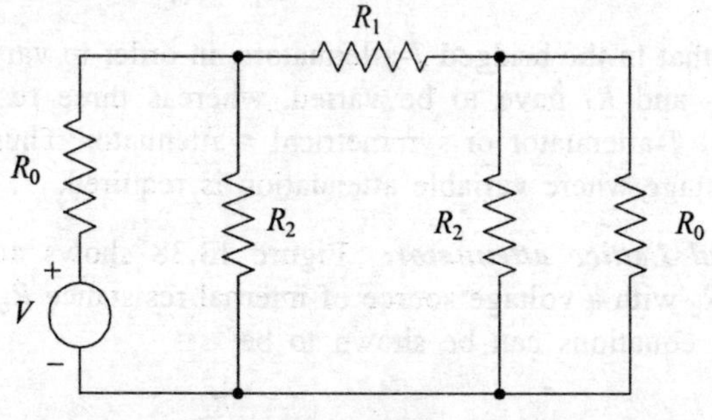

Fig. 13.36

The design of the attenuator consists in finding R_1 and R_2, given R_0 and the propagation constant γ. Here also $Z_0 = R_0$ and $\gamma = \alpha$.

It can be shown that

$$R_1 = R_0 \frac{N^2-1}{2N} \tag{13.56}$$

and

$$R_2 = R_0 \frac{N+1}{N-1} \tag{13.57}$$

Symmetrical bridged T-attenuator: Figure 13.37 shows an attenuator with a voltage source of internal resistance R_0 and with a load resistance R_0 at the output port. The

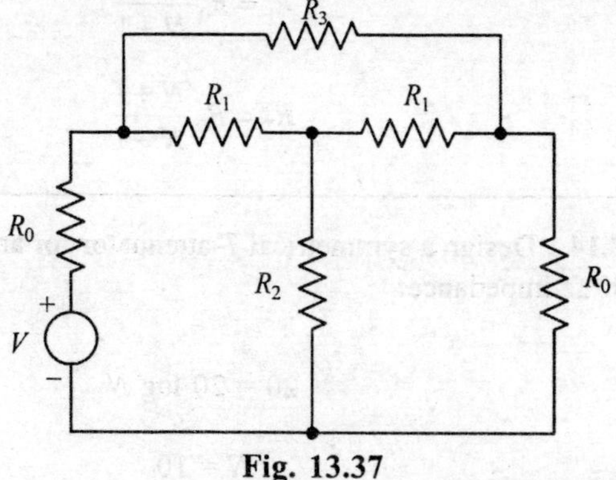

Fig. 13.37

bridged T-network may be designed to have any characteristic resistance R_0 and any attenuation by making $R_3R_2 = R_1^2 = R_0^2$. The design equations for the symmetrical bridged T-attenuator are:

$$N = 1 + \frac{R_0}{R_2}$$

$$R_3 = R_0\,(N - 1) \tag{13.58}$$

$$R_2 = \frac{R_0}{N-1} \tag{13.59}$$

$$R_1 = R_0 \tag{13.60}$$

Note that in the bridged T-attenuators, in order to vary the attenuation N only two resistors R_2 and R_3 have to be varied, whereas three resistors have to be varied in symmetrical T-attenuator or symmetrical π-attenuator. Thus, bridged T-attenuator is of great advantage where variable attenuation is required.

Symmetrical Lattice attenuator: Figure 13.38 shows attenuator with characteristic resistance R_0 with a voltage source of internal resistance R_0 and a load resistance of R_0. The design equations can be shown to be

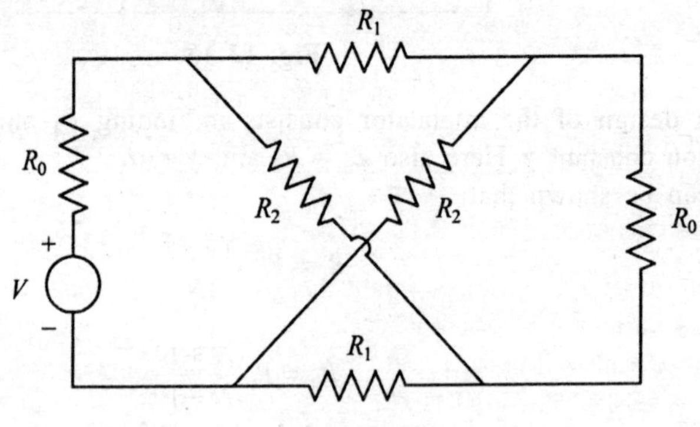

Fig. 13.38

$$R_1 = R_0 \frac{N-1}{N+1} \tag{13.61}$$

and

$$R_2 = R_0 \frac{N+1}{N-1} \tag{13.62}$$

EXAMPLE 13.14 Design a symmetrical T-attenuator for an attenuation of 20 dB to work in a line of 600 Ω impedance.

SOLUTION:

$$20 = 20 \log N$$

which gives

$$N = 10$$

The resistances are given by

$$R_1 = R_0 \frac{N-1}{N+1} = 600 \times \frac{10-1}{10+1} = 490.90\ \Omega$$

See Eqs. (13.54) and (13.55)

$$R_2 = R_0 \frac{2N}{N^2-1} = 600 \times \frac{2 \times 10}{10^2-1} = 121.21\ \Omega$$

The parameters are shown in Fig. 13.38a.

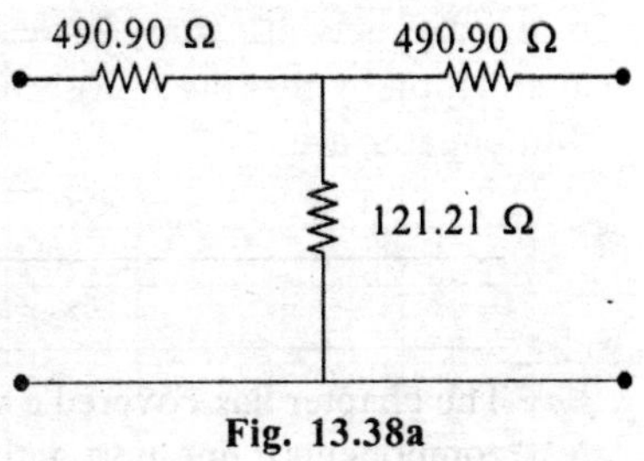

Fig. 13.38a

EXAMPLE 13.15 Design a symmetrical π-attenuator with an attenuation of 20 dB and design impedance of 600 Ω.

SOLUTION:

$$20 = 20 \log N$$

$$\therefore \quad N = 10$$

The series impedance is written as:

$$R_1 = R_0 \frac{N^2-1}{2N} = 600 \frac{10^2-1}{2 \times 10} = 2970\ \Omega$$

and

$$R_2 = R_0 \frac{N+1}{N-1} = 600 \frac{10+1}{10-1} = 733.33\ \Omega$$

For the circuit diagram, see Fig. 13.36. **Ans.**

13.12.2 Asymmetrical Attenuators

Asymmetrical attenuators are used when the attenuator is to work between two unequal impedances. The attenuator is then designed on the basis that the two unequal impedances form image impedances. Figure 13.39 shows an asymmetrical T-attenuator.

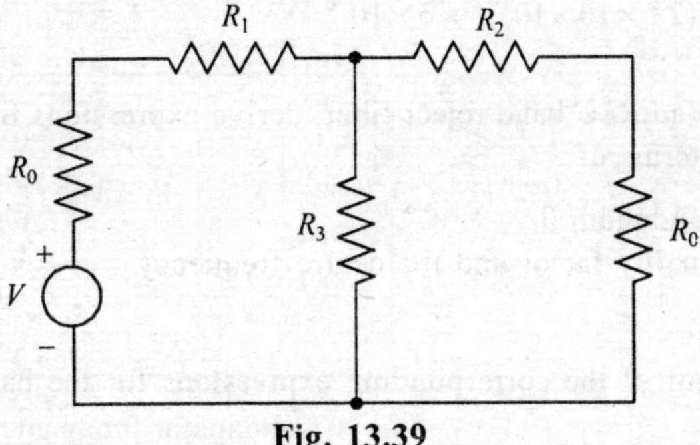

Fig. 13.39

Asymmetrical L-attenuator: This attenuator uses only two resistances. It is useful as an impedance matching circuits between two unequal impedances equal to the image impedances of the attenuator. The design equations are

$$R_1 = \sqrt{R_{i1}(R_{i1} - R_{i2})} \tag{13.63}$$

$$R_2 = \sqrt{\frac{R_{i1}R_{i2}^2}{(R_{i1} - R_{i2})}} \tag{13.64}$$

SUMMARY

The chapter has covered a wide range of wave filters—one-port passive filters (with *RLC* components), one-port active filters (using *RC* + op amp), two-port reactive filters (*LC*) and attenuators (*R* only). The design equations for the commonly used *T*-section and π-section filters have been derived and their applications illustrated with examples and problems. In other cases the equations are stated without proof. One-port filter is treated in detail in Reference 4. The subject of two-port filters including attenuators has been elaborated in Reference 3.

Solved Problems

P13.1 Compute the values for *R* and *L* for a band pass filter with centre frequency of 10 kHz and a bandwidth of 400 Hz using a 5 μF capacitor. (See Fig. 13.8)

SOLUTION:

$$\beta = \frac{1}{RC}$$

$\therefore$ $$R = \frac{1}{2\pi \times 400 \times 5 \times 10^{-6}} = 79.57\ \Omega \qquad \textbf{Ans.}$$

Now, $$\omega_0 = \sqrt{1/(LC)}$$

Hence $$L = \frac{1}{(2\pi \times 10 \times 10^3)^2 \times 5 \times 10^{-6}} = 5.06 \times 10^{-5}\ \text{H} \qquad \textbf{Ans.}$$

P13.2 For an *RLC* band reject filter, derive expressions for the cut off frequencies ω_{c1} and ω_{c2} in terms of

(a) Bandwidth β
(b) Quality factor and the centre frequency

SOLUTION:
See the proof of the corresponding expressions for the band pass filter.

P13.3 Calculate the resistance values required for the low pass filter in the circuit of Fig. P13.1 to have the transfer function

$$H(s) = \frac{-10000}{s + 5000}$$

Use a 10 μF capacitor.

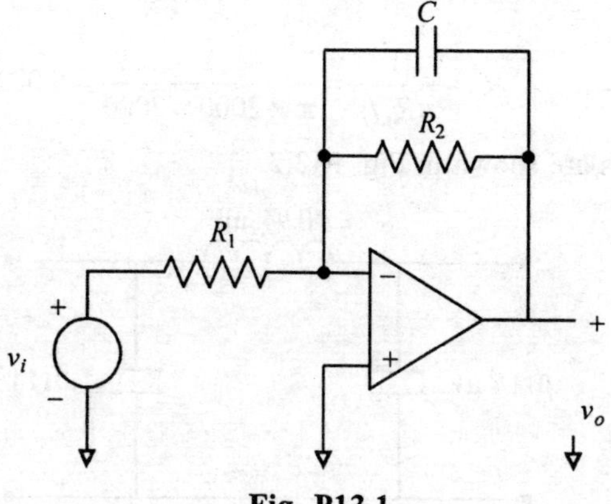

Fig. P13.1

SOLUTION:

This expression for $H(s)$ is in the standard form as given in Eq. (13.26) where

$$K = R_2/R_1, \qquad \omega_c = 1/(R_2C)$$

With a 10 μF capacitor,

$$5000 = \frac{1}{R_2 \times 10 \times 10^{-6}}$$

or $$R_2 = 20\ \Omega \qquad \textbf{Ans.}$$

Now, $$K = \frac{20}{R_1} = \frac{10000}{5000}$$

or $$2 = \frac{20}{R_1}$$

∴ $$R_1 = 10\ \Omega \qquad \textbf{Ans.}$$

P13.4 Compute the value for R_2 and C that yield a high pass filter with a gain of 1 and a cut off frequency of 1 rad/s if R_1 is 1 Ω.

SOLUTION: $$R_2 = KR_1 = 1 \times 1 = 1\ \Omega \qquad \textbf{Ans.}$$

From Eq. (13.26),

$$C = \frac{1}{1 \times 1} = 1\ \text{F} \qquad \textbf{Ans.}$$

This is the prototype high pass filter.

P13.5 Design a π section constant k low pass filter having cut off frequency of 7000 Hz and nominal characteristic impedance $R_0 = 2000\ \Omega$.

SOLUTION:

$$L = \frac{R_0}{\pi f_c} = \frac{2000}{\pi \times 7000} = 90.94\ \text{mH}$$

and $$C=\frac{1}{\pi R_0 f_c}=\frac{1}{\pi\times 2000\times 7000}=0.0227\ \mu\text{F}$$

The parameters are shown in Fig. P13.2.

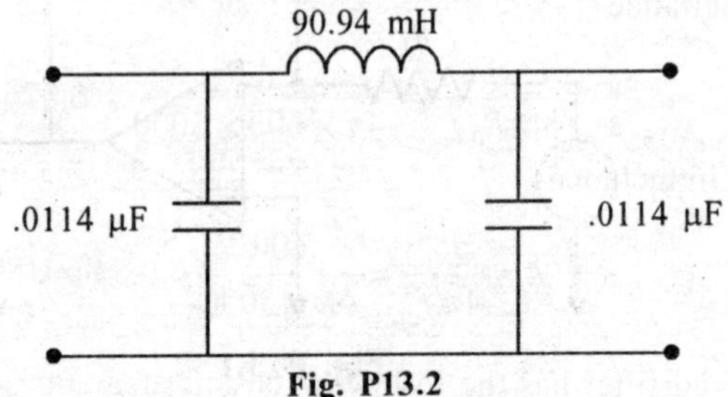

Fig. P13.2

P13.6 Design a *T*-section constant k high pass filter having cut off frequency $f_c = 4$ kHz and nominal characteristic impedance $R_0 = 600\ \Omega$. Also find its characteristic impedance at frequency of 8 kHz.

SOLUTION:

$$L=\frac{R_0}{4\pi f_c}=\frac{600}{4\pi\times 4\times 10^3}=11.94\text{ mH}$$

$$C=\frac{1}{4\pi R_0 f_c}=\frac{1}{4\pi\times 600\times 4\times 10^3}=3.31\times 10^{-8}\text{ F}$$ **Ans.**

$$Z_{0T}=R_0\sqrt{1-\left(\frac{f_c}{f}\right)^2}$$

and $$=600\times\sqrt{1-\left(\frac{4\times 10^3}{8\times 10^3}\right)^2}=519.6\ \Omega$$

The parameters are shown in Fig. P13.3.

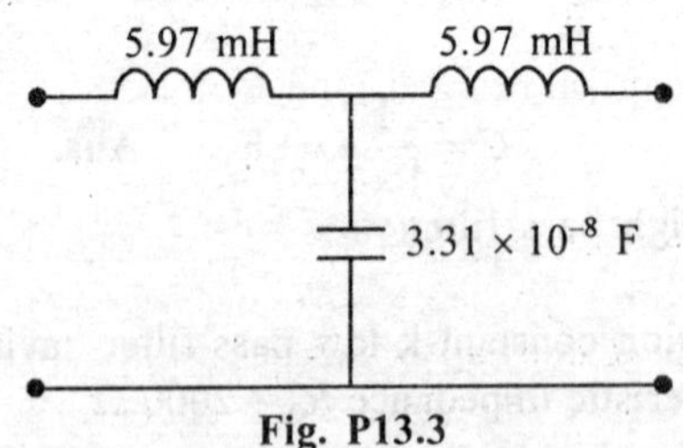

Fig. P13.3

P13.7 Design an *m*-derived π-section high pass filter having characteristic impedance 600 Ω, cut off frequency 5 kHz and infinite attenuation at 4 kHz.

SOLUTION:

$$m = \sqrt{1 - \left(\frac{f_\infty}{f_c}\right)^2} = \sqrt{1 - \left(\frac{4000}{5000}\right)^2} = 0.6$$

Series capacitance

$$C = \frac{1}{4\pi R_0 f_c} = \frac{1}{4\pi \times 600 \times 5000} = 0.026 \text{ μF}$$

Shunt arm inductance

$$L = \frac{R_0}{4\pi f_c} = \frac{600}{4\pi \times 5000} = 9.54 \text{ mH}$$

π-section m-derived filter has the following circuit elements:

$$\frac{2L}{m} = \frac{2 \times 9.54 \times 10^{-3}}{0.6} = 31.8 \text{ mH}$$

$$\frac{C}{m} = \frac{0.026 \times 10^{-6}}{0.6} = 0.043 \text{ μF}$$

$$\frac{4m}{1 - m^2} L = \frac{4 \times 0.6}{1 - (0.6)^2} \times 9.54 \times 10^{-3} = 35.77 \text{ mH}$$

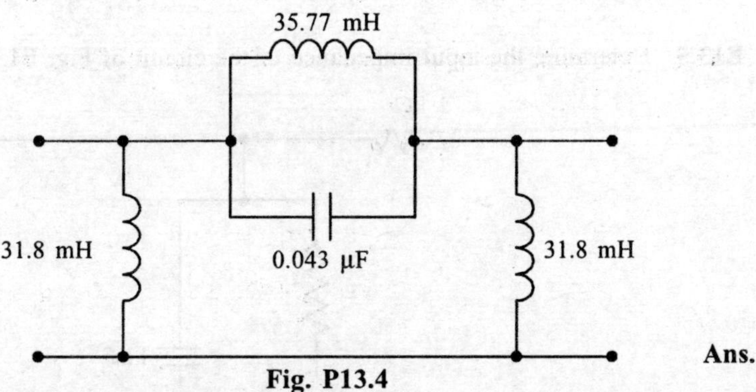

Ans.

Fig. P13.4

P13.8 Design a symmetrical bridged *T*-network with an attenuation of 40 dB and design impedance 600 Ω:

$$40 = 20 \log N$$

∴ $N = 100$

Now, $R_1 = R_0 = 600 \ \Omega$

$$R_2 = \frac{R_0}{N-1} = \frac{600}{100-1} = 6.06 \ \Omega$$

and $R_3 = R_0 (N-1) = 600 (100-1) = 59.4 \text{ kΩ}$

See Fig. 13.31 **Ans.**

Exercises

E13.1 A series *RC* high pass filter has $C = 1\ \mu$F, compute the cut off frequency for (a) $R = 100\ \Omega$ and (b) $R = 30$ kΩ.

[*Ans:* (a) 10 k rad/s (b) 33.33 rad/s]

E13.2 (a) Specify the cut off frequencies for a band pass filter in terms of its centre frequency and bandwidth.

(b) Cut off frequencies in terms of quality factor Q and the centre frequency.

[**Hint:**

(a) Derive Eqs. (13.14) and (13.15) from Eqs. (13.9) and (13.10).

(b) Derive Eqs. (13.16) and (13.17) from Eqs. (13.14) and (13.15).]

E13.3 For a series *RLC* band reject filter, derive expressions for the cut off frequencies, bandwidth β and quality factor Q for a band reject filter.

[**Hint:** Start with the equations for the transfer function magnitude and the phase angle.]

E13.4 Show that a second order delay equalizer can be realized by the function:

$$\frac{V_o}{V_i} = \frac{s^2 - as + b}{s^2 + as + b}$$

[**Hint:** The gain of the given function may be shown to be 0 dB. Such a function which has a flat gain of unity at all frequencies is referred to as all pass function.]

E13.5 Determine the input impedance of the circuit of Fig. E13.1 at resonance.

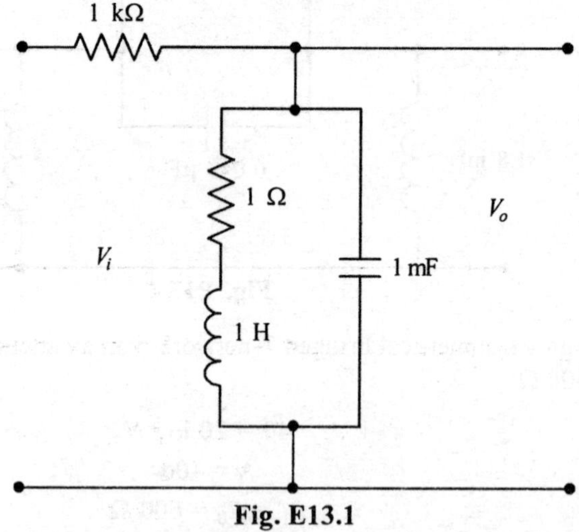

Fig. E13.1

[**Hint:**

$$\omega_r = \frac{1}{\sqrt{(LC)}} = 31.6 \text{ rad/s}$$

$$Q_s = 31.6$$

The impedance of the parallel combination

$$Z = Q_s^2 R = 1000\ \Omega$$

and
$$Z_{in} = 2000\ \Omega$$

E13.6 A filter circuit using an op amp is shown in Fig. E13.2. If $Z_1 = R_1$ and Z_{C2} corresponds to an *RC* parallel circuit with R_2 and *C*. Find *H*(*s*).

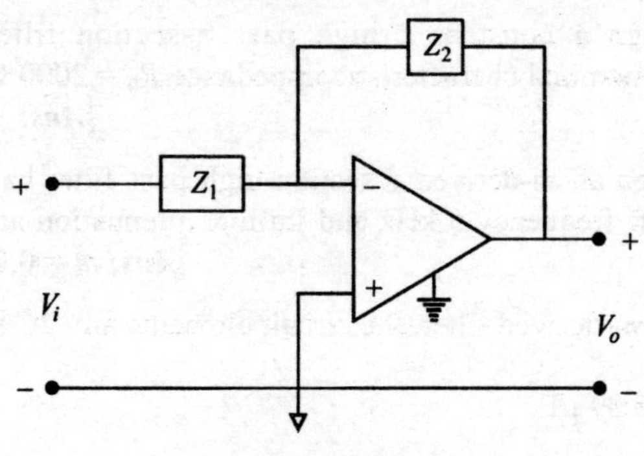

Fig. E13.2

[Hint:

$$\frac{V_o}{V_i} = \frac{Z_2}{Z_1}$$

$$Z_2 = R_2 \frac{1/(sC)}{R_2 + 1/(sC)} = \frac{R_2}{1 + sCR_2}$$

and
$$\frac{V_o}{V_i} = \frac{R_2/R_1}{1 + sCR_2}$$

E13.7 Obtain the filter characteristics of the circuit of Problem E13.1.

[Hint:
$$\frac{V_o}{V_i} = \frac{R}{1 + sL}$$

where $R = \dfrac{R_2}{R_1}$ and $CR_2 = L$.

In the frequency domain,

$$\frac{V_o}{V_i} = \frac{R}{1 + j\omega L}$$

$$V_o = V_i \frac{R}{\sqrt{(R^2 + X_L^2)}}$$

which shows that the circuit is a low pass filter.

The cut off frequency is given by

$$(0.707)^2 = \frac{R^2}{R^2 + X_L^2}$$

or

$$R = X_L$$

$$f_c = \frac{R}{2\pi L} \text{ Hz}$$

E13.8 Design a constant k high pass π-section filter having cut off frequency 7000 Hz and nominal characteristic impedance $R_0 = 2000\ \Omega$.

[***Ans:*** $L = 22.73$ mH, $C = 0.0056\ \mu$F]

E13.9 Design an m-derived *T*-section high pass filter having characteristic impedance 600 Ω, cut off frequency 5 kHz and infinite attenuation at 4 kHz.

[***Ans:*** $m = 0.6$, $C = 0.026\ \mu$F, $L = 9.54$ mH]

For the m-derived filter, the circuit elements are: $2C/m = 0.088\ \mu$F, $L/m = 15.9$ mH,

$$\frac{4m}{1 - m^2}\ C = 0.09\ \mu\text{F}$$

E13.10 Design a symmetrical lattice attenuator to have characteristic impedance of 600 Ω and attenuation of 40 dB.

[***Ans:*** $N = 100$, $R_1 = 588.1\ \Omega$, $R_2 = 612.12\ \Omega$]

Fourier Series and Fourier Transforms

14

CHAPTER CONTENTS

Waveform analysis using Fourier techniques is explained in this chapter. The circuits in previous chapters were assumed to have excitations with constant, sinusoidal or exponential waveforms. In engineering practice one comes across signals which are much broader in scope than simple dc or ac signals.

A general *periodic function* is defined by

$$f(t \pm T) = f(t) \qquad -\infty < t < \infty \tag{14.1}$$

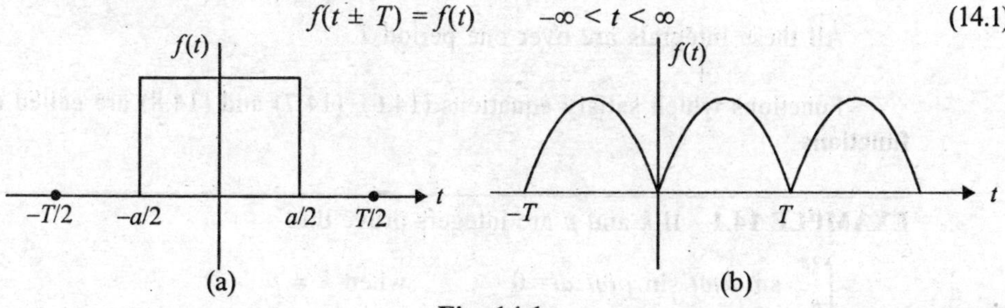

Fig. 14.1

Two examples of such periodic functions are shown in Fig. 14.1. The problem in network analysis is to find the zero-state response of a circuit when the input is such a periodic waveform. Laplace transformation could be used but the procedure is rather involved. It is in this context that the use of *Fourier series* in periodic waveforms and that of *Fourier transforms* in aperiodic waveforms, have to be viewed.

FOURIER SERIES

14.1 THE FOURIER SERIES

Consider a function $f(t)$ defined by Eq. (14.1). The frequency of $f(t)$ is given by

$$f = 1/T$$

French mathematician Fourier proved that, subject to certain conditions, such a periodic function can be expanded into the infinite trigonometric series.

$$\begin{aligned} f(t) = a_0/2 + a_1 \cos \omega t + a_2 \cos 2\omega t + \cdots \\ + b_1 \sin \omega t + b_2 \sin 2\omega t + \cdots \end{aligned} \tag{14.2}$$

$f(t)$ has been written as a sum of pure sinusoids in the fundamental frequency,

$$\omega = 2\pi/T = 2\pi f \tag{14.3}$$

and in its integral multiples, $k\omega$, called **harmonics**.

The Fourier coefficients in Eq. (14.2) for a given $f(t)$ are calculated on the basis of the following identities known as **Euler relations**.

$$\int^T .\sin k\omega t \, dt = 0 \qquad \text{for all integer } k \tag{14.4}$$

$$\int^T \cos k\omega t \, dt = 0 \qquad \text{when } t \neq 0 \tag{14.5}$$

$$\int^T \sin k\omega t \cos p\omega t \, dt = 0 \qquad \text{for all } k, p \tag{14.6}$$

$$\begin{aligned} \int^T \sin k\omega t \sin p\omega t \, dt &= 0 \qquad \text{when } k \neq p \\ &= T/2 \qquad k = p \end{aligned} \tag{14.7}$$

$$\begin{aligned} \int^T \cos k\omega t \cos p\omega t \, dt &= 0 \qquad \text{when } k \neq p \\ &= T/2 \qquad k = p \end{aligned} \tag{14.8}$$

All these integrals are over one period T.

Functions which satisfy equations (14.6), (14.7) and (14.8) are called **orthogonal functions**.

EXAMPLE 14.1 If k and p are integers prove that

$$\begin{aligned} \int_0^{2\pi} \sin k\omega t \sin p\omega t \, dt &= 0 \qquad \text{when } k \neq p \\ &= \pi \qquad k = p \end{aligned}$$

SOLUTION:

The integrand may be written as $1/2\{\cos (k - p)\omega t - \cos (k + p)\, \omega t\}$ and, therefore, if $k = p$ the given integral equals

$$1/2 \int_0^{2\pi} (1 - \cos 2p\omega t)\, dt = \left[\frac{1}{2}t - \frac{\sin 2p\omega t}{2p\omega}\right]_0^{2\pi} = \pi$$

If $k \neq p$ then

$$\int_0^{2\pi} \sin k\omega t \sin p\omega t\, dt = \frac{1}{2}\left[\frac{\sin(k-p)\omega t}{(k-p)\omega} - \frac{\sin(k+p)\omega t}{(k+p)\omega}\right]_0^{2\pi} = 0$$

The other Euler relations are easily proved on similar lines.

Calculation of the constant term $a_0/2$ in the Fourier series: Integrating both sides of Eq. (14.2) over one period,

$$\int^T f(t)\, dt = \int^T a_0/2\, dt + \int^T a_1 \cos \omega t\, dt + \cdots + \int^T b_1 \sin \omega t + \cdots$$

All the integrals on the right side, except the first one, vanish due to Eqs. (14.4) and (14.5).

Hence
$$\int^T f(t)\, dt = \frac{a_0}{2} T$$

or
$$\frac{a_0}{2} = \frac{1}{T}\int^T f(t)\, dt \tag{14.9}$$

This integral represents the net area under one cycle divided by T and the constant term is, therefore, the *average value* of $f(t)$, it is also called the **dc term**.

Calculation of a_k and b_k: Multiplying both sides of Eq. (14.2) by $\cos \omega t$ and then integrating over T,

$$\int^T f(t) \cos \omega t\, dt = \int^T a_0/2 \cos \omega t\, dt + \int^T a_1 \cos^2 \omega t\, dt + \int^T a_2 \cos 2\omega t \cos \omega t\, dt + \cdots$$

$$+ \int^T b_1 \sin \omega t \cos \omega t\, dt + \cdots$$

All the integrals on the right side, except the second one, vanish due to the Euler relations—Eqs. (14.4) to (14.8).

Hence

$$\int^T f(t) \cos \omega t\, dt = \frac{a_1 T}{2}$$

or

$$a_1 = \frac{2}{T}\int^T f(t) \cos \omega t\, dt \tag{14.10}$$

In general,

$$a_k = 2\int^{T} \frac{f(t)}{T} \cos k\omega t \, dt \qquad k = 0, 1, 2, \dots \tag{14.11}$$

Similarly,

$$b_k = 2/T\int^{T} f(t) \sin k\omega t \, dt \qquad k = 1, 2, 3, \dots \tag{14.12}$$

$a_0/2$ is introduced in the series in Eq. (14.9) so that Eq. (14.11) giving the general term a_k will be applicable for $k = 0$ as well. An alternate form of the evaluation integrals, with the variable $\theta = \omega t$ and the corresponding period 2π radians is:

$$a_k = 1/\pi \int_0^{2\pi} f(\theta/\omega) \cos k\theta \, d\theta \tag{14.11a}$$

$$b_k = 1/\pi \int_0^{2\pi} f(\theta/\omega) \sin k\theta \, d\theta \tag{14.12a}$$

14.1.1 Convergence of the Fourier Series—Dirichlet Conditions

There are certain conditions to be satisfied by the periodic function $f(t)$, if it is to be expressed as a Fourier series.

- $f(t)$ may have at most, a finite number of discontinuities over one period.
- $f(t)$ may have, at most, a finite number of maximum and minimum over one period.
- $f(t)$ must be absolutely integrable over one period that is to say

$$\int^{T} |f(t)| \, dt = M \qquad M < \infty$$

Most engineering functions satisfy these Dirichlet conditions. The error in approximating a periodic function by a finite Fourier series is considered in Reference 5.

EXAMPLE 14.2 Find the Fourier series for the saw tooth waveform shown in Fig. 14.2.

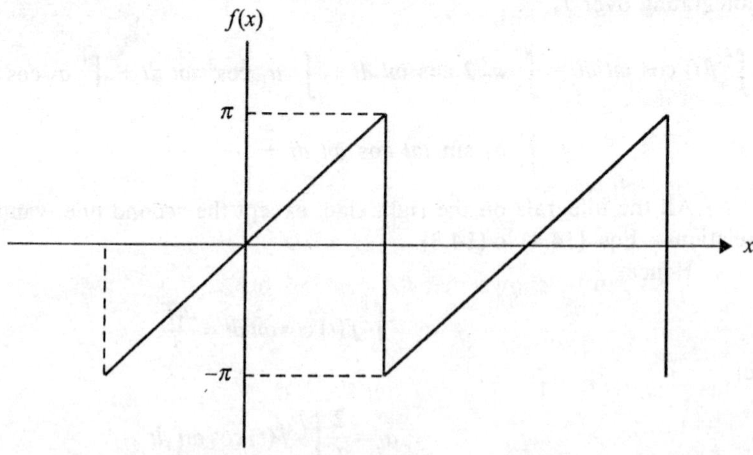

Fig. 14.2

SOLUTION:

$$f(x) = x \qquad -\pi < x < \pi$$

and
$$f(x + 2\pi) = f(x)$$

By inspection, $a_0/2 = 0$

$$a_k = \frac{2}{2\pi}\int_{-\pi}^{\pi} x \cos kx\, dx = \frac{1}{\pi}\left[\frac{x \sin kx}{k} + \frac{\cos kx}{k^2}\right]_{-\pi}^{\pi} = 0$$

$$b_k = \frac{2}{2\pi}\int_{-\pi}^{\pi} x \sin kx\, dx = \frac{1}{\pi}\left[\frac{\sin kx}{k^2} - \frac{x \cos kx}{k}\right]_{-\pi}^{\pi} = \frac{2(-1)^{k+1}}{k}$$

In particular $b_1 = \dfrac{1}{\pi}\left[\dfrac{\pi}{1} + \dfrac{\pi}{1}\right] = 2$

$$b_2 = \frac{1}{\pi}\left[\frac{-\pi}{2} - \frac{\pi}{2}\right] = -1$$

$$b_3 = \frac{1}{\pi}\left[\frac{\pi}{3} + \frac{\pi}{3}\right] = \frac{2}{3}$$

and $f(x) = 2 \sin x - \sin 2x + 2/3 \sin 3x + \cdots$

$$= 2\left[\sin x - \frac{\sin 2x}{2} + \frac{\sin 3x}{3} - \cdots\right]$$ **Ans.**

14.1.2 Symmetry Conditions

When $f(t)$ satisfies certain symmetry conditions, the calculation of Fourier coefficients is considerably simplified.

1. If $f(t)$ is an even function

$$f(t) = f(-t) \tag{14.13}$$

   It can be shown that $b_k = 0$.

2. If $f(t)$ is an odd function,

$$f(t) = -f(-t)$$

   It can be shown that $a_k = 0$, as in Example 14.2. (14.14)

3. Half wave symmetry,

$$f(t \pm T/2) = -f(t) \qquad \text{(as in Fig. 14.3)}$$

   It can be shown that $f(t)$ contains only odd harmonic terms, that is,

$$a_k = b_k = 0, \quad \text{when } k \text{ is even.}$$

and
$$a_k = 4/T \int_0^{T/2} f(t) \cos k\omega t\, dt$$

$$b_k = 4/T \int_0^{T/2} f(t) \sin k\omega t\, dt \qquad \text{when } k \text{ is odd.} \tag{14.15}$$

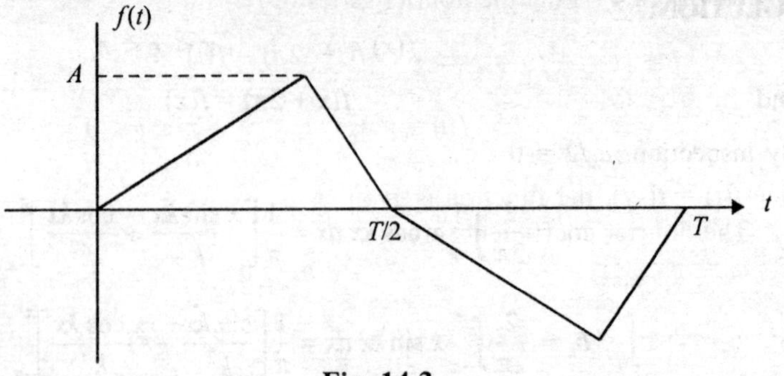

Fig. 14.3

EXAMPLE 14.3 Find the Fourier series that represents the numerical value of $|\sin t|$.

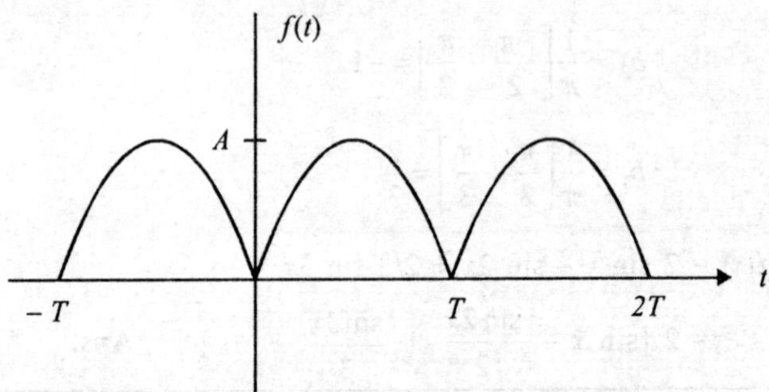

Fig. 14.4

SOLUTION:

The signal may be represented by

$$f(t) = A\,|\sin t|$$

The period $T = \pi$ and the fundamental frequency is $\omega = 2$.

$$b_k = \frac{2}{\pi}\int_0^{\pi} f(t)\sin 2kt\,dt = 0$$

$$a_0 = 2\frac{A}{\pi}\int_0^{\pi}\sin t\,dt = \frac{4A}{\pi}$$

$$a_k = \frac{2}{\pi}\int_0^{\pi} f(t)\cos 2kt\,dt$$

$$= \frac{1}{1-4k^2}\frac{4A}{\pi}$$

Thus, the Fourier series for the rectified sine wave is:

$$f(t) = \frac{4A}{2\pi}\left[2 + \sum_{k=1}^{\infty}\frac{2}{1-4k^2}\cos 2kt\right] \qquad \textbf{Ans.}$$

EXAMPLE 14.4 Find the Fourier series of $f(t)$ if

$$f(t + 2\pi) = f(t)$$

and

$$f(t) = \pi + t \qquad -\pi \le t < 0$$
$$= \pi - t \qquad 0 \le t < \pi$$

Since $f(t) = f(-t)$, the function is even.

The Fourier coefficients are:

$a_0 = \pi \qquad b_k = 0$

$$a_k = 2/\pi \int_0^{\pi} (\pi - t) \cos kt \, dt = \frac{2}{\pi k^2} (1 - \cos k\pi)$$

$$= \begin{cases} 0 & n \text{ even} \\ \dfrac{4}{\pi k^2} & n \text{ odd} \end{cases}$$

and the Fourier series of $f(t)$ is:

$$\pi/2 + 4/\pi \sum \frac{1}{k^2} \cos kx \qquad \textbf{Ans.}$$

where k is odd.

14.1.3 Cosine Form of the Fourier Series

The sine and cosine terms of like frequencies in Eq. (14.2) can be combined into a single cosine (or sine) term with a phase angle.

$$a_k \cos k\omega t + b_k \sin k\omega t = \sqrt{a_k^2 + b_k^2}\left(\frac{a_k \cos k\omega t}{\sqrt{a_k^2 + b_k^2}} + \frac{b_k \sin k\omega t}{\sqrt{a_k^2 + b_k^2}}\right)$$

$$= A_k \{\cos \theta_k \cos k\omega t + \sin \theta_k \sin k\omega t\}$$

$$= A_k \{\cos (k\omega t - \theta_k)\} \tag{14.16}$$

where $A_k = \sqrt{a_k^2 + b_k^2}$, is the amplitude and $\theta_k = \tan^{-1} \dfrac{b_k}{a_k}$ is the phase angle, as indicated in Fig. 14.5.

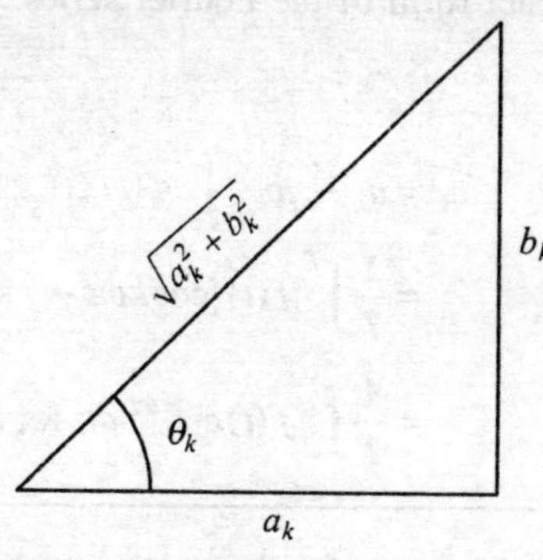

Fig. 14.5

The Fourier cosine series is then

$$f(t) = A_0 + A_1 \cos(\omega t - \theta_1) + A_1 \cos(2\omega t - \theta_2) + \cdots + A_k \cos(k\omega t - \theta_k) + \cdots \quad (14.17)$$

The dc term is $A_0 = a_0/2$.

From Eq. (14.16), the sine form of the Fourier series is:

$$f(t) = A_0 + \sum_{k=1}^{\infty} \infty\, A_k \sin(k\omega t + \pi/2 - \theta_k)$$

14.2 EXPONENTIAL FORM OF FOURIER SERIES

Expressing cos $k\omega t$ and sin $k\omega t$ in terms of complex exponentials, using Euler's identity, the Fourier series can be written as:

$$f(t) = \frac{a_0}{2} + \sum_{k=1}^{\infty} \left[a_k \frac{(e^{jk\omega t} + e^{-jk\omega t})}{2} + b_k \frac{(e^{jk\omega t} - e^{-jk\omega t})}{2j} \right]$$

$$= \frac{a_0}{2} + \sum_{k=1}^{\infty} \left[\frac{(a_k - jb_k)}{2} e^{jk\omega t} + \frac{(a_k + jb_k)}{2} e^{-jk\omega t} \right]$$

Define $\quad c_k = \dfrac{a_k - jb_k}{2} \qquad |c_k| = \dfrac{1}{2}\sqrt{a_k^2 + b_k^2} = \dfrac{1}{2} A_k$

$$c_{-k} = \frac{a_k + jb_k}{.2}$$

$$c_0 = a_0/2$$

The complex form of the Fourier series is then

$$f(t) = c_0 + \sum_{k=1}^{\infty} [c_k e^{jk\omega t} + c_{-k} e^{-jk\omega t}] \quad (14.18)$$

or

$$f(t) = \sum_{k=-\infty}^{\infty} c_k e^{jk\omega t}$$

which is a very compact form of the Fourier series.

Explicit form of c_k:

$$c_k = a_k - jb_k$$

$$= \frac{1}{T}\int^{T} f(t)\,[\cos k\omega t - j \sin k\omega t]\, dt$$

$$= \frac{1}{T}\int^{T} f(t) e^{-jk\omega t}\, dt \quad \text{for all } k \quad (14.19)$$

EXAMPLE 14.5 Determine c_k for the square wave shown in Fig. 14.6.

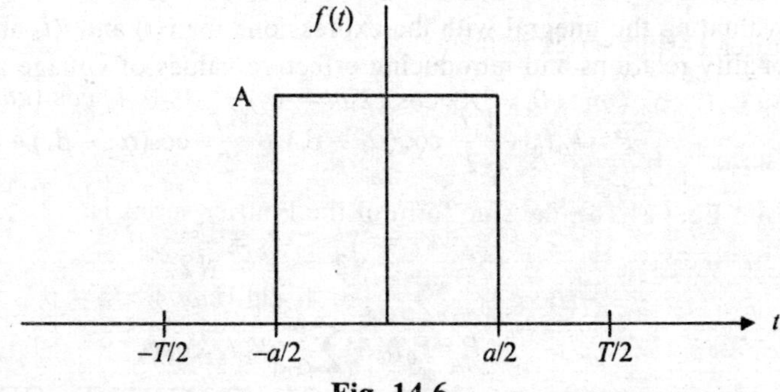

Fig. 14.6

$f(t)$ is an even function with $b_k = 0$. Hence

$$c_k = a_k/2$$

or
$$c_k = \frac{A}{2}\frac{\sin k\pi/2}{k\pi/2}$$

Now,

$$c_0 = \frac{A}{2}\frac{\sin k\omega/2}{k\omega/2}\bigg|_{k=0} = \frac{A}{2}\cdot\frac{\pi}{2}\frac{\cos k\pi/2}{\pi/2}\bigg|_{k=0} = \frac{A}{2}$$

by L'Hospital rule for indeterminate forms

$$c_1 = \frac{A}{2}\frac{\sin \pi/2}{\pi/2} = \frac{A}{\pi}$$

$$c_2 = \frac{A}{2}\frac{\sin \pi}{\pi} = 0$$

$$c_3 = \frac{A}{2}\frac{\sin 3\pi/2}{3\pi/2} = \frac{A}{2}\frac{2}{3\pi}(-1) = \frac{-A}{3\pi}$$

Similarly, the other coefficients may be determined.

14.3 POWER RELATIONSHIPS

Power in a network when voltage and current are periodic and expressed in their Fourier series. Using the cosine form,

$$v(t) = V_0 + V_{m1} \cos(\omega t - \alpha_1) + V_{m2} \cos(2\omega t - \alpha_2) + \cdots$$
$$i(t) = I_0 + I_{m1} \cos(\omega t - \beta_1) + I_{m2} \cos(2\omega t - \beta_2) + \cdots$$

Instantaneous power $p(t) = v(t)\, i(t)$

Average power $P = \frac{1}{T}\int_0^T p(t)\,dt$

$$= \frac{1}{T}\int_0^T v(t)\,i(t)\,dt$$

Evaluating the integral with the expressions for $v(t)$ and $i(t)$ and making use of the orthogonality relations and introducing effective values of voltage and current, we have

$$P = V_0 I_0 + \frac{V_1 I_1}{2} \cos(\alpha_1 - \beta_1) + \frac{V_2 I_2}{2} \cos(\alpha_2 - \beta_2) + \cdots$$

where
$$V_k = \frac{V_{mk}}{\sqrt{2}} \qquad I_k = \frac{I_{mk}}{\sqrt{2}}$$

Hence
$$P = V_0 I_0 + \sum_{k=1}^{\infty} V_k I_k \cos\phi_k \tag{14.20}$$
$$= P_0 + \sum_{k=1}^{\infty} P_k$$

where P_k is the average power of the kth harmonic and $\cos \phi_k = \cos(\alpha_k - \beta_k)$ is the power factor of the kth harmonic.

Thus, the average power equals the sum of the average powers of all the harmonics. It may be noted *different* harmonics of voltage and current do not contribute to the average power.

14.4 FOURIER SERIES IN CIRCUIT ANALYSIS

Response of LTI systems by Fourier series: If the input to a linear network is periodic, superposition and phasors may be used to calculate the steady state response.

EXAMPLE 14.6 A periodic current source is applied to a parallel RC circuit shown in Fig. 14.7. Calculate the response $v(t)$.

$$i(t) = 5 + 3\cos(100t + 45°) + 2\cos(200t - 10°)$$

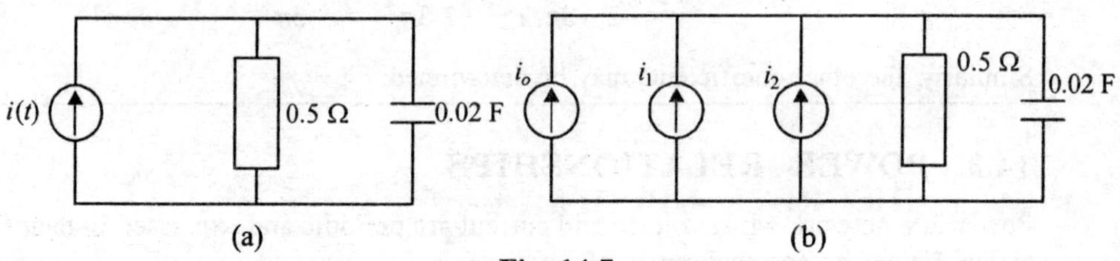

Fig. 14.7

SOLUTION:
Each source is sinusoidal and hence

$$V = ZI$$

where V is the phasor voltage, I is the phasor current, and Z is the driving point impedance of the RC circuit, which is given by

$$Z = \frac{1}{Y} = \frac{1}{2 + j\omega(0.02)}.$$

With the dc source,

$$V_0 = \frac{1}{2}.5 = 2.5$$

With the first harmonic source,

$$\omega = 100,\ I_1 = 3 \angle 45°$$

and
$$V_1 = \frac{1}{2+2j} 3 \angle 45° = \frac{3 \angle 45°}{2.828 \angle 45°} = 1.06 \angle 0°$$

Thus,

$$v_1(t) = 1.06 \cos 100t$$

With the second harmonic,

$$\omega = 200,\quad I_2 = 2 \angle -10°$$

and

$$V_2 = \frac{1}{2+j4} 2 \angle -10° = \frac{2 \angle -10°}{4.472 \angle 63°} = 0.447 \angle -73°$$

Thus,

$$v_2\,(t) = 0.447 \cos\,(200t - 73°)$$

By superposition, the output voltage is: $v(t) = 2.5 + 1.06 \cos 100t + 0.447 \cos (200t - 73°)$

Note that it is periodic with same period as the input.

EXAMPLE 14.7 Find the average power for the voltage and current given in Example 14.6.

SOLUTION:

$$\begin{aligned} P_{av} &= v(t)\, i(t) \\ &= 5 \times 2.5 + 3/\sqrt{2} \times 1.06/\sqrt{2} \cos 45° + 2/\sqrt{2} \times 0.477/\sqrt{2} \cos 63° \\ &= 12.5 + 1.59 \cos 45° + 0.477 \cos 63° \\ &= 13.84 \text{ W} \quad \textbf{Ans.} \end{aligned}$$

14.5 DISCRETE SPECTRA

Consider the cosine form of the Fourier series:

$$f(t) = A_0 + \sum_{k=1}^{\infty} A_k \cos\,(k\omega t - \theta_k)$$

The amplitudes A_0, A_1, ..., and the angles θ_0, θ_1, ..., of the various harmonics can be plotted versus $k\omega$. These plots are called the *discrete amplitude spectrum* and the ***discrete phase spectrum*** for $f(t)$.

The discrete magnitude spectrum and the phase angle spectrum are defined also for the exponential form of the Fourier series.

EXAMPLE 14.8 Find the amplitude and phase spectra of the square wave shown in Fig. 14.8(a).

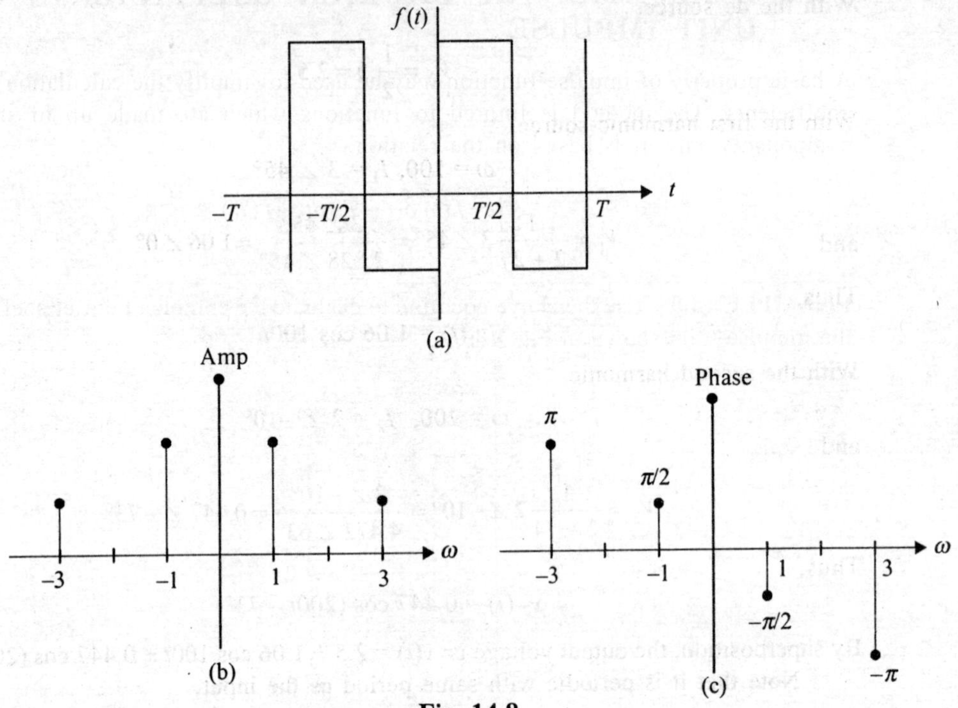

Fig. 14.8

SOLUTION:

$f(t)$ is an odd function and $f(t - T/2) = -f(t)$. The series has therefore only odd harmonics.

From Eq. (14.19), the coefficients of the complex Fourier series are:

$$c_k = \frac{1}{T}\int_0^{T/2} Ae^{-k\omega t}dt - \frac{1}{T}\int_{T/2}^{T} A\, e^{-jk\omega t}\, dt$$

$$= \frac{A}{jk\omega T}\left(1 - 2e^{-jk\omega T/2} + e^{-jk\omega T}\right)$$

Since $k\omega T = k2\pi$,

$$c_k = \frac{A}{j2k\pi}(1 - 2e^{-jk\pi} + e^{-j2k\pi})$$

$$= \frac{2A}{jk\pi} \qquad k \text{ odd}$$

$$= 0 \qquad k \text{ even}$$

The amplitude spectrum and the phase spectrum of the square wave are given in Figs. 14.8(b) and (c). **Ans.**

14.6 EVALUATION OF FOURIER COEFFICIENTS USING UNIT IMPULSE

A basic property of impulse function may be used to simplify the calculation of Fourier coefficients. The method is limited to functions which are made up of straight-line components only. It is based on the relation:

$$\int_{-\infty}^{\infty} f(t)\,\delta(t-T_1)\,dt = f(T_1)$$

EXAMPLE 14.9 Use the above equation to evaluate the complex Fourier coefficients for the impulse train shown in Fig. 14.9.

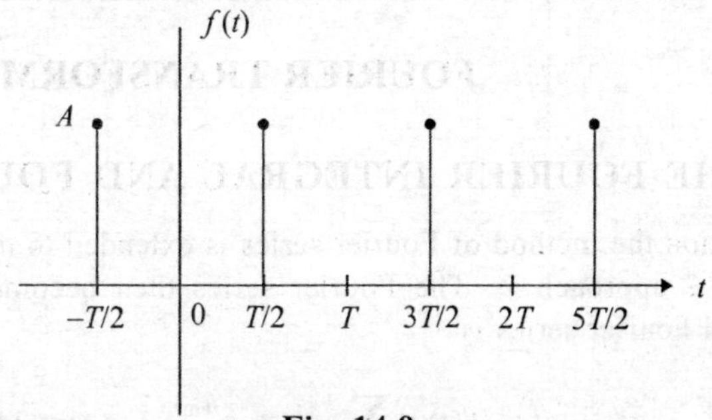

Fig. 14.9

SOLUTION:

$$c_k = \frac{A}{T}\int_0^T \delta(t - T/2)e^{-jk\omega t}dt = \frac{A}{T}e^{-jk\omega T/2} \qquad \textbf{Ans.}$$

It is seen that the complex Fourier coefficients for impulse functions are obtained by simply substituting the time at which the impulses occur into the expression $e^{-jk\omega t}$.

Inverting the Fourier series: The process of obtaining time function $f(t)$ from discrete spectra is illustrated in Example 14.10.

EXAMPLE 14.10 The amplitude and the phase spectrum of a periodic signal $f(t)$ are shown in Fig. 14.10. Write an expression for $f(t)$ in terms of trigonometric functions.

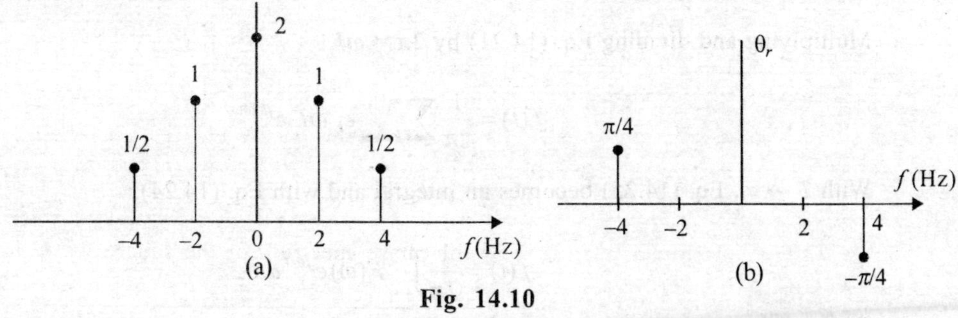

Fig. 14.10

SOLUTION:

The function $$f(t)=\sum_{k=-\infty}^{\infty} c_k e^{jk\omega t}, t'<t<t'+T$$

where $$\omega = 2\pi f = 2\pi \times 2$$

The Fourier coefficients c_k are:

$$c_{-4} = ½e^{j45},\quad c_{-2} = 1e^{j0},\quad c_0 = 2,\quad c_2 = 1.e^{j0}, \text{ and } c_4 = ½e^{j(-45)}$$

Therefore,

$$f(t) = ½\ e^{j45}\ e^{-j2\pi.4t} + 1\ e^{j0}\ e^{-j2\pi 2t} + 2 + 1e^{j2\pi 2t} + ½e^{-j45}\ e^{2\pi 4t}$$

$$= 2 + 2\cos 4\pi t + \cos(8\pi t - \pi/4) \qquad \textbf{Ans.}$$

FOURIER TRANSFORMS

14.7 THE FOURIER INTEGRAL AND FOURIER TRANSFORM

In this section the method of Fourier series is extended to *aperiodic functions* by letting the period T approach ∞. The Fourier series then becomes the Fourier integral. The exponential Fourier series is:

$$f(t)=\sum_{k=-\infty}^{\infty} c_k e^{jk\omega t} \qquad \omega = 2\pi/T \tag{14.21}$$

and

$$c_k = \frac{1}{T}\int_{-T/2}^{T/2} f(t)\, e^{-jk\omega t}\, dt \tag{14.22}$$

It follows

$$c_k T = \int_{-T/2}^{T/2} f(t)\, e^{-jk\omega t}\, dt \tag{14.23}$$

The amplitude and phase spectra have been found to consist of discrete lines. As period T becomes larger, the spacing between the adjacent harmonic lines in the spectrum becomes smaller. For aperiodic signals, if $T \to \infty$, the discrete spectrum becomes *continuous* and the integral in Eq. (14.23) becomes

$$\int_{-\infty}^{\infty} f(t)e^{-j\omega t}\, dt = F(\omega) \tag{14.24}$$

Multiplying and dividing Eq. (14.21) by $2\pi = \omega T$,

$$f(t)=\frac{1}{2\pi}\sum_{k=-\infty}^{\infty} c_k\, \omega T\, e^{jk\omega t} \tag{14.25}$$

With $T \to \infty$, Eq. (14.25) becomes an integral and with Eq. (14.24),

$$f(t)=\frac{1}{2\pi}\int_{-\infty}^{\infty} F(\omega)\, e^{j\omega t}\, d\omega \tag{14.26}$$

Equation (14.24) defines the **Fourier integral or transform** of $f(t)$ written as:

$$F\, f(t) = F(\omega)$$

and Eq. (14.26) is the **inverse Fourier transform** of $F(\omega)$ written as:

$$F^{-1}\, F(\omega) = f(t)$$

$F(\omega)$ is seen to be the extension of Eq. (14.22) for aperiodic function.

The heuristic approach in this derivation is considered to be quite satisfactory. As in the case of Fourier series, Dirichlet conditions (see section 14.1.1) apply for Fourier transforms.

Continuous spectra:

$$F(\omega) = R(\omega) + j\, X(\omega)$$

or

$$F(\omega) = |\, F(\omega)\, |\, e^{j\phi(\omega)}$$

where $|\, F(\omega)\, |$ is the *continuous amplitude spectrum* of $f(t)$ and $\phi(\omega)$ is the *continuous phase spectrum* of $f(t)$.

EXAMPLE 14.11 Find the continuous amplitude spectrum and the continuous phase spectrum of $f(t) = e^{-at}u(t)$.

Its Fourier transform is:

$$F(\omega) = \int_{-\infty}^{\infty} e^{-at}u(t)e^{-j\omega t}dt = \int_{0}^{\infty} e^{-at}\, e^{-j\omega t}dt = \frac{1}{a + j\omega}$$

SOLUTION:

Its magnitude spectrum is:

$$|F(\omega)| = \frac{1}{\sqrt{a^2 + \omega^2}}$$

and its phase spectrum is:

$$\phi\,(\omega) = -\tan^{-1} \omega/a$$

The plots are given in Fig. 14.11.

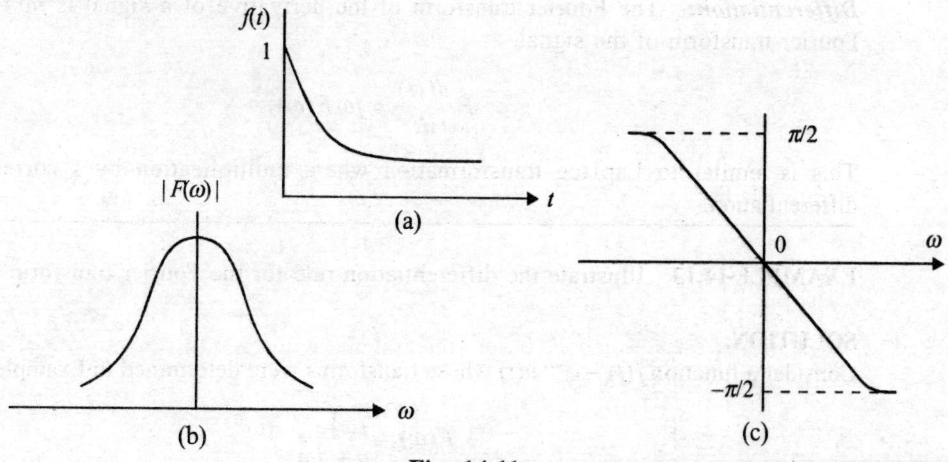

Fig. 14.11

EXAMPLE 14.12 Find the amplitude spectrum of $f(t) = A\delta(t)$.

SOLUTION:

$$F(\omega) = \frac{1}{2\pi}\int_{-\infty}^{\infty} A\,\delta(t)\,e^{-j\omega t}\,dt = Ae^{-j\omega t}\Big|_{\omega=0} = 1$$

Using the sampling property of the impulse function, the amplitude spectrum is

$A(\omega) = A$ (See Fig. 14.12) **Ans.**

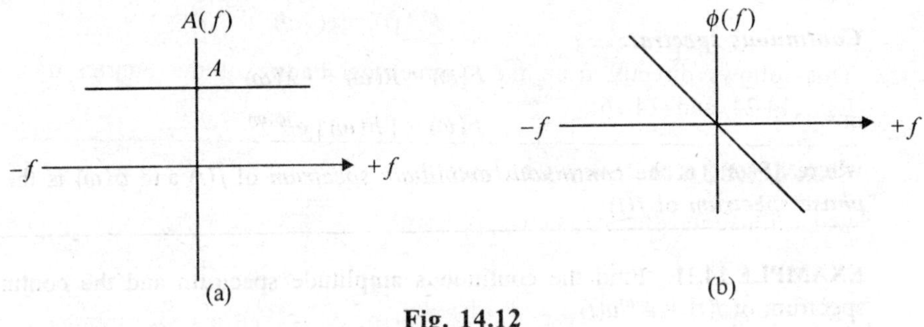

Fig. 14.12

14.7.1 Properties of the Fourier Transform

Several properties of Fourier transform are similar to those of Laplace transforms. Some of them are listed as follows:

Linearity:

$$F[a_1 f_1(t) + a_2 f_2(t)] = a_1F_1(\omega) + a_2F_2(\omega)$$

In words, the linearity property states that the Fourier transform of a sum of two signals is the sum of their individual transforms. The proof is based on the defining integral $F(\omega)$ [see Eq. (14.24)].

Differentiation: The Fourier transform of the derivative of a signal is $j\omega$ times the Fourier transform of the signal.

$$F\frac{df(t)}{dt} = j\omega F(\omega)$$

This is similar to Laplace transformation where multiplication by s corresponds to differentiation.

EXAMPLE 14.13 Illustrate the differentiation rule for the Fourier transform.

SOLUTION:

Consider a function $f(t) = e^{-at}\,u(t)$ whose transforms were determined in Example 14.11 as:

$$F(\omega) = \frac{1}{a + j\omega}$$

The derivative of $f(t)$ is:

$$f'(t) = \delta(t) - ae^{-at}u(t)$$

Its Fourier transform is:

$$Ff'(t) = 1 - \frac{a}{a + j\omega} = \frac{j\omega}{a + j\omega} = j\omega\, F(\omega)$$

Symmetry: The symmetry property states that if

$$f(xt) = X(\omega)$$

Then

$$F\, X(t) = x(-\omega)$$

This follows directly from the symmetrical nature of the Fourier transform pair in Eqs. (14.24) and (14.26).

EXAMPLE 14.14 Show that F rect t

which illustrates the symmetry property.

$$\text{rect } t = \begin{cases} 1 & |t| < 1/2 \\ 0 & |t| > 1/2 \end{cases}$$

SOLUTION:

$$F(\omega) = \int_{-\infty}^{\infty} f(t)\, e^{-j\omega t}\, dt = \int_{-1/2}^{1/2} 1\, e^{-j\omega t}\, dt = \frac{\sin \pi f}{\pi f}$$

which when plotted gives the waveform in Fig. 14.13(b) called the sinc waveform. Hence

$$F \text{ rect } t = \text{sinc } f$$

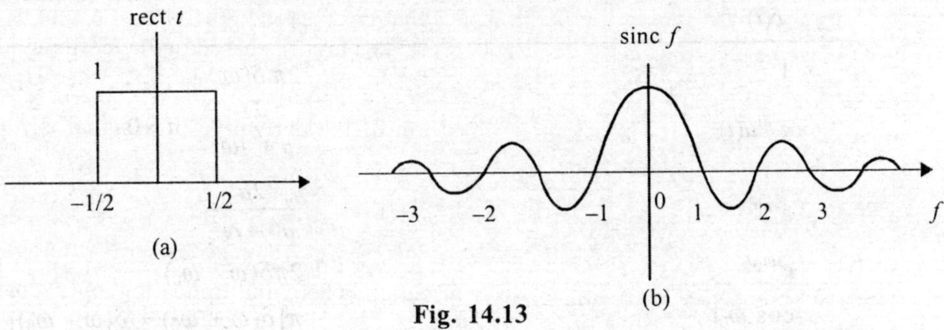

Fig. 14.13

Time shifting: If a signal is delayed by t_0 in the time domain, the transform of the undelayed signal is multiplied by $e^{-j\omega t}$

or
$$F\, f(t - t_0) = e^{-j\omega t}\, F(\omega)$$

Frequency shifting (modulation): If a Fourier transform is shifted in frequency by ω_0 then $Ff(t)\, e^{j\omega t} = F(\omega - \omega_0)$

Important properties are summarized in Table 14.1.

Table 14.1 Fourier Transform—Properties and Operations

$f(t)$	$F(\omega)$
Linearity $a_1 f_1(t) + a_2 f_2(t)$	$a_1 F_1(\omega) + a_2 F_2(\omega)$
Time differentiation $\dfrac{df(t)}{dt}$	$j\omega F(\omega)$
$\dfrac{d^n f(t)}{dt^n}$	$(j\omega)^n F(\omega)$
Time integration $\int_{-\infty}^{t} f(x)\, dx$	$\dfrac{1}{j\omega} F(\omega) + \pi F(0)\delta(\omega)$
Time shifting $f(t - t_0)$	$e^{-j\omega t} F(\omega)$
Modulation $f(t)\, e^{j\omega t}$	$F(\omega - \omega_0)$
Time scaling $f(at)$	$\dfrac{1}{\lvert a \rvert} F(\omega/a)$
Time convolution $\int_{-\infty}^{\infty} f_1(\lambda) f_2(t - \lambda)\, d\lambda$	$F_1(\omega) F_2(\omega)$
Frequency convolution $f_1(t)\, f_2(t)$	$1/2\pi \int_{-\infty}^{\infty} F_1(\omega - \lambda) F_2(\lambda)\, d\lambda$

A short list of Fourier transform pairs is given in Table 14.2.

Table 14.2 Typical Transform Pairs

$\delta(t)$	1
1	$2\pi\, \delta(\omega)$
$e^{-at} u(t)$	$\dfrac{1}{a + j\omega} \quad a > 0$
$e^{-a\lvert t \rvert}$	$\dfrac{2a}{a^2 + \omega^2}$
$e^{j\omega_0 t}$	$2\pi\delta(\omega - \omega_0)$
$\cos \omega_0 t$	$\pi[\delta(\omega + \omega_0) + \delta(\omega - \omega_0)]$
$\sin \omega_0 t$	$j\pi[\delta(\omega + \omega_0) - \delta(\omega - \omega_0)]$
$u(t)$	$\pi\delta(\omega) + \dfrac{1}{j\omega}$
Signum (t)	$\dfrac{2}{j\omega}$

14.8 PARSEVAL'S THEOREM

An important theorem which relates energy in the time and frequency domains is Parseval's theorem, which states

$$\int_{-\infty}^{\infty} f_1(t)\, f_2(t)\, dt = \frac{1}{2\pi}\int_{-\infty}^{\infty} F_1(\omega)F_2(-\omega)\, d\omega$$

The proof is obtained by making use of the frequency convolution relationship (see Table 14.1)

Let
$$f_1(t) = f_2(t) = f(t)$$

Then
$$\int_{-\infty}^{\infty} f^2(t) = \frac{1}{2\pi}\int_{-\infty}^{\infty} |\, F(\omega)\, |^2\, d\omega$$

If $f(t)$ is the current through or the voltage across, a 1 Ω resistance, the integral on the left side is the total energy in the signal $f(t)$. The total energy is also equal to [the area under the curve $F^2(\omega)]/2\pi$, (see Example 14.15). Hence $|\, F(\omega)\, |^2$ is called **energy density** or **energy spectral density** of $f(t)$. The energy of a periodic signal can therefore be determined directly from $F(\omega)$.

EXAMPLE 14.15 Prove that if $f(t)$ is real, its Fourier transforms satisfies $F(-\omega) = F^*(\omega)$.

SOLUTION:

By definition,

$$F(\omega) = \int_{-\infty}^{\infty} f(t)\, e^{-j\omega t}\, dt = \int_{-\infty}^{\infty} f(t)\,(\cos \omega t - j \sin \omega t)\, dt, \qquad \text{(Using Euler's identity)}$$

Putting $F(\omega)$ into its rectangular form:

$$F(\omega) = R(\omega) + j\,X(\omega)$$

$$= \int_{-\infty}^{\infty} f(t) \cos \omega t\; dt - j\int_{-\infty}^{\infty} f(t) \sin\omega t\; dt$$

Now,
$$F(-\omega) = \int_{-\infty}^{\infty} f(t) \cos(-\omega t)\; dt - j\int_{-\infty}^{\infty} f(t) \sin(-\omega t)\; dt$$

$$= \int_{-\infty}^{\infty} f(t) \cos \omega t\; dt + j\int_{-\infty}^{\infty} f(t) \sin \omega t\; dt$$

$$F(-\omega) = F^*(\omega)$$

Hence the result. **Ans.**

14.9 THE FOURIER TRANSFORM AND THE LAPLACE TRANSFORM

From F(s) to F(ω): If $f(t)$ is causal [$f(t) = 0$ for $t < 0$] and all the poles of $F(s)$ are inside the left half plane ($\sigma < 0$), then Fourier transform is a special case of Laplace transform, with $s = j\omega$.

or
$$F(\omega) = F(s)\big|_{s=j\omega} \tag{14.27}$$

The restriction on the poles of $F(s)$ ensures that $f(t)$ is absolutely integrable and ensures that the Fourier integral converges. Note that if $F(s)$ has poles in the RH plane, there is no $F(\omega)$ for $f(t)$.

From F(ω) to F(s): Under certain conditions the reverse process of Eq. (14.27) may be written as:

$$F(s) = F(\omega)\big|_{\omega=s}$$

EXAMPLE 14.16 If $f(t) = Ke^{-at}u(t)$ and $a > 0$, find the Fourier transform directly and compare the value obtained from the Laplace transform.

SOLUTION:

On putting $s = j\omega$,

$$F(\omega) = K\int_0^{\infty} e^{-at}\, e^{-j\omega t}\, dt = \frac{K}{j\omega + a}$$

Laplace transform $F(s) = \dfrac{K}{s + a}$

With $s = j\omega$, the Laplace transform becomes $\dfrac{K}{j\omega + a}$ which is the same as $F(\omega)$ obtained directly. **Ans.**

SUMMARY

The chapter brings out the significance of Fourier series for *periodic* waveforms and Fourier transforms for *aperiodic* waveforms. Both types of waveforms occur frequently in networks. Fourier techniques have proved to be very valuable in network analysis. The Laplace transform is a generalization of Fourier transform. Hence the former has wider applicability to problems like finding the current in a circuit with a square pulse input. The chapter ends with a quick comparison of Fourier transform and Laplace transform.

Solved Problems

P14.1 Find the Fourier series defined as:

$$f(\theta) = \begin{cases} 0 & -\pi \le \theta < 0 \\ 1 & 0 \le \theta < \pi \end{cases}$$

and

$$f(\theta + 2\pi) = f(\theta)$$

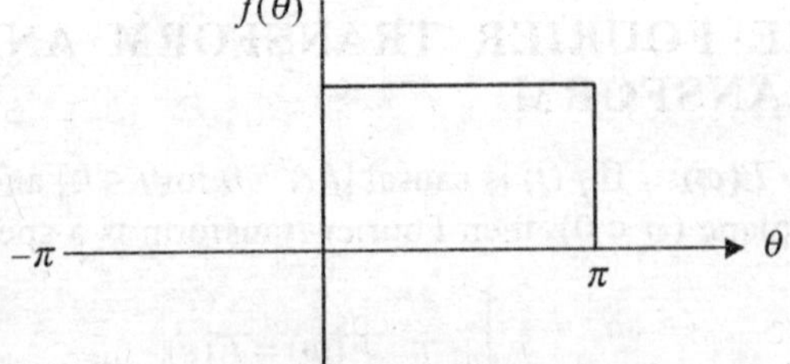

SOLUTION:

Using the Euler relation Eqs. (14.9), (14.11) and (14.12),

$$\frac{a_0}{2} = \frac{1}{2\pi}\int_0^{\pi} d\theta = \frac{1}{2}$$

$$a_k = \frac{1}{\pi}\int_0^{\pi} \cos k\theta \, d\theta$$

$$= \frac{1}{\pi}\left[\frac{\sin k\theta}{k}\right]_0^{\pi} = 0$$

$$b_k = \frac{1}{\pi}\int_0^{\pi} \sin k\theta \, d\theta$$

$$= \frac{1}{\pi}\left[\frac{-\cos k\theta}{k}\right]_0^{\pi} = \frac{1-(-1)^k}{k\pi} = 0 \qquad \text{for } k \text{ even}$$

$$= \frac{2}{k\pi} \qquad \text{for } k \text{ odd}$$

The Fourier series for $f(\theta)$ is:

$$\frac{1}{2} + \frac{2}{\pi}\frac{\Sigma \sin k\theta}{k}$$

where k is odd. **Ans.**

P14.2 Find the Fourier series for the waveform shown in Fig. P14.1.

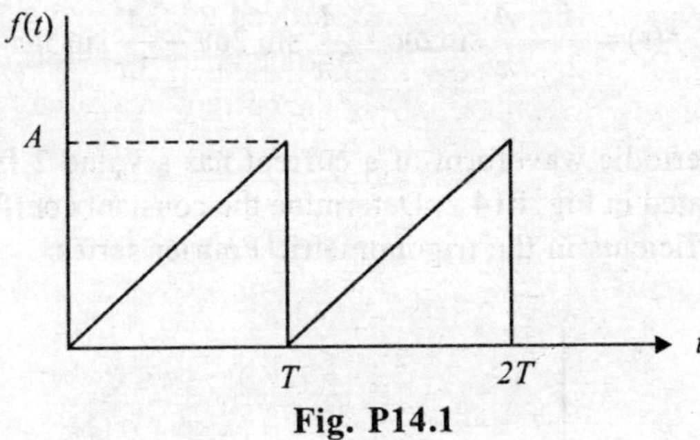

Fig. P14.1

SOLUTION:

The waveform is periodic with period T. It is continuous for $0 < t < T$ and is given by

$$f(t) = \frac{At}{T}$$

$$a_0 = \frac{A}{2}$$

$$a_k = \frac{2}{T}\int_0^{T} \frac{At}{T} \cos k\omega t \, dt$$

$$= \frac{2A}{T^2}\left[\frac{t \sin k\omega t}{k\omega}\right]_0^T - \frac{2A}{T^2}\int_0^T \frac{\sin k\omega t}{k\omega}\,dt$$

$$= 0 + \frac{2A}{T^2k^2\omega^2}\left[\cos k\frac{2\pi t}{T}\right]_0^T$$

$$= 0 + \frac{2A}{T^2k^2\omega^2}[\cos k\,2\pi - \cos 0] = 0$$

$$b_k = \frac{2}{T}\int_0^T \frac{At}{T}\sin k\omega t\,dt$$

$$= \frac{2A}{T^2}\left[\frac{-t\cos k\omega t}{k\omega}\right]_0^T - \frac{2A}{T^2}\int_0^T - \frac{\cos k\omega t\,dt}{k\omega}$$

$$= \frac{2A}{T^2}\frac{(-T)}{k\omega} - \frac{2A}{T^2}\left[\frac{-\sin k\omega t}{(k\omega)^2}\right]_0^T$$

$$= \frac{-2AT^2}{k2\pi}$$

$$= -\frac{A}{k\pi}$$

The Fourier series is:

$$f(t) = \frac{A}{2} - \frac{A}{\pi}\sin\omega t - \frac{A}{2\pi}\sin 2\omega t - \frac{A}{3\pi}\sin 3\omega t - \cdots \qquad \textbf{Ans.}$$

P14.3 The periodic waveform of a current has a value 2 from 0 to $\pi/2$ and –4 from π to 2π as illustrated in Fig. P14.2. Determine the constant coefficient, fundamental and third harmonic coefficients in the trigonometric Fourier series.

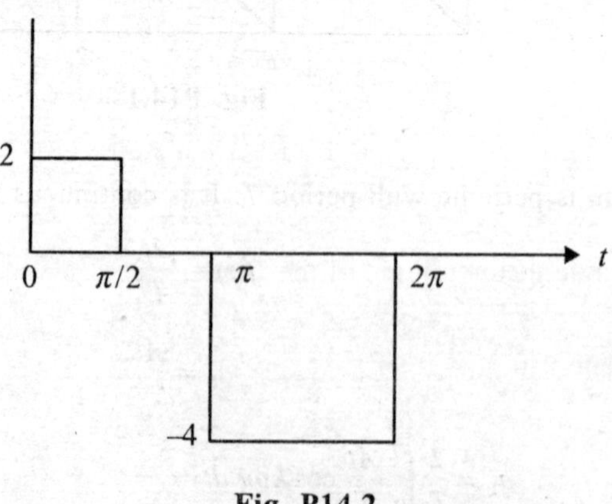

Fig. P14.2

SOLUTION:

$$\frac{a_0}{2} = \frac{1}{2\pi}\left[\frac{2\pi}{2} - 4\pi\right] = -1.5 \qquad \textbf{Ans.}$$

$$a_1 = \frac{1}{\pi}\int_0^{\pi/2} 2 \cos \omega t + \frac{1}{\pi}\int_{\pi}^{2\pi} - 4 \cos \omega t \, dt$$

$$= \frac{2}{\pi}[\sin \omega t]_0^{\pi/2} - \frac{4}{\pi}[\sin \omega t]_{\pi}^{2\pi}$$

$$= \frac{2}{\pi} - \frac{4}{\pi} = -0.637$$

$$b_1 = \frac{1}{\pi}\int_0^{\pi/2} 2 \sin \omega t + \frac{1}{\pi}\int_{\pi}^{2\pi} - 4 \sin \omega t \, dt$$

$$= -\frac{2}{\pi}[\cos \omega t]_0^{\pi/2} + \frac{4}{\pi}[\cos \omega t]_{\pi}^{2\pi}$$

$$= \frac{2}{\pi} + \frac{4}{\pi} 2 = 3.183 \qquad \textbf{Ans.}$$

$$a_3 = \frac{1}{\pi}\int_0^{\pi/2} 2 \cos 3\omega t \, dt + \frac{1}{\pi}\int_{\pi}^{2\pi} - 4 \cos 3\omega t \, dt$$

$$= \frac{1}{\pi}\cdot\frac{2}{3}[\sin 3\omega t]_0^{\pi/2} - \frac{4}{3}\cdot\frac{1}{\pi}[\sin 3\omega t]_{\pi}^{2\pi}$$

$$= -\frac{2}{3\pi} = -0.212$$

$$b_3 = \frac{1}{\pi}\int_0^{\pi/2} 2 \sin 3\omega t \, dt + \frac{1}{\pi}\int_{\pi}^{2\pi} -4 \sin 3\omega t \, dt$$

$$= \frac{-1}{\pi}\cdot\frac{2}{3}.[\cos 3\omega t]_0^{\pi/2} + \frac{4}{3\pi}[\cos 3\, \omega t]_{\pi}^{2\pi}$$

$$= \frac{2}{3\pi}\cdot\frac{4}{3\pi} 2 = 1.06 \qquad \textbf{Ans.}$$

P14.4 Determine the amplitude of the fundamental current in Problem P14.3.

SOLUTION:

From the Problem P14.3,

$$a_1 = -0.637$$

$$b_1 = 3.183$$

Hence

$$A_1 = \sqrt{a_1^2 + b_1^2}$$

$$= \sqrt{(-0.637)^2 + (3.183)^2}$$

$$= 3.246 \text{ A} \quad \textbf{Ans.}$$

P14.5 Find the trigonometric Fourier series for the output $f(t)$ of a half-wave rectifier given by

$$f(t) = \begin{cases} 0 & \text{for } -T/2 \le t < 0 \\ V \sin 2\pi t/T & 0 \le t < T/2 \end{cases}$$

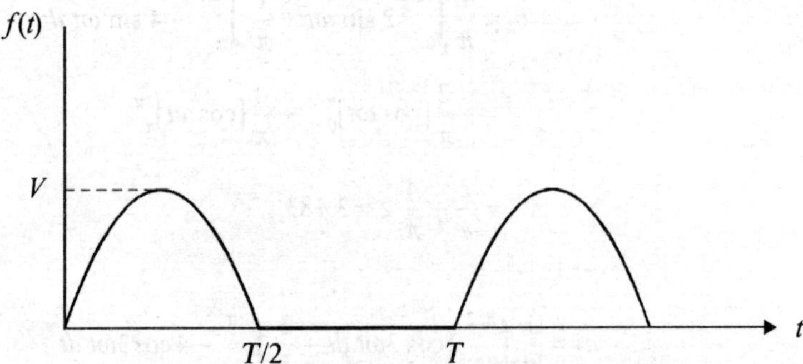

Fig. P14.3 Output $f(t)$ of a half-wave rectifier.

SOLUTION:

$$\frac{a_0}{2} = \frac{1}{\pi}\int_0^{\pi} V \sin \omega t \, dt = \frac{V}{\pi}$$

$$a_k = \frac{1}{\pi}\int_0^{\pi} V \sin \omega t \cos k\omega t \, dt$$

$$= \frac{2V}{\pi(1-k^2)} \qquad \text{with } k \text{ even}$$

$$= 0 \qquad \text{with } k \text{ odd.}$$

The value for a_1 being found separately because the expression for a_k is indeterminate for $k = 1$.

and $\quad b_k = 0$

$b_1 = V/2$ (Evaluated separately?)

The Fourier series is:

$$f(t) = \frac{V}{\pi}\left(1 + \frac{\pi}{2}\sin \omega t - \frac{2}{3}\cos 2\omega t - \frac{2}{15}\cos 4\omega t - \cdots\right) \qquad \textbf{Ans.}$$

P14.6 Find the exponential Fourier series for the waveform in Problem P14.2 and derive a_k and b_k of the trigonometric series.

SOLUTION:

$$c_k = \frac{1}{T}\int_0^T \frac{At}{T} e^{-jk\omega t} dt$$

$$= \frac{A}{T^2}\left[\frac{te^{-jk\omega t}}{-jk\omega}\right]_0^T - \int_0^T \frac{e^{-jk\omega t}}{(-jk\omega)^2} dt$$

$$= \frac{A}{T^2}\left[\frac{T.1}{-jk\omega} - \left(\frac{-1}{k^2\omega^2} + \frac{1}{k^2\omega^2}\right)\right]$$

$$= \frac{jA}{k2\pi}$$

The trigonometric series coefficients are:

$$a_k = 2 \text{ Re } c_k = 0 \qquad \textbf{Ans.}$$

$$b_k = -2 \text{ Im } c_k = \frac{-A}{\pi k} \qquad \text{(see Section 14.2)} \qquad \textbf{Ans.}$$

and
$$f(t) = \frac{A}{2} - \frac{A \sin \omega t}{\pi} - \frac{A}{2\pi} \sin 2\omega t - \cdots \qquad \textbf{Ans.}$$

which agrees with the series in Problem P14.2.

P14.7 Find the Fourier transform and plot the continuous spectra of the rectangular function $f(t)$ as shown in Fig. P14.4 with $A = 1$ pulse width a.

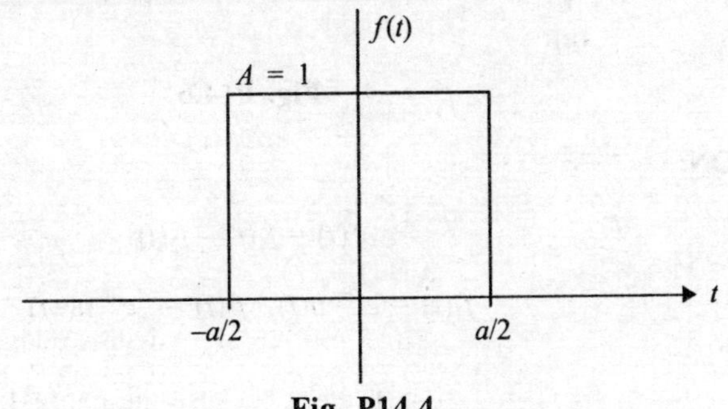

Fig. P14.4

SOLUTION:

$$F(\omega) = \int_{-\infty}^{+\infty} f(t)e^{-j\omega t} dt = \int_{-a/2}^{a/2} Ae^{-j\omega t} dt$$

$$= 2A\,a/2 \frac{\sin \omega\, a/2}{\omega\, a/2}$$

$$= \frac{a \sin \omega a/2}{\omega\, a/2}$$

The plot of sin ω/ω is called sinc t which is the continuous amplitude spectrum of the rectangular function f_{ct} as shown in Fig. P14.5.

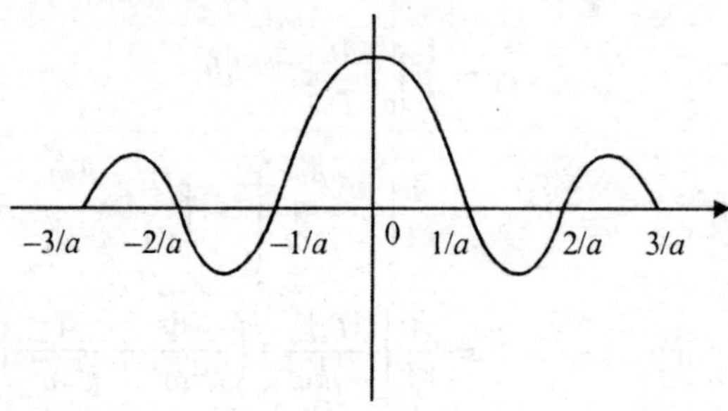

Fig. P14.5

Thus, sinc t is the inverse transform of rect x. **Ans.**

P14.8 Find the spectrum of the function $f(t)$ as shown in Fig. P14.6(a).

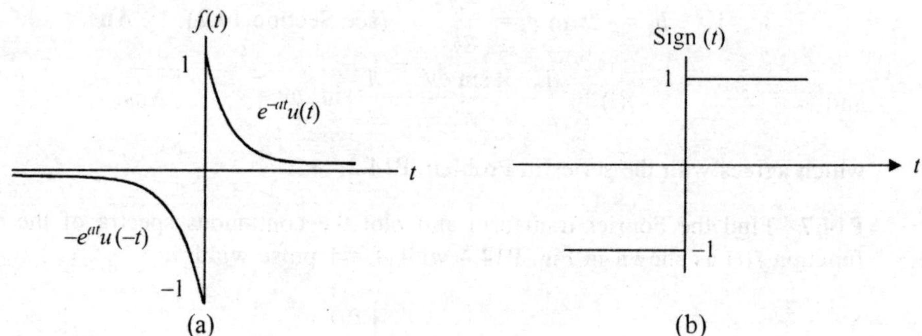

Fig. P14.6

SOLUTION:

$$f(t) = f_1(t) - f_2(t)$$

and
$$f_1(t) = e^{-at}\,u(t),\;\; f_2(t) = -e^{at}\,u(-t)$$

It follows

$$F_1(\omega) = \frac{1}{a + j\omega} \quad \text{and} \quad F_2(\omega) = \frac{1}{a - j\omega}$$

Hence
$$F(\omega) = F_1(\omega) - F_2(\omega)$$

$$= \frac{-j2\omega}{a^2 + \omega^2}$$

Note: If $a \to 0$, $f(t)$ becomes signum (t) as shown in Fig. P.14.6(b).

P14.9 Find the function $v(t)$ corresponding to the function $V(f)$ in Fig. P14.7.

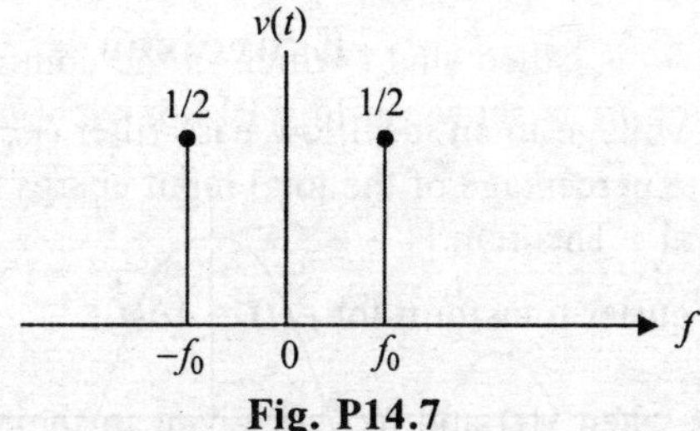

Fig. P14.7

SOLUTION:
Using the defining integral for the inverse Fourier transform,

$$v(t) = \int_{-\infty}^{\infty} \left[\frac{1}{2}\delta(f - f_0) + \frac{1}{2}\delta(f + f_0)\right] e^{j\omega t}\, df$$

$$= \frac{1}{2}e^{j2\pi ft} + \frac{1}{2}e^{-j2\pi ft}$$

$$= \cos 2\pi f_0 t \qquad \textbf{Ans.}$$

P14.10 A series RL circuit has an applied voltage $v = 10 + 5 \sin \omega t + 2 \sin 3\omega t$ with $\omega = 200$ rad/s. Find the current and the average power.

SOLUTION:

At $\omega = 0$, $\quad Z_0 = R = 1.0\ \Omega$

$\therefore \quad I_0 = V_0/R = 10/1.0 = 10$ A

At $\omega = 250$, $\quad Z_1 = 1 + j250 \times 10^{-2} = 1 + j\,2.5 = 2.69 \angle 68.2°$

$\therefore \quad i_1 = \dfrac{5}{7.69} \sin (\omega t - 68.2°) = 1.86 \sin (\omega t - 68.2°)$

At $\omega = 750$, $\quad Z_3 = 1 + j750 \times 10^{-2} = 1 + j7.5 = 7.57 \angle 82.5°$

$i_3 = \dfrac{2}{7.57} \sin (3\omega t - 82.5°) = 0.26 \sin (3\omega t - 82.5°)$

Then

Total current $i = 10 + 1.86 \sin (\omega t - 68.2°) + 0.26 \sin (3\omega t - 82.5°)$

Effective value of the current is given by

$$I^2 = 10^2 + (1.86/\sqrt{2})^2 + (0.26/\sqrt{2})^2 = 101.75$$

and $\qquad I = 10.09$ A $\qquad$ **Ans.**

Thus,

Average power $P = I^2R = 101.75 \times 1.0 = 101.75$ W $\qquad$ **Ans.**

Exercises

E14.1 The input voltage to an ideal low pass filter (ω_c = rad/s, $k = 1$) is $f(t) = 25\, e^{-10t}$ $u(t)$ V. Calculate the percentage of the total input energy available at the output terminal.
[**Hint:** Use Parseval's Theorem.] **Ans:** 50%

E14.2 Find the Fourier transform for $f(t) = A\delta(t)$.

E14.3 Show that when $v(t)$ and $i(t)$ are given in their exponential Fourier series, the average power is $P = \sum_{-\infty}^{\infty} c_{vk}\, c_{ik}^*$ where c_{vk} and c_{ik} are the coefficients in the exponential Fourier series for v and i respectively.
[**Hint:** See Section 14.3]

E14.4 Find the trigonometric Fourier series for the half-wave rectified sine wave shown in Fig. E14.1.

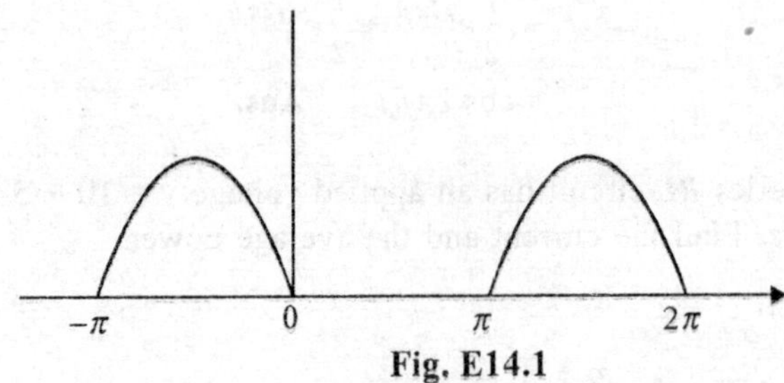

Fig. E14.1

[**Hint:** The function $f(t) = -V \sin \omega t$ in the interval $-\pi$ to 0. The average value is:

$$\frac{1}{2} a_0 = V/\pi \text{ (see Problem P14.5)}$$

and
$$a_k = \frac{1}{\pi}\int_{-\pi}^{0} -V \sin \omega t \cos k\omega t\, dt = \frac{V}{\pi}\frac{(1+\cos k\pi)}{(1-k^2)}$$

For k even $a_k = \dfrac{2V}{\pi(1-k^2)}$

For k odd $a_k = 0$ except a_1 which is also zero.

Now, $b_k = 0$ but $b_1 = -V/2$.

$$\therefore f(t) = \frac{V}{\pi}\left[1 - \frac{\pi}{2}\sin \omega t - \frac{2}{3}\cos 2\omega t - \frac{2}{15}\cos 4\omega t - \cdots\right]$$

Compare this series with that in Problem P14.5.

E14.5 The amplitude and phase spectrum of a periodic signal $f(t)$ are shown in Fig. E14.2. Find $f(t)$ in terms of trigonometric functions.

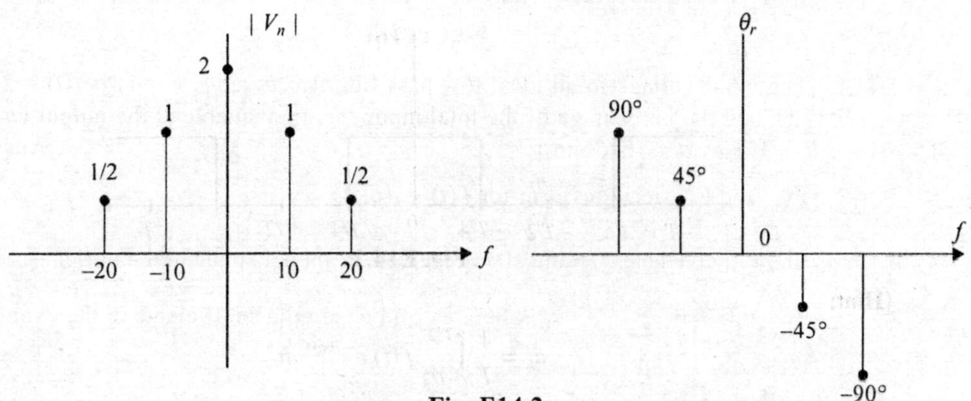

Fig. E14.2

[**Hint:** The function $f(t)$ is given by

$$f(t) = \sum_{k=-\infty}^{\infty} F_k e^{jk\omega t}$$

where $\omega = 2\pi f = 2\pi.\ 10$

From Fig. E14.2, the values of F_k are:

$F_{-2} = ½\ e^{j90}$, $F_{-1} = 1e^{j45}$, $F_0 = 2$, $F_1 = 1e^{-j,45}$, $F_2 = ½e^{-j90}$, and $F_k = 0$ for all other k.

$f(t)$ is then given by $½\ e^{j90}e^{-j2\pi 20t} + 1e^{j45}\ e^{-j2\pi 10t} + 2 + 1.e^{-j45}\ e^{j2\pi 10t} + ½e^{-j90}e^{j2\pi 20t}$
$= 2 + 2\cos(20\pi t - 45°) + \cos(40\pi t - 90°)$]

E14.6 Represent the function e^t over the interval $(0 < t < 1)$ by the exponential Fourier series.

[**Hint:**

$$F_k = \frac{1}{T}\int_0^T f(t)e^{-k\omega t}dt$$

$$F_k = \int_0^1 e^t e^{-jk\omega t}dt$$

$$= \frac{e^{(1-jk2\pi)} - 1}{1 - jk2\pi}$$

Note that F_k represents a periodic function with $T = 1$ although $f(t)$ is not periodic.

E14.7 Find the Fourier coefficients c_k for the periodic square wave $f(t)$ shown in Fig. E14.3.

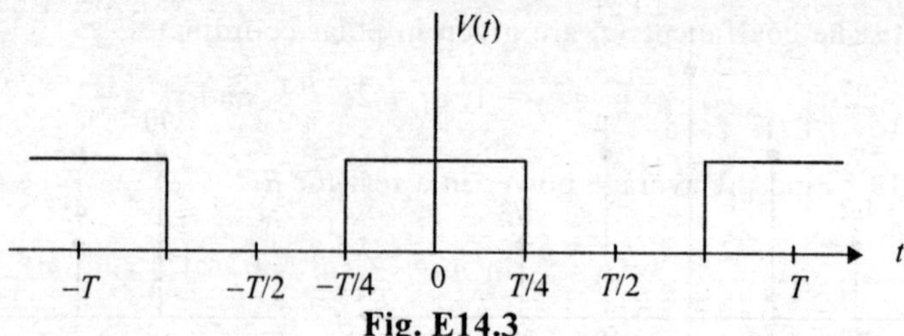

Fig. E14.3

[**Hint:**

$$c_k = \frac{1}{T}\int_{-T/2}^{T/2} f(t)\, e^{-jk\omega t}\, dt$$

The fundamental frequency $\omega = \dfrac{2\pi}{T}$

$$c_k = \frac{1}{T}\int_{-T/2}^{T/4} Ae^{-jk\omega 2\pi}\, dt$$

$\therefore$

$$= \frac{A}{2}\frac{\sin k\pi/2}{k\pi/2}$$

E14.8 If the periodic square wave of Exercise (E14.7) is shifted to the right by $T/4$, find the new Fourier coefficients.

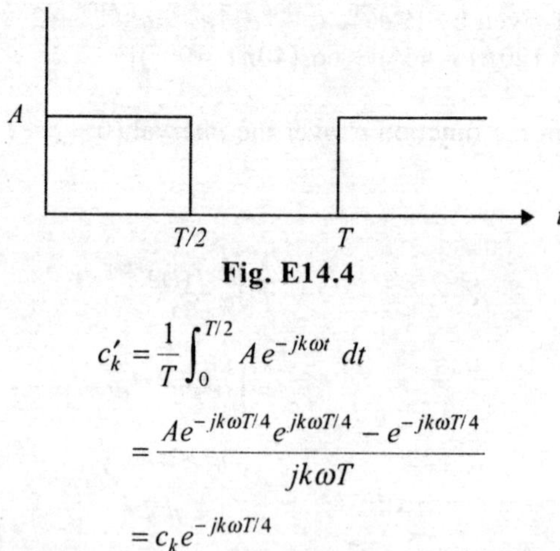

Fig. E14.4

$$c_k' = \frac{1}{T}\int_0^{T/2} Ae^{-jk\omega t}\, dt$$

$$= \frac{Ae^{-jk\omega T/4}e^{jk\omega T/4} - e^{-jk\omega T/4}}{jk\omega T}$$

$$= c_k e^{-jk\omega T/4}$$

E14.9 Find the time function $v(n)$ whose transform $V(k)$ is shown in Fig. E.14.5.

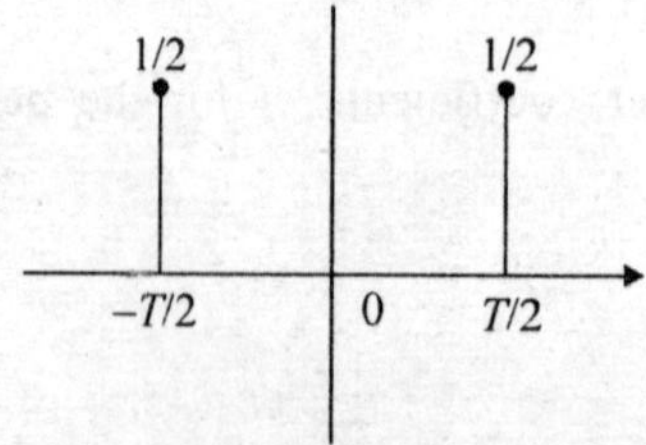

Fig. E14.5

[**Hint:** The coefficients c_k are given in polar coordinates.

$$\therefore \qquad c_0 = 1,\ c_1 = 2e^{-j1.1} \text{ and } c_2 = -3$$

E14.10 Find the average power in a resistor $R = 17.5\ \Omega$, if the current has the Fourier series

$$i = 5 \sin \omega t + 3 \sin 3\omega t + 1.0 \sin 5\omega t$$

[**Hint:** Effective $I^2 = (5/\sqrt{2})^2 + (3/\sqrt{2})^2 + (1/\sqrt{2})^2$

Average power $P = I^2 R$

$= 306.25$ W]

Network Analysis using PSpice and MATLAB

15

CHAPTER CONTENTS

A system analyst used to be faced with the drudgery of tedious and difficult problem solving involved in his work. But, thanks to the high-speed computer, he is now in the happy position of finding quick and easy solutions to the diverse problems occurring in network analysis, design and optimization.

The use of electronic digital computer for numerical solutions to problems has become routine, for instance, in the solution of simultaneous equations by matrix inversion (or by other methods). However, two of the notable tools which revolutionized problem solving in network analysis are SPICE and MATLAB.

The University and the Industry quickly realised the potential of these tools. The former in every undergraduate curriculum revision in Electrical, Electronics and related subjects is placing greater and greater emphasis on the new approach and the industry also has began to use it in a big way.

Simulation Program with Integrated Circuit Emphasis (SPICE) was announced by University of California in 1975. SPICE is a computer aided simulation program that enables the design of a circuit and thereafter to simulate the design on a computer. Many software versions are available that implement SPICE on personal computers, such as AllSpice, IS-SPICE, Z-SPICE, DSPICE, PSpice. The most popular version is PSpice (from OrCad and MicroSim). Various factors such as user friendly interface, possibility to support digital circuits and cost effectiveness have contributed to the acceptance of SPICE.

MATLAB stands for *Matrix Laboratory*. MATLAB was originally written to provide easy access to matrix software, developed by the LINPACK and EISPACK projects. Today, MATLAB represents the state-of-the-art in software for matrix computation.

MATLAB is an interactive system whose basic data element is an array that does not require dimensioning. This allows the solution of many technical computing problems, especially those with matrix and vector formulations to be done in much lesser time than it would take to write a program in a scalar non-interactive language such as C or Fortran.

Its uses include mathematical computations, algorithm development, modelling, simulation, prototyping, data analysis, exploration, visualization, scientific and engineering graphics. Many *toolboxes* are available which help in solving specialized classes of problems which include signal processing, control systems, neural networks, fuzzy logic, wavelets, and many others.

This chapter introduces the elementary principles of PSpice and MATLAB and illustrates them with the solution of simple networks. The aim is to highlight the versatility and simplicity of the techniques presented.

15.1 PSpice

PSpice enables the simulation of various electric circuits either by drawing the circuit using *schematic capture* front-end program as schematic (.sch) file, or by typing the commands for identifying circuit components and their attributes, in a circuit (.cir) file.

A schematic file stores the circuit as interconnected circuit symbols as in the actual circuit. This is translated into a *Netlist*, which is a collection of statements that represent the components, its attributes, topology, etc. and a circuit file for the PSpice to understand it. On simulation the output is stored as the output (.out) file.

15.1.1 Basics of PSpice

The elements in a circuit are called the **parts**. They are represented by names and have many attributes like value, units and reference. For example, a sinusoidal voltage source may have attributes like ac, dc, offset voltage, amplitude, frequency and phase.

The element value may have two suffixes: the scale suffix and the unit suffix; these are shown in Tables 15.1(a)and (b).

Table 15.1 Scale Suffixes and Unit Suffixes

Suffix	*Scale*	
F	1E-15	femto
P	1E-12	pico
N	1E-9	nano
U	1E-6	micro
MIL	25.4E-6	—
M	1E-3	milli
K	1E3	kilo
MEG	1E6	mega
G	1E9	giga
T	1E12	tera

(a)

Suffix	*Unit*
V	volt
A	amp
Hz	hertz
Ohm	ohm
H	henry
F	farad
Deg	degree

(b)

PSpice is not case sensitive hence the suffixes can be written in both upper and lower cases. The unit suffix for an element is not necessary as the PSpice always puts the default unit for each element. The floating point values without scale suffixes recognized by PSpice are:

2   2.   2.0   2E-3   2e-3   2.0E-3   2E+6   2E6

The element's name must start with a letter for symbol corresponding to that element.

Table 15.2 First Letter to be Used for Naming Circuit Elements

First letter	*Circuit elements*
C	Capacitor
D	Diode
E	Voltage-controlled voltage source
F	Current-controlled current source
G	Voltage-controlled current source
H	Current-controlled voltage source
I	Independent current source
J	JFET
K	Mutual inductors (Transformers)
L	Inductor
Q	BJT
R	Resistor
S	Voltage-controlled switch
T	Transmission line
V	Independent voltage source
W	Current-controlled switch

In PSpice each element is connected between two *nodes* and is identified using these. No node can be left floating. Nodes can be designated either by numbers or letters. An important point to be taken care of during drawing a schematic is that a ground/earth must be provided before simulation, without which the circuit would not be simulated. For example, a resistor named Rload of 15 kΩ connected between nodes 3 and 4 is represented as:

Rload 3 4 15 K

The attributes of some commonly used elements are

*V*dc = dc voltage
*I*dc = dc current
*V*sin = dc, ac, offset voltage, peak voltage, frequency, time delay, damping factor, phase
*I*sin = dc, ac, offset current, current amplitude, frequency, time delay, damping factor, phase
C = capacitance, initial condition
L = inductance, initial condition

15.1.2 Types of Analysis

After drawing the circuit the computer must be told how to analyze it, this is done by specifying the analysis type. The type of analysis can be specified from the Analysis setup window. The main methods of analysis are DC, AC and transient.

DC analysis: This is used for circuits with constant sources. The inductances and capacitances present in the circuit are considered as short-circuit and open-circuit, respectively, as the frequency during analysis is taken as zero. The common types are:

DC sweep. The input variable (voltage source, current source, temperature, model parameter, etc.) is varied over a range of values in a linear manner or swept logarithmically in octave or decades. The values can also be taken from a list. The syntax is:

.DC <type of sweep> <variable name> <initial value> <final value> <step size>

DC operating point or bias point details. This prints the dc operating point of all independent sources. This is enabled in the netlist with the command **.OP**.

Small signal transfer function. This calculates the small signal dc gain, output and input resistances of the circuit. The Syntax is:

.TF <output variable> <input variable>

AC analysis: This is done for analysis of circuits with variable frequencies. The frequency responses of the circuit elements for node voltages, currents with their magnitude and phase angles can be easily found. The AC sweep analysis covers a range of frequencies in linear or logarithmic scales. Its syntax is:

.AC <type of sweep> < total points> <start frequency> <end frequency>

Transient analysis: The transient behaviour of energy storing elements, ac-dc sources, etc. are analyzed using this. It gives the variation in node voltage, branch currents over an interval of time. The responses to various analog or digital inputs or stimulus can be found. Its syntax is:

.TRAN <print steps> <final time value>

Other methods: Other advanced types of analysis are also available such as *Parametric*, in which dc sources, global parameters, temperatures can be swept. Monte Carlo/Worst case analysis, sensitivity are statistical methods.

15.1.3 Simulating a Circuit

When the circuit file is typed in from the keyboard as a series of statements describing the elements and the type of analysis, it is executed by opening this file from PSpice executable file.

When the circuit is described using schematic capture front-end program, the netlist (.net) and circuit (.cir) files are created by selecting the "create netlist" option from "Analysis" menu. The output file and other outputs such as plots, probe are created when the schematic file is simulated.

The circuit is drawn using the *Schematics Capture* tool. The view of Schematics is shown in Fig. 15.1. Take a New file from the FILE menu and draw the circuit in the

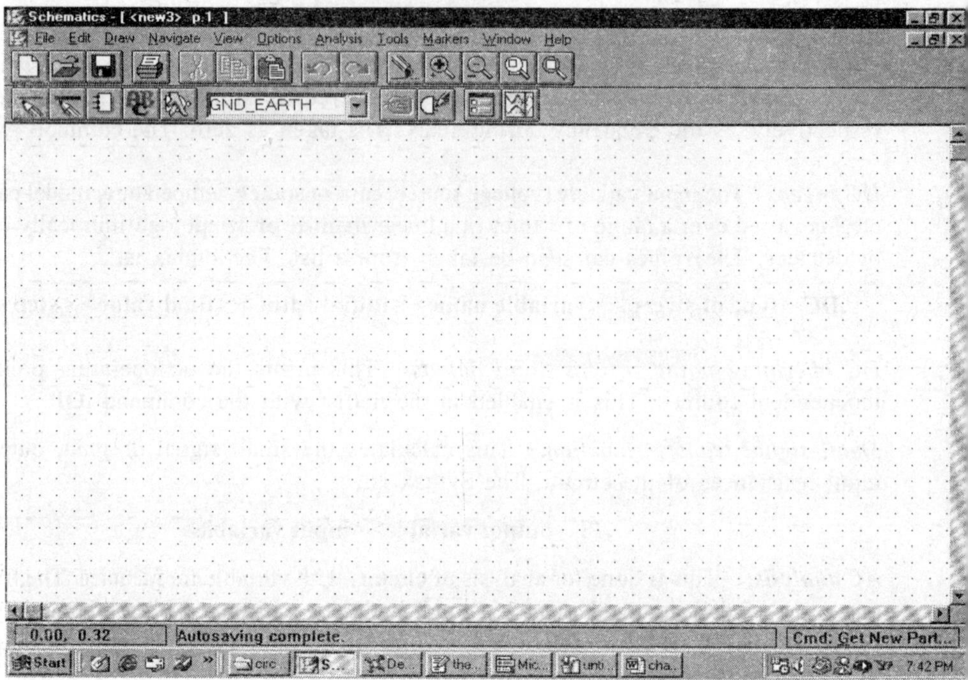

Fig. 15.1

dotted region. Drawing is done by placing parts and connecting them. This is done using the DRAW menu. Parts are chosen from the Part Browser shown in Fig. 15.2. The properties of the parts can be altered by double clicking the respective part or selecting *Attributes*...from EDIT menu.

A sample circuit is shown in Fig. 15.3(a). After drawing, specify the analysis to be performed using Analysis Setup dialog box Fig. 15.3(b) selected from ANALYSIS menu. The file is then saved. Now select the *create netlist* and then *simulate* option from this Analysis menu to simulate the circuit. The simulation result is available in the PSpice dialog box Fig. 15.3(c). The output may be a text file, a plot, etc. The text output in the text editor is displayed in Fig. 15.3(d).

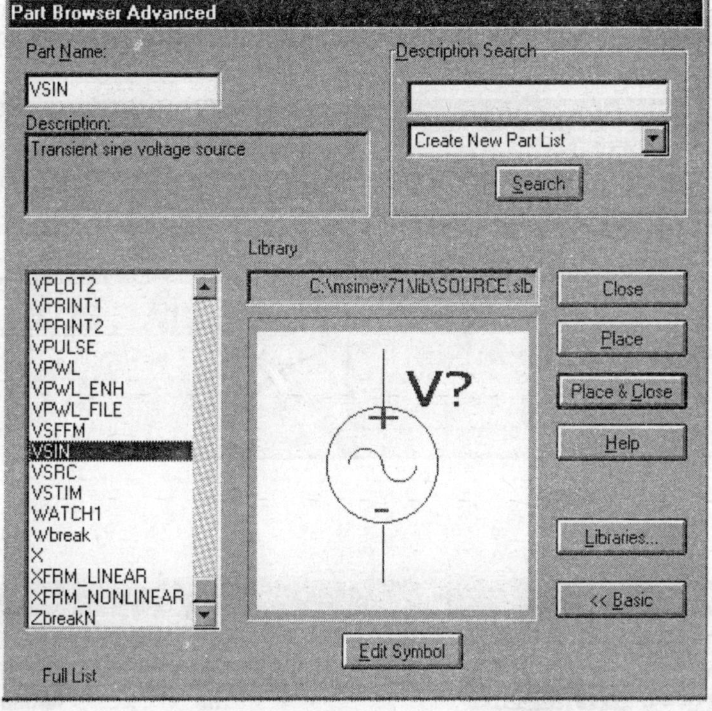
Part Browser Advanced
Part Name:
VSIN
Description:
Transient sine voltage source
Description Search
Create New Part List
Search
Library
C:\msimev71\lib\SOURCE.slb
VPLOT2
VPRINT1
VPRINT2
VPULSE
VPWL
VPWL_ENH
VPWL_FILE
VSFFM
VSIN
VSRC
VSTIM
WATCH1
Wbreak
X
XFRM_LINEAR
XFRM_NONLINEAR
ZbreakN
V?
Close
Place
Place & Close
Help
Libraries...
<< Basic
Edit Symbol
Full List

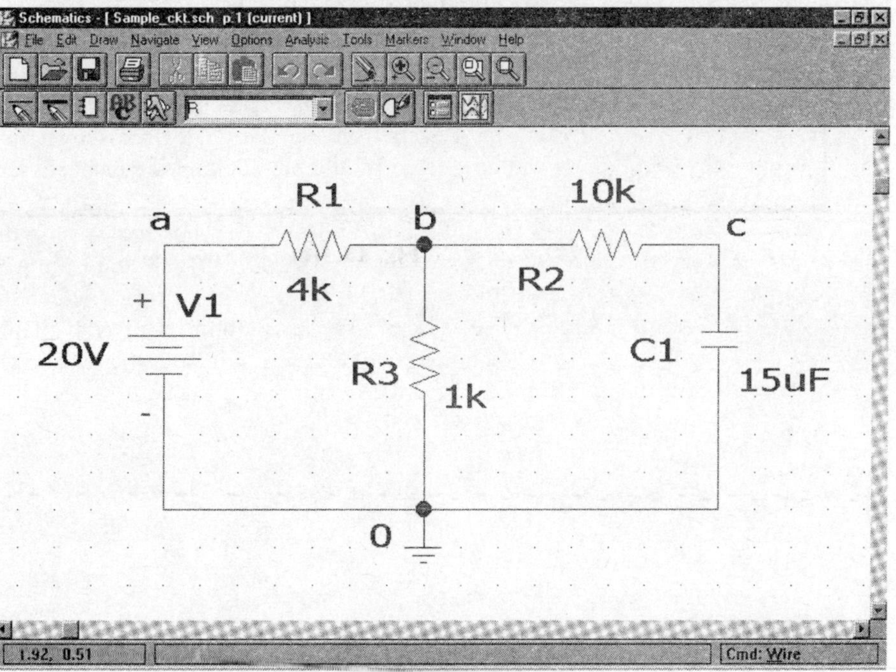
Schematics - [Sample_ckt.sch p.1 (current)]
File Edit Draw Navigate View Options Analysis Tools Markers Window Help
R1
a
b
10k
c
4k
R2
+
V1
20V
C1
R3
15uF
1k
-
0
1.92, 0.51
Cmd: Wire

Fig. 15.3(a)

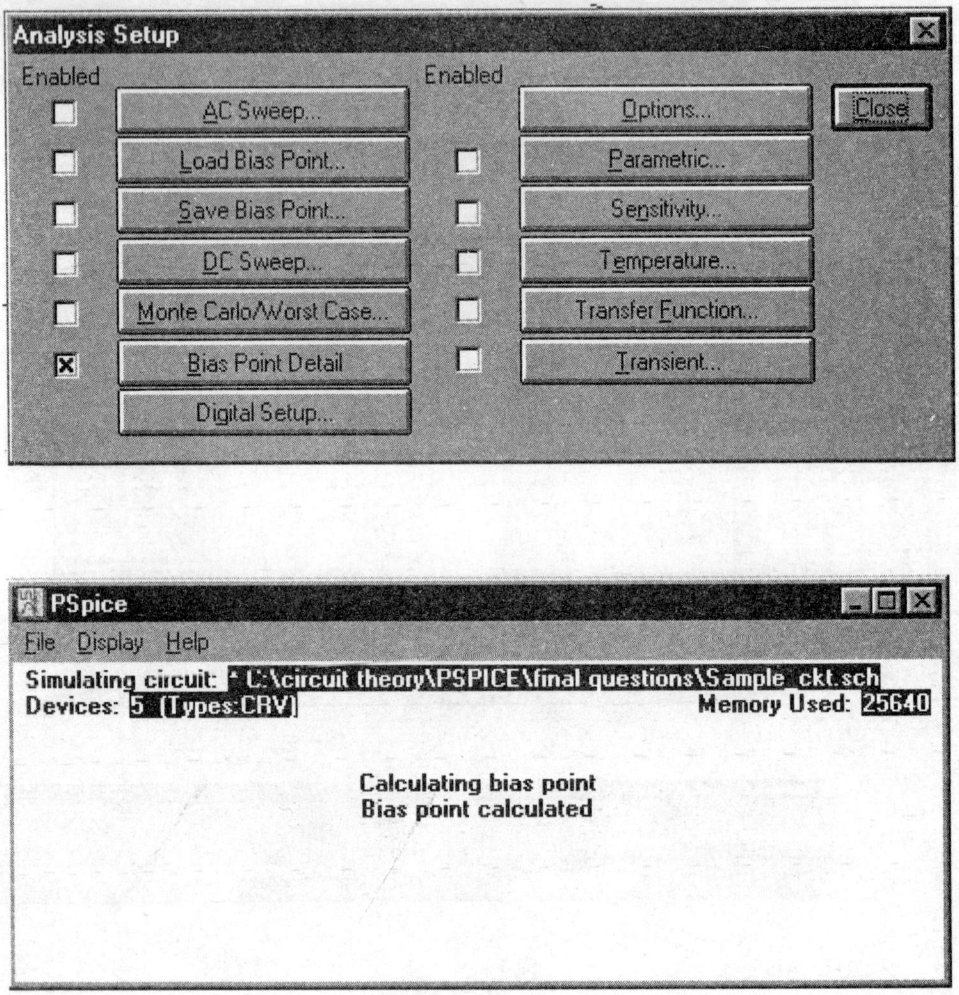

Analysis Setup
Enabled
AC Sweep...
Load Bias Point...
Save Bias Point...
DC Sweep...
Monte Carlo/Worst Case...
Bias Point Detail
Digital Setup...
Enabled
Options...
Parametric...
Sensitivity...
Temperature...
Transfer Function...
Transient...
Close

PSpice
File Display Help
Simulating circuit: * C:\circuit theory\PSPICE\final questions\Sample_ckt.sch
Devices: 5 (Types:CRV)
Memory Used: 25640
Calculating bias point
Bias point calculated

Fig. 15.3(c)

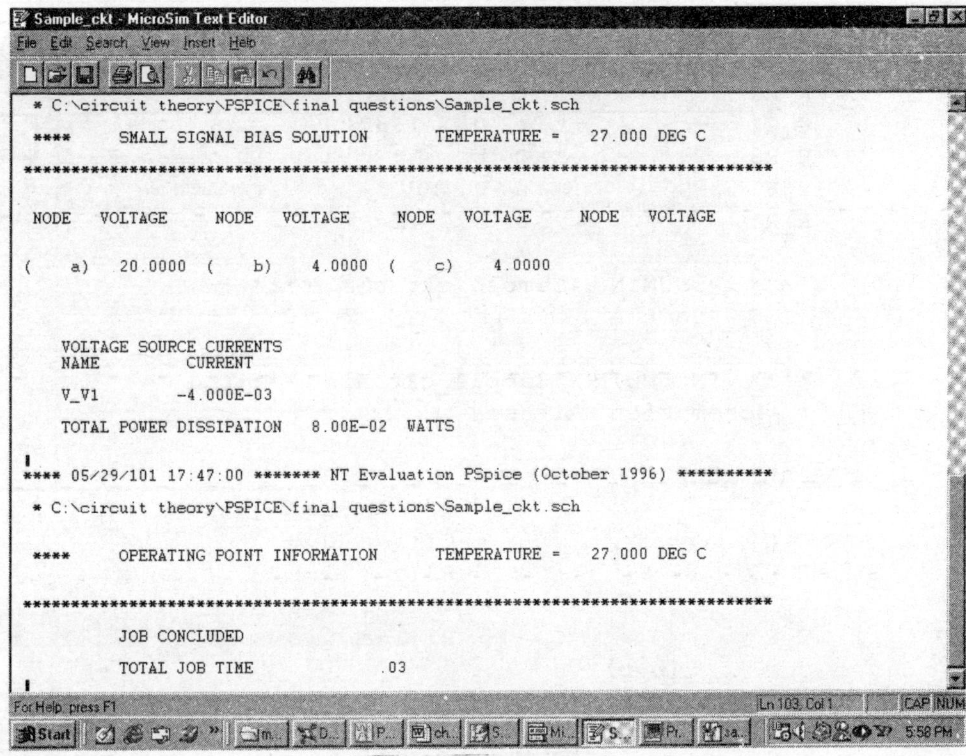

```
Sample_ckt - MicroSim Text Editor
File  Edit  Search  View  Insert  Help

 * C:\circuit theory\PSPICE\final questions\Sample_ckt.sch

 ****     SMALL SIGNAL BIAS SOLUTION       TEMPERATURE =   27.000 DEG C

******************************************************************************

 NODE   VOLTAGE     NODE   VOLTAGE     NODE   VOLTAGE     NODE   VOLTAGE

(    a)   20.0000  (    b)    4.0000  (    c)    4.0000

    VOLTAGE SOURCE CURRENTS
    NAME         CURRENT

    V_V1        -4.000E-03

    TOTAL POWER DISSIPATION   8.00E-02  WATTS

**** 05/29/101 17:47:00 ******* NT Evaluation PSpice (October 1996) **********
 * C:\circuit theory\PSPICE\final questions\Sample_ckt.sch

 ****     OPERATING POINT INFORMATION      TEMPERATURE =   27.000 DEG C

******************************************************************************

          JOB CONCLUDED

          TOTAL JOB TIME            .03

For Help, press F1                                   Ln 103, Col 1    CAP NUM
```

Fig. 15.3(d)

The contents of the output file for the aforementioned example are given as follows:

```
* C:\circuit theory\PSpice\final questions\Sample_ckt.sch

  ****        CIRCUIT DESCRIPTION
*********************************************************************
* Tue May 29 17:47:00 2001

** Analysis setup **
.OP


* From [SCHEMATICS NETLIST] section of msim.ini:
.lib "nom.lib"
.lib "nom.lib"
.lib "C:\msimev71\lib\Nom.lib"

.INC "Sample_ckt.net"

**** INCLUDING Sample_ckt.net ****
```

```
* Schematics Netlist *

R_R2            c b   10k
C_C1            c 0   15uF
R_R3            0 b   1k
V_V1            a 0  20V
R_R1            a b   4k

**** RESUMING Sample_ckt.cir ****
.INC "Sample_ckt.als"

**** INCLUDING Sample_ckt.als ****
* Schematics Aliases *

.ALIASES
R_R2                    R2(1=c  2=b )
C_C1                    C1(1=c  2=0 )
R_R3                    R3(1=0  2=b )
V_V1                    V1(+=a  -=0 )
R_R1                    R1(1=a  2=b )
_      _(b=b)
_      _(c=c)
_      _(a=a)
.ENDALIASES

**** RESUMING Sample_ckt.cir ****

.probe

.END

* C:\circuit theory\PSpice\final questions\Sample_ckt.sch

 **   SMALL SIGNAL BIAS SOLUTION          TEMPERATURE =
27.000 DEG C

******************************************************************
```

NODE	VOLTAGE	NODE	VOLTAGE	NODE	VOLTAGE	NODE	VOLTAGE
(a)	20.0000	(b)	4.0000	(c)	4.0000		

VOLTAGE SOURCE CURRENTS

NAME	CURRENT
V_V1	-4.000E-03

```
TOTAL POWER DISSIPATION 8.00E-02  WATTS

* C:\circuit theory\PSpice\final questions\Sample_ckt.sch

** OPERATING POINT INFORMATION TEMPERATURE = 27.000 DEG C

*****************************************************************
      JOB CONCLUDED

      TOTAL JOB TIME            .57
```

The output file has the contents of the circuit file (.cir file), net list, analysis types, the output which includes the voltages across various nodes, the current in the voltage source and the power dissipated by the circuit.

If there is any plot or graph, output can be viewed on the **PROBE**. Probe is the *waveform analyzer* for PSpice simulations. It can display complex arithmetic expressions involving voltages and currents and the Fourier transforms of these expressions other than simple voltages and currents. Many waveforms (traces) can be simultaneously displayed using this. The command for invoking this in the netlist is .Probe.

PLOT statement is used to plot the waves in the form of line printer plots in the output file. The plot is drawn using characters. The netlist statement is:

.PLOT DC/TRAN <output variable> (lower limit, upper limit)

.PLOT AC <magnitude> <phase>

Format of circuit file: When the circuit is simulated without the schematic capture tool (graphical method), a circuit file is typed describing the circuit. The structure of this file is as follows:

1. The title – this is the first line and may contain any text, as this is not compiled.
2. Circuit description – contains statements defining the circuit elements.
3. Analysis description – defines the analysis to be performed.
4. Output description – defines the mode of the output.
5. End of file statement – the last line is a .END command.

The order of the parts 2, 3 and 4 of the circuit file may be altered.

The statements are not case sensitive. The statements can be extended to next line by using a plus (+) sign. The comments line is preceded by an asterisk(*), within a statement it is preceded by a semicolon (,).

In the following examples, the Analysis setup and the Schematics Netlist together constitute the source file for PSpice simulation without drawing the circuit.

DC Analysis

EXAMPLE 15. 1 For the circuit in Fig. 15.4(a), find the voltage across each element.

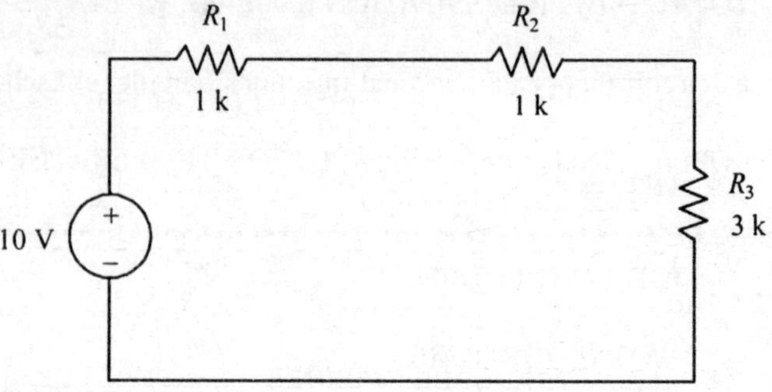

Fig. 15.4(a)

SOLUTION:

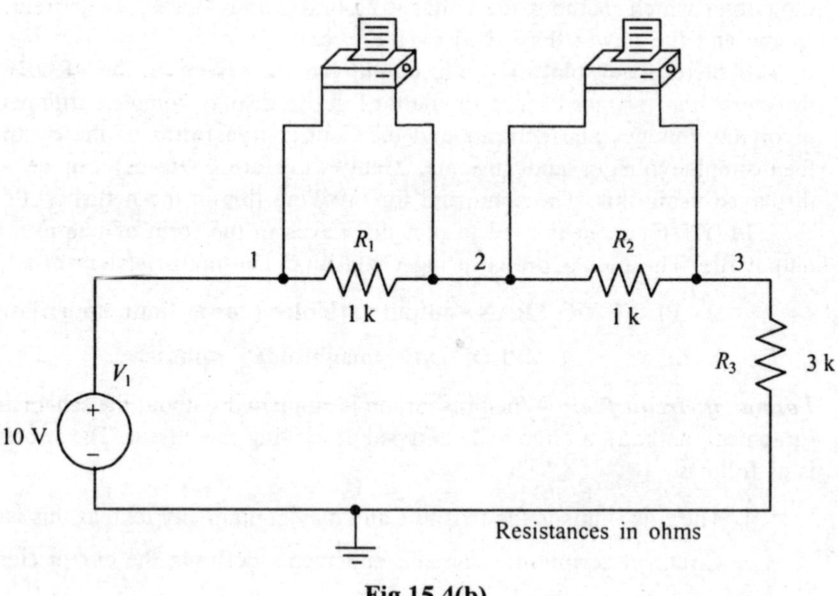

Fig 15.4(b)

The voltages across resistances R_1 and R_2 between nodes 1-2 and 2-3 [see Fig. 15.4(b)] is determined using VPRINT2 parts.

Note: While using VPRINT2, the attribute DC must be given a value *Y*. Always put the ground node in all the ciruits.

The output file is:

```
***  CIRCUIT DESCRIPTION

**  Analysis  setup  **
.DC  LIN  V_V1  10  10  1
.OP

*  Schematics  Netlist  *
```

```
R_R1                1 2    1k
R_R2                2 3    1k

.PRINT                  DC  V([2],[3])

.PRINT                  DC  V([1],[2])
R_R3                0 3    3k
V_V1                1 0  DC  10
.End

**   DC TRANSFER CURVES TEMPERATURE  =  27.000  DEG  C
*****************************************************************

     V_V1              V(2,3)

     1.000E+01      2.000E+00

     V_V1              V(1,2)

     1.000E+01      2.000E+00

 ** SMALL SIGNAL BIAS SOLUTION TEMPERATURE = 27.000 DEG C
*****************************************************************

 NODE VOLTAGE NODE VOLTAGE NODE VOLTAGE NODE VOLTAGE

  (1)    10.0000  (2)  8.0000    (3)    6.0000

      VOLTAGE  SOURCE  CURRENTS
      NAME               CURRENT

      V_V1               -2.000E-03

      TOTAL  POWER  DISSIPATION     2.00E-02    WATTS
```

The `.PRINT` command or `IPRINT/VPRINT` parts give the voltage across or current through an element for various values of the parameter (voltage, current…) swept using the `DC sweep` command. **Ans.**

EXAMPLE 15.2 In Fig. 15.5(a), find the voltage across nodes A-B and current through R_1.

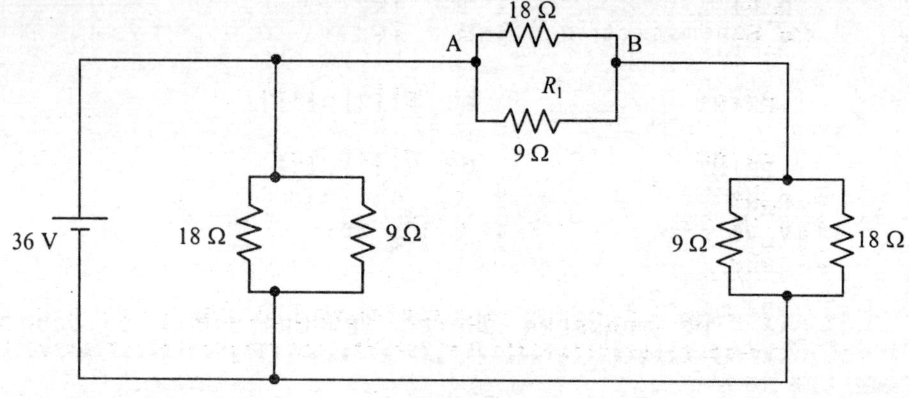

Fig. 15.5(a)

SOLUTION:

In the circuit [see Fig. 15.5(b), a 0 V dummy voltage source V_2 is added to get the current through R_1, as PSpice gives the current of all voltage sources in the circuit. Else a `IPRINT` part may be used.

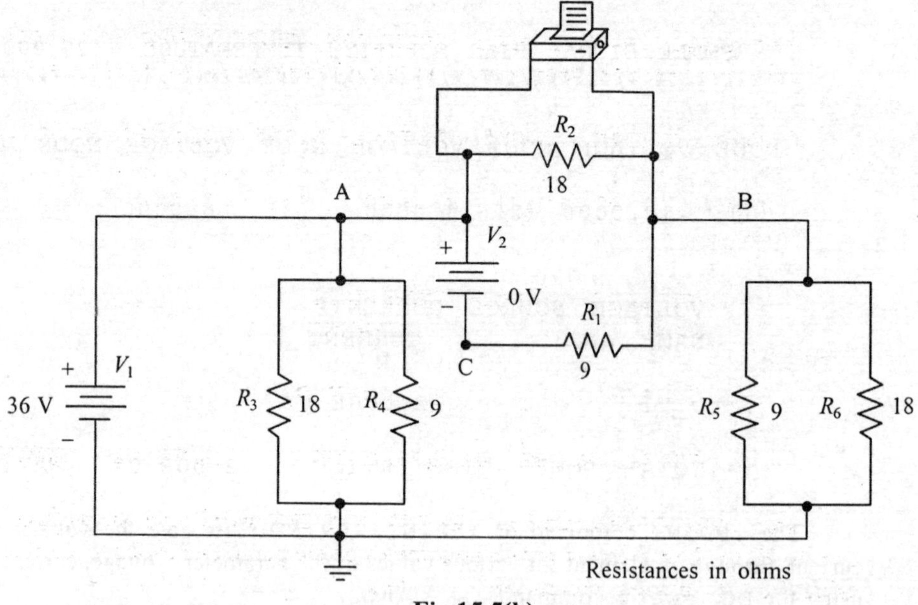

Fig 15.5(b)

```
**    CIRCUIT DESCRIPTION

**Analysis  setup  **
.DC  LIN  V_V1  36  36  18
.OP
```

```
* Schematics Netlist *

R_R5            0 B   9
V_V1            A 0 36V
R_R3            0 A   18
R_R4            0 A   9
R_R2            A B   18

.PRINT            DC V([A],[B])
R_R1            C B   9
R_R6            0 B   18
V_V2            A C 0V
.probe

.END


DC TRANSFER CURVES          TEMPERATURE = 27.000 DEG C
***************************************************************

     V_V1              V(A,B)

   3.600E+01     1.800E+01

SMALL SIGNAL BIAS SOLUTION TEMPERATURE = 27.000 DEG C

 NODE VOLTAGE NODE VOLTAGE   NODE VOLTAGE

(A)     36.0000    (B)     18.0000    (C)     36.0000

      VOLTAGE SOURCE CURRENTS
      NAME               CURRENT

      V_V1              -9.000E+00
      V_V2               2.000E+00

      TOTAL POWER DISSIPATION 3.24E+02 WATTS
```

The voltage across *AB* is 18 V and current across R_1 is 2 A. **Ans.**

EXAMPLE 15.3 Find the current through R_5 and R_2, and the voltages across each resistor in the circuit of Fig. 15.6(a).

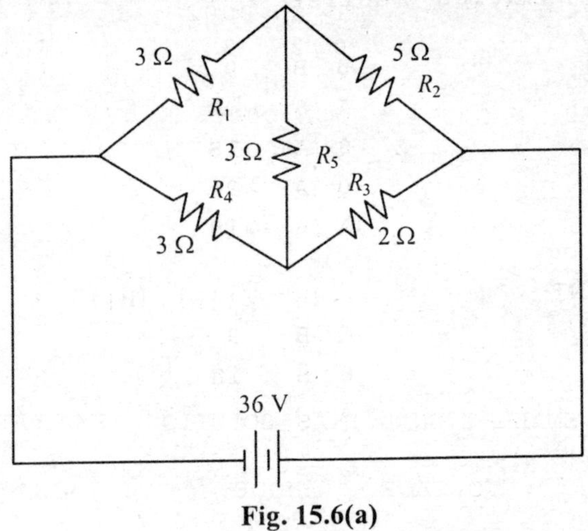

Fig. 15.6(a)

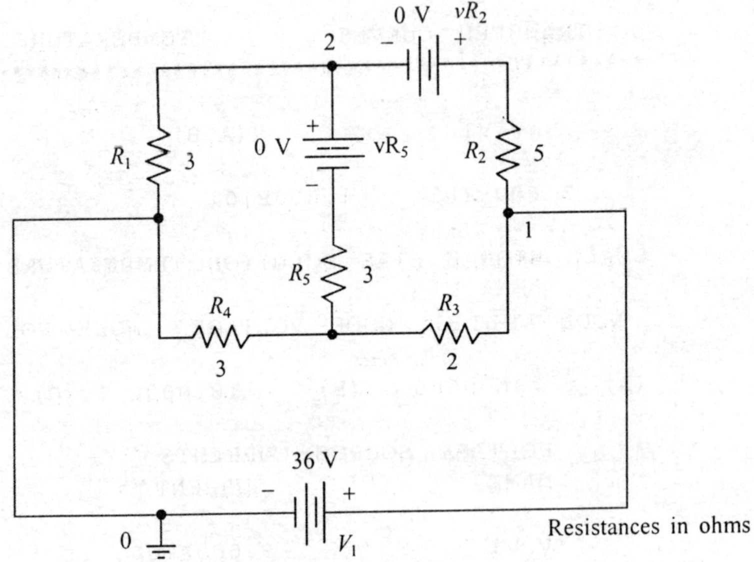

Fig. 15.6(b)

SOLUTION:

```
*** CIRCUIT DESCRIPTION

** Analysis setup **
.DC LIN V_V1 0 36 36
.OP

* Schematics Netlist *

R_R4                0  $N_0001   3
```

```
R_R3          $N_0001 1   2
R_R2          1  $N_0002   5
R_R1          0  2   3
V_vR5          2  $N_0003  0V
V_vR2          $N_0002  2  0V
R_R5          $N_0001  $N_0003   3
V_V1          1  0  36V

.probe

.END
```

```
*** SMALL SIGNAL BIAS SOLUTION TEMPERATURE = 27.000 DEG C

NODE     VOLTAGE    NODE     VOLTAGE      NODE    VOLTAGE

(1)     36.0000    (2)     16.0000    ($N_0001)     20.0000

($N_0002)     16.0000                  ($N_0003)     16.0000


     VOLTAGE  SOURCE  CURRENTS
     NAME                CURRENT

     V_vR5              -1.333E+00
     V_vR2               4.000E+00
     V_V1               -1.200E+01

     TOTAL  POWER  DISSIPATION     4.32E+02    WATTS
```

EXAMPLE 15.4 Find the currents in and voltage across each branch of the circuit of Fig. 15.7(a). Check the answers with those in Example 6.1.

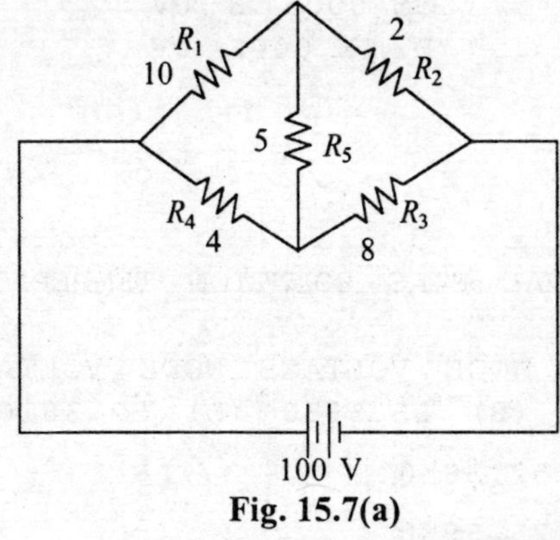

Fig. 15.7(a)

SOLUTION:

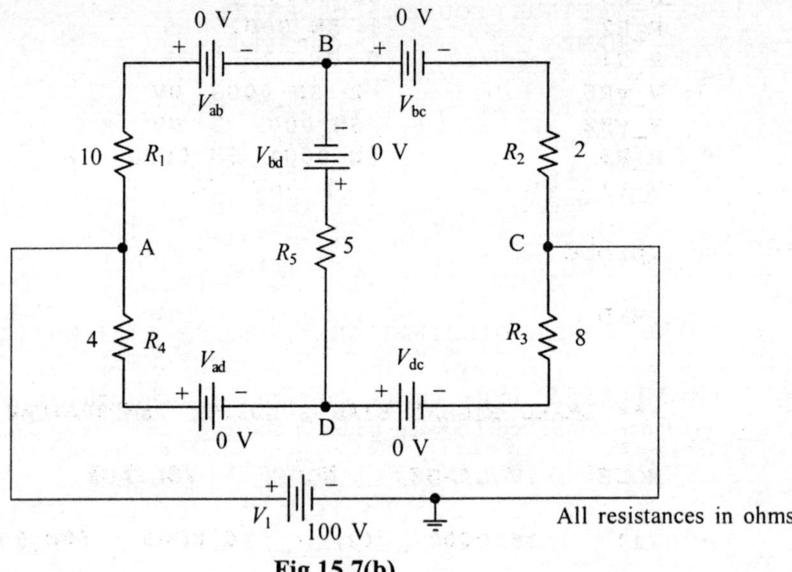

Fig 15.7(b)

```
** Analysis setup **
.OP

Schematics Netlist *

R_R2              0  $N_0001   2
R_R3              $N_0002  0   8
R_R4              $N_0003  A   4
R_R1              A  $N_0004   10
R_R5              D  $N_0005   5
V_V1              A  0  100V
V_Vab              $N_0004  B  0V
V_Vbc              B  $N_0001  0V
V_Vad              $N_0003  D  0V
V_Vbd              $N_0005  B  0V
V_Vdc              D  $N_0002  0V

.probe

.END

** SMALL SIGNAL BIAS SOLUTION TEMPERATURE = 27.000 DEG C

NODE VOLTAGE NODE VOLTAGE NODE VOLTAGE NODE VOLTAGE
(A) 100.0000 (B) 25.5950 (D) 52.3810 ($N_0001) 25.5950

($N_0002)    52.3810              ($N_0003)    52.3810

($N_0004)    25.5950              ($N_0005)    25.5950
```

```
    VOLTAGE  SOURCE  CURRENTS
    NAME            CURRENT

    V_V1            -1.935E+01
    V_Vab            7.440E+00
    V_Vbc            1.280E+01
    V_Vad            1.190E+01
    V_Vbd            5.357E+00
    V_Vdc            6.548E+00

TOTAL  POWER  DISSIPATION      1.93E+03    WATTS
```

EXAMPLE 15.5 In the circuit shown in Fig. 15.8(a), find the current through the 4 Ω resistance connected between nodes n and 0.

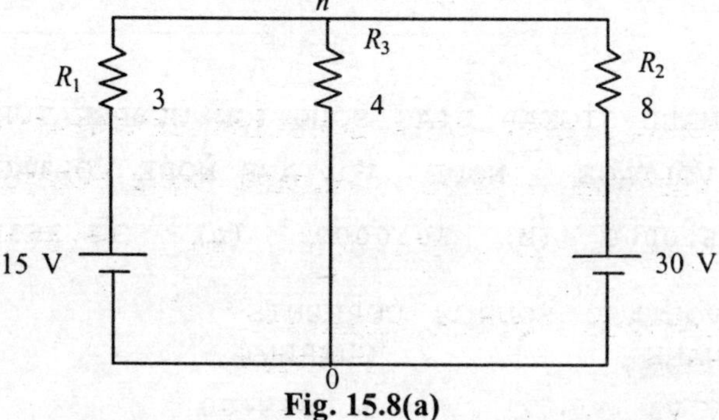

Fig. 15.8(a)

SOLUTION:

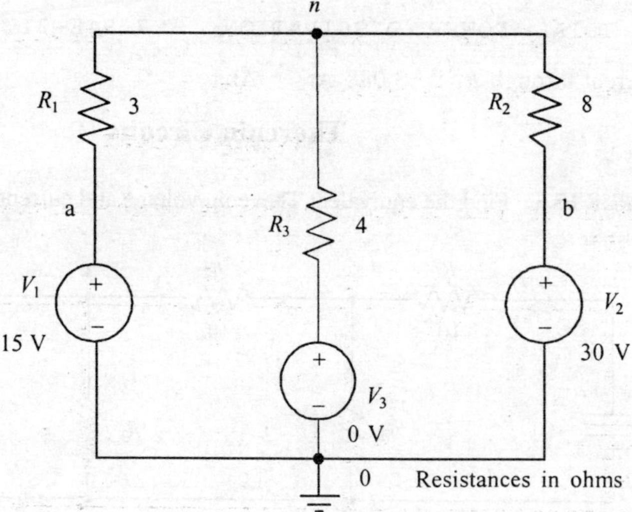

Fig 15.8(b)

The schematic is drawn as shown in Fig. 15.8(b) and a 0V DC source is added between the nodes n and 0 V. The current through this voltage source is the current through the resistance R_3. Output file is shown as follows:

```
**  Analysis  setup  **
.OP

*  Schematics  Netlist  *

V_V3                $N_0001  0  DC  0v  AC  0
R_R1                a  n    3
V_V1                a  0  DC  15v
R_R2                b  n    8
V_V2                b  0  DC  30v
R_R3                $N_0001  n    4

.END

****  SMALL SIGNAL BIAS SOLUTION TEMPERATURE  =  27.000 DEG C

NODE    VOLTAGE    NODE    VOLTAGE NODE VOLTAGE NODE     VOLTAGE

(a)   15.0000   (b)   30.0000   (n)    12.3530  ($N_0001)   0.00

    VOLTAGE  SOURCE  CURRENTS
    NAME                CURRENT

    V_V3                  3.088E+00
    V_V1                 -8.824E-01
    V_V2                 -2.206E+00

    TOTAL  POWER  DISSIPATION     7.94E+01    WATTS
```

The current through R_3 is 3.088 A. **Ans.**

Thevenin Circuits

EXAMPLE 15.6 Find the equivalent Thevenin voltage and current for the circuit illustrated in Fig. 15.9.

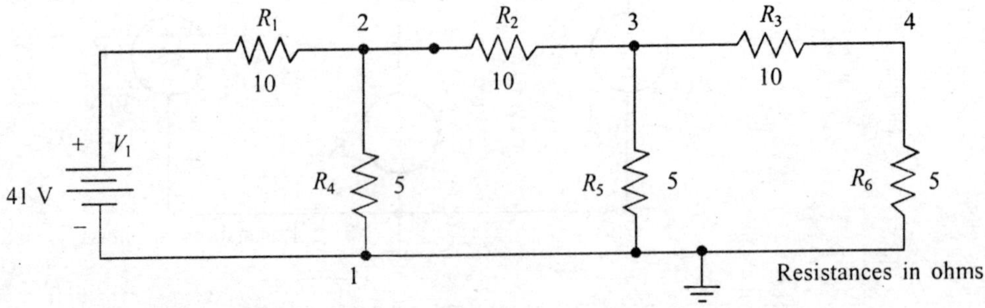

Fig. 15.9

SOLUTION:

Transfer function analysis (.TF) is used for getting the Thevenin resistances. The output resistance of the transfer function will be the R_{TH}. The voltage across R_1 will give the V_{TH}.

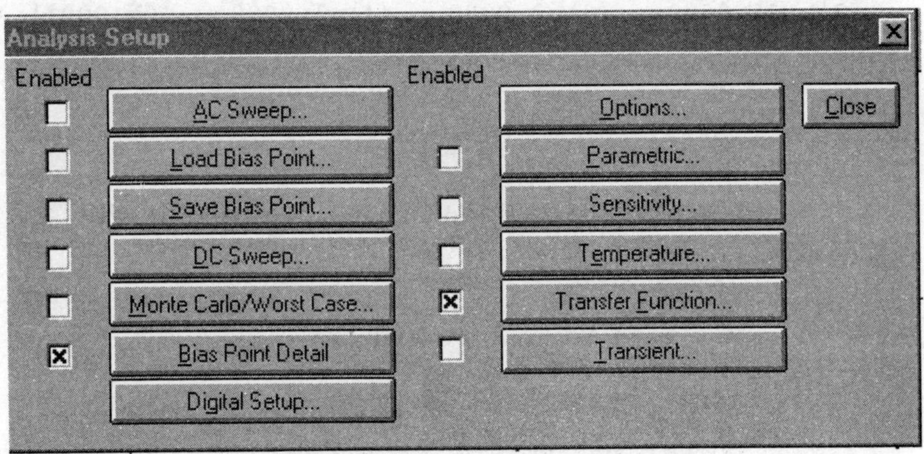

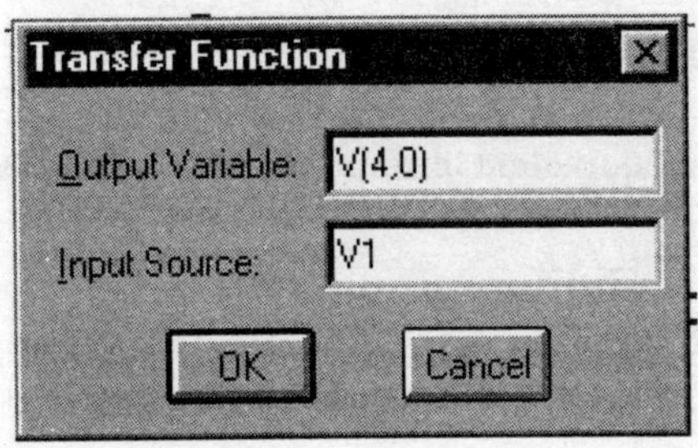


```
** Analysis setup **
.TF V([4],[0]) V_V1
.OP
* Schematics Netlist *

R_R1           $N_0001 2    10
R_R2           2 3    10
R_R3           3 4    10
R_R5           0 3    5
R_R4           0 2    5
R_R6           0 4    5
V_V1           $N_0001 0  41V

.probe

.END
```

```
*** SMALL SIGNAL BIAS SOLUTION TEMPERATURE = 27.000 DEG C

NODE    VOLTAGE    NODE   VOLTAGE     NODE     VOLTAGE

(2)   11.0000    (3) 3.0000    (4) 1.0000    ($N_0001) 41.0000

    VOLTAGE SOURCE CURRENTS
    NAME               CURRENT

    V_V1              -3.000E+00

    TOTAL POWER DISSIPATION      1.23E+02    WATTS

      *SMALL-SIGNAL CHARACTERISTICS

      V(4,0)/V_V1  =    2.439E-02

      INPUT RESISTANCE AT V_V1  =    1.367E+01

      OUTPUT RESISTANCE AT V(4,0)  =    3.659E+00
```

Hence the V_{TH} is 1 V and R_{TH} is 3.659 Ω. **Ans.**

EXAMPLE 15.7 Find the Thevenin's equivalent circuit for the circuit shown in Fig. 15.10(a).

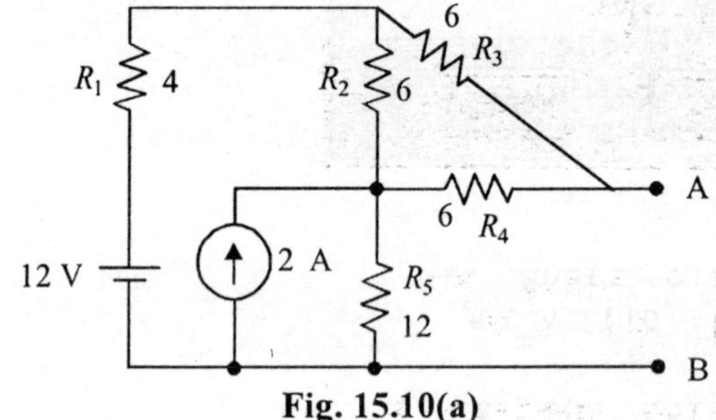

Fig. 15.10(a)

SOLUTION:

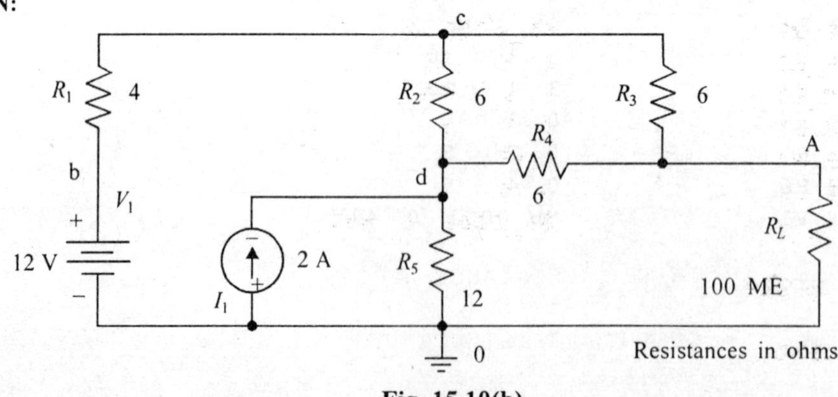

Fig. 15.10(b)

Note: The R_L for 100 M ohm is used to get an open-circuit at the output as nodes cannot be left floating [Refer Fig. 15.10(b)]

```
*   CIRCUIT DESCRIPTION
**  Analysis  setup  **

.TF  v([A],[0])  V_V1
.OP

*  Schematics  Netlist  *

V_V1              b  0  12V
R_R2              d  c   6
R_R3              A  c   6
I_I1              0  d  DC  2A
R_RL              0  A   100MEG
R_R1              b  c   4
R_R5              0  d   12
R_R4              d  A   6

.probe

.END

**  SMALL  SIGNAL  BIAS  SOLUTION  TEMPERATURE  =  27.000  DEG  C

 NODE   VOLTAGE   NODE  VOLTAGE   NODE  VOLTAGE   NODE   VOLTAGE

(A)   15.6000    (b)   12.0000   (c)   14.4000    (d)  16.8000

     VOLTAGE  SOURCE  CURRENTS
     NAME              CURRENT

     V_V1              6.000E-01

     TOTAL  POWER  DISSIPATION   -7.20E+00   WATTS

 ****          SMALL-SIGNAL  CHARACTERISTICS

          V(A,0)/V_V1  =   7.000E-01

          INPUT  RESISTANCE  AT  V_V1  =   2.000E+01

          OUTPUT  RESISTANCE  AT  V(A,0)  =   6.200E+00
```

Here the output resistance gives the R_{Th} = 6.2 Ω, and voltage across (*A*,0) is V_{Th} = 15.6 V. **Ans.**

EXAMPLE 15.8 Find the Thevenin equivalent of the circuit depicted in Fig. 15.11(a) at (*d*, 0). Also find the current through the 15 Ω resistor.

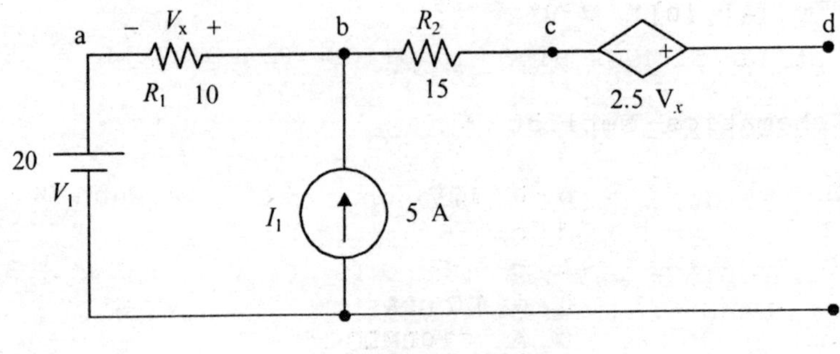

Fig. 15.11(a)

SOLUTION:

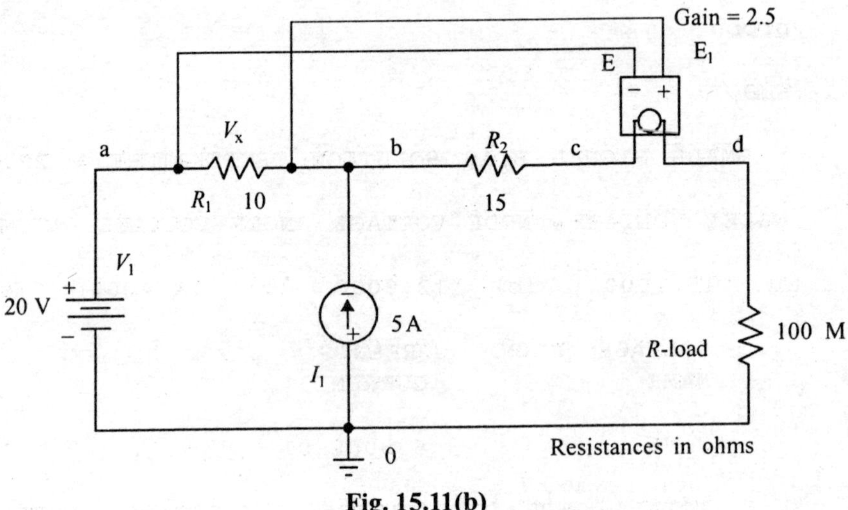

Fig. 15.11(b)

For finding the Thevenin equivalent circuit, the 15 Ω resistor may be replaced by a high resistance (100 M), the voltage-controlled voltage source is represented by the part *E*, the two nodes are attached across the controlling voltage (R_1) [Refer Fig. 15.11(b)]

```
** Analysis setup **
.TF V([d],[0]) V_V1
.OP

* Schematics Netlist *
```

```
R_R1          a b   10
R_R2          b c   15
I_I1          0 b DC  5A
V_V1          a 0  20V
E_E1          d c b a 2.5
R_R_load         0 d   100 meg

** SMALL SIGNAL BIAS SOLUTION TEMPERATURE = 27.000 DEG C

NODE  VOLTAGE  NODE  VOLTAGE   NODE  VOLTAGE  NODE  VOLTAGE

(a)   20.0000  (b)   70.0000   (c)   70.0000  (d)  195.0000

     VOLTAGE SOURCE CURRENTS
     NAME           CURRENT

     V_V1           5.000E+00

     TOTAL POWER DISSIPATION   -1.00E+02   WATTS

**** VOLTAGE-CONTROLLED VOLTAGE SOURCES

NAME             E_E1
V-SOURCE       1.250E+02
I-SOURCE      -1.950E-05

 ****        SMALL-SIGNAL CHARACTERISTICS

       V(d,0)/V_V1  =  1.000E+00

       INPUT RESISTANCE AT V_V1  =   1.000E+07

       OUTPUT RESISTANCE AT V(d,0)  =   5.000E+01
```

Output resistance R_{Th} = 5.0 ohm
Voltage across (d,0) V_{Th} = 195 V **Ans.**

AC Analysis

Using this the frequency response of various circuits can be found, also the response for the circuit with different values of an element can be simulated by making it a global parameter. The values of the passive circuit elements can be swept in different intervals over a range.

EXAMPLE 15.9 Plot the variation of current and voltage across the resistance with frequency for different values of resistance in the *RLC* circuit shown in Fig. 15.12(a).

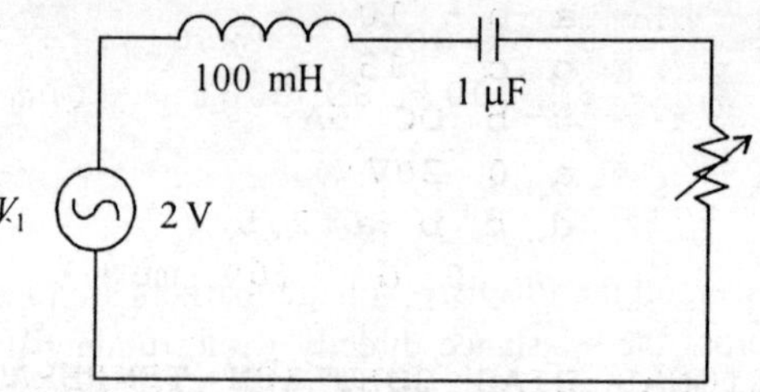

Fig. 15.12(a)

SOLUTION:

The circuit is drawn as shown in Fig. 15.12(b) and AC Sweep analysis is performed by specifying the start and end frequencies, the number of points. The resistance R_1 is made variable by using the command .PARAM or part PARAMETER. The element to be made variable is named inside braces "{ }" and it is given a value in the part PARAMETER. The different values to be swept can be specified as global parameters in the *parametric analysis* option. Voltage markers are used across the "rval" to directly get the voltage across it.

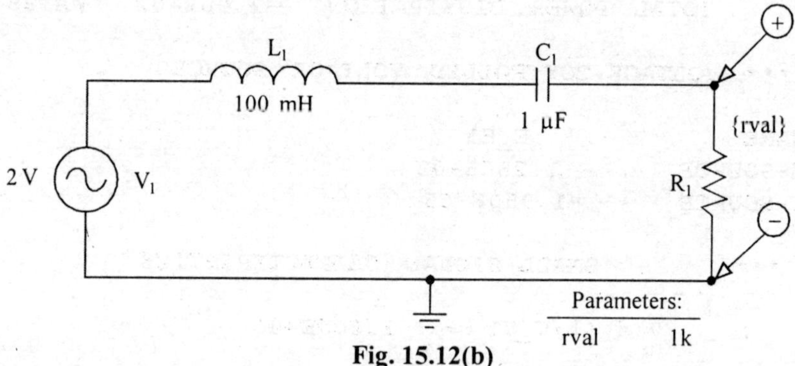

Fig. 15.12(b)

```
*  CIRCUIT DESCRIPTION

.PARAM               rval=1k

**  Analysis  setup  **

.ac  DEC  1001  10  1meg
.STEP   PARAM  rval  LIST
+  500,1k,2k
.OP

*  Schematics  Netlist  *

C_C1                $N_0001  $N_0002    1u
V_V1                $N_0003  0    AC  2
```

```
+SIN 0 2V 491.25 0 0 0
R_R1                    0 $N_0002    {rval}
L_L1                    $N_0003 $N_0001   100m
.probe
.END
```

The plot is seen on the PROBE. Voltage markers are placed across the resistance R_1 to get the voltage across the resistance directly while running the probe. The plot for current is selected from the Trace menu. Figure 15.12(c) shows the current for different values of R.

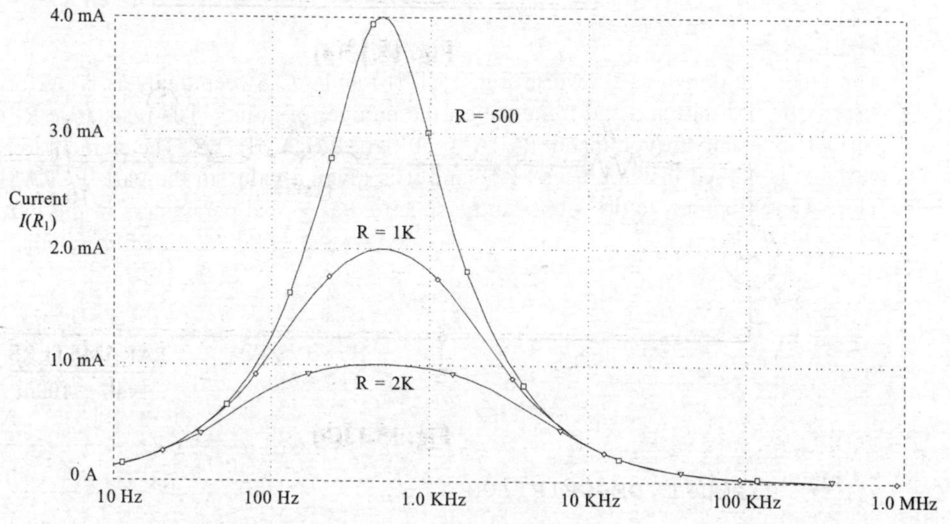

Fig. 15.12(c)

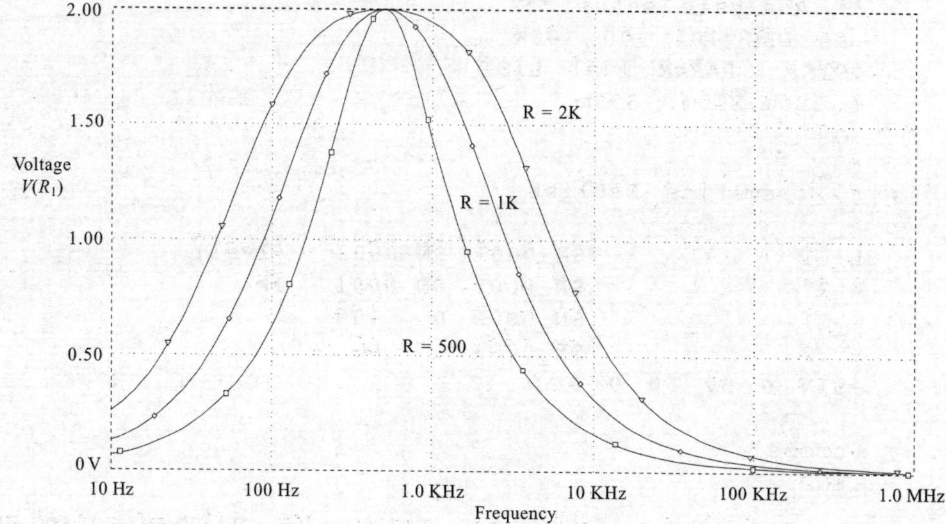

Fig. 15.12(d)

Note: The Phase frequency plot may be obtained making use of the trace menu.

EXAMPLE 15.10 Plot the amplitude frequency response for three different values of inductance in the circuit illustrated in Fig. 15.13(a).

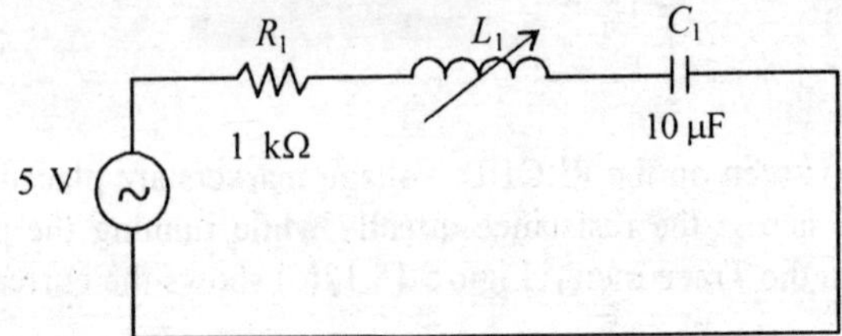

Fig. 15.13(a)

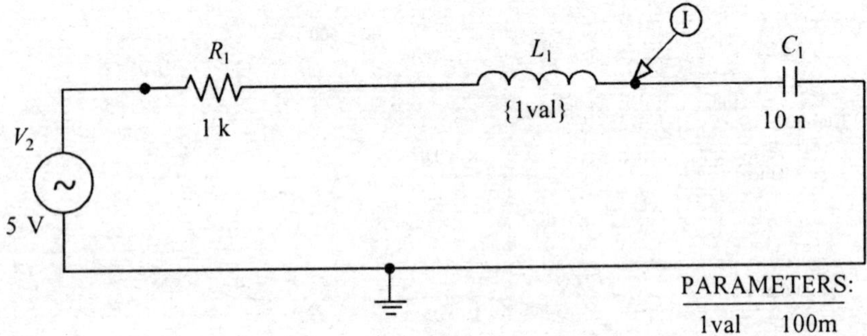

Fig. 15.13(b)

```
 ** CIRCUIT DESCRIPTION

.PARAM              lval=100m

** Analysis setup **
.ac DEC 101 100 100k
.STEP  PARAM lval LIST
+ 100m,250m, 500m
.OP

* Schematics Netlist *

L_L1           $N_0001 $N_0002   {lval}
R_R1           $N_0003 $N_0001   1k
C_C1           $N_0002 0    10n
V_V2           $N_0003 0    AC 2
+SIN 0 5v 50 0 0 0

.probe
.END
```

Here the three values of inductances are taken as 100 mH, 250 mH and 500 mH. Figure 15.13(b) shows the schematic for the circuit and the current–frequency response is shown in Fig. 15.13(c).

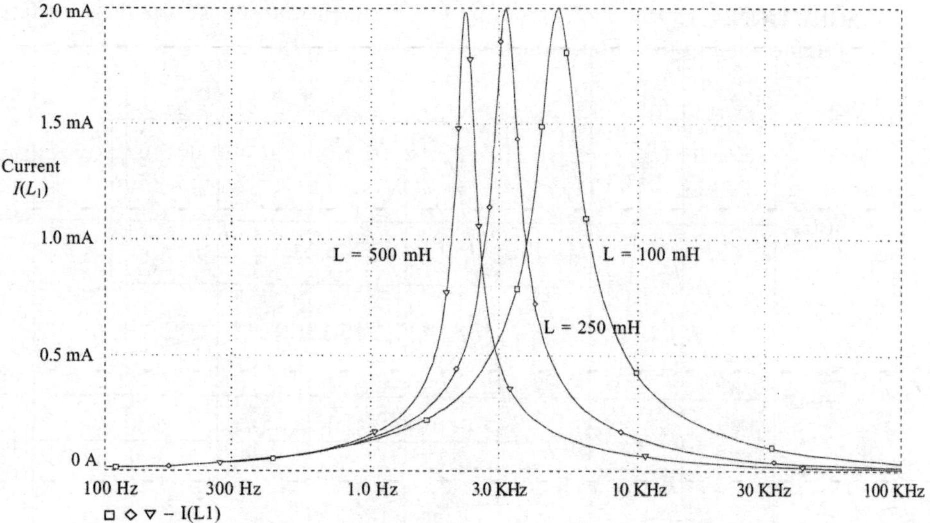

Fig. 15.13(c)

EXAMPLE 15.11 Obtain the resonant frequencies of the circuit given below:

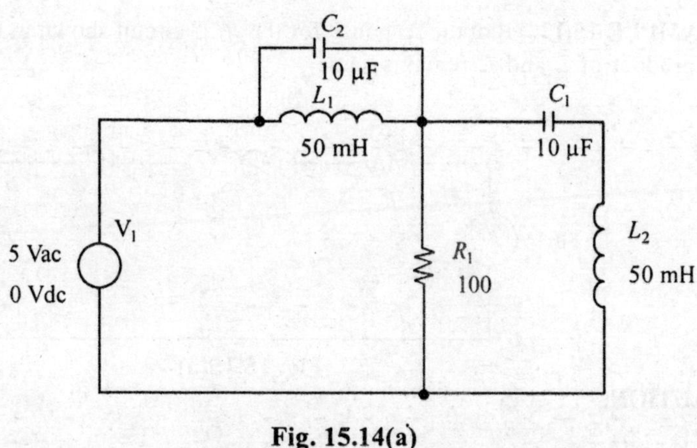

Fig. 15.14(a)

Plot the current wave form using the current marker. Perform the ac analysis and examine the wave form.

The plot reaches the maximum at frequencies 135 Hz and 375 Hz and reaches the minimum at 225 Hz [see Fig. 15.14(b). Hence the resonant frequencies are 135 Hz, 225 Hz and 375 Hz. **Ans.**

SOLUTION:

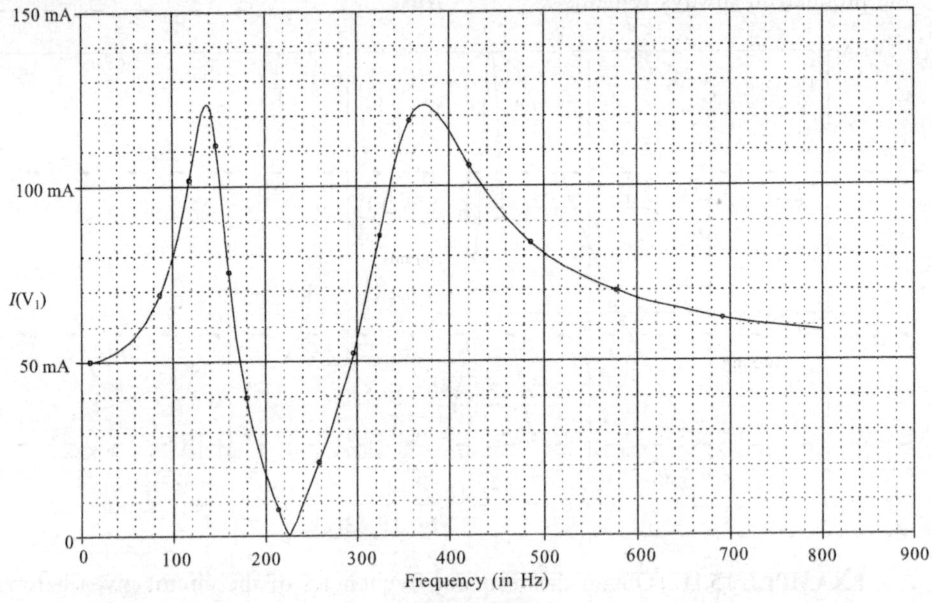

Fig 15.14(b)

EXAMPLE 15.12 Plot the response for the *RLC* circuit shown in Fig. 15.15(a) such that the product of *L* and *C* remains same.

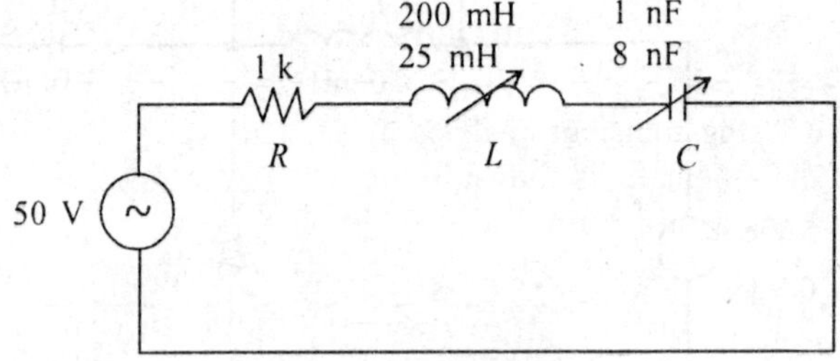

Fig. 15.15(a)

SOLUTION:

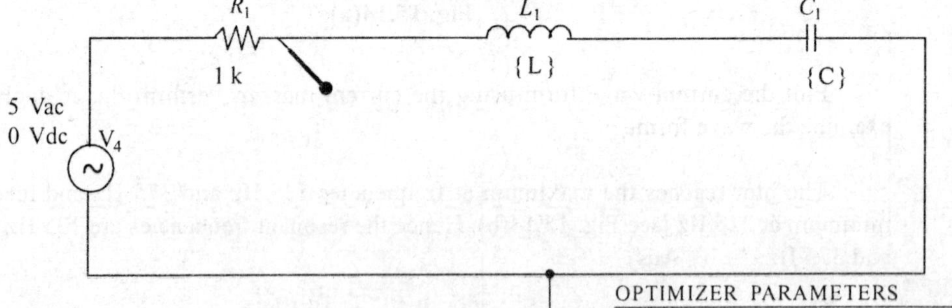

Fig. 15.15(b)

The schematic using optimizer parameters is shown in Fig. 15.15(b), and its response is shown in Fig. 15.15(c).

The values taken here are L = 200 mH, C = 1 nF and L = 25 mH, C = 8 nF. Hence the product LC always remains 2^{-10}. **Ans.**

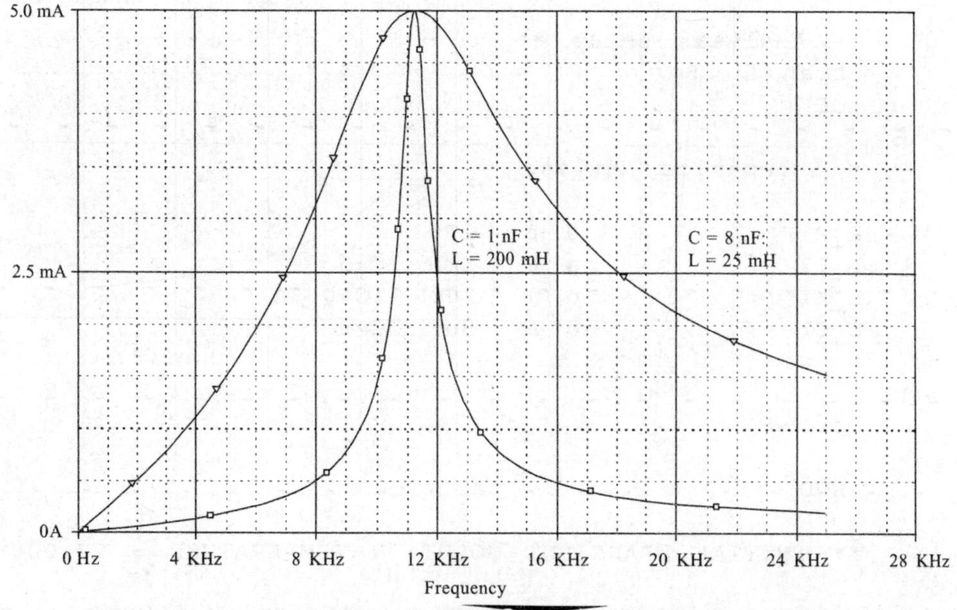

Fig. 15.15(c)

Transient Analysis

EXAMPLE 15.13 Using transient analysis plot the voltage across the resistance in the circuit, for which the schematic is shown in Fig. 15.16(a) during the interval $0 < t < 8$ ms. The inductance L_1 has a initial current of 0.8 A.

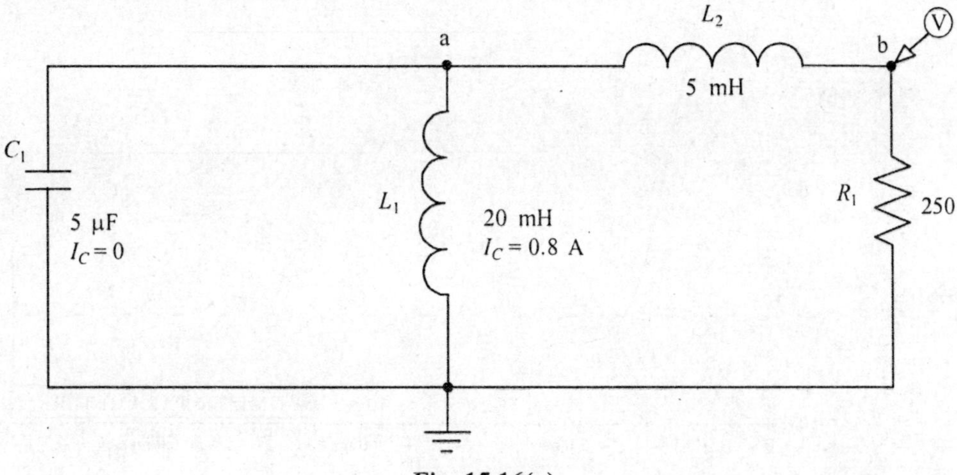

Fig. 15.16(a)

From the analysis setup menu (Fig. 15.3(b)). Transient analysis is selected with print step 5 μs and final time as 8 ms. The plot is depicted in Fig. 15.16(b).

```
** Analysis setup **
.tran 5u 8m
.OP

* Schematics Netlist *

R_R1          0 b   250
L_L2          a b   5mH  IC=0
L_L1          0 a   20mH  IC=0.8A
C_C1          0 a   5uF  IC=0


.probe

.END

** INITIAL TRANSIENT SOLUTION TEMPERATURE = 27.000 DEG C

NODE    VOLTAGE      NODE    VOLTAGE      NODE    VOLTAGE
NODE    VOLTAGE

(a)        .0016    (b)  400.0E-12

TOTAL POWER DISSIPATION    0.00E+00    WATTS
```

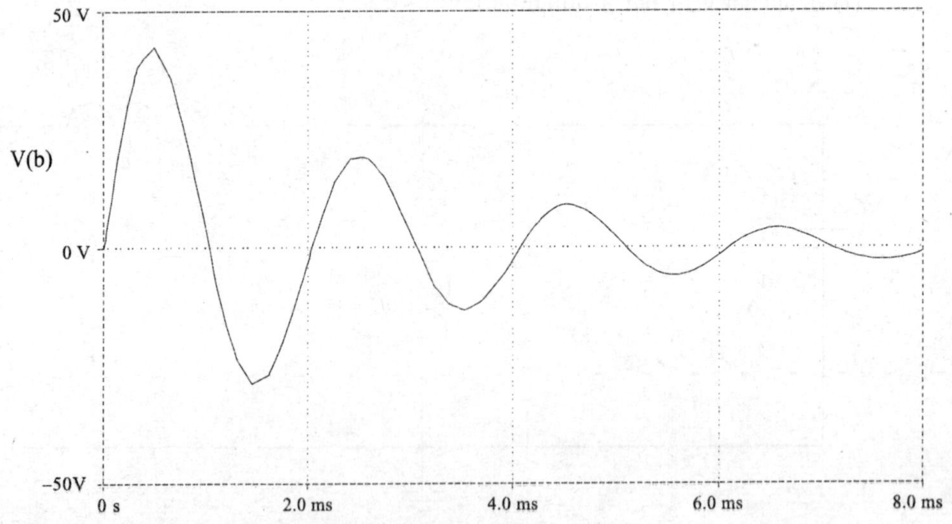

Fig. 15.16(b)

For transient analysis an input stimulus can be given to get various responses for a circuit. The stimuli may be a pulse, sin, PWL, digital clock, etc. A *Stimulus Editor* is used to create such analog and digital input signals or stimuli for use in simulation. The stimulus is first created and saved, this is then implemented in the required circuit.

EXAMPLE 15.14 Plot the response of the given circuit [see Fig. 15.17(a)] to a pulse input.

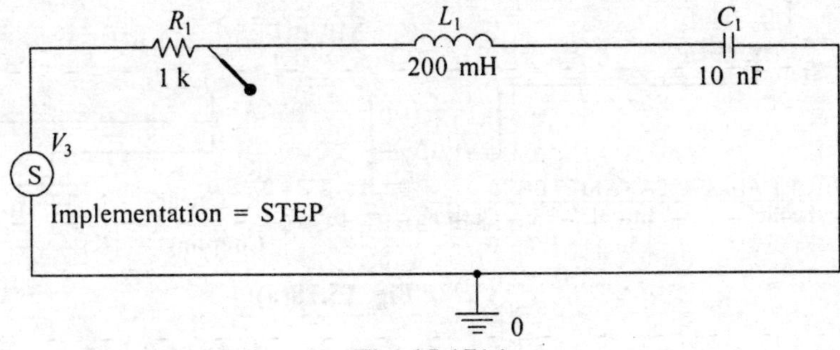

Fig. 15.17(a)

```
** Analysis setup **
.TRAN/OP   0 1ms  0  .001ms  SKIPBP
.PROBE

* Schematics Netlist *

L_L1             N02 N03   200mH
C_C1             N03 0   10n
R_R1             N01 N02   1K
V_V3             N01 0      STIMULUS=STEP

.STIMULUS step PULSE( 0 5 1ns 1ns 1ns 10 10 )
.END
```

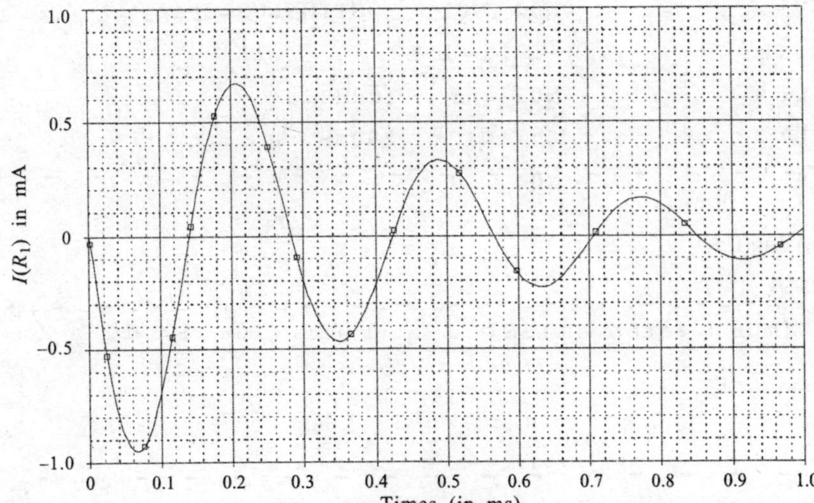

Fig. 15.17(b)

EXAMPLE 15.15 For the coupled circuit, whose schematic is given in Fig. 15.18(a), show the variation of output voltage for different values of the coupling coefficient *K*.

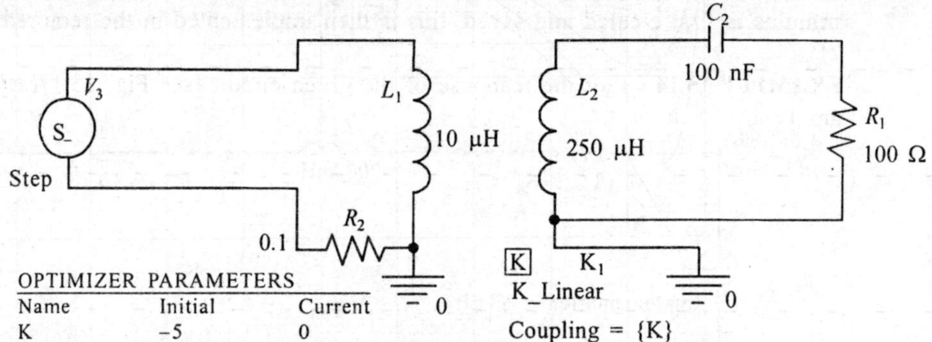

Fig. 15.18(a)

The part K_Linear is used to define the coupling coefficient. The name of the inductances between which the coupling is present is also to be specified. The part *K* can be varied by using optimizer parameters.

Here *K* is varied from 0.25 to 1 in steps of 0.25. The input is a pulse stimulus.

```
*  Analysis  setup  *
.TRAN    0  50us  0  10ns  SKIPBP
.STEP  LIN  PARAM  K  .25   1   .25
.PROBE

*  Schematics  Netlist  *
C_C2              N12   N13   100n
R_R1              0   N13   100
V_V3              N02  N01      STIMULUS=step
Kn_K1              L_L1  L_L2          {K}
L_L1              N02    0    10uH
L_L2              0   N12   250uH
R_R2              N01    0    .1

.param  K  =  .25
.END
.STIMULUS  step  PULSE(  0  5  0  1ns  1ns  10  10  )
```

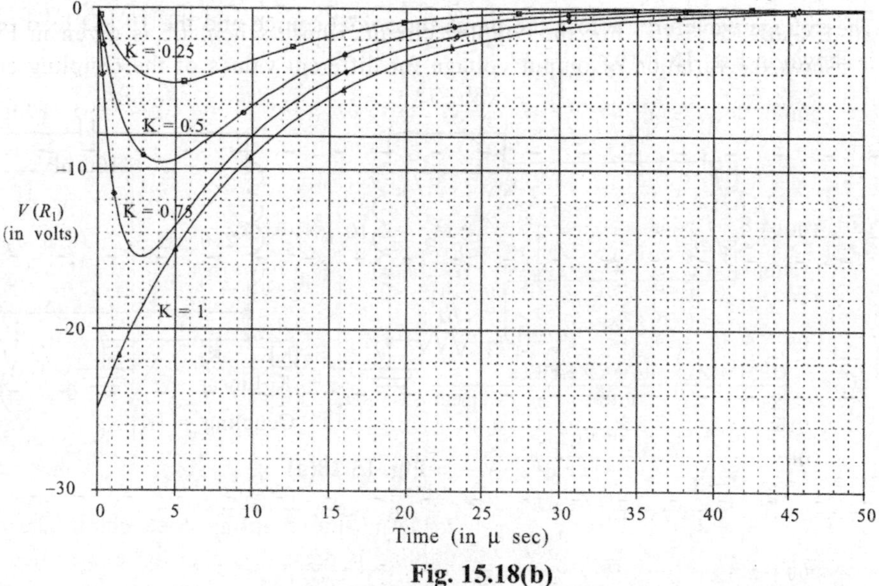

Fig. 15.18(b)

EXAMPLE 15.16 Simulate a Schmitt trigger and show the square wave generated. Also perform the Fourier analysis of the wave. See Fig. 15.19(a).

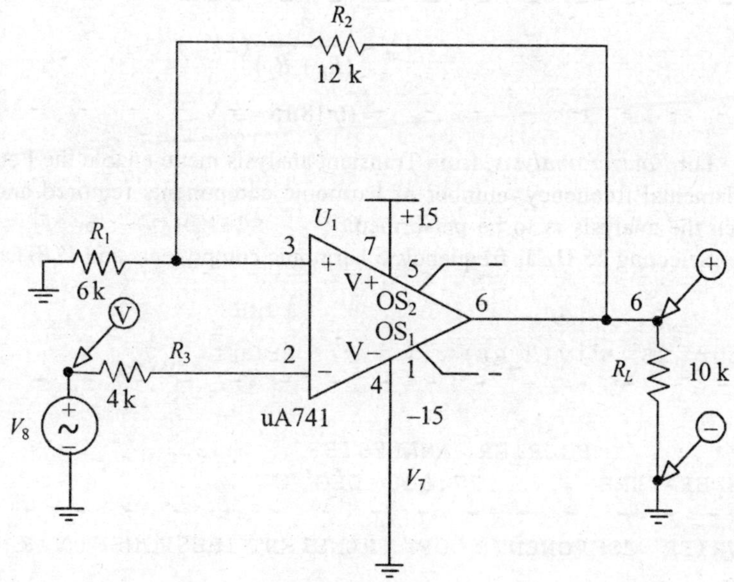

Fig. 15.19(a)

The voltage waveform across the input sinusoidal source and the 10 kΩ load resistance is shown in Fig. 15.19(b).

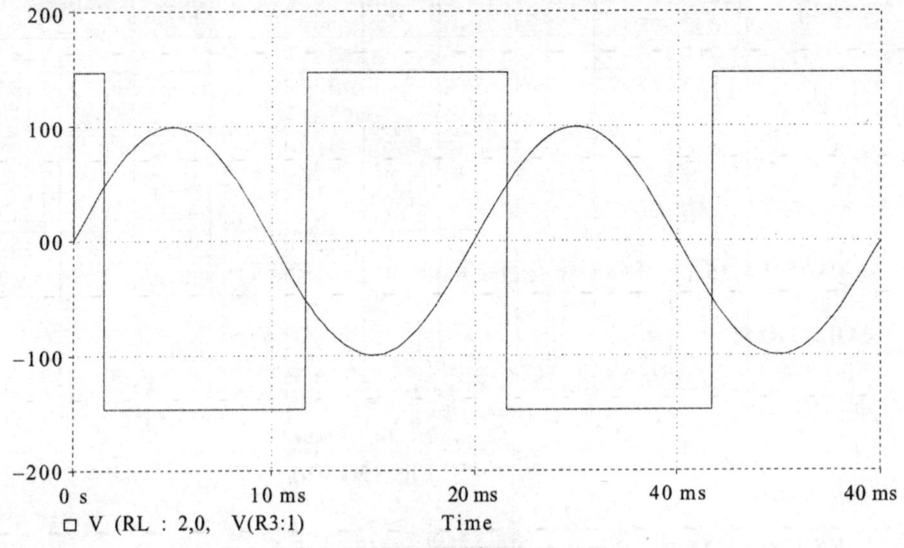

Fig. 15.19(b)

For a Schmitt Trigger,

$$V_{ut} = \frac{R_1}{(R_1 + R_2)}(\pm V_{\text{sat}})$$

$$= (6/18)15 = 5 \text{ V}$$

For *Fourier analysis*, from Transient analysis menu enable the Fourier and enter the fundamental frequency, number of harmonic components required and the variable for which the analysis is to be performed.

Selecting 25 Hz as frequency, 5 harmonic components and $V(R_L)$ as the variable, we get

```
.four 25 5 V(R_RL)


****        FOURIER ANALYSIS
TEMPERATURE =       27.000 DEG C

FOURIER COMPONENTS OF TRANSIENT RESPONSE V(R_RL)

DC COMPONENT =      1.182856E-02
```

HARMONIC NO	FREQUENCY (HZ)	FOURIER COMPONENT	NORMALIZED COMPONENT	PHASE (DEG)	NORMALIZED PHASE(DEG)
1	2.500E+01	1.904E-01	1.000E+00	-1.678E+01	0.000E+00
2	5.000E+01	1.847E+01	9.698E+01	-2.869E+01	-1.192E+01
3	7.500E+01	4.827E-01	2.535E+00	1.465E+02	1.633E+02
4	1.000E+02	2.727E-01	1.432E+00	1.465E+02	1.632E+02
5	1.250E+02	1.968E-01	1.033E+00	-1.700E+02	-1.532E+02

TOTAL HARMONIC DISTORTION = 9.702460E+03 PERCENT

EXAMPLE 15.17 Draw an astable multivibrator using Op-amp and show its waveforms.

SOLUTION:

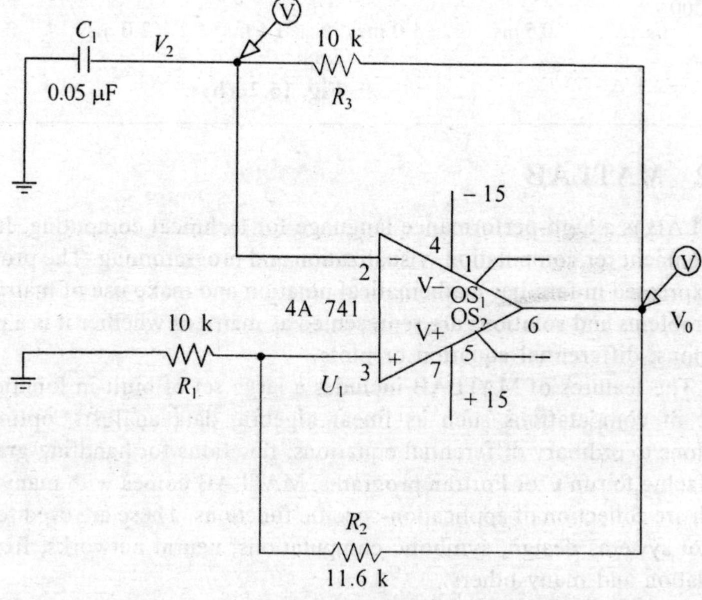

Fig. 15.20(a)

The voltages at nodes V_2 and V_0 are shown in Fig. 15.20(b).

Fig. 15.20(b)

15.2 MATLAB

MATLAB is a high-performance language for technical computing. It has an easy-to-use environment for computation, visualization and programming. The problems and solutions are expressed in familiar mathematical notation and make use of matrix methods. Most of the problems and solutions are represented as matrices whether it is a polynomial, transfer functions, differential equation or plots.

The features of MATLAB includes a large set of built-in functions used for a wide range of computations such as linear algebra, data analysis, optimization, numerical solutions to ordinary differential equations, functions for handling graphics and external interfacing to run C or Fortran programs. MATLAB comes with many optional *Toolboxes* which are collection of application-specific functions. These are used for signal processing, control systems design, symbolic computations, neural networks, fuzzy logic, wavelets, simulation and many others.

15.2.1 MATLAB Basics

The MATLAB Desktop (see Fig. 15.21) has many tools such as Command Window, Command History, Launch Pad, Help Browser, Current Directory Browser, Workspace Browser, Array Editor, Editor/Debugger and Graphics Window. The functions and commands are written in the *Command Window*. The previous command lines entered in the command window are stored in the Command History. The *Launch Pad* provides an easy access to the other tools. The workspace consists of the set of variables (named arrays) built up during a MATLAB session and stored in memory. The *browsers* are used to browse through the corresponding details. The *Array Editor* is used to view and edit a visual representation of one- or two-dimensional numeric arrays, strings and cell arrays of strings that are in the workspace. The *Editor/Debugger or Edit Window* is used to create and debug M-files, which are programs written to run MATLAB functions. The Editor/ Debugger provides a graphical user interface for basic text editing, as well as for M-file debugging. The graphical output are shown in the *Graphics/Figure window*.

File types: Information in MATLAB are stored as three types of files:

1. *M-Files* (.m extension). These are ASCII text files, created by users or buit-in function. These are of two types: Script file and Function file.
2. *Mat-Files* (.mat extension). These are binary data files which are created when data is saved using *save* command. Mat-files are loaded by using *load* command.
3. *Mex-Files* (.mex extension). These are Fortran or C programs callable by MATLAB.

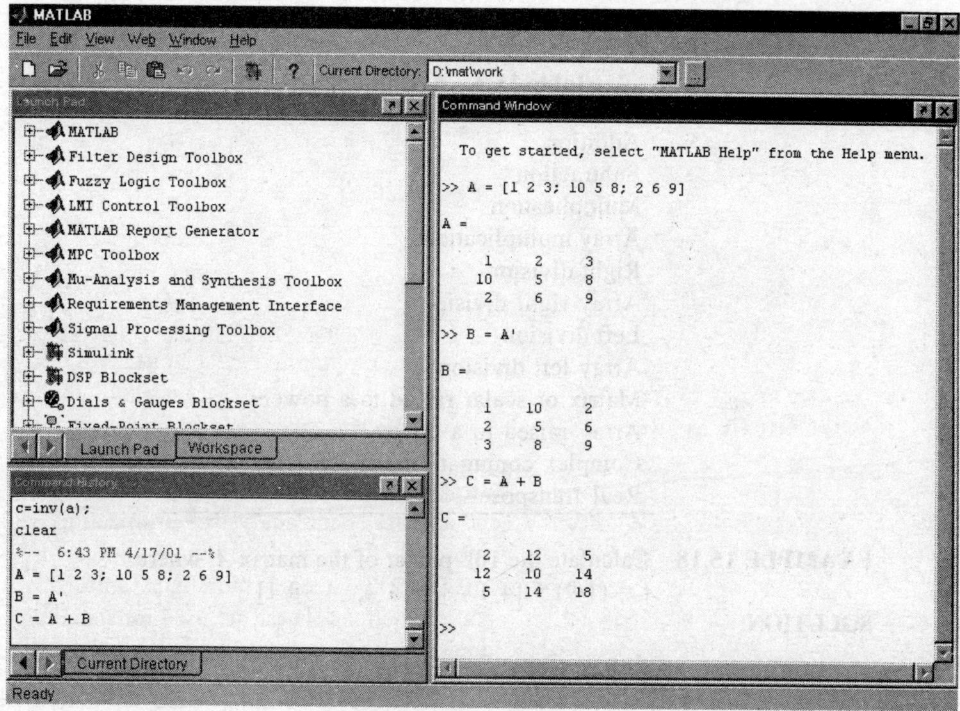

Fig. 15.21

15.2.2 Matrix

Matrix is the basic building block of MATLAB. The fundamental data type is the *array*.

The commands are written against the *command line prompt* ">>" in the command window. The variables can be given any name. MATLAB is case sensitive and hence a variable/matrix named "*A*" and "*a*" will be regarded as two different matrices. The functions are written in lower case. A matrix is defined by writing the elements in the rows separated by blank spaces and each row separated by semi colon within square brackets "[ ]". The "Enter or Return" key is used to show the output of the current statement.

>> A = [1 2 3; 10 5 8; 2 6 9]

On pressing 'Enter'

A =

1	2	3
10	5	8
2	6	9

If the screen output is to be suppressed, a semi-colon can be used at the end of the command statement. All outputs except graphics and on-line help commands are suppressed.

Some of the *arithmetic operators* are shown in Table 15.3. These can be used for matrix operations or matrix-scalar operations. For matrix operations, the order of the matrices must be compatible.

Table 15.3 Arithmetic Operations

Addition	+
Subtraction	−
Multiplication	*
Array multiplication	.*
Right division	/
Array right division	./
Left division	\
Array left division	.\
Matrix or scalar raised to a power	^
Array raised to a power	.^
Complex conjugate transpose	'
Real transpose	.'

EXAMPLE 15.18 Calculate the 10th power of the matrix A, where $A = (1/9) * [4\ 1\ -8;\ 7\ 4\ 4;\ 4\ -8\ 1]$

SOLUTION:

```
>> A = (1/9)*[4  1  -8;  7  4  4  ;  4  -8  1]

A =

        0.4444        0.1111       -0.8889
        0.7778        0.4444        0.4444
        0.4444       -0.8889        0.1111

>> a1= A^10

a1 =

       -0.1111        0.8889       -0.4444
        0.8889       -0.1111       -0.4444
       -0.4444       -0.4444       -0.7778
```

To get the result in a better form, the result has been multiplied by 9.

```
>> a2  =  a1*9

a2  =

    -1.0000         8.0000         -4.0000
     8.0000        -1.0000         -4.0000
    -4.0000        -4.0000         -7.0000
```

EXAMPLE 15.19 For the given matrix *A*, find the transpose and add it to the matrix *A*.

$$A = \begin{bmatrix} 1 & 2 & 3 \\ 10 & 5 & 8 \\ 2 & 6 & 9 \end{bmatrix}$$

SOLUTION:

```
>>  A  =  [1  2  3;  10  5  8;  2  6  9];
>>  B  =  A'

B  =

        1        10        2
        2         5        6
        3         8        9

>>  C  =  A  +  B

C  =

        2        12        5
       12        10       14
        5        14       18
```

The variables defined in the command window retain their value until they are cleared using the "*clear*" command.

EXAMPLE 15.20 Find the product of the *A* and *B* matrices in the previous Example and then find its inverse.

SOLUTION:

```
>>  A  =  [1  2  3;  10  5  8;  2  6  9];
>>  B  =  A' ;
>>  M  =  A*B
M  =

            14        44        41
            44       189       122
            41       122       121
```

Note: Consider a matrix $AX = B$.

Then

$$X = A\backslash B$$

which is same as X = (Inverse of A) * B

The functions used to do some basic matrix operation, are listed in Table 15.4. Let A be a square matrix.

Table 15.4 Matrix Functions

$D = \det(A)$	D returns the determinant of a matrix A
$I = \text{inv}(A)$	I returns the inverse of matrix A
$R = \text{rank}(A)$	R gives the rank of matrix A
$E = \text{eig}(A)$	E gives the eigen values of matrix A in column form
$[V,D] = \text{eig}(A)$	D matrix is the canonical form of A with the eigen values as the main diagonal elements and V is the Modal matrix gives the eigen vectors as columns
eye (n)	Creates a $n \times n$ unit matrix
zeros(m,n)	Creates an $m \times n$ matrix of zeros
ones(m,n)	Creates an $m \times n$ matrix of ones
rand(m,n)	Creates an $m \times n$ matrix of random numbers
diag(v)	Generates a diagonal matrix with vector v on the diagonal

Note: The variables used above may be replaced by any other variable.

EXAMPLE 15.21 Find the inverse of the given matrix D.

SOLUTION:

```
>> D = [3/4 -7/12 2/3;  -1/4  5/12 -1/3;  -3/4 1/4 0]

D =

      0.7500    -0.5833        0.6667
     -0.2500     0.4167       -0.3333
     -0.7500     0.2500             0

>>  inv(D)

ans =

      1.0000     2.0000       -1.0000
      3.0000     6.0000        1.0000
      3.0000     3.0000        2.0000
```

Note: The matrix elements can be defined as fractions also.

Matrix indexing and vectors: The elements of an existing matrix can be specified by their row and column indices. $A(i, j)$ refers to element of matrix A in ith row and jth column. The statement `A(m:n, p:q)` refers to rows m to n and columns p to q of matrix A. To specify a full row or column in a matrix, only the colon (:) as the row or column index. Thus, $P(:, 2:5)$ refers to elements in column 2, 3, 4 and 5 and all row of matrix P.

A vector of numbers can be created over a range of values by the following command:

```
v = Initial value : increment : Final value
```

By default the increment is 1.

Other commonly used vector declarations are:

`linspace(a,b,n)` this generates n linearly spaced vectors from a to b.
`logspace(a,b,n)` this generates n logarithmically spaced vectors from 10^a to 10^b.

15.2.3 User-written Programs

The user can write programs in *m*-files. These are of two types:

Script file: A script file is a user created file with a set of MATLAB commands. This is written in the Edit Window which is taken from the FILE menu by selecting the NEW + M-File. It is saved with **.m extension** and executed by simply typing the name of the file (without extension) on the command prompt.

It is normally started using comment lines. The variables defined here are global variables. To view the code of the script file, use

```
>> type <script file name>
```

A script file to compute the loop currents and total current through AB in Fig. 15.22 for different values of voltage V and resistance R is shown below. (The statement followed by % is considered as comment line.)

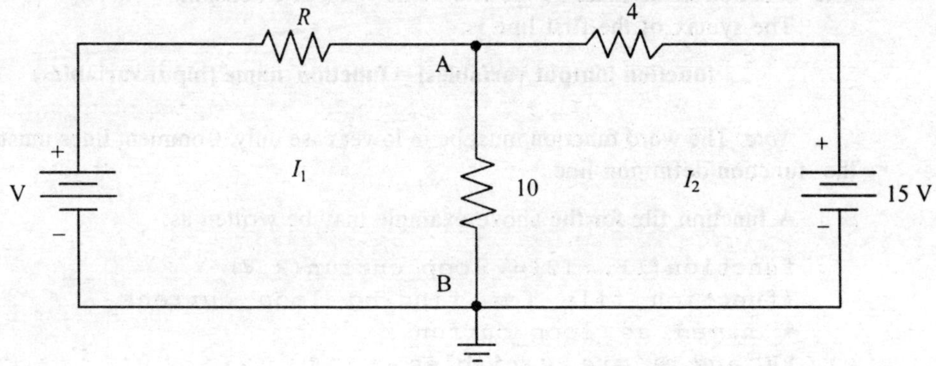

Fig. 15.22

```
%script file for finding loop current
% named as loop_cur.m
%V and R are variables.
Resis =[R+10   -10 ; -10   4+10];
Volt  =  [V ;  -15 ];
```

```
loopCur = Resis \ Volt; % for inv(Resis)
I1 = loopCur(1,1)
I2 = loopCur(2,1)
Cur_AB = I1-I2
```

The statements in the command window are:

```
>> R=8; V=35;
>> loop_cur

I1 =

    2.2368

I2 =

    0.5263

Cur_AB =

    1.7105
```

Note: The script file must not be named same as the variables used in it. The name must start with an alphabet.

Function file: It is like a script file except that the variables are all local. The function file begins with the *function definition line* which defines the list of input and outputs. The function name must be the file name without extension.

The syntax of the first line is:

function [output variables] = function_name (input variables)

Note: The word function must be in lower case only. Comment lines must start after the function definition line.

A function file for the above example may be written as:

```
function[I1, I2]= loop_curfun(R,V)
%function file for finding loop current
% named as loop_curfun.m
%V and R are variables.
Resis =[R+10  -10 ; -10  4+10];
Volt = [V ; -15 ];
loopCur = Resis \ Volt; % for inv(Resis)
I1 = loopCur(1,1);
I2 = loopCur(2,1);
Cur_AB = I1-I2
```

In the command window, it is executed as:

```
>>  [P,Q]=  loop_curfun(8,32)

Cur_AB  =

         1.6316
P  =

         1.9605
Q  =
         0.3289
```

Note: The name of the output variables has no significance.

15.2.4 Graphics

MATLAB can display vectors and matrices as graphs, annotate and print these graphs. We will discuss some of the methods for doing so.

`plot` is one of the most basic command and can be used in many ways. Its basic syntax is:

plot(x-values, y-values, 'style-options')

x-values are the vectors specifying the *x*-coordinate values and *y*-values are for *y*-coordinate. Style-option is an optional argument which may be used to specify the style in which the plots should appear in the output (e.g. Solid line, dashed, starred, dotted...). If only one argument is specified e.g. `plot(y)` then plot against *y* and its index is drawn if *y* is real. If *y* is complex then it plots between real(*y*) and img(*Y*).

Multiple plots can also be drawn by simply adding more arguments. e.g. Plot(x_1,y_1,x_2,y_2)

`fplot` plots functions of a single variable. The function must be specified in terms of *x*.

fplot('function of x', [x_min x_max])

The plots may be annotated by using command like `xlabel`, `ylabel`, `title`, `text`.

EXAMPLE 15.22 Plot two graphs for functions $y = e^{t/10}$ sin *t* for *t* = 0 to 20 s and *y* = *t* cos *t* for *t* = 0 to 6π with labels.

SOLUTION:

```
>>  x=  0:0.1:20;

>>  y=exp(0.1*x).*sin(x);
>>  [k,s]=fplot('x.*cos(x)',  [0  6*pi]);   %defining vectors
k  and  s
>>  plot(x,y,'*',k,s)
>> title('SAMPLE PLOT'); xlabel('TIME t'); ylabel('Y OUTPUT ');
>>  legend('e^t^/^1^0  sint','tcost')
```

The output is shown in Fig. 15.23.

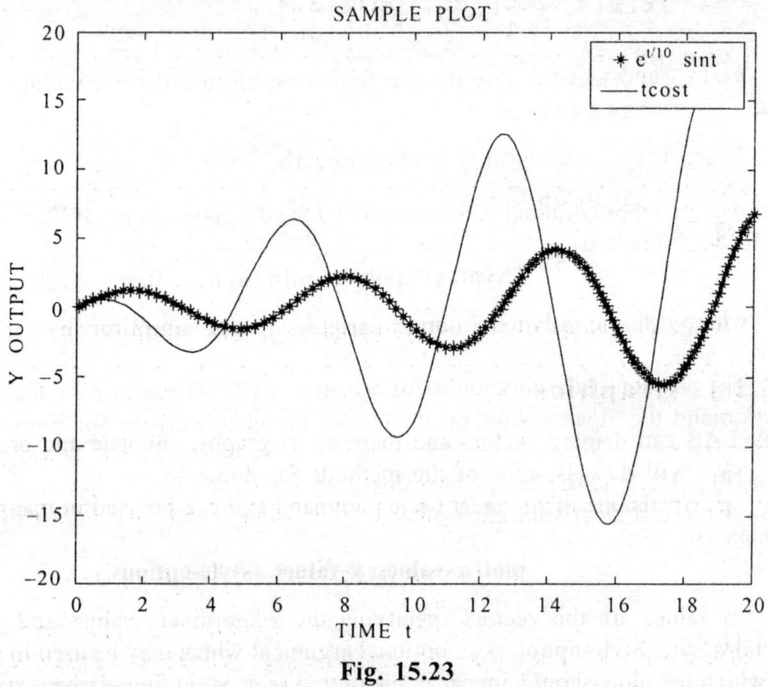

Fig. 15.23

Other functions are area, bar, comet, contour, hist, pie, semilogx and semilogy. `plot3, mesh, surf, pie3,` are some of the 3-dimensional graphic functions. The reader may use help command to know more about these and other MATLAB commands.

>>help <function name>

Advanced applications: MATLAB represents polynomials as row vectors containing coefficients ordered by

descending powers. For example, the polynomial

$$x^3 - 2x - 5$$

can be entered into MATLAB as:

```
p = [1 0 -2 -5];
```

The `roots` function calculates the roots of a polynomial.

```
r = roots(p)
r =

      2.0946
     -1.0473 + 1.1359i
     -1.0473 - 1.1359i
```

By convention, MATLAB stores roots in column vectors. The function `poly` returns to the polynomial coefficients.

```
p2 = poly(r)
p2 =
          1        8.8818e-16         -2         -5
```

`poly` and `roots` are inverse functions, up to ordering, scaling, and round off error.

Note: The second term is tending to zero.

Any given polynomial can be evaluated for different values of the variable by using `polyval` function

Syntax: polyval(p,[n_1 n_2 n_3 ...]),

where p is the polynomial and n_1, n_2, n_3 are the values for the variable.

The characteristic polynomial of a matrix $(\lambda I - A)$ can also be found using `poly` function and the eigen values by `roots` of the characteristic polynomial.

EXAMPLE 15.23 Find the eigen values, vectors of the matrix B and check it with the roots of the characteristic equation.

$$B = \begin{bmatrix} 5 & 1 \\ -2 & 2 \end{bmatrix}$$

SOLUTION:

```
>>  B=[ 5   1 ;   -2   2 ]
B =
        5         1
       -2         2

>>  [v,d]= eig(B)
v =
        0.7071     -0.4472
       -0.7071      0.8944
d =
         4         0
         0         3

>>  Chs_Poly  =  poly(B)

Chs_Poly  =

        1       -7       12

>>  r  =  roots(Chs_Poly)

r  =
        4
        3
```

EXAMPLE 15.24 Evaluate the determinant and the inverse of the matrix:

$$\begin{bmatrix} 3+2i & 0+2i \\ 0+2i & 12+8i \end{bmatrix}$$

SOLUTION:

```
>> c1=[3+2j   2j  ;  2j 12+8j]

c1 =
   3.0000  + 2.0000i                      0  + 2.0000i
        0  + 2.0000i            12.0000  + 8.0000i

>> det(c1)

ans =

   24.0000 +48.0000i

>> inv(c1)

ans =
    0.2333  -  0.1333i                     -0.0333  -  0.0167i
   -0.0333  -  0.0167i                      0.0583  -  0.0333i
```

EXAMPLE 15.25 Solve Example 6.1 using MATLAB.

SOLUTION:

```
>> r=[12  -4  -8  ;  -4  19  -5;  -8  -5  15]

r =
        12        -4        -8
        -4        19        -5
        -8        -5        15

>> ri=  inv(r)

ri =
        0.1935      0.0744      0.1280
        0.0744      0.0863      0.0685
        0.1280      0.0685      0.1577

>> i=ri*[100;0;0]

i =

     19.3452
      7.4405
     12.7976
```

Network functions: The network functions can be represented in transfer function, state space or zero pole gain forms. Consider a transfer function:

$$t = \frac{s+1}{s^2+2s+2}$$

Transfer functions are written using `tf` function.

Syntax : tf(num , den)

The num (numerator) may be directly given or may be represented as a matrix polynomial as shown below.

```
>>  a=[1  1];
>>  b=[1  2  2];
>>  c=  tf(a,b)

Transfer  function:
   s  +  1
----------------
s^2  +  2s  +  2

OR
This  can  be  represented  as:

>>  t=  tf([1  1],[1  2  2])

Transfer  function:
   s  +  1
----------------
s^2  +  2s  +  2
```

The other functions are:
`zpk` is used to create zero-pole-gain models (ZPK objects) or to convert TF or SS models to zero-pole-gain form.

Syntax: zpk(z,p,k)

`ss` is used to create real- or complex-valued *state-space models* (SS objects) or to convert transfer function or zero-pole-gain models to state space. Here *a,b,c* and *d* are the four matrices.

Syntax: ss(a,b,c,d)

EXAMPLE 15.26 For the given transfer function c find the `zpk` and state space form:

$$C = (s + 1)/s^2 + 2s + 2$$

SOLUTION:

```
>>  c=  tf([1  1],[1  2  2]);

Transfer  function:
    s  +  1
---------------
s^2  +  2s  +  2


>>  zpk(c)

Zero/pole/gain:
       (s+1)
--------------
(s^2  +  2s  +  2)


>>  ss(c)

a  =
                    x1       x2
          x1        -2       -1
          x2         2        0


b  =
              u1
       x1      1
       x2      0


c  =
            x1          x2
       y1    1          0.5


d  =
          u1
       y1  0

Continuous-time  model.
```

Thus, here $A = \begin{bmatrix} -2 & -1 \\ 2 & 0 \end{bmatrix}$ $B = \begin{bmatrix} 1 \\ 0 \end{bmatrix}$ $C = [1 \quad 0.5]$ $D = [0]$ **Ans.**

`pole` computes the poles p of the SISO or MIMO LTI model sys. Here t is a transfer function in any form(tf, zpk or ss).

Syntax: p= pole(t)

`zero` computes the zeros of SISO systems and the transmission zeros of MIMO systems.

Syntax: z = zero(t)

`pzmap` Compute the pole-zero map of one/n LTI model.

Syntax: **pzmap(t)**
pzmap(t1,t2,...,tn)

`rlocus` computes the Evans root locus of a SISO open-loop model. The root locus gives the closed-loop pole trajectories as a function of the feedback gain K (assuming negative feedback).

Syntax: **rlocus(t)**
rlocus(t,k)

`bode` computes the magnitude and phase of the frequency response of LTI models. It produces a Bode plot on the screen. The magnitude is plotted in decibels (dB), and the phase in degrees.

Syntax: bode(t)

EXAMPLE 15.27 Find the `zpk` form of the given system, its poles and zeroes, then plot its pole-zero map, the root locus and Bode plot.

$$s = \frac{-10s^2 + 20s}{s^5 + 7s^4 + 20s^3 + 28s^2 + 19s + 5}$$

SOLUTION:

```
>> s = tf([-10 20 0],[1 7 20 28 19 5])

Transfer function:
                 -10 s^2 + 20 s
----------------------------------------
s^5 + 7 s^4 + 20 s^3 + 28 s^2 + 19 s + 5

>> zpk(s)

Zero/pole/gain:

     -10 s (s-2)
----------------------
(s+1)^3 (s^2 + 4s + 5)
```

```
>> pole(s)

ans =

   -2.0000 + 1.0000i
   -2.0000 - 1.0000i
   -1.0000
   -1.0000 + 0.0000i
   -1.0000 - 0.0000i

>> zero(s)

ans =

     0
     2

>> pzmap(s)
>> rlocus(s)
>> bode(s)
```

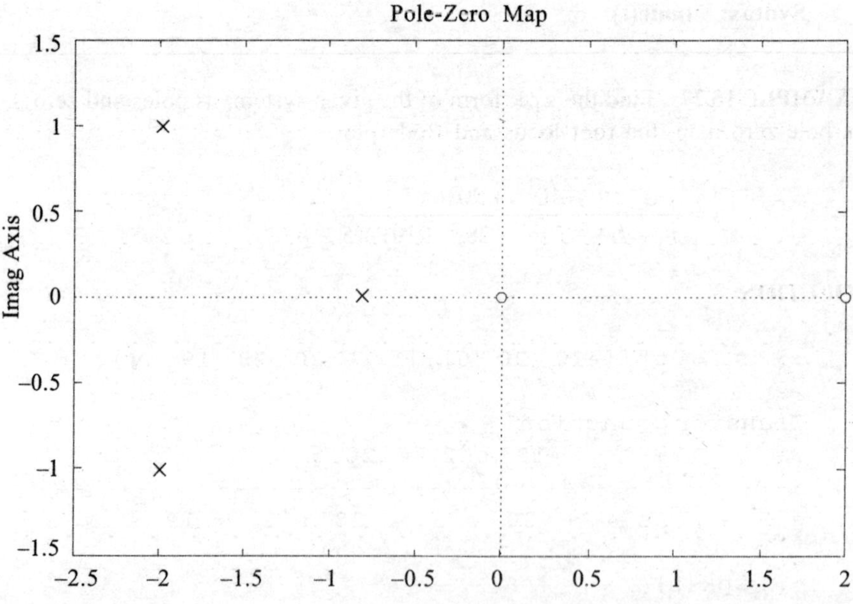

Fig. 15.24(a) Pole-Zero map.

The Fig. 15.24a, b and c shows respectively the pole-zero map. Root locus and the Bode plot for the system.

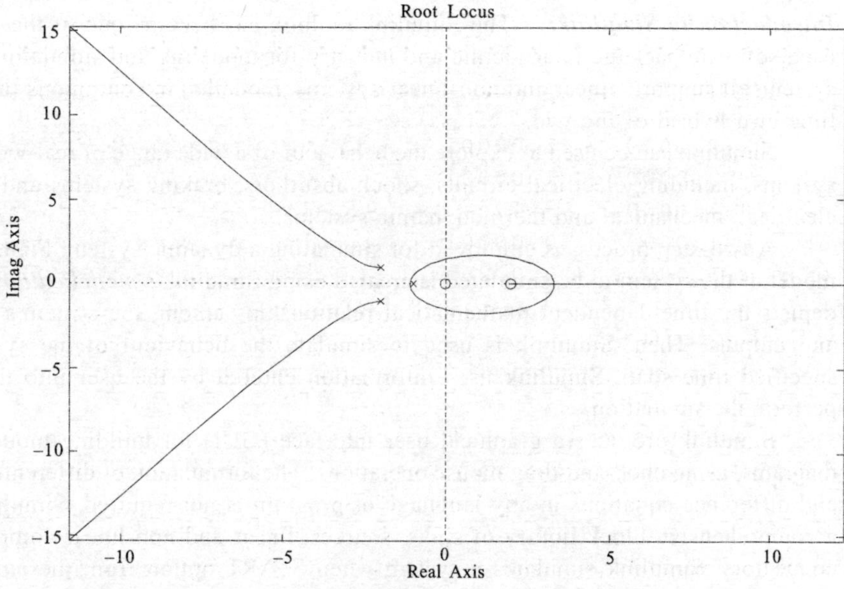

Fig. 15.24(b) Root locus.

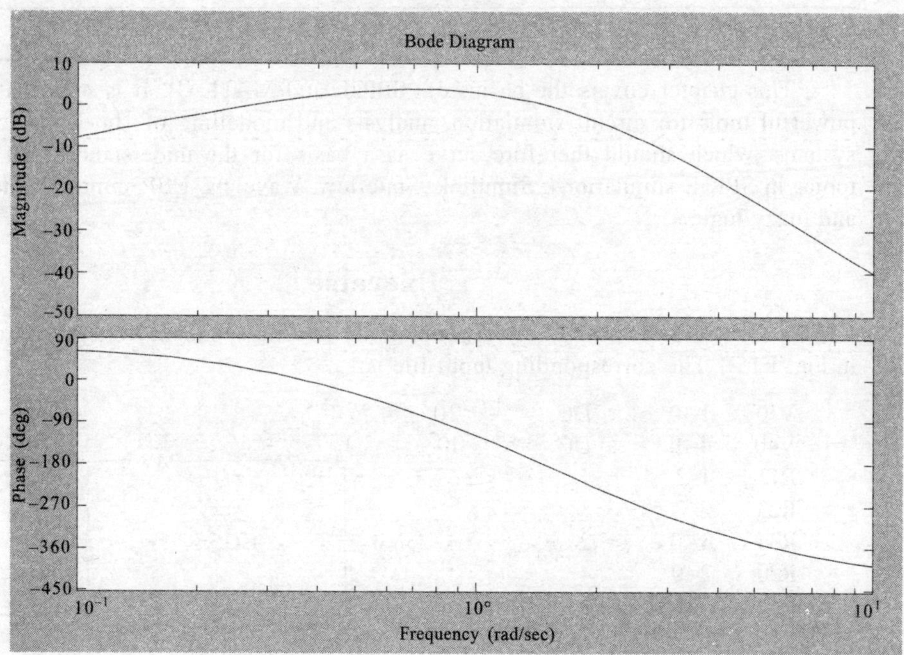

Fig. 15.24(c) Bode plot.

Introduction to Simulink: The Simulink toolbox has become one of the most widely used software package in academic and industry for modeling and simulating dynamical systems. It supports linear and non-linear systems, modelled in continuous time, sampled time or a hybrid of the two.

Simulink can be used to explore the behaviour of a wide range of real-world dynamic systems, including electrical circuits, shock absorbers, braking systems and many other electrical, mechanical and thermodynamic systems.

A two-step process is employed for simulating a dynamic system. First, a graphical model of the system to be simulated is created using Simulink's *model editor*. The model depicts the time-dependent mathematical relationships among the system's input states and outputs. Then, Simulink is used to simulate the behaviour of the system over a specified time span. Simulink uses information entered by the user into the model to perform the simulation.

Simulink provides a graphical user interface (GUI) for building models as block diagrams, using click-and-drag mouse operations. The formulation of differential equations and difference equations in any language or program is not required. Simulink includes a comprehensive block library of sinks, sources, linear and non-linear components, and connectors. Simulink simulates a system when START option from the *model editor's* Simulation menu is chosen, with the system's model open. Detailed study of Simulink is out of the scope of this book.

SUMMARY

This chapter covers the basics of SPICE and MATLAB. It is seen that these are powerful tools for circuit simulation, analysis and modelling of linear and non-linear systems, which should therefore serve as a basis for the understanding of advanced topics in SPICE simulations, Simulink, Stateflow, Wavelets, DSP, control system design and fuzzy logic.

Exercise

E15.1 Consider the simple circuit containing two independent voltage sources shown in Fig. E15.1 The corresponding input file is:

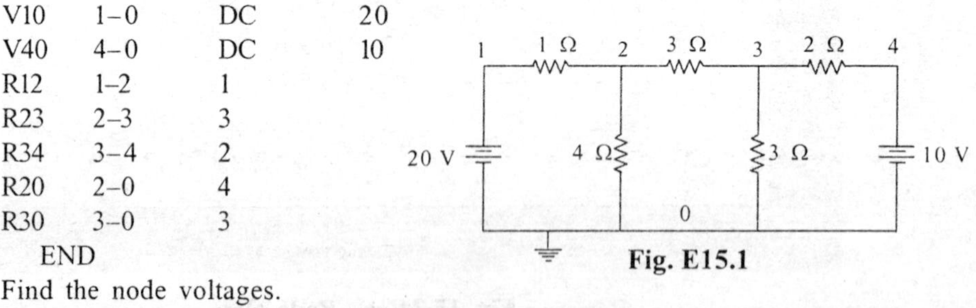

```
V10   1–0   DC   20
V40   4–0   DC   10
R12   1–2   1
R23   2–3   3
R34   3–4   2
R20   2–0   4
R30   3–0   3
 END
```


Fig. E15.1

Find the node voltages.

Ans: NODE VOLTAGE	NODE VOLTAGE	NODE VOLTAGE	NODE VOLTAGE
(1) 20.000	(2) 14.4000	(3) 8.4000	(4) 10.0000

Part 3

Advanced Topics in Network Analysis

State Variable Analysis

16

CHAPTER CONTENTS

Analysis and design of physical systems involve the following six stages:

Identification—measurement of system variables
Representation—analytical form or a mathematical model
Solution—determination of system response
Stability
Design
Optimization

The two common methods available for analysis are the *conventional* approach and the *modern* approach. The former is based on complex variables whereas the latter is based on state variables.

The conventional approach concentrates on the use of *high order differential equations* in representation and pole zeros in analysis. The main mathematical tool is Laplace transformation; and the *analog* computer facilitates computation.

The state variable approach relies on a set of *first order differential equations* in representation and linear transformation in analysis. The main mathematical tool is matrix theory; and the *digital* computer facilitates computation.

The more important advantages of state variables are: (1) the set of first order differential equations is quite suitable for solution by digital computers and (2) the technique is applicable to time-varying and non-linear systems.

This chapter is solely devoted to analysis in state space. State equations for simple systems (networks) may be written by inspection, but more complicated networks require the use of network topology. (See Chapter 7)

FORMULATION OF STATE MODEL

The state of a system refers to the past, present and future conditions of the system. It is convenient to define a set of state variables and state equations to represent systems.

16.1 DEFINITION OF STATE VARIABLES

The state variables are defined as the minimal set of variables $x_1, x_2, \ldots, x_n$ such that the knowledge of these at any point of time t, plus information on the input excitation subsequently applied is sufficient to determine the state at any future time. These definitions are applicable to electrical as well as mechanical systems.

The state variables are directly related to the energy storage elements of the network. By assigning v_c and i_L as the state variables, a complete description of the past history (initial values) and present and future states of the network is made available.

EXAMPLE 16.1 Determine the state equations for the network in Fig. 16.1

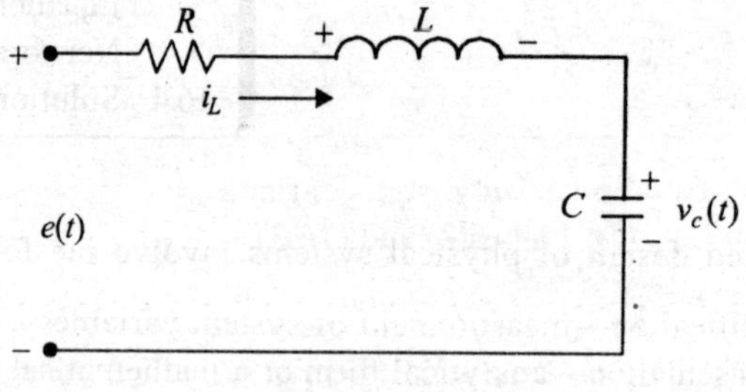

Fig. 16.1

SOLUTION:

The conventional method is to write KVL for the circuit leading to an integro-differential equation which can be reduced to a second order differential equation. In state space, the current in C is given by

$$C\frac{dv_c}{dt} = i_L$$

and the voltage across L is:

$$L\frac{di_L}{dt} = -v_c - Ri_L + e$$

In matrix form,

$$\begin{bmatrix} \dfrac{dv_c}{dt} \\ \dfrac{di_L}{dt} \end{bmatrix} = \begin{bmatrix} 0 & 1/C \\ 1/L & -R/L \end{bmatrix} \cdot \begin{bmatrix} v_c \\ i_L \end{bmatrix} + \begin{bmatrix} 0 \\ 1/L \end{bmatrix} e \qquad \textbf{Ans.}$$

EXAMPLE 16.2 Write the state equations for the circuit shown in Fig. 16.2.

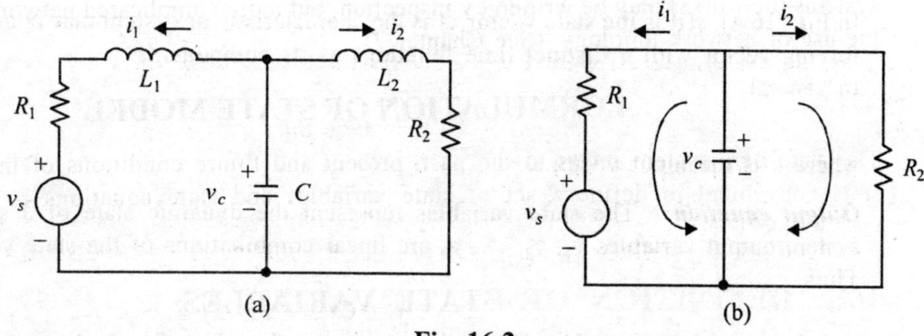

Fig. 16.2

Select a tree with a capacitor only, and none of the inductors. Use the tree-branch capacitor voltage v_c and the link inductor currents i_1 and i_2 as the state variables.

KCL for the junction is

$$C\frac{dv_c}{dt} + i_1 + i_2 = 0 \tag{16.1}$$

KVL for the fundamental loops are:

$$L_1\frac{di_1}{dt} - i_1R_1 + v_s - v_c = 0 \tag{16.2}$$

$$L_2\frac{di_2}{dt} + i_2R_2 - v_c = 0 \tag{16.3}$$

In these equations, only state variables occur and hence no rearrangement is required. From Eqs. (16.1), (16.2) and (16.3),

$$\frac{dv_c}{dt} = -\frac{1}{C}i_1 - \frac{1}{C}i_2$$

$$\frac{di_1}{dt} = \frac{1}{L_1}v_c - \frac{R_1i_1}{L_1} - \frac{v_s}{L_1}$$

$$\frac{di_2}{dt} = \frac{1v_c}{L_2} - \frac{R_2i_2}{L_2}$$

In matrix form,

$$\frac{d}{dt}\begin{bmatrix} v_c \\ i_1 \\ i_2 \end{bmatrix} = \begin{bmatrix} 0 & -1/C & -1/C \\ 1/L_1 & -R_1/L_1 & 0 \\ 1/L_2 & 0 & -R_2/L_2 \end{bmatrix} \cdot \begin{bmatrix} v_c \\ i_1 \\ i_2 \end{bmatrix} + \begin{bmatrix} 0 \\ -v_s/L_1 \\ 0 \end{bmatrix}$$

In general,

$$\dot{x} = Ax + f(t) \tag{16.4}$$

Equation (16.4) represents linear systems having more than one forcing function.

16.2 STATE EQUATIONS FOR LINEAR SYSTEMS

In Eq. (16.4), $x(t)$ is the state vector, A is the characteristic or system matrix and $f(t)$ is the forcing vector with n distinct time functions as its components.
In general,

$$\dot{x} = Ax + Bu \tag{16.5}$$

where u is the input vector.

Output equation: The state variables represent the dynamic state of a system. The system output variables $y_1, y_2, \ldots, y_p$ are linear combinations of the state variables. Thus,

$$y = Cx \tag{16.6}$$

and C is a $p \times n$ matrix. Very often the output is a function of both the state vector and the force vector. In this case,

$$y = Cx + Du \tag{16.7}$$

where D is a $p \times m$ matrix.

This is the output equation of the system. (for linear systems only)

State model: The state equation of a system determines its dynamic states and the output equation gives the output at any time $t > 0$. These two together form the *state model* of the system. Thus,

$$\begin{aligned} \dot{x} &= Ax + Bu \\ y &= Cx + Du \end{aligned} \tag{16.8}$$

The solution of the state equations will yield the zero input response ($u = 0$), the natural frequency of the network and the zero state response [$x(0-) = 0$].

The state variables determine completely all the other variables in the network or system.

Derivation of the state model. The state model may be obtained for a system which is governed by the differential equation:

$$\frac{d^n y}{dt^n} + a_{n-1}\frac{d^{n-1} y}{dt^{n-1}} + \cdots + a_2 \frac{d^2 y}{dt^2} + a_1 \frac{dy}{dt} + a_0 y = f(t) \tag{16.9}$$

By setting $y = x_1, \dfrac{dy}{dt} = x_2, \dfrac{d^2 y}{dt^2} = x_3, \cdots, \dfrac{d^{n-1} y}{dt^{n-1}} = x_n$ and using the dot notation for time derivative,

$$\begin{aligned}
\dot{x}_1 &= \quad + x_2 \\
\dot{x}_2 &= \quad\quad + x_3 \\
&\text{M} \\
\dot{x}_{n-1} &= \quad\quad\quad x_n \\
\dot{x}_n &= -a_0 x_1 - a_1 x_2 \ldots\ldots\ldots\ldots - a_{n-1} x_n + f(t)
\end{aligned}$$

The above equations result in the state equation.

$$\begin{bmatrix} \dot{x}_1 \\ \dot{x}_2 \\ \vdots \\ \dot{x}_{n-1} \\ \dot{x}_n \end{bmatrix} = \begin{bmatrix} 0 & 1 & 0 & .. & 0 \\ 0 & 0 & 1 & .. & 0 \\ . & . & . & .. & . \\ 0 & 0 & 0 & .. & 1 \\ -a_0 & -a_1 & -a_2 & .. & -a_{n-1} \end{bmatrix} \begin{bmatrix} x_1 \\ x_2 \\ .. \\ x_{n-1} \\ x_n \end{bmatrix} + \begin{bmatrix} 0 \\ 0 \\ \vdots \\ 0 \\ b_0 \end{bmatrix} u$$

or

$$\dot{x} = Ax + Bu$$

Note that matrix A has a very special form. The upper off diagonal terms are 1 and the last row is the negative of the coefficients of the differential equation (16.9); further all other elements are zero. This form of matrix A is known as **Bush form**.

16.3 SYSTEMATIC FORMULATION OF STATE EQUATIONS FOR LINEAR, CONSTANT, PASSIVE NETWORKS

Combined loop and node method is illustrated in Examples 16.3 and 16.4.

EXAMPLE 16.3 Write the state equations for the circuit shown in Fig. 16.3.

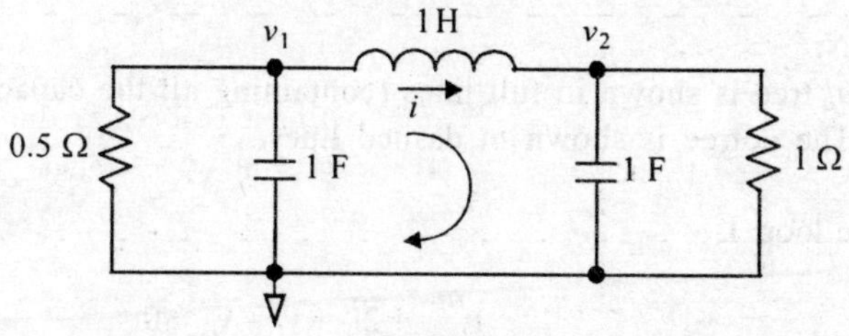

Fig. 16.3

SOLUTION:
Two node equations are:

$$1\frac{dv_1}{dt} + 2v_1 + i = 0 \tag{16.10}$$

$$1\frac{dv_2}{dt} + 1v_2 - i = 0 \tag{16.11}$$

One loop equation is:

$$1\frac{di}{dt} + v_2 - v_1 = 0 \tag{16.12}$$

In matrix form, these equations become

$$d/dt \begin{bmatrix} v_1 \\ v_2 \\ i \end{bmatrix} = \begin{bmatrix} -2 & 0 & -2 \\ 0 & -1 & 1 \\ 1 & -1 & 0 \end{bmatrix} \begin{bmatrix} v_1 \\ v_2 \\ i \end{bmatrix} \quad \textbf{Ans.}$$

The formulation of state equations for complicated networks requires the use of network topology. The concepts of node, branch, path, loop, mesh, tree, cotree and link have been introduced in Chapter 7. The procedure is illustrated through Example 16.4.

EXAMPLE 16.4 Write the state equations for the circuit shown in Fig. 16.4.

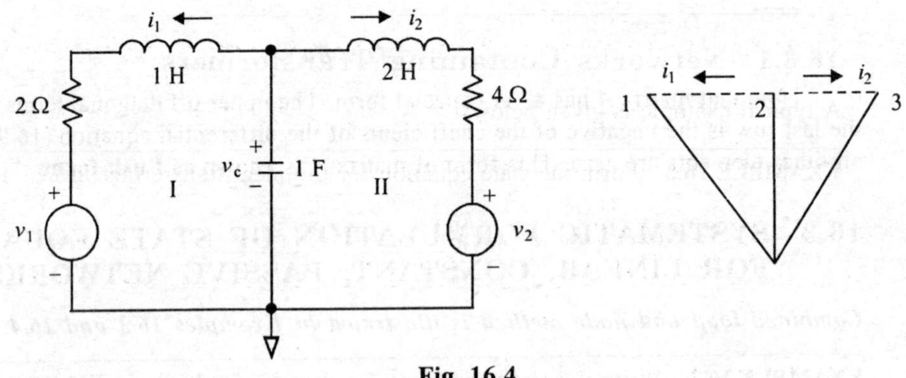

Fig. 16.4

SOLUTION:

The *normal* tree is shown in full lines (containing all the capacitors and all the voltage sources). The cotree is shown in dashed lines.

By KVL in loop 1,

$$1\frac{di_1}{dt} + 2i_1 + v_1 - v_c = 0$$

or

$$\frac{di_1}{dt} = -2i_1 + v_c - v_1$$

and by KVL in loop 2,

$$2\frac{di_2}{dt} + 4i_2 + v_2 - v_c = 0$$

or

$$\frac{di_2}{dt} = -2i_2 + 0.5v_c - 0.5v_2$$

By KCL at node 2,

$$1\frac{dv_c}{dt} + i_1 + i_2 = 0$$

or

$$\frac{dv_c}{dt} = -i_1 - i_2$$

In matrix form,

$$\frac{d}{dt}\begin{bmatrix} i_1 \\ i_2 \\ v_c \end{bmatrix} = \begin{bmatrix} -2 & 0 & 1 \\ 0 & -2 & 0.5 \\ -1 & -1 & 0 \end{bmatrix}\begin{bmatrix} i_1 \\ i_2 \\ v_c \end{bmatrix} + \begin{bmatrix} -v_1 \\ -0.5v_2 \\ 0 \end{bmatrix}$$

Note that the components of the forcing vector are not the forcing functions v_1 and v_2, but are linear combinations of these functions.

16.3.1 Networks Containing Transformers

A typical example is given below:

EXAMPLE 16.5 Form the state equation for the ideal transformer in Fig. 16.5.

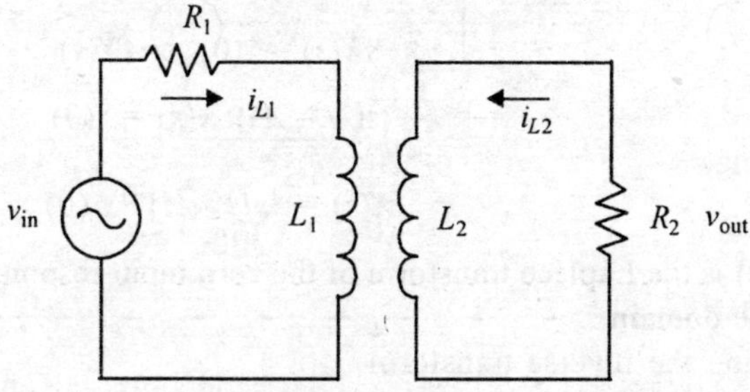

Fig. 16.5

SOLUTION:

The state variables are i_{L1} and i_{L2} and the two loop equations are:

$$R_1 i_{L1} + L_1 i_{L1} + M i_{L2} = v_i$$

$$L_2 i_{L2} + M i_{L1} + R_2 i_{L2} = 0$$

Solving for i_{L1} and i_{L2},

$$\frac{d}{dt}\begin{bmatrix} i_{L1} \\ i_{L2} \end{bmatrix} = \begin{bmatrix} \dfrac{-R_1 L_2}{\Delta} & \dfrac{R_2 M}{\Delta} \\ \dfrac{R_1 M}{\Delta} & \dfrac{-R_2 L_1}{\Delta} \end{bmatrix} \cdot \begin{bmatrix} i_{L1} \\ i_{L2} \end{bmatrix} + \begin{bmatrix} L_2/\Delta \\ -M/\Delta \end{bmatrix} v_i$$

where $\Delta = L_1 L_2 - M^2 \neq 0$

The output equation becomes

$$v_{out} = -R_2 i_{L2} \qquad \textbf{Ans.}$$

16.4 SOLUTION OF STATE EQUATIONS

Solving a differential equation means finding a relation between the variables involved, free of derivatives, which satisfies the equation identically. Thus, the solution of a set of first order differential equations is a certain set of algebraic equations. In the first part of this section, solution techniques for linear equations with constant coefficients, making use of Laplace transformation, will be presented. In this connection the *transition matrix* will be introduced. A classical approach to the same equations will be presented subsequently.

LAPLACE TRANSFORMATION METHOD

16.4.1 Free Motion or Zero Input Response

A set of homogeneous differential equations with constant coefficients is a mathematical model of a time-invariant system. The problem is to find the zero input response of a system which is governed by the set of linear differential equations.

$$\dot{x} = Ax \tag{16.13}$$

where A is a square constant matrix, x is a column matrix and $x(0) = x_0$ (specified). Taking Laplace transforms on Eq. (16.13),

$$sX(s) - x(0) = AX(s)$$

or

$$\{s(U - A)\} X(s) = x(0)$$

Rearranging,

$$X(s) = \{sU - A\}^{-1} x(0)$$

where $X(s)$ is the Laplace transform of the zero input response and $x(0)$ is the initial vector in the time domain.

Taking the inverse transform

$$x(t) = L^{-1} \{sU - A\}^{-1} x(0)$$

where $x(t)$ is the zero input response of the system.

EXAMPLE 16.6 Obtain the time response of a system governed by the differential equations.

$$\frac{d}{dt}\begin{bmatrix} x_1 \\ x_2 \end{bmatrix} = \begin{bmatrix} 1 & 0 \\ 1 & 1 \end{bmatrix}\begin{bmatrix} x_1 \\ x_2 \end{bmatrix}$$

SOLUTION:

$$[sU - A] = \begin{bmatrix} s & 0 \\ 0 & s \end{bmatrix} - \begin{bmatrix} 1 & 0 \\ 1 & 1 \end{bmatrix} = \begin{bmatrix} s-1 & 0 \\ -1 & s-1 \end{bmatrix}$$

$$\det sU - A = (s - 1)^2$$

$$\text{adj } (sU - A) = \begin{bmatrix} s-1 & 0 \\ 1 & s-1 \end{bmatrix}$$

$$\phi(s) = (sU - A)^{-1}$$

$$= 1/(s-1)^2 \begin{bmatrix} s-1 & 0 \\ 1 & s-1 \end{bmatrix}$$

$$= \begin{bmatrix} 1/(s-1) & 0 \\ 1/(s-1)^2 & 1/(s-1) \end{bmatrix}$$

and $$\phi(t) = \begin{bmatrix} e^t & 0 \\ te^t & e^t \end{bmatrix} = e^{At}$$

The general form of the solution of Eq. (16.13) may be written as:

$$x(t) = \phi(t)\, x(0) \qquad \textbf{Ans.}$$

where $\phi(t)$ is called the *transition matrix*, $x(0)$ the initial value and $x(t)$ the state.

From a geometrical point of view, $x(t)$ is a point in the n-dimensional space which defines the state of the system at time t, $x\,(0)$ is the initial state of the system. If, therefore, $x(0)$ is specified, the future behaviour of $x(t)$ is completely determined. $\phi(t)$ serves as an operator changing the system from $x(0)$ to $x(t)$; $\phi(t)$ is, therefore, designated as the transition matrix.

16.4.2 Forced Motion—Zero State Response

If a forcing function is applied to a system with certain initial conditions given, the response can be considered in two parts, namely the *zero input response* and the *zero state response*. The former is the effect of the initial conditions only and the latter arises from the forcing function with zero initial conditions.

To find the zero state response, consider the system:

$$\dot{x} = Ax + B \tag{16.14}$$

Laplace transformation gives

$$\{sU - A\}\, x(s) = x(0) + B(s)$$

$$x(s) = (sU - A)^{-1}\, x(0) + (sU - A)^{-1}\, B(s)$$

or

$$x(s) = \phi(s)\, x(0) + \phi(s)\, B(s)$$

The complete solution is obtained by inverse transformation.

$$x(t) = \phi(t)\, x(0) + L^{-1}\, \phi(s)\, B(s) \tag{16.15}$$

The second term on the right hand side is the zero state response.
Note that when $x(0) = 0$, $x(t)$ is determined by only the last term of Eq. (16.15).

EXAMPLE 16.7 The differential equation of a system is specified as:

$$\frac{d^2y}{dt^2} + \frac{4dy}{dt} + 3y = u(t)$$

Obtain the complete response of the system to a unit step input with initial conditions:

$$\frac{dy}{dt}(0) = 2$$

SOLUTION:
Choosing state variables $x_1 = y$, $x_2 = y$ and using the dot notation:

$$\frac{d}{dt}\begin{bmatrix} x_1 \\ x_2 \end{bmatrix} = \begin{bmatrix} 0 & 1 \\ -3 & -4 \end{bmatrix}\begin{bmatrix} x_1 \\ x_2 \end{bmatrix} + \begin{bmatrix} 0 \\ u(t) \end{bmatrix}$$

$$\dot{x} = Ax + f$$

with the initial conditions,

$$x(0) = \begin{bmatrix} 0 \\ 1 \end{bmatrix}$$

(see Section 16.2)

$$(sU - A)^{-1} = \begin{bmatrix} s & -1 \\ 3 & s+4 \end{bmatrix}^{-1}$$

$$= 1/\{(s+3)(s+1)\} \begin{bmatrix} s+4 & 1 \\ -3 & s \end{bmatrix}$$

$$= \begin{bmatrix} \dfrac{s+4}{(s+3)(s+1)} & \dfrac{1}{(s+3)(s+1)} \\ \dfrac{-3}{(s+3)(s+1)} & \dfrac{s}{(s+3)(s+1)} \end{bmatrix}$$

and $\phi(t) = L^{-1}\{(sU - A)^{-1}\}$

$$= \begin{bmatrix} (3/2)e^{-t} - (1/2)e^{-3t} & (1/2)e^{-t} - (1/2)e^{-3t} \\ -(3/2)e^{-t} + (3/2)e^{-3t} & -(1/2)e^{-t} + (3/2)e^{-3t} \end{bmatrix}$$

The solution to the state equation is given by

$$x(t) = \phi(t)\, x(0) + \int_0^t \phi(t - \tau)\, f(\tau)\, d\tau$$

Now,

$$x_1(t) = y(t)$$

$$= [(3/2)e^{-t} - (½)e^{-3t} \;\; (½)e^{-t} - (½)e^{-3t}] \begin{bmatrix} 0 \\ 1 \end{bmatrix}$$

$$+ \int_0^t [(3/2)e^{-(t-\tau)} - (½)e^{-3(t-\tau)} \;\; (½)e^{-(t-\tau)} - (½)e^{-3(t-\tau)}] \begin{bmatrix} 0 \\ u(\tau) \end{bmatrix} d\tau$$

$$= (½)e^{-t} - (½)e^{-3t} + \int_0^t (½)e^{-(t-\tau)} - (½)e^{-3(t-\tau)}\, 1 dt$$

$$= 1/3 - (1/3)e^{-3t}$$ **Ans.**

Properties of the transition matrix:

1. Multiplicative property

$$\phi(t)\ \phi(\tau) = \phi(t + \tau)$$

2. Commutative property

$$\phi(\tau + t) = \phi(\tau)\ \phi(t)$$

This may be written

$$\phi(t + \tau) = \phi(\tau)\ \phi(t)$$

Thus,

$$\phi(\tau)\ \phi(t) = \phi(t)\ \phi(\tau)$$

which is the commutative property.

3. $\phi(t)$ is the solution of the differential equation:

$$\phi(t) = A\phi(t) \qquad \phi(0) = U$$

where A is the constant square matrix.

CLASSICAL METHOD

In the classical theory the transition matrix is known as the *matrizant*. Hence the solution consists of deriving the matrizant from a set of first order differential equations. The principle underlying the classical approach is illustrated through the following example:

EXAMPLE 16.8 Solve the differential equation:

$$dx/dt = x, \quad x = x(0) = x_0$$

SOLUTION:
Direct integration with respect to t gives

$$x(t) = x_0 + \int_0^t x(t)\ dt \tag{16.16}$$

Equation (16.16) may be written as:

$$x(t) = x_0 + \int_0^t x(\tau_1)\ d\tau_1 \tag{16.17}$$

where τ_1 is a dummy variable.
For $t = \tau_1$, Eq. (16.17) becomes

$$x(\tau_1) = x_0 + \int_0^{\tau_1} x(\tau_1)\ d\tau_1 \tag{16.18}$$

The dummy variable can be changed again giving,

$$x(\tau_1) = x_0 + \int_0^{\tau_1} x(\tau_2)\ d\tau_2 \tag{16.19}$$

Making use of Eq. (16.19) in Eq. (16.17),

$$x(t) = x_0 + \int_0^t \{x_0 + \int_0^{\tau_1} x(\tau_2)\, d\tau_2\}\, d\tau_1 \tag{16.20}$$

$$= x_0 + x_0 \int_0^t d\tau_1 + \int_0^t \int_0^{\tau_1} x(\tau_2)\, d\tau_2\, d\tau_1 \tag{16.21}$$

$$= \{1 + Q(x) + Q(x)\, Q(x) + \Lambda\ \}\, x_0$$

where

$$Q = \int_0^t (\quad) d\tau$$

In the given example,

$$dx = x\, dt$$

$$x = x_0 + \int_0^t x\, dt$$

$$= x_0 + x_0 \int_0^t d\tau_1 + x_0 \int_0^t \int_0^\tau d\tau_2\, d\tau_1 + \Lambda$$

$$= (1 + t + \frac{t^2}{2!} + \Lambda\ )\, x_0 = x_0\, e^t \quad \textbf{Ans.}$$

Note: Extension of this technique to a matrix differential equation gives

$$\frac{d}{dx} x = Ax \qquad x(0) = x_0$$

$$x = (U + QA + QA\, QA + \Lambda\ )\, x_0$$

The matrizant of A is written as:

$$\phi(t) = U + QA + QA\, QA + \Lambda$$

LINEAR TRANSFORMATION METHOD

A state model with a diagonal matrix A can be of great advantage in the time domain solution because each component state variable equation is a first order differential equation and is decoupled from all other component equations. A matrix may be diagonalised by the use of eigen vectors (See Reference 18).

EXAMPLE 16.9 Diagonalise the matrix:

$$A = \begin{bmatrix} 0 & 1 \\ -2 & -3 \end{bmatrix}$$

SOLUTION:
The eigen values are given by

$$\det (A - \lambda U) = 0$$

or
$$= \begin{vmatrix} -\lambda & 1 \\ -2 & -3-\lambda \end{vmatrix} = 0$$

or
$$(\lambda + 1)\,(\lambda + 2) = 0$$

The eigen values are: $\lambda = -1, -2$.

The eigen vector associated with $\lambda = -1$ is:

$$\left\{\begin{bmatrix} 0 & 1 \\ -2 & -3 \end{bmatrix} - \begin{bmatrix} -1 & 0 \\ 0 & -1 \end{bmatrix}\right\}\begin{bmatrix} x_1 \\ x_2 \end{bmatrix} = \begin{bmatrix} 0 \\ 0 \end{bmatrix}$$

or
$$\begin{bmatrix} 1 & 1 \\ -2 & -2 \end{bmatrix}\begin{bmatrix} x_1 \\ x_2 \end{bmatrix} = \begin{bmatrix} 0 \\ 0 \end{bmatrix}$$

from which
$$\begin{bmatrix} x_1 \\ x_2 \end{bmatrix} = \begin{bmatrix} \alpha \\ -\alpha \end{bmatrix}$$

for any value of α.

Similarly, the eigen vector with $\lambda = -2$ is:

$$\left\{\begin{bmatrix} 0 & 1 \\ -2 & -3 \end{bmatrix} - \begin{bmatrix} -2 & 0 \\ 0 & -2 \end{bmatrix}\right\}\begin{bmatrix} x_1 \\ x_2 \end{bmatrix} = \begin{bmatrix} 0 \\ 0 \end{bmatrix}$$

or
$$\begin{bmatrix} 2 & 1 \\ -2 & -1 \end{bmatrix}\begin{bmatrix} x_1 \\ x_2 \end{bmatrix} = \begin{bmatrix} 0 \\ 0 \end{bmatrix}$$

from which
$$\begin{bmatrix} x_1 \\ x_2 \end{bmatrix} = \begin{bmatrix} \beta \\ -2\beta \end{bmatrix}$$

for any value of β.

Thus, the eigen vector matrix becomes

$$M = \begin{bmatrix} \alpha & \beta \\ -\alpha & -2\beta \end{bmatrix}$$

Assuming $\alpha = 1$, $\beta = 1$, the matrix reduces to

$$M = \begin{bmatrix} 1 & 1 \\ -1 & -2 \end{bmatrix}$$

M is also called **modal** matrix.

It is easy to show that the equation:

$$(A - \lambda U)x = 0$$

may be written as:

$$M^{-1} AM = \Lambda$$

which is the required diagonal matrix. **Ans.**

Similarity:

Definition. Two square matrices A and B of the same order are *similar* if a non-singular matrix M exists such that

$$M^{-1} AM = B$$

Theorem. An important special similarity relation is the similarity of A to a diagonal matrix. If A and Λ are similar, the diagonal elements of this Λ are the eigen values of A. Also it can be shown that the matrix M which diagonalises A is the eigen vector matrix of A.

EXAMPLE 16.10 Given

$$A = \begin{bmatrix} 0 & 1 \\ -2 & -3 \end{bmatrix} \qquad M = \begin{bmatrix} \alpha & \beta \\ -\alpha & -2\beta \end{bmatrix}$$

For $\alpha = \beta = 1$,

$$M = \begin{bmatrix} 1 & 1 \\ -1 & -2 \end{bmatrix}$$

Show that $M^{-1}AM = \begin{bmatrix} -1 & 0 \\ 0 & -2 \end{bmatrix}$.

SOLUTION:
Taking inverse of M,

$$M^{-1} = \begin{bmatrix} 2 & 1 \\ -1 & -1 \end{bmatrix}$$

Then

$$M^{-1}AM = \begin{bmatrix} 2 & 1 \\ -1 & -1 \end{bmatrix} \begin{bmatrix} 0 & 1 \\ -2 & -3 \end{bmatrix} \begin{bmatrix} 1 & 1 \\ -1 & -2 \end{bmatrix}$$

$$= \begin{bmatrix} -1 & 0 \\ 0 & -2 \end{bmatrix}$$

Hence proved.

SUMMARY

- The chapter has covered the analysis of dynamic systems using state variable technique, concentrating on formulation of state equation from classical representations, as well as the solution for the state equations. The topological approach has been used in the transformation of the circuit equations defining a system to state variable representation. The concepts presented are related to systems which are linear.
- Techniques for non-linear systems such as a feedback control system involving non-linearity in the forward loop are described in textbooks on control systems (See Reference 18).
- Methods of solution presented in this chapter cover the following:
  1. Laplace transformation
  2. The classical method using the matrizant.
  3. Linear transformation method based on eigen values and diagonalisation of matrices.
- The techniques of *flow graphs* which is best suited for systems characterized by a set of non-linear equations as well as numerical solution (Runge Kutta method) are described in references 2 and 3.

Solved Problems

P16.1 Write the state equation for the circuit of Fig. P16.1.

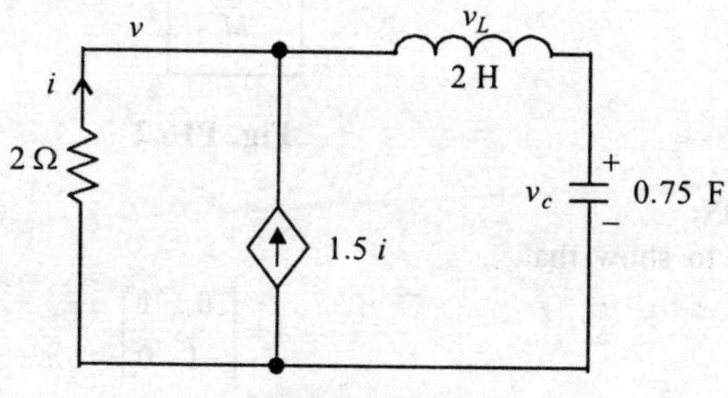

Fig. P16.1

SOLUTION:

$$i_L = i + 1.5i = 2.5i$$

and
$$-i = v/2 = \frac{v_L + v_c}{2}$$

Now,
$$v_L = 2\frac{di_L}{dt}$$

and
$$\frac{di_L}{dt} = -0.4i_L - 0.5v_c$$

Then
$$0.75\frac{dv_c}{dt} = i_L$$

or
$$\frac{dv_c}{dt} = (4/3)i_L$$

Therefore,
$$\begin{bmatrix} \dfrac{di_L}{dt} \\ \dfrac{dv_c}{dt} \end{bmatrix} = \begin{bmatrix} -0.4 & -0.5 \\ 4/3 & 0 \end{bmatrix} \begin{bmatrix} i_L \\ v_c \end{bmatrix} \quad \textbf{Ans.}$$

P16.2 Determine $\phi(t)$ for the system of Fig. P16.2 given the state equation as:

$$\frac{d}{dt}\begin{bmatrix} x_1 \\ x_2 \end{bmatrix} = \begin{bmatrix} 0 & 1 \\ -1 & 0 \end{bmatrix} \begin{bmatrix} x_1 \\ x_2 \end{bmatrix}$$

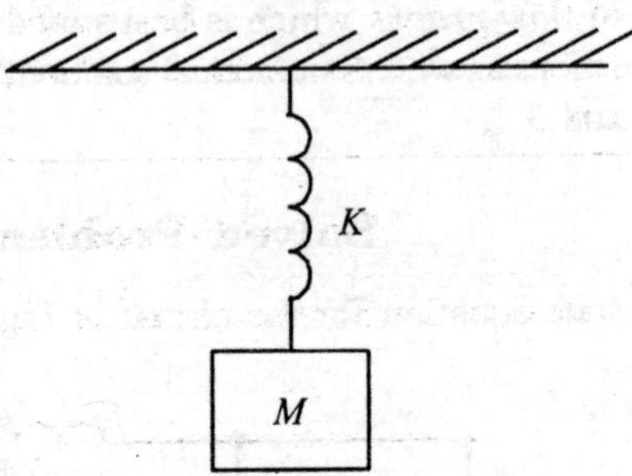

Fig. P16.2

SOLUTION:
It is easy to show that
$$A = \begin{bmatrix} 0 & 1 \\ -1 & 0 \end{bmatrix}$$

Then
$$\phi(t) = e^{At} = \Sigma_{n=0}^{\infty} \frac{1}{n!} A^n t^n$$

$$\therefore \phi(t) = \begin{bmatrix} 1 & 0 \\ 0 & 1 \end{bmatrix} + \begin{bmatrix} 0 & 1 \\ -1 & 0 \end{bmatrix} t + \begin{bmatrix} -1 & 0 \\ 0 & -1 \end{bmatrix} \frac{t^2}{2!} + \begin{bmatrix} 0 & -1 \\ 1 & 0 \end{bmatrix} \frac{t^3}{3!} + \cdots$$

$$= \begin{bmatrix} 1 & 0 \\ 0 & 1 \end{bmatrix} + \begin{bmatrix} 0 & t \\ -t & 0 \end{bmatrix} + \begin{bmatrix} -t^2/2! & 0 \\ 0 & -t^2/2! \end{bmatrix} + \begin{bmatrix} 0 & -t^3/3! \\ t^3/3! & 0 \end{bmatrix} + \cdots$$

$$= \begin{bmatrix} 1 - t^2/2! + \cdots & t - t^3/3! + \cdots \\ -t + t^3/3! - \cdots & 1 - t^2/2! + \cdots \end{bmatrix} + \cdots$$

$$= \begin{bmatrix} \cos t & \sin t \\ -\sin t & \cos t \end{bmatrix} \quad \textbf{Ans.}$$

P16.3 Write the matrix state equation for the circuit shown in Fig. P16.3.

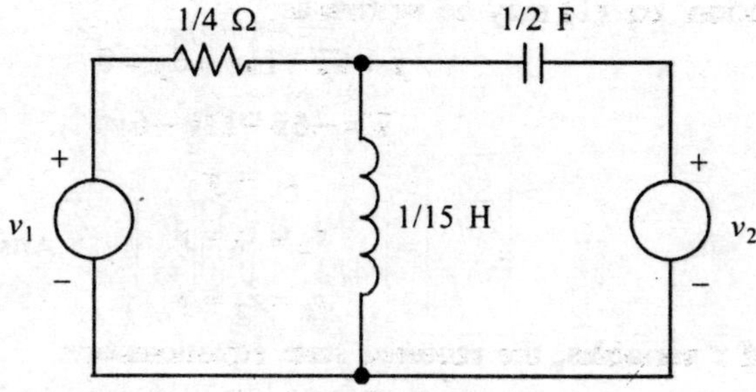

Fig. P16.3

SOLUTION:

$$\frac{1}{2}\frac{dv_c}{dt} = i_2$$

$$v_1 = \frac{1}{4}i_1 + \frac{1}{15}\frac{di_L}{dt}$$

$$v_2 - v_c - \frac{1}{15}\frac{di_L}{dt} = 0$$

Now,

$$i_L = i_1 + i_2$$

Then

$$\frac{dv_c}{dt} = -8v_c + 2i_L + 8v_2 - 8v_1$$

and

$$\frac{di_L}{dt} = -15v_c + 15v_2$$

In matrix form,

$$\begin{bmatrix} \frac{dv_c}{dt} \\ \frac{di_L}{dt} \end{bmatrix} = \begin{bmatrix} -8 & 2 \\ -15 & 0 \end{bmatrix}\begin{bmatrix} v_c \\ i_L \end{bmatrix} + \begin{bmatrix} 8v_2 - 8v_1 \\ 15v_2 \end{bmatrix} \qquad \textbf{Ans.}$$

P16.4 A mechanical system is described by the differential equation:

$$\frac{d^3y}{dx^3} + 6\frac{d^2y}{dx^2} + 11\frac{dy}{dx} + 6y = 0 \qquad (1)$$

Find the characteristic matrix of the system.

SOLUTION:

In dot notation, eq. (1) may be written as:

$$\dddot{y}+6\ddot{y}+11\dot{y}+6y=0$$

or

$$\dddot{y}=-6\ddot{y}-11\dot{y}-6y$$

Let

$$x_1 = y$$

$$x_2 = \dot{x}_1 = \dot{y}$$

and

$$x_3 = \dot{x}_2 = \ddot{y}$$

In terms of x variables, the required state equations are:

$$\dot{x}_1 = x_2$$

$$\dot{x}_2 = x_3$$

$$\dot{x}_3 = -6x_1 - 11x_2 - 6x_3$$

In matrix form,

$$\begin{bmatrix} x_1 \\ x_2 \\ x_3 \end{bmatrix} = \begin{bmatrix} 0 & 1 & 0 \\ 0 & 0 & 1 \\ -1 & -11 & -6 \end{bmatrix}\begin{bmatrix} x_1 \\ x_2 \\ x_3 \end{bmatrix}$$ **Ans.**

P16.5 Solve the state equations:

$$\frac{d}{dt}\begin{bmatrix} i \\ v \end{bmatrix} = \begin{bmatrix} -2 & 0 \\ 0 & -4 \end{bmatrix}\begin{bmatrix} i \\ v \end{bmatrix} + \begin{bmatrix} e^{-t} \\ 2e^{-t} \end{bmatrix}$$

with initial conditions

$$\begin{bmatrix} i_0 \\ v_0 \end{bmatrix} = \begin{bmatrix} 1 \\ 1 \end{bmatrix}$$

SOLUTION:

$$A = \begin{bmatrix} -2 & 0 \\ 0 & -4 \end{bmatrix}$$

$$(sU - A)^{-1} = \left\{\begin{bmatrix} s+2 & 0 \\ 0 & s+4 \end{bmatrix}\right\}^{-1}$$

$$= \begin{bmatrix} 1/(s+2) & 0 \\ 0 & 1/(s+4) \end{bmatrix}$$

and

$$\phi(t) = L^{-1}\{(sU - A)^{-1}\}$$

$$= \begin{bmatrix} e^{-2t} & 0 \\ 0 & e^{-4t} \end{bmatrix}$$

Hence

$$\begin{bmatrix} i \\ v \end{bmatrix} = \begin{bmatrix} e^{-2t} & 0 \\ 0 & e^{-4t} \end{bmatrix}\begin{bmatrix} 1 \\ 1 \end{bmatrix} + \int_0^t \left\{ \begin{bmatrix} e^{-2(t-\tau)} & 0 \\ 0 & e^{-4(t-\tau)} \end{bmatrix}\begin{bmatrix} e^{-t} \\ 2e^{-t} \end{bmatrix} d\tau \right\}$$

$$= \begin{bmatrix} e^{-2t} \\ e^{-4t} \end{bmatrix} + \begin{bmatrix} e^{-2t}(e^{t}-1) \\ 2/3e^{-4t}(e^{3t}-1) \end{bmatrix}$$

$$= \begin{bmatrix} e^{-t} \\ \dfrac{e^{-4t}}{3} + \dfrac{2e^{-t}}{3} \end{bmatrix} \qquad \textbf{Ans.}$$

Exercises

E16.1 Obtain the state equations for the circuit shown in Fig. E16.1.

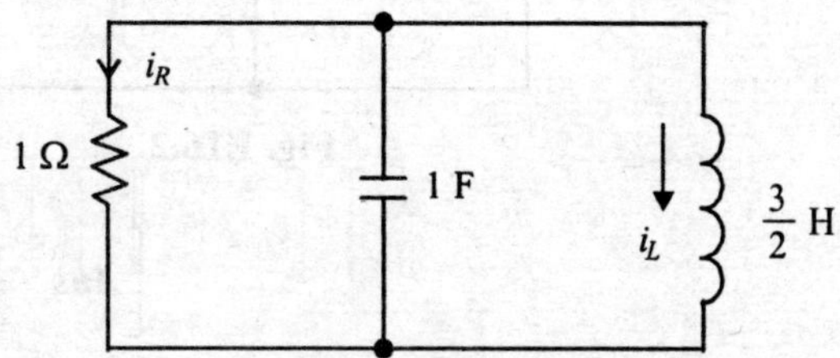

Fig. E16.1

[**Hint:**

$$v_c = \frac{3}{2}\frac{di_L}{dt} \tag{1}$$

$$i_R = \frac{v_c}{1}$$

and
$$i_c = \frac{dv_c}{dt}$$

Since
$$i_R + i_L + i_c = 0$$

or
$$i_c = \frac{dv_c}{dt} = -i_L - \frac{v_c}{1} \tag{2}$$

From eqs. (1) and (2),

$$\begin{bmatrix} \dfrac{di_L}{dt} \\ \dfrac{dv_c}{dt} \end{bmatrix} = \begin{bmatrix} 0 & 2/3 \\ -1 & -1 \end{bmatrix}\begin{bmatrix} i_L \\ v_c \end{bmatrix}$$

E16.2 Find the eigenvalues for the circuit of Fig. E16.1.

[**Hint:** The characteristic equation

$$\det(sU - A) = 0$$

or

$$\begin{vmatrix} s & -2/3 \\ 1 & s+1 \end{vmatrix} = 0$$

or

$$3s^2 + 3s + 2 = 0$$

The eigen values are $-0.5 + 0.65j, -0.5 - 0.65j$]

E16.3 Write state equation in matrix form for the circuit in Fig. E16.2.

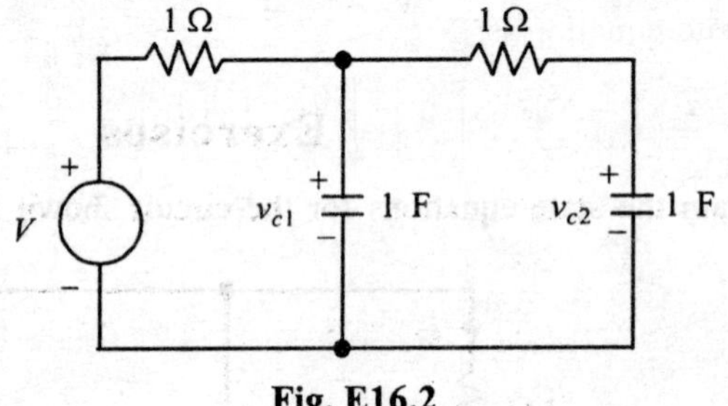

Fig. E16.2

$$\left[\textbf{\textit{Ans}}: \frac{d}{dt}\begin{bmatrix} v_{c1} \\ v_{c2} \end{bmatrix} = \begin{bmatrix} -2 & 1 \\ 1 & -1 \end{bmatrix}\begin{bmatrix} v_{c1} \\ v_{c2} \end{bmatrix} + \begin{bmatrix} v \\ 0 \end{bmatrix} \right]$$

E16.4 For the bridge network shown in Fig. E16.3 obtain the characteristic matrix using state variables:

$$x_1 = v_c \qquad x_2 = i_L$$

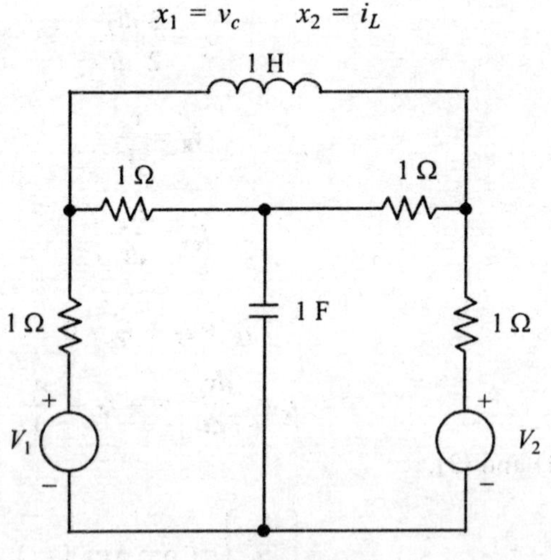

Fig. E16.3

$$\left[\textbf{\textit{Ans}}: A = \begin{bmatrix} -1 & 0 \\ 0 & -1 \end{bmatrix} \right]$$

E16.5 Find the response of the circuit shown in Fig. E16.4.

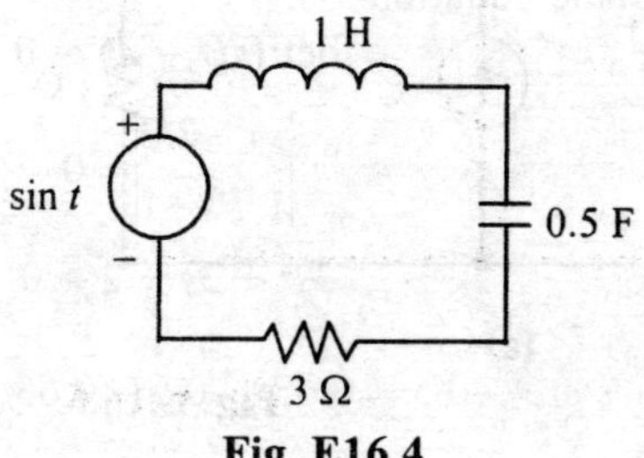

Fig. E16.4

[**Hint:** The state equation is:

$$\begin{bmatrix} \dot{x}_1 \\ \dot{x}_2 \end{bmatrix} = \begin{bmatrix} 0 & 1 \\ -2 & -3 \end{bmatrix} \begin{bmatrix} x_1 \\ x_2 \end{bmatrix} + \begin{bmatrix} 0 \\ \sin t \end{bmatrix}$$

Proceed to show that the total response is:

$$\begin{bmatrix} \dot{x}_1 \\ \dot{x}_2 \end{bmatrix} = \begin{bmatrix} \dfrac{s+3}{s^2+3s+2} & \dfrac{1}{s^2+3s+2} \\ \dfrac{-2}{s^2+3s+2} & \dfrac{s}{s^2+3s+2} \end{bmatrix} \begin{bmatrix} x_1(0) \\ x_2(0) \end{bmatrix} + \begin{bmatrix} \dfrac{s+3}{s^2+3s+2} & \dfrac{1}{s^2+3s+2} \\ \dfrac{-2}{s^2+3s+2} & \dfrac{s}{s^2+3s+2} \end{bmatrix} \begin{bmatrix} 0 \\ \dfrac{1}{s^2+1} \end{bmatrix}$$

The first term on the RHS is the zero input response in the s domain and the second term on the RHS is the zero state response.]

E16.6 For the network shown in Fig. E16.5, obtain the state equation.

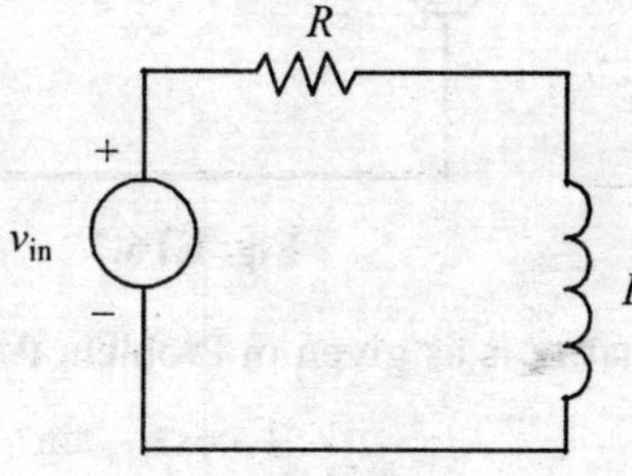

Fig. E16.5

$$\left[\textbf{\textit{Ans:}}\ \frac{di_L}{dt} = -\frac{Ri_L}{L} + v_{in} \right]$$

E16.7 Write the state equations for the circuit illustrated in Fig. E16.6.

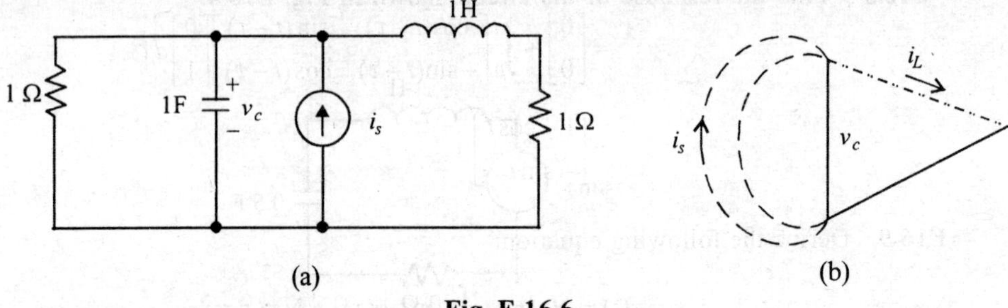

Fig. E.16.6

[**Hint:** Figure shows the normal tree with the capacitor in the tree and the inductor in the cotree.

$$\left[\textbf{\textit{Ans:}}\ \frac{d}{dt}\begin{bmatrix} i_L \\ v_c \end{bmatrix} = \begin{bmatrix} -1 & 1 \\ -1 & -1 \end{bmatrix}\begin{bmatrix} i_L \\ v_c \end{bmatrix} + \begin{bmatrix} 0 \\ i_s \end{bmatrix} \right]$$

E16.8 Find the state response of the system shown in Fig. E16.7.

$$\begin{bmatrix} x_1 \\ x_2 \end{bmatrix} = \begin{bmatrix} 0 & 1 \\ -1 & 0 \end{bmatrix}\begin{bmatrix} x_1 \\ x_2 \end{bmatrix} + \begin{bmatrix} 0 \\ 1 \end{bmatrix}$$

$$\begin{bmatrix} x_1(0) \\ x_2(0) \end{bmatrix} = \begin{bmatrix} 0 \\ 0 \end{bmatrix}$$

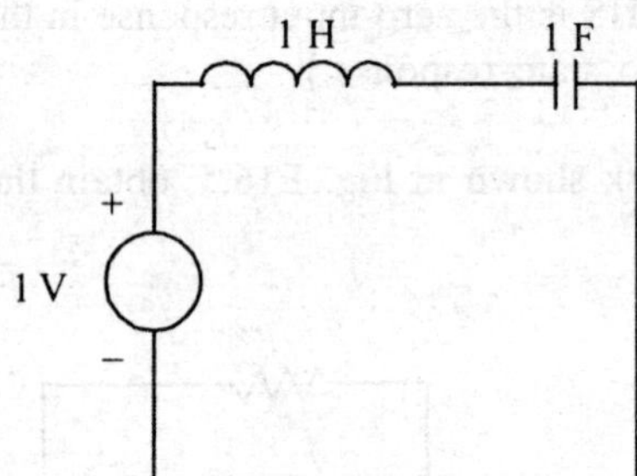

Fig. E16.7

[**Hint:** The transition matrix is as given in Problem P16.2:

$$\phi(t) = \begin{bmatrix} \cos t & \sin t \\ -\sin t & \cos t \end{bmatrix}$$

Show that

$$x(t) = \phi(t)\, x_0 + \int_0^t \{\phi(t)\, \phi(\tau)\}^{-1}\, b(\tau)\, d\tau$$

$$= \begin{bmatrix} \cos t & \sin t \\ -\sin t & \cos t \end{bmatrix}\begin{bmatrix} 0 \\ 0 \end{bmatrix} + \int \{\phi(t)\, \phi(\tau)\}^{-1}\, b(\tau)\, d\tau$$

$$= \begin{bmatrix} 0 \\ 0 \end{bmatrix} + \int_0^t \begin{bmatrix} \cos(t-\tau) & \sin(t-\tau) \\ -\sin(t-\tau) & \cos(t-\tau) \end{bmatrix} \begin{bmatrix} 0 \\ 1 \end{bmatrix} d\tau$$

$$= \begin{bmatrix} 1-\cos t \\ \sin t \end{bmatrix}$$

E16.9 Derive the following equation:

$$x(t) = \phi(t)x_0 + \int_0^t \{\phi(t)\phi(\tau)\}^{-1} b(\tau)\, d\tau$$

Digital Systems and z-Transforms

17

CHAPTER CONTENTS

Digital systems have variables that are digital in nature and can assume only a finite number of values (often only two values, 0 and 1). This is in comparison with analog systems where the variables can assume an infinite number of values.

A well-known example of digital network is the digital computer. The knowledge of digital circuits brought about by the computer paved the way for the giant strides taken by present day digital signal processing (DSP). This chapter introduces the concept of *difference equation* as models of digital systems. The first and the second order digital systems are considered briefly.

17.1 DIFFERENCE EQUATIONS

A discrete time system (digital) may be represented by a difference equation. A difference equation is a function of a variable which is restricted to a set of integers. Compare the analog equation:

$$x(t) = 4t - 10 \qquad 0 \le t \le 3$$

and the difference equation

$$x(n) = 4n - 10 \qquad n = 0, 1, 2, 3$$

$x(t)$ is an analog signal with an infinite number of values between $t = 0$ and $t - 3$ whereas $x(n)$ can have only four values (−10, −6, −2, 2).

A discrete time system is constructed of elements that operate on a discrete time signal to produce another discrete time signal according to a given rule. In analog systems, the elements are resistors, inductors, capacitors and energy sources and the operation each of these performs on an input signal resulting in an output signal are well defined. For example, if the input signal to an inductor is $i(t)$, the output is:

$$v(t) = L\, di/dt$$

17.1.1 Digital Signals

A discrete time signal is a sequence of numbers as shown in Fig. 17.1.

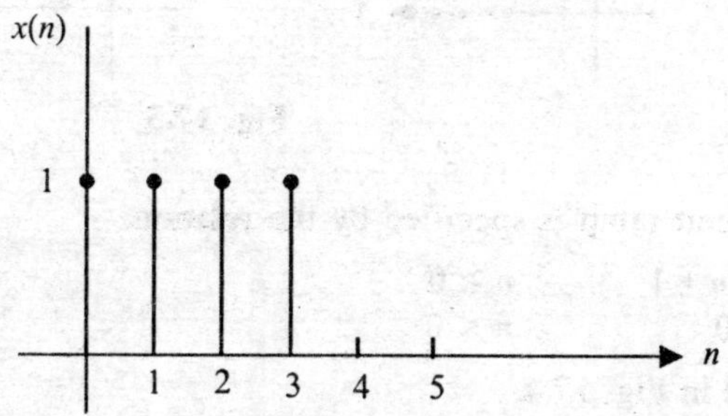

Fig. 17.1

A *sequence* is a function whose domain is a set of integers. On the other hand, a *series* of n terms is of the form $r_1 + r_2 + \cdots + r_n$ where each term is formed by some definite rule. In the sequence shown in Fig. 17.1,

$$S_1 = 2$$
$$S_2 = 3$$
and $$S_3 = 4$$

Thus, the two terms, sequence and series are quite distinct. A sequence is a function, while a series is a single number, the sum of the terms in a sequence.

There are several digital signals that occur frequently in digital systems. They are the *unit step*, *delta* and *ramp* functions. To distinguish between analog and digital versions, the analog unit step is designated by $u(t)$ while digital unit step is designated by $u(n)$.

The digital unit step function in Fig. 17.2 is defined by

$$u(n) = 1 \quad n \geq 0 \qquad (17.1)$$
$$= 0 \quad n < 0$$

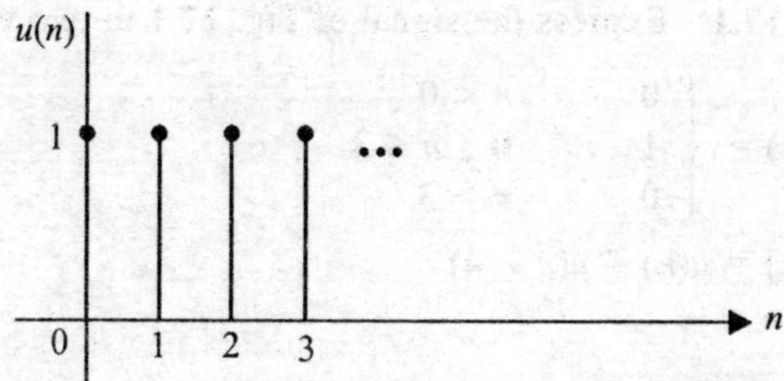

Fig. 17.2

The digital delta function is shown in Fig. 17.3. It is specified by

$$\delta(n) = 1 \qquad n = 0$$
$$= 0 \qquad \text{otherwise} \tag{17.2}$$

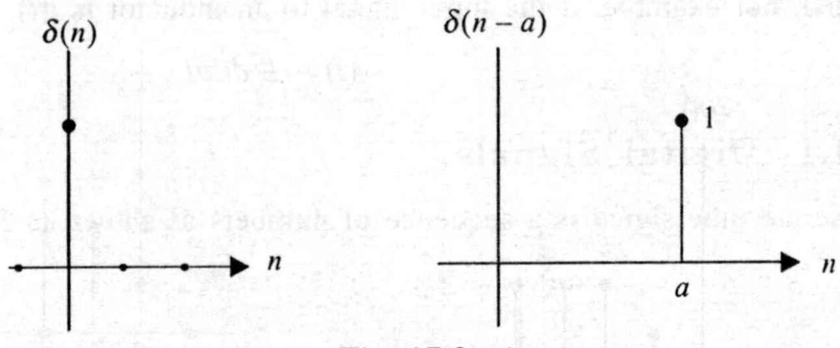

Fig. 17.3

The digital unit ramp is specified by the relation:

$$r(n) = n + 1 \qquad n \geq 0$$
$$= 0 \qquad n < 0 \tag{17.3}$$

as illustrated in Fig. 17.4.

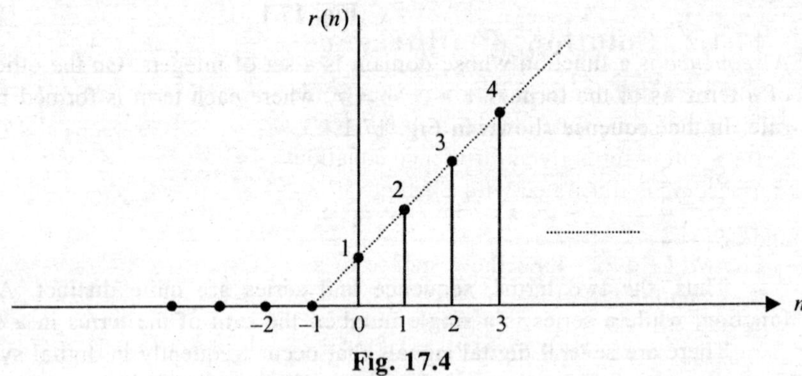

Fig. 17.4

Composite signals may be specified either directly as in Eq. (17.3) or in terms of basic signals such as the step and ramp as in equation:

$$r(n) = (n + 1)\, u(n)$$

EXAMPLE 17.1 Express the signal of Fig. 17.1 in two ways:

(a)
$$x(n) = \begin{cases} 0 & n < 0 \\ 1 & 0 \leq n \leq 3 \\ 0 & n > 3 \end{cases}$$

(b) $x(n) = u(n) - u(n - 4)$

SOLUTION:

The truncated ramp in Fig. 17.5 may be expressed as:

$$x(n) = \begin{cases} 0 & n < 0 \\ 2(n+1) & 0 \le n \le 2 \\ 6 & n > 2 \end{cases}$$

or in terms of two ramp functions as shown in Fig. 17.5(b).

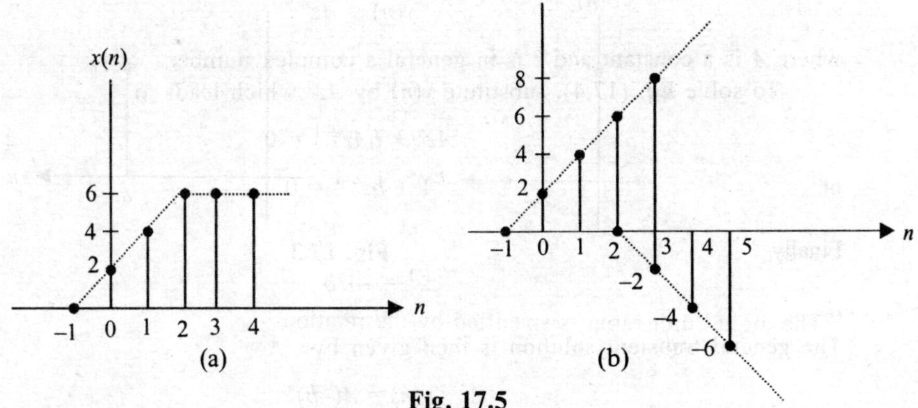

Fig. 17.5

17.1.2 Solution of Difference Equation

Solving a difference equation means to find that sequence which satisfies the equation, (a discrete time function). A step by step (iterative) procedure may be adopted to obtain the solution for a given difference equation with known input and initial conditions. The procedure is illustrated with Example 17.2.

EXAMPLE 17.2 Find the response to a discrete unit step of the system described by the equation

$$y(n) = 1.5x(n) - 1.5x(n-1) + 1.9y(n-1) \text{ with } y(-1) = 0$$

SOLUTION:

Use the step by step method.
The input sequence $x(n)$ is given by …, 0,0,0, 1,1,1,1, …,
where the first 1 occurs at $n = 0$.
The first value of the response can be found as:

$$y(0) = 1.5x(0) - 1.5x(-1) + 1.9y(-1) = 1.5$$

$$y(1) = 1.5x(1) - 1.5x(0) + 1.9y(0) = 2.85$$

$$y(2) = 1.5x(2) - 1.5x(1) + 1.9y(1) = 5.415$$

A general rule for calculating $y(n)$ can be derived from the above results.

$$y(n) = 1.5\,(1.9)^n \qquad n > 0 \qquad \textbf{Ans.}$$

17.2 FIRST ORDER DIGITAL SYSTEMS

Source-free solution: For a first order system, the difference equation reduces to

$$y(n) + by(n-1) = 0 \tag{17.4}$$

Assume a solution

$$y(n) = Az^n$$

where A is a constant and z is in general a complex number.

To solve Eq. (17.4), substitute $y(n)$ by Az^n which leads to

$$Az^n + bAz^{n-1} = 0$$

or

$$1 + bz^{n-1} = 0 \tag{17.5}$$

Finally,

$$z^{-1} = -1/b$$

or

$$z = -b$$

The general transient solution is then given by

$$y(n) = A(-b)^n$$

For example, find the solution for the first order difference equation:

$$y(n) + 1.5y(n-1) = 0$$

The transient solution can be readily written as:

$$y(n) = A(-1.5)^n$$

Equation (17.5) is called the **characteristic equation** for the system and $z = -b$ is called a **root** of the characteristic equation. $z = -b$ is also called a *natural frequency* of the system. For a first order system, the natural frequency will always be a real number. For a higher order systems, there will be as many natural frequencies as the order of the system and these roots may be complex numbers.

EXAMPLE 17.3 Obtain the transient solution for the difference equation:

$$y(n) = x(n) + 0.7y(n-1)$$

SOLUTION:

Assuming a solution:

$$y(n) = Az^n$$

$$Az^n - 0.7z^{-1}Az^n = 0$$

The characteristic equation is therefore,

$$1 - 0.7z^{-1} = 0$$

Solving for z,

$z = 0.7$, which is the natural frequency.

Thus, the transient solution is:

$$y_{tr}(n) = A(0.7)^n$$

A plot of the solution is shown in Fig. 17.6. Note the exponential nature of the solution.

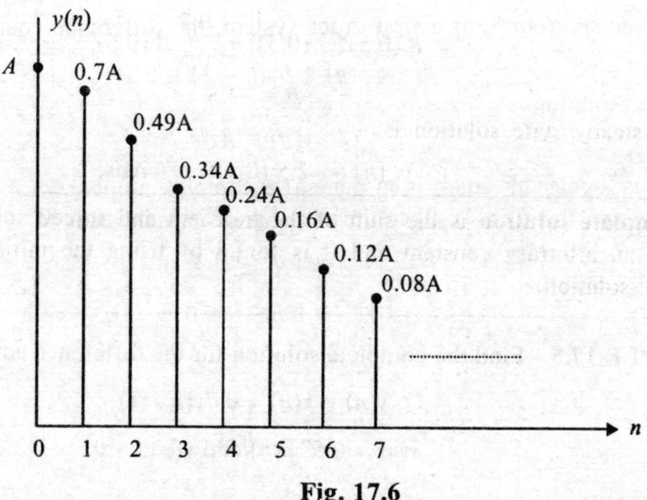

Fig. 17.6

The forced response: If the forcing function is one of the forms a^n, e^{bn}, n^p, sin dn or sums or products of these forms, the forced (steady state) solution is found by the method of undetermined coefficients. For example, if $y(n) = a^n$,
Then

$$y(n+1) = a^{n+1} = a.a^n$$
$$y(n+m) = a^m.a^n$$

Thus, the form a^n is retained.

For the cosine terms,

If $$y(n) = \cos dn$$

Then

$$y(n+m) = \cos d(n+m) = \cos dm \cos dn - \sin dm \sin dn$$

Thus, the sin dn and cos dn forms are retained.

If the forcing function is a member of one of the group a^n, n^p, cos dn, sin dn then the steady state solution will also be a member of this group. It remains only to solve for the undetermined coefficients in the solution. The procedure is illustrated in Example 17.4.

EXAMPLE 17.4 Find the steady state solution for the difference equation:

$$y(n) = x(n) + 0.7y(n-1) \tag{17.6}$$

where $$x(n) = 0.5^n\, u(n) \quad \text{and} \quad y(-1) = 0$$

SOLUTION:

Assuming the solution is of the form:

$$y_{ss}(n-1) = B(0.5)^{n-1} = \frac{B}{0.5}(0.5)^n$$

From Eq. (17.6),

$$B\,(0.5)^n = (0.5)^n + \frac{0.7}{0.5}\,B\,(0.5)^n$$

or

$$B = -2.5$$

and the steady state solution is:

$$y_{ss}(n) = -2.5\,(0.5)^n \quad \textbf{Ans.}$$

The complete solution is the sum of the transient and forced solutions. The solution contains an arbitrary constant and it is found by using the initial condition into the complete solution.

EXAMPLE 17.5 Find the complete solution for the difference equation:

$$y(n) = x(n) + 0.7y(n-1)$$

where

$$x(n) = 0.5^n\,u(n) \text{ and } y(-1) = 0.$$

SOLUTION:
The complete solution is:

$$\begin{aligned} y(n) &= y_{tr}(n) + y_{ss}(n) \\ &= A(0.7)^n - 2.5(0.5)^n \end{aligned} \tag{17.7}$$

From Eq. (17.7),

$$\begin{aligned} y(0) &= x(0) + 0.7y(-1) \\ &= 1 + 0 = 1 \end{aligned}$$

Substituting this in Eq. (17.7),

$$y(0) = A\,(0.7)^0 - 2.5\,(0.5)^n$$

or

$$1 = A - 2.5$$

$$A = 3.5$$

and the complete solution is:

$$y(n) = 3.5\,(0.7)^n - 2.5\,(0.5)^n \quad \textbf{Ans.}$$

17.3 SECOND ORDER DIGITAL SYSTEMS

Second order system can be put in the form:

$$b_0y(n) + b_1y(n-1) + b_2y(n-2) = a_0x(n) + a_1x(n-1) + a_2x(n-2)$$

The solution is outlined as follows:

EXAMPLE 17.6 Find the forced response for the system:

$$y(n) - 0.7y(n-1) + 0.25y(n-2) = (0.8)^n \tag{17.8}$$

If the forcing function is $x(n) = (0.8)^n$.

SOLUTION:
Assuming the steady state solution as:

$$y_{ss}(n) = B(0.8)^n$$

It follows

$$y_{ss}(n-1) = B(0.8)^{n-1} = \frac{B}{0.8}(0.8)^n$$

and
$$y_{ss}(n-2) = B(0.8)^{n-2} = \frac{B}{0.64}(0.8)^n$$

Substituting these values in Eq. (17.8),

$$B(0.8)^n - 0.7\frac{B}{0.8}(0.8)^n + 0.25\frac{B}{0.64}(0.8)^n = (0.8)^n$$

or
$$B = 1.94$$

The particular solution is then

$$y_{ss}(n) = 1.94\,(0.8)^n \qquad \textbf{Ans.}$$

As in the case of first order system the complete solution is the sum of the transient solution and the particular solution, with the initial conditions taken into account.

17.4 z-TRANSFORM

Definition: The z-transform is an operation performed on a discrete time sequence $v(n)$ to produce a function $V(z)$, where z is a complex variable. There are two forms of the z-transform, the bilateral form:

$$V(z) = \sum_{n=-\infty}^{\infty} v(n)z^{-n}$$

and the unilateral form:

$$V(z) = \sum_{n=0}^{\infty} v(n)z^{-n}$$

Note that the only difference is in the lower limit on the summation.

The inverse z-transform is given by

$$v(n) = \frac{1}{2\pi j}\oint V(z)z^{n-1}dz \tag{17.9}$$

If $v(n)$ is to have a z-transform the magnitude of $V(z)$ must be finite. The set of all z for which this condition holds is called the **region of convergence** in the complex plane.

The z-transform is a power series. Both negative and positive powers of z are involved in bilateral transform; hence it is necessary to split the equation into two parts.

$$V(z) = \sum_{n=-\infty}^{-1} v(n)z^{-n} + \sum_{n=0}^{\infty} v(n)z^{-n}$$

Theorem: A very useful theorem for power series is

(a) Every power series

$$\sum_{n=0}^{\infty} v(n)\,(z-z_0)^n$$

converges in a circle of radius R_1 centred at z_0, that is, $|z - z_0| < R_1$.

The value of R_1 is given by

$$R_1 = \lim_{n\to\infty}\left|\frac{v(n)}{v(n+1)}\right| \tag{17.9a}$$

if this limit exists.

(b) Every power series

$$\sum_{n=0}^{\infty} v(n)\,(z - z_0)^{-n}$$

converges outside a circle of radius R_2 centred at z_0, that is, $|z - z_0| > R_2$

The value of R_2 is given by

$$R_2 = \lim_{n\to\infty} \left|\frac{v(n+1)}{v(n)}\right| \tag{17.9b}$$

if this limit exists.

A discrete time function where $v(n) = 0$, $n > 0$, is called **negative-time function**. When $v(n) = 0$, $n < 0$ then $v(n)$ is called a **positive time function.**

EVALUATING THE z-TRANSFORMS

EXAMPLE 17.7 Evaluate the z-transform of $v(n)$ consisting of two values as shown in Fig. 17.7.

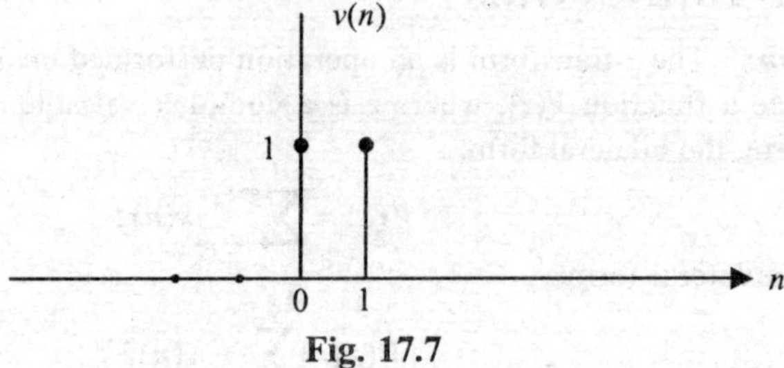

Fig. 17.7

SOLUTION:

From the defining equation for $V(z)$,

$$V(z) = \sum_{n=0}^{1} z^{-n} = 1 + 1/z = \frac{z+1}{z} \qquad z \neq 0$$

At $z = 0$, the series does not converge. **Ans.**

EXAMPLE 17.8 Find the z-transform of unit step function shown in Fig. 17.8.

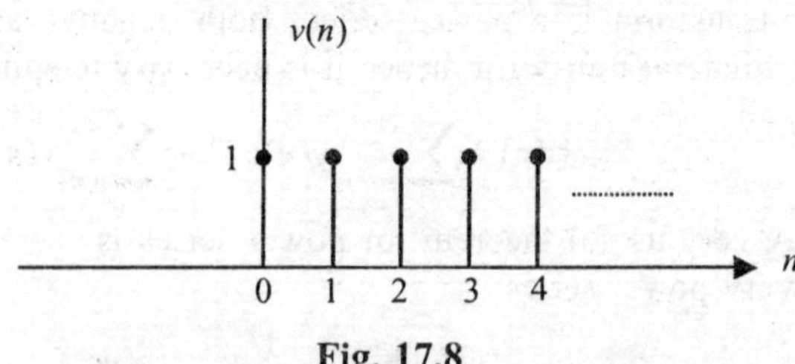

Fig. 17.8

SOLUTION:

$$V(z) = \sum_{n=0}^{\infty} z^{-n} = 1 + 1/z + (1/z)^2 + \cdots \qquad |z| > 1$$

$$= \frac{1}{1 - 1/z} = \frac{z}{z-1} \qquad \textbf{Ans.}$$

17.4.1 The Inverse z-transform

The inverse z-transform may be obtained by any of the following methods:

(a) By use of the definition Eq. (17.9).
(b) By referring to a table of transform pairs in Tables 17.1 and 17.2. Partial fraction expansions will be useful here.
(c) Power series method.

The first method involves the evaluation of complex integrals.

SOLUTION BY PARTIAL FRACTION METHOD

EXAMPLE 17.9 Find the inverse transform of

$$V(z) = \frac{z(z-2)}{(z^2 - 7z + 12)} \qquad z > 4$$

SOLUTION:
By cover up rule,

$$\frac{V(z)}{z} = \frac{-1}{z-3} + \frac{2}{z-4}$$

or

$$V(z) = \frac{-z}{z-3} + \frac{2z}{z-4}$$

Then

$$v(n) = [-(3)^n + 2(4)^n]\, u(n) \quad \textbf{Ans.} \tag{17.10}$$

Solution by power series
By long division,

$$
\begin{array}{rl}
z^2 - 7z + 12 \,\big)\, z^2 - 2z & \big(\, 1 + 5z^{-1} + 23z^{-2} + 101z^{-3} \\
\underline{z^2 - 7z + 12} & \\
5z - 12 & \\
\underline{5z - 35 + 60z^{-1}} & \\
23 - 60z^{-1} & \\
\underline{23 - 161z^{-1} + 276z^{-2}} & \\
101z^{-1} - 276z^{-2} & \\
\underline{101z^{-1} - 707z^{-2} + 1212z^{-3}} & \\
431z^{-2} - 1212z^{-3} & \quad \textbf{Ans.}
\end{array}
$$

This result checks with the expression for $v(n)$ in Eq. (17.10).

Table 17.1 Transforms of Common Sequences

S. No	$v(n)\ n \geq 0$	$V(z)$	*Region of convergence*
1	1	$\frac{z}{z-1}$	$1 < \lvert z \rvert$
2	a^n	$\frac{z}{z-a}$	$a < \lvert z \rvert$
3	n	$\frac{z}{(z-1)^2}$	$1 < \lvert z \rvert$
4	na^{n-1}	$\frac{z}{(z-a)^2}$	$a < \lvert z \rvert$
5	$\frac{a^n}{n!}$	e^{az-1}	$0 < \lvert z \rvert$
6	$\sin \omega n$	$\frac{z \sin \omega}{z^2 - 2z \cos \omega + 1}$	$1 < \lvert z \rvert$
7	$\cos \omega n$	$\frac{z^2 - z \cos \omega}{z^2 - 2z \cos \omega + 1}$	$1 < \lvert z \rvert$
8	$a^n \cos \pi n$	$\frac{z}{z+a}$	$a < \lvert z \rvert$

Table 17.2 Transforms of Common Sampled Time Functions

S. No. $v(t)$	$v(nT)\ n \geq 0$	$V(z)$	*Region of convergence*
1 $u(t)$	1	$\frac{z}{z-1}$	$1 < \lvert z \rvert$
2 $tu(t)$	nT	$\frac{Tz}{(z-1)^2}$	$1 < \lvert z \rvert$
3 $e^{-at} u(t)$	e^{-anT}	$\frac{z}{z - e^{-aT}}$	$e^{-at} < \lvert z \rvert$
4 $\sin(\omega t)\, u(t)$	$\sin(n\omega T)$	$\frac{z \sin \omega T}{z^2 - 2z \cos \omega T + 1}$	$1 < \lvert z \rvert$
5 $\cos(\omega t)\, u(t)$	$\cos(n\omega T)$	$\frac{z^2 - z \cos \omega T}{z^2 - 2z \cos \omega T + 1}$	$1 < \lvert z \rvert$

17.4.2 Properties of z-transforms

Some of the more important properties of the *unilateral* z-transform are following:

Superposition:

$$av_1(n) + bv_2(n) \leftrightarrow aV_1(z) + bV_2(z)$$

Scaling:

$$a^n v(n) \leftrightarrow V(z/a)$$

Delay:

$$v(n-k) \leftrightarrow z^{-k}\, V(z) \qquad k \geq 0$$

Advance:

$$v(n+1) \leftrightarrow zV(z) - v(0)$$

Differentiation:

$$nv\,(n) \leftrightarrow \quad \frac{dV(z)}{dz}$$

Summation:

$$\Sigma\, v(k) \leftrightarrow \frac{z}{z-1}\, V(z)$$

Initial value:

$$v(0) = \lim_{z\to\infty} V(z)$$

Final value:

$$\lim_{n\to\infty} v(n) = \lim_{z\to 1}(z-1)\, V(z)$$

EXAMPLE 17.10 Find the z-transform of the signal $v(n)$ shown in Fig. 17.9.

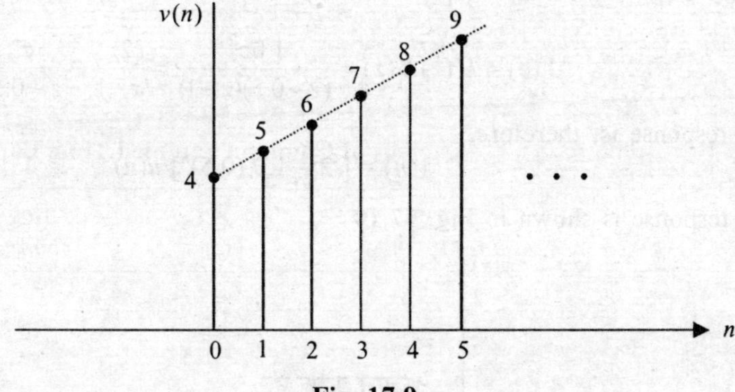

Fig. 17.9

SOLUTION:

Figure shows

$$v(n) = 4u(n) + n \quad \text{(the sum of a step and a ramp)}$$

By the superposition property,

$$V(z) = \frac{4z}{z-1} + \frac{z}{(z-1)^2}$$

$$= \frac{4z^2 - 3z}{(z-1)^2} \qquad |\, z\,| > 1 \qquad \textbf{Ans.}$$

(See entries 1 and 3 in Table 17.1 for the transforms of the step and ramp.)

17.4.3 Response by *z*-transform

The procedure to be followed in finding the response of a LTI discrete-time system by the transform method is:

(a) Find the system transfer function.

(b) Find the transform of the input signal.

(c) Obtain the transform of the product and find its inverse.

This is the desired response.

EXAMPLE 17.11 A unit step is applied to the system described by the difference equation:

$$y(n) = 0.5y(n-1) + 1.0x(n)$$

Find the response by *z*-transform.

SOLUTION:
The system transfer function

$$H(z) = \frac{Y(z)}{X(z)} = \frac{1.0z}{z-0.5}$$

The transform of the input is:

$$u(n) \leftrightarrow \frac{z}{z-1}$$

Hence

$$Y(z) = H(z)X(z) = \frac{1.0z^2}{(z-0.5)(z-1)} = \frac{2z}{z-1} - \frac{z}{z-0.5}$$

The response is, therefore,

$$y(n) = [\,2 - 1.0\,(0.5)^n]\,u(n)$$

The response is shown in Fig. 17.10

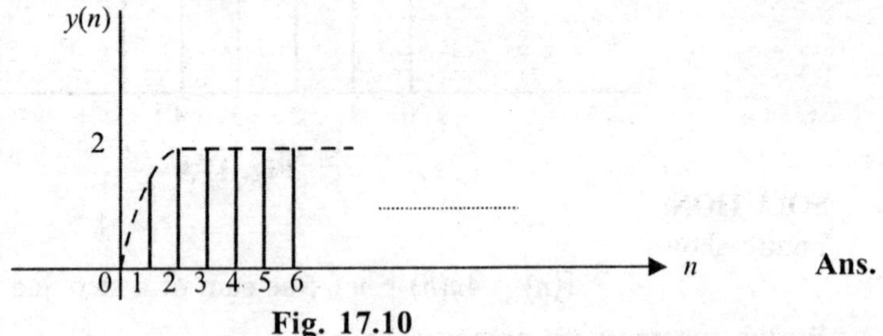

Ans.

Fig. 17.10

SUMMARY

An introduction to the theory underlying digital signal processing has been the aim of this chapter. The concept of difference equation and models of elements of digital circuits are introduced. Solution of difference equation, in particular first order, is explained.

Solved Problems

P17.1 Determine the region of convergence for the series:

(a) $v(n) = 2u(n)$

(b) $v(n) = \begin{cases} 2^n & n \geq 0 \\ 1 & n < 0 \end{cases}$

SOLUTION:

(a) R_2 in Eq. (17.9b) is given by

$$R_2 = \lim_{n\to\infty}\left|\frac{2}{2}\right| = 1$$

Hence $|z| > 1$ is the region of convergence. **Ans.**

(b) $$R_2 = \lim_{n\to\infty}\frac{2^{n+1}}{2^n} = 2$$

$$R_1 = \lim_{n\to\infty}\frac{1}{1} = 1$$

Since $R_2 > R_1$, there is no region of convergence. **Ans.**

P17.2 Use *z*-transform to solve the difference equation:

$$y(n+1) - 0.5\,y(n) = 0.75^n, \qquad y(0) = 1$$

SOLUTION:

Taking the *z*-transform of each term and applying the advance property to $y(n + 1)$,

$$y(n+1) \leftrightarrow zY(z) - zy(0)$$

Hence

$$zY(z) - 1.z - 0.5Y(z) = \frac{z}{z-0.75}$$

Solving,

$$Y(z) = \frac{z}{z-0.5} + \frac{z}{(z-0.5)(z-0.75)}$$

$$= \frac{z}{z-0.5} + \frac{-4z}{z-0.5} + \frac{4z}{z-0.75}$$

$$= \frac{4z}{z-0.75} - \frac{3z}{z-0.5}$$

From Table 17.1 the inverse transform:

$$y(n) = 4(0.75)^n - 3(0.5)^n \qquad \textbf{Ans.}$$

P17.3 Find the initial and final values of the time function $v(n)$ if

$$V(z) = \frac{z}{(z-1)(z-0.2)} \quad |z|>1$$

Verify by finding $v(n)$.

SOLUTION:

Initial value is given by

$$v(0) = \lim_{z\to\infty} V(z) = 0$$

Final value is given by

$$\lim_{n\to\infty} v(n) = \lim_{z\to 1} (z-1)\, V(z) = 5/4$$

Check:

$$v(n) = 5/4 - 5/4\,(0.2)^n$$

Initial value $\lim_{n\to 0} v(0) = 5/4 - 5/4 = 0$

Final value $\lim_{n\to\infty} v(n) = 5/4$ **Ans.**

P17.4 Find the inverse transform of

$$V(z) = \frac{z(z-1)}{(z-3)(z-4)} \quad 3<|z|<4$$

SOLUTION:

$$\frac{V(z)}{z} = \frac{z-1}{(z-3)(z-4)}$$

$$= \frac{3}{z-4} - \frac{2}{z-3}$$

or $$V(z) = \frac{3z}{z-4} - \frac{2z}{z-3}$$

$\therefore\ v(n) = [3(4)^n - 2(3)^n]\, u(n) \quad |z| > 4$ **Ans.**

Exercises

E17.1 Show that the truncated ramp function $x(n)$ in Fig. E17.1 may be expressed as the sum of two ramp functions.

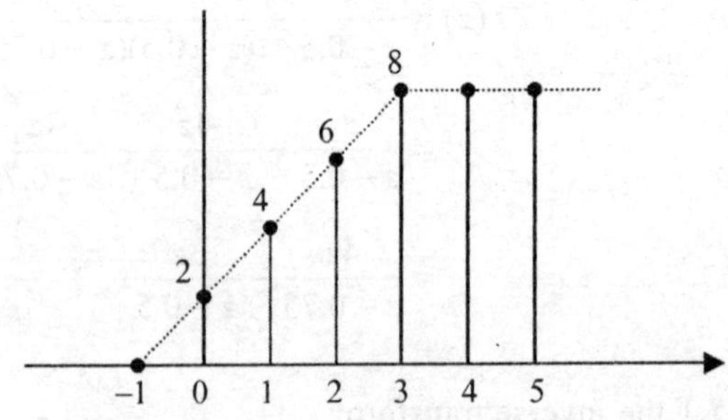

Fig. E17.1

[*Ans:* $x(n) = 2(n+1)u(n) - 2[(n+1) - 4]u(n-4)$]

E17.2 Express the signal shown in Fig. E17.2 in two ways.

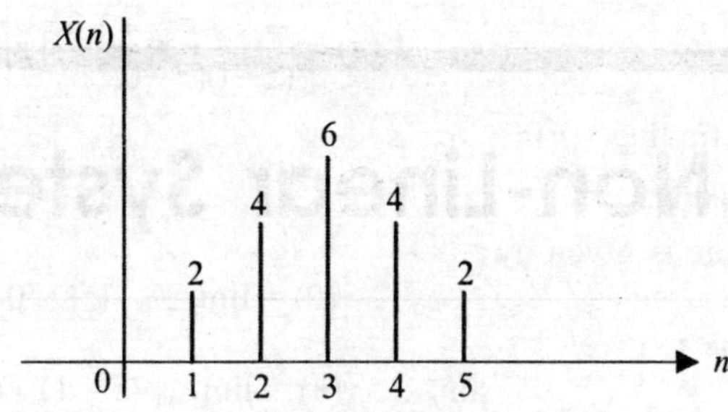

Fig. E17.2

$$[\textit{Ans:}\quad x(n) = \begin{cases} 0 & n < 0 \\ 2n & 0 < n \le 3 \\ 2(6-n) & 3 < n \le 6 \\ 0 & n > 6 \end{cases}$$

or $x(n) = 2r(n)\,u(n) - 4r(n-3)\,u(n-3) + 2r(n-6)\,u(n-6)]$

E17.3 Calculate the z-transform and specify the region of convergence for $v(n) = nu(n)$.

$$[\textit{Ans:}\quad V(z) = \frac{z}{(z-1)^2} \quad 1 < |z|]$$

E17.4 Calculate the inverse z-transform of

$$\frac{z}{z-3} \quad 3 > |z|$$

[*Ans:* $3^n\,u(n)$]

Non-Linear Systems

18

CHAPTER CONTENTS

The methods of network analysis described in previous chapters were based on the assumption of linearity and time invariance. Actual circuit elements and devices however are mostly non-linear.

Exact analysis of non-linear systems is rather difficult and, therefore, approximate methods have been developed to get solutions within acceptable limits of accuracy.

18.1 NON-LINEAR ELEMENTS

Non-linear elements which occur in practical networks include

- Non-linear resistor
- Semiconductor diode
- Iron cored inductor

Non-linear resistor: A linear resistor has a potential across it, which is proportional to the current through it; but in practical resistors the resistance increases with current as shown in the *v-i* characteristic of Fig. 18.1.

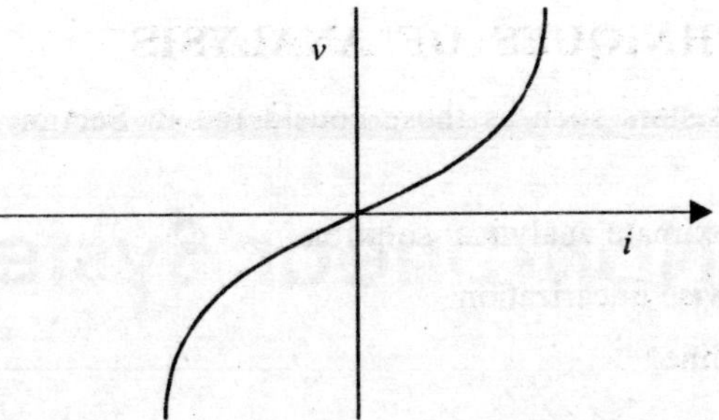

Fig. 18.1 Non-linear resistor characteristics.

Semiconductor diode: An ideal diode has zero resistance to current passing through it in one direction but infinite resistance in the other direction. Figure 18.2 shows a typical *i-v* characteristics of a practical diode.

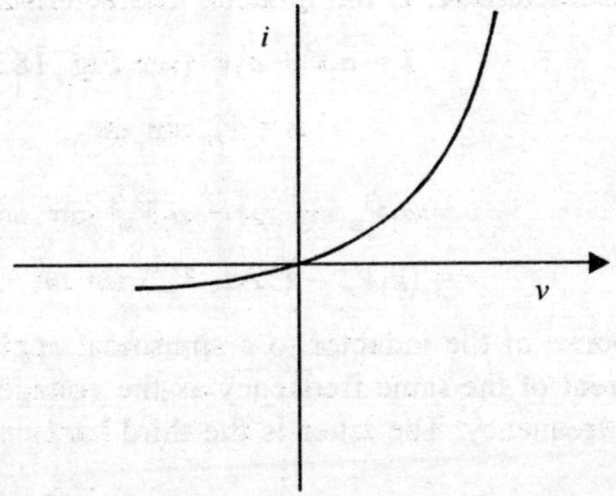

Fig. 18.2 Semiconductor diode characteristic.

Iron-cored inductor: Linear inductors have air-cored coils. In order to achieve higher inductance, iron cored inductors are used, which however results in non-linear BH characteristics and they have *v-i* characteristic similar to that shown in Fig. 18.3.

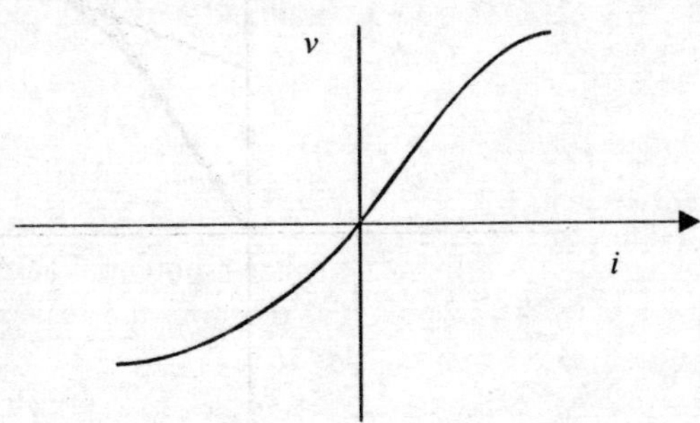

Fig. 18.3 Iron-cored inductor characteristic.

18.2 TECHNIQUES OF ANALYSIS

Non-linear resistors such as those considered in Section 18.1 may be analyzed using techniques of

- Approximate analytical solution
- Piecewise linearization
- Load line

18.2.1 Approximate Analytical Solution

It is assumed that the *v-i* characteristic of a non-linear device is available in a mathematical form (power series).

$$i = a_0 + a_1 v + a_2 v^2 + \cdots$$

If powers of *v* higher than one are neglected then this is a linear approximation of the non-linear characteristic. If the inductor characteristic is taken as:

$$i = a_1 v + a_3 v^3 \text{ (see Fig. 18.3)}$$

and

$$v = V_m \sin \omega t$$

Then

$$i = a_1 V_m \sin \omega t + a_3 V_m^3 \sin^3 \omega t$$

$$= (a_1 V_m - (3/4) a_3 V_m^3) \sin \omega t + (1/4) a_3 V_m^3 \sin 3\omega t$$

The response of the inductor to a sinusoidal applied voltage is seen to produce a sinusoidal current of the same frequency as the voltage, with an additional current of three times that frequency. The latter is the third harmonic current taken by iron-cored inductor.

18.2.2 Piecewise Linearization

A non-linear characteristic, as shown in Fig. 18.4, may be split into sections each represented by a straight line. Below point *A* the characteristics is that of a linear element.

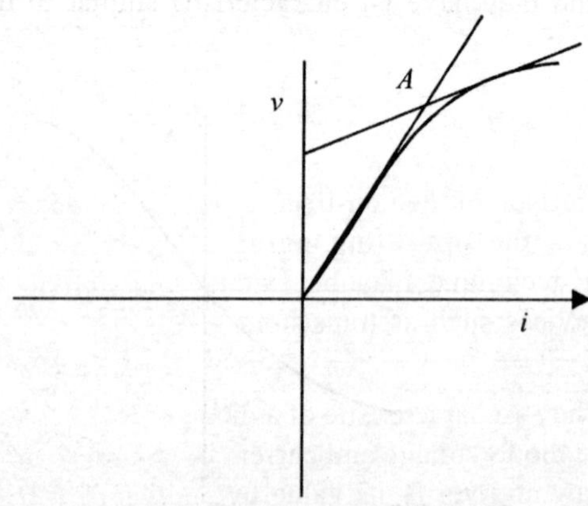

Fig. 18.4 Piecewise linearization of a non-linear characteristic.

18.2.3 Load Line

A load line provides a powerful technique when in a circuit there is only one non-linearity. For example, Fig. 18.5 shows the Thevenin equivalent of a one-port network with linear resistors and a load resistance which is non-linear.

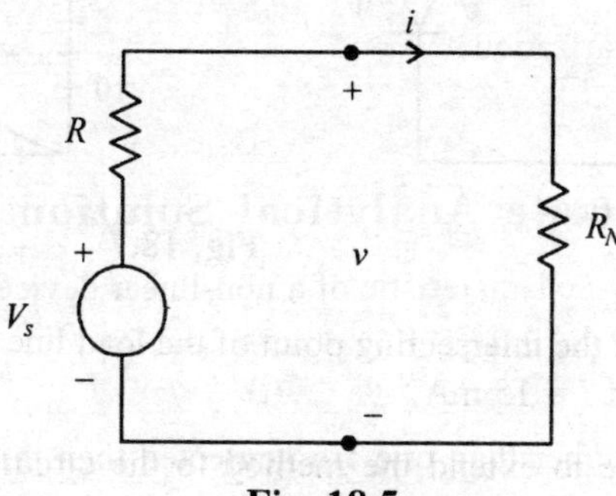

Fig. 18.5

Terminal voltage $v = V_s - Ri$
where i is the current supplied to the load circuit.

This is the equation of a straight line with v as the ordinate as shown in Fig. 18.6. The straight line representing the v-i relationship is called a **load line** and the method is referred to as the load line analysis. Note that the intercepts of the load line on the v and i axes are V_s and V_s/R and its slope is $-R$, as shown in Fig. 18.6.

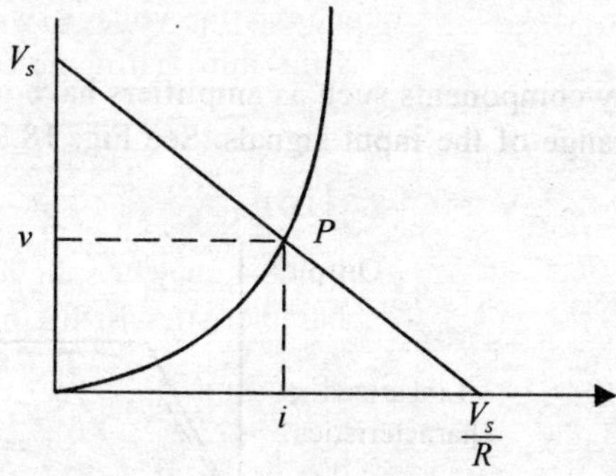

Fig. 18.6

The v-i characteristic of the non-linear resistance is also shown in the Fig. 18.6. The point P is known as the **operating point** where both characteristics are satisfied simultaneously. The technique may be extended to determine the operating point of multiple terminal devices such as transistors.

EXAMPLE 18.1 The i-v characteristic of a diode in Fig. 18.7(a) is shown in Fig. 18.7(b). Find the value of the diode voltage and current by drawing the load line. Find the resistor voltage by Ohm's law and verify its value by applying KVL.

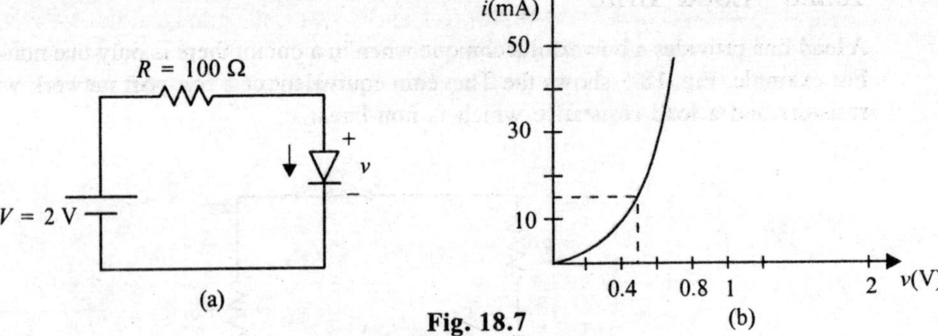

Fig. 18.7

It is seen that the intersecting point of the load line with the non-linear characteristics is at $V = 0.5$ V and $I = 15$ mA.

Note: It is possible to extend the method to the circuits in which the source voltage is not constant.

18.3 COMMON PHYSICAL NON-LINEARITIES

The non-linearities present in a system can be classified as *incidental* and *intentional*. Incidental non-linearities are those which are inherently present in the system. Examples of incidental non-linearities are saturation, dead-zone and backlash. On the other hand intentional non-linearities are those which are deliberately introduced to the system to modify the system characteristics. The most common example of this type of non-linearity is a relay.

Saturation: Many components such as amplifiers have output proportional to the input only in a limited range of the input signals. See Fig. 18.8.

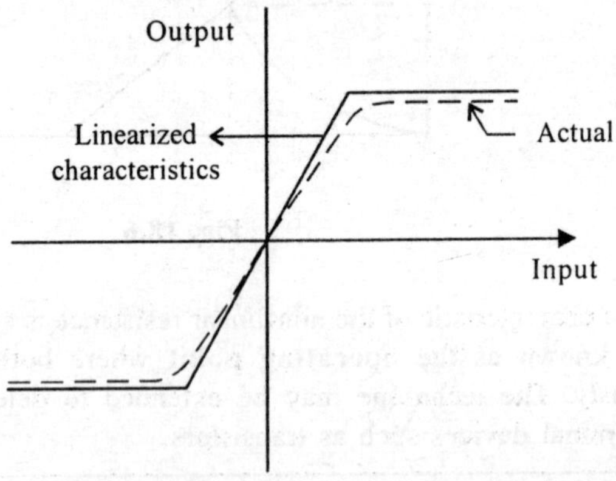

Fig. 18.8 Piecewise linearization of saturation.

Friction: Retarding frictional forces exist whenever mechanical surfaces come in sliding contact. Viscous friction is proportional to the relative velocity of sliding surfaces.

Viscous friction is thus linear in nature. Fig. 18.9(a) is the graphical representation of such friction. In addition to viscous friction, there is Coulomb friction and stiction. The former is a constant retarding force always opposing relative motion and the latter is the force required to initiate motion.

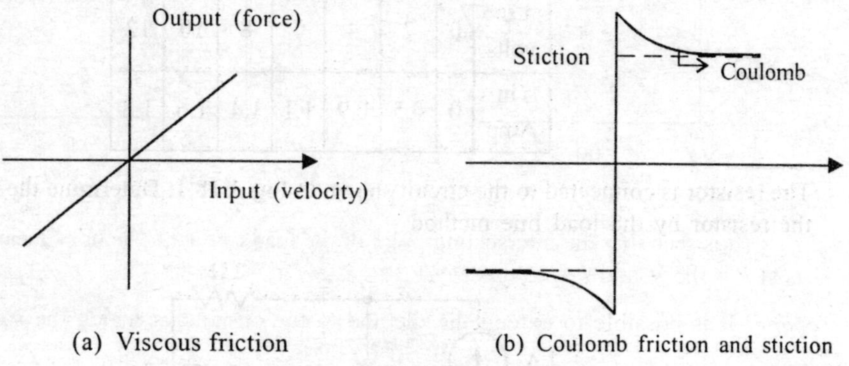

(a) Viscous friction (b) Coulomb friction and stiction

Fig. 18.9

Backlash: Another important non-linearity occurring in physical systems is hysteresis in mechanical transmission, such as gear trains and linkages. In a servo system, gear backlash may cause sustained oscillations which might lead to instability.

Dead-zone: Figure 18.10 shows the dead-zone when gears with backlash drive a spring load.

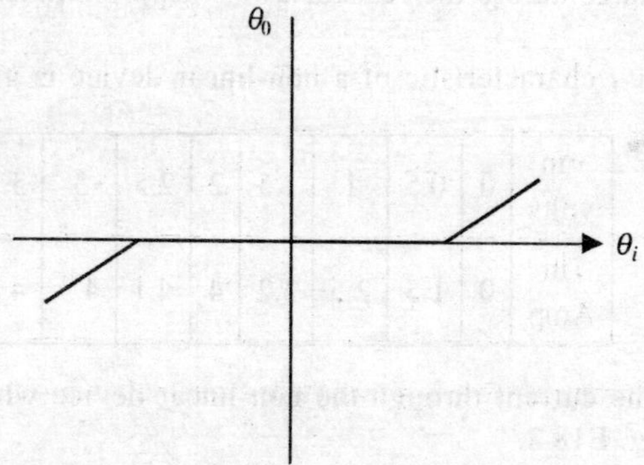

Fig. 18.10 Dead-zone non-linearity.

Relay: A relay is a non-linear power amplifier which can provide large amplification inexpensively and is therefore very commonly used in physical systems, such as control systems.

SUMMARY

The chapter is intended only to serve as an introduction to non-linear systems. There are powerful techniques which facilitate the analysis of non-linearities occurring in control systems; these are elaborated in Reference 18.

Exercises

E18.1 The *v-i* characteristic of a non-linear resistor is given below:

v in volts	0	2	4	6	8	10	12
i in Amp	0	0.5	0.9	1.1	1.4	1.6	1.8

The resistor is connected to the circuit shown in Fig. E18.1. Determine the voltage across the resistor by the load line method.

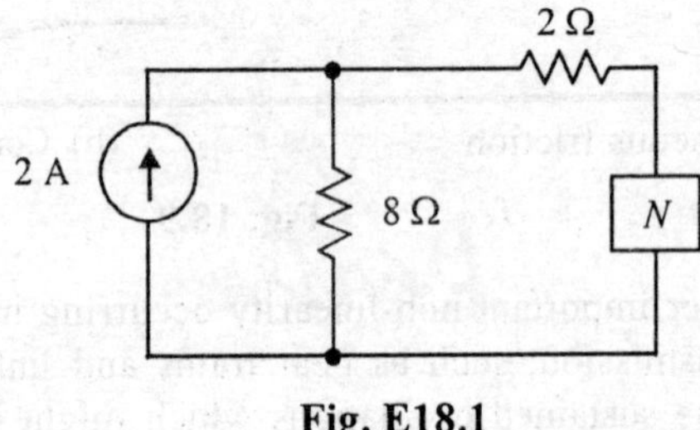

Fig. E18.1

[*Ans:* The intersection of the load line and the non-linear characteristics gives the required voltage across the resistor as 5 V approximately.]

E18.2 The *v-i* characteristic of a non-linear device is given below:

v in volts	0	0.5	1	1.5	2	2.5	3	3.5	4	4.5	5
i in Amp	0	1.5	2.5	3.2	4	4.1	4.3	4.4	4.5	4.6	4.7

Determine the current through the non-linear device when it is connected in the circuit shown in Fig. E18.2.

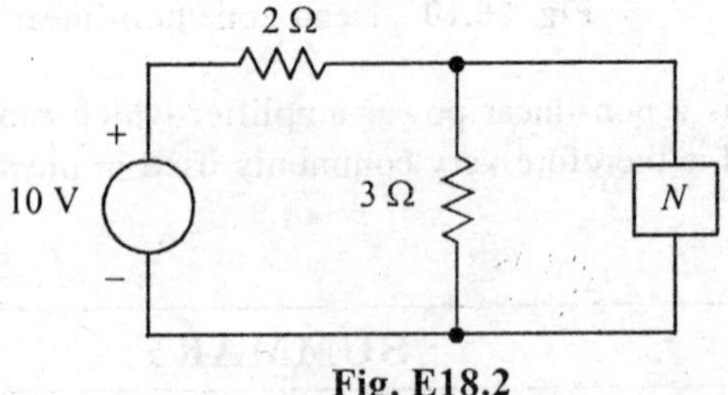

Fig. E18.2

[*Ans:* The intersection of the load line and the non-linear characteristics give the required current as 3.5 A approximately.]

Part 4

Elements of Network Synthesis

Network Synthesis

19

CHAPTER CONTENTS

Network synthesis is a branch of network theory that deals with the problem of realization of Systems that satisfy prescribed input-output characteristics (See Chapter 1). Elements of realizability are presented in this Chapter.

19.1 GENERAL BACKGROUND

Impedance functions: Laplace transform techniques are eminently suitable for evaluating the properties of a network having zero initial conditions. The driving point impedance function, for example, is equal to

$$\frac{V(s)}{I(s)} = R + sL + \frac{1}{sC}$$

See Fig. 19.1.

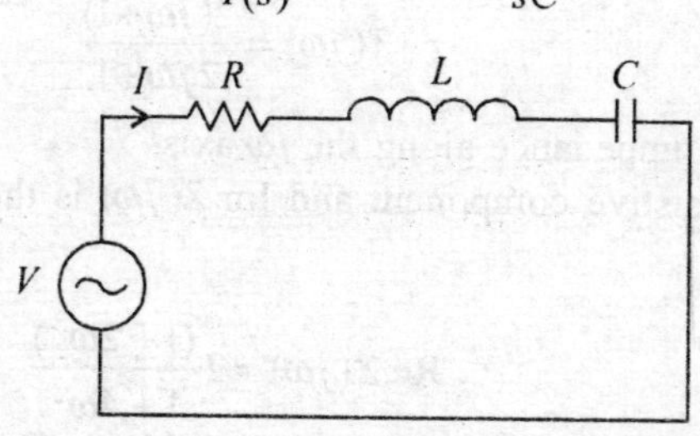

Fig. 19.1

EXAMPLE 19.1 The network in Fig. 19.1 with $R = 1\ \Omega$, $L = 1$ H and $C = 1$ F and zero initial conditions, has the driving point impedance function:

$$Z(s) = 1 + s + 1/s$$

EXAMPLE 19.2 $Z(s)$ for the circuits in Fig. 19.2 are:

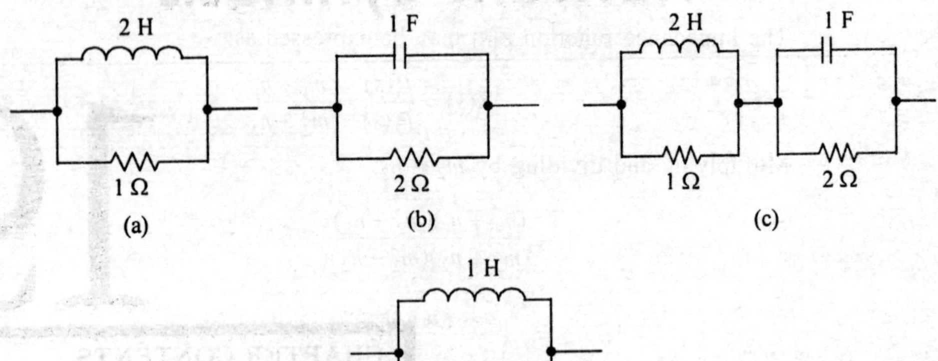

Fig. 19.2

(a) $Z(s) = \dfrac{2s}{2s+1}$

(b) $Z(s) = \dfrac{2/s}{2+1/s} = \dfrac{2}{2s+1}$

(c) $Z(s) = \dfrac{2s}{2s+1} + \dfrac{2}{2s+1} = \dfrac{2(s+1)}{2s+1}$ (19.1)

(d) $Z(s) = \dfrac{s.1/s}{s+1/s} = \dfrac{s}{s^2+1}$

Note that the circuit of Fig. 19.2(c) is a tandem connection of Fig. 19.2(a) and 19.2(b).

If s is set at $j\omega$, $Z(s)$ becomes an ordinary impedance.

$$Z(j\omega) = \frac{2(j\omega+1)}{2j\omega+1} \tag{19.2}$$

$Z(j\omega)$ is called the impedance along the $j\omega$ axis.

Re $Z(j\omega)$ is the resistive component and Im $Z(j\omega)$ is the reactive component. Thus, in Eq. (19.2),

$$\text{Re } Z(j\omega) = 2\frac{(1+2\omega^2)}{1+4\omega^2}$$

$$\text{Im } Z(j\omega) = \frac{-2\omega}{1+4\omega^2}$$

In general, $Z(s)$ is a ratio of two rational polynomials of s. A polynomial $P(s)$ can be put in terms of its even and odd parts as:

$$P(s) = m + n = \text{Ev } P(s) + \text{Od } P(s) = \frac{P(s)+P(-s)}{2} + \frac{P(s)-P(-s)}{2} \tag{19.3}$$

The impedance function $Z(s)$ may be expressed as:

$$Z(s) = \frac{P_1(s)}{P_2(s)} = \frac{m_1 + n_1}{m_2 + n_2}$$

Multiplying and dividing by $m_2 - n_2$,

$$= \frac{(m_1 + n_1)(m_2 - n_2)}{(m_2 + n_2)(m_2 - n_2)}$$

$$= \frac{(m_1 m_2 - n_1 n_2) + (n_1 m_2 - n_2 m_1)}{m_2^2 - n_2^2}$$

The first term on the right side is an even function of s and is denoted by

$$\text{Ev } Z(s) = \frac{m_1 m_2 - n_1 n_2}{m_2^2 - n_2^2}$$

Similarly,

$$\text{Od } Z(s) = \frac{n_1 m_2 - n_2 m_1}{m_2^2 - n_2^2}$$

EXAMPLE 19.3 Find the even and odd parts of $Z(s) = \dfrac{2(s+1)}{2s+1}$ of Eq. (19.1).

SOLUTION:

$$m_1 = 2, \quad m_2 = 1, \quad n_1 = 2s, \quad n_2 = 2s$$

Therefore $\quad \text{Ev } Z(s) = \dfrac{2-4s^2}{1-4s^2}$ **Ans.**

and $\quad \text{Od } Z(s) = \dfrac{2s-4s}{1-4s^2} = \dfrac{-2s}{1-4s^2}$ **Ans.**

Along the $j\omega$ axis Ev $Z(s)$ is made up of real quantities only, therefore it is a real function of ω.

Od $Z(s)$ is made up of imaginary components; therefore it is an imaginary function of ω, i.e.,

Along the $j\omega$ axis, Ev $Z(s)$ equals Re $Z(s)$

Along the $j\omega$ axis Od $Z(s)$ equals j Im $Z(s)$

In Example 19.3

$$\operatorname{Re} Z(s)\Big|_{s=j\omega} = \frac{2-4s^2}{1-4s^2}\Big|_{s=j\omega}$$

$$= \frac{2+4\omega^2}{1+4\omega^2}$$

and

$$j \operatorname{Im} Z(s)\Big|_{s=j\omega} = \frac{-2s}{1-4s^2}\Big|_{s=j\omega}$$

$$= \frac{-2j\omega}{1+4\omega^2}$$

It may be noted that the Re $Z(s)$ is never negative. The conclusion is that the impedance function $Z(s)$ of Example 19.3 may be realized in terms of *RLC* circuits.

EXAMPLE 19.4 Show that Re $Z(s) \geq 0$ for Re $s = 0$ if

$$Z(s) = \frac{s^2 + s + 2}{2s^2 + s + 1\delta} \tag{19.4}$$

SOLUTION:

$m_1 = s^2 + 2, \quad m_2 = 2s^2 + 1, \quad n_1 = s, \quad n_2 = s$

$$\therefore \qquad \operatorname{Ev} Z(s) = \frac{(s^2+2)(2s^2+1) - s.s}{(2s^2+1)^2 - s^2}$$

Both numerator and denominator are easily seen to be $Z(s) \geq 0$ for Re $s = 0$.

It may be verified that both Eq. (19.1) and Eq. (19.4) satisfy the constraints for a positive real function stated in Section 19.2.

19.2 POSITIVE REAL FUNCTIONS

Definition: A rational polynomial function is called a **positive real function** if it satisfies the constraints.

$Z(s)$ = real when s = real
Re $Z(s) \geq 0$ when Re $s = 0$
Re $Z(s) > 0$ when Re $s > 0$

Rational functions of s satisfying these constraints are called **positive real functions** (prf). Positive real functions play a very important role in network synthesis. It can be shown that all *RLC* networks yield positive real functions. It is also true that all positive real functions are realizable as *RLC* networks.

Geometrically $Z(s)$ is said to map the right half s plane into a portion of the right half s plane. That is to say for any point s_0 in the RH s plane the corresponding $Z(s_0)$ is located in the RH $Z(s)$ plane.

Theorem I: A positive real function $Z(s)$ cannot have any poles or zeros in the RH s plane. Poles of $Z(s)$ and $1/Z(s)$ must be simple with real positive residues.

An important corollary is that if $Z(s)$ is prf, the degree of the numerator cannot differ from that of the denominator by more than unity.

Analytic function: $Z(s)$ is analytic if the derivative of $Z(s)$ exists throughout a certain domain D and it is unique irrespective of $d\omega/d\sigma$, for $s = \sigma + j\omega$.

19.3 TEST FOR POSITIVE REAL FUNCTIONS

It has been shown (see Section 19.2) that if $Z(s)$ is positive real function, it has no zeros or poles in the RHS plane. $Z(s)$ is then said to be analytic in the RHS plane. This leads to three simple tests for positive realness.

1. Test for analyticity of $Z(s)$ in RHS plane. Polynomials with all their zeros in the LHS plane are called **Hurwitz polynomial**; and hence both numerator and denominator of $Z(s)$ must be Hurwitz polynomials.
2. Verify whether Re $Z(s) \geq 0$ on the j-axis.
3. Ensure that all j-axis poles are simple with positive residues.

The necessary and sufficient conditions for a :ational function with real coefficients $Z(s)$ to be positive real are:

(a) $Z(s)$ must have no poles in RHS plane.
(b) $Z(s)$ may have only simple poles on the $j\omega$ axis with real and positive residues.
(c) Re $Z(j\omega) \geq 0$ for all ω.

These conditions are more convenient to use than the basic conditions stated in Section 19.2.

The conditions that must be fulfilled in order for

$$Z(s) = \frac{s+a}{s^2+bs+c}$$

to be positive real can be shown to be

$$a,\ b,\ c \geq 0 \qquad b > a \tag{19.5}$$

For example,

$$\frac{s+3}{s^2+4s+6} \text{ is positive real}$$

$$\frac{s+4}{s^2+3s} \text{ is } \textit{not} \text{ positive real.}$$

EXAMPLE 19.5 Check the following function for positive realness:

$$Z(s) = \frac{s+2}{s^2+3s+2}$$

SOLUTION:
Making use of Eq. (19.5), $Z(s)$ is prf

Aliter
The poles are $s = -1, -2$
Zeros are $s = -2$
All in the LHS plane.
On the $j\omega$-axis

$$Z(j\omega) = \frac{1}{j\omega + 1} = \frac{1 - j\omega}{1 + \omega^2}$$

Re $Z(j\omega)$ is positive. Thus, both tests 1 and 2 are satisfied and $Z(s)$ is prf. **Ans.**

19.4 PROPERTIES OF POSITIVE REAL FUNCTIONS

1. If $F(s)$ is positive real then $1/F(s)$ is also positive real.
2. The sum of positive real functions is positive real; the difference is not necessarily positive real. For example, $s - 1/s$ is not positive real.
3. Poles and zeros cannot have positive real parts.
4. For only simple poles with real positive residues can exist on the j-axis.
5. Poles and zeros of a prf are real or occur in conjugate pairs.
6. The highest powers of the numerator and denominator polynomials may differ at most by unity. This condition prohibits multiple poles and zeros at $s = \infty$.
7. The lowest powers of numerator and denominator polynomials may differ by at most unity. This condition prevents the possibility of multiple poles or zeros at $s = 0$.

For example, if

$$F(s) = \frac{3s^2 + 5}{s(s^2 + 1)}$$

Using partial fractions,

$$F(s) = \frac{-2s}{s^2 + 1} + \frac{5}{s}$$

The residue of the poles at $s = \pm j$ is negative. Therefore $F(s)$ is **not** prf.

19.5 HURWITZ POLYNOMIALS

One condition for a function to be positive real is that all the poles and zeros lie in the LHS plane. For polynomials having all their zeros in the LHS plane are called **Hurwitz polynomials**.

Theorem II: If P is a Hurwitz polynomial then the ratio of its even and odd parts is a positive real function.

EXAMPLE 19.6 Check whether $s^3 + s^2 - 4s + 6$ is Hurwitz.

SOLUTION:

$$m/n = \frac{s^3 - 4s}{s^2 + 6}$$

Poles are $s = \pm\sqrt{6}$
Zeros $s = 0$, $s = 4$

Then

$$Z(j\omega) = \frac{-j\omega^3 - 4j\omega}{-\omega^2 + 6}$$

Re $Z(j\omega) = 0$ and m/n is prf. It follows $P(s)$ is Hurwitz. **Ans.**

19.6 ELEMENTARY SYNTHESIS PROCEDURE

The basic idea behind the synthesis of driving point functions is to breakup a positive real function $Z(s)$ into a sum of simpler positive real functions $Z_1(s)$, $Z_2(s)$, ..., and then to synthesize these individual functions as elements of the overall network whose driving point impedance is:

$$Z(s) = Z_1(s) + Z_2(s) + \cdots + Z_n(s) \tag{19.6}$$

For example, consider the positive real function:

$$Z(s) = \frac{s^2 + 2s + 8}{s(s+4)}$$

A partial fraction expansion gives

$$Z(s) = \frac{2}{s} + \frac{s}{s+4} = Z_1(s) + Z_2(s)$$

$Z_1(s) = 2/s$ is positive real, because when s is real, $Z_1(s)$ is real, further when Re $s = \sigma >$ 0,

$$\text{Re } Z_1(s) = \frac{2\sigma}{\sigma^2 + \omega^2} > 0$$

Therefore, $Z_1(s)$ is positive real.

$Z_2(s)$ satisfies the following positive real criteria.

1. $Z_2(s)$ must have no poles in the right half plane.
2. Poles of $Z_2(s)$ on the imaginary axis must be simple and their residues must be real and positive. Re $|Z_2(j\omega)| \geq 0$ for all ω.

Hence $Z_2(s)$ is positive real.

Since both $Z_1(s)$ and $Z_2(s)$ are positive real, their sum can be synthesized into the network shown in Fig. 19.3.

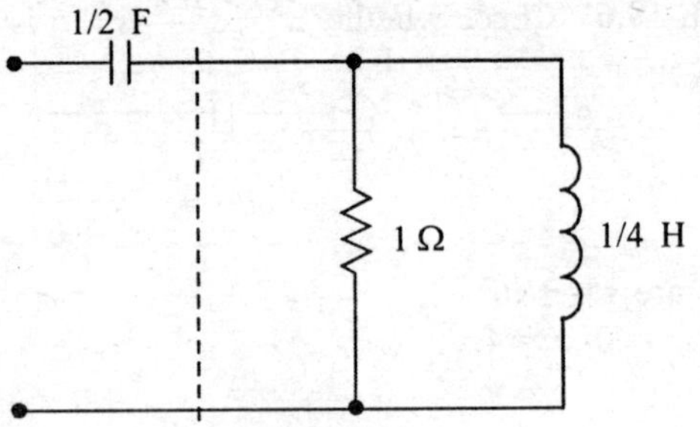

Fig. 19.3

Note: The basic passive impedance (and admittance) are positive real (pr) functions. Similarly, the admittances K, Ks, K/s are pr if K is real and positive.

The following properties of LC impedance or admittance functions may be noted:

1. $Z_{LC}(s)$ or $Y_{LC}(s)$ is the ratio of odd to even or even to odd polynomials.
2. The poles and zeros are simple and lie on the $j\omega$ axis.
3. The poles and zeros interlace on the $j\omega$ axis.
4. The highest powers of numerator must differ by unity; the lowest power also differ by unity.
5. There must be either a zero or a pole at origin and infinity.

EXAMPLE 19.7 Test whether the function

$$Z(s) = \frac{s(s^2+4)}{(s^2+1)(s^2+3)} \text{ is } LC.$$

SOLUTION:
The poles are $\pm j$, $\pm j\sqrt{3}$ and the zeros are $0, \pm j2$. Evidently the poles and zeros do not interlace on the $j\omega$ axis and hence $Z(s)$ is *not LC*.

19.6.1 Foster Type Realization

The LC networks synthesized by *partial fraction* expansion are called **Foster type networks.** The impedance form is called a **Foster series network** and the admittance form is a **Foster parallel network**.

EXAMPLE 19.8 Synthesize the LC function

$$Z(s) = \frac{(s^2+1)(s^2+9)}{s(s^2+4)}$$

SOLUTION:
A partial fraction expansion gives

$$Z(s) = s + \frac{9/4}{s} + \frac{s(15/4)}{s^2+4}$$

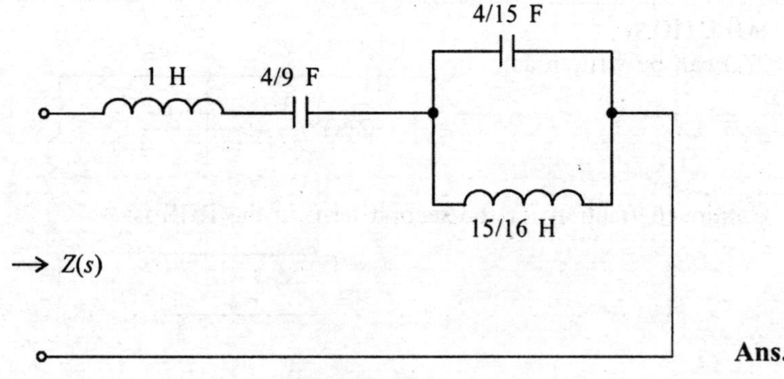

Fig. 19.4

The synthesized network is obtained as shown in Fig. 19.4.
The last term in the partial fraction may be verified to be the Laplace transform of parallel combination of the capacitance and inductance.

EXAMPLE 19.9 Synthesize the admittance function

$$Y(s) = \frac{s(s^2+2)(s^2+4)}{(s^2+1)(s^2+3)}$$

SOLUTION:
The partial fraction expansion of $Y(s)$ is expressed as:

$$Y(s) = s + \frac{(1/2)s}{s^2+3} + \frac{(3/2)s}{s^2+1}$$

from which the network can be synthesized as shown in Fig. 19.5.

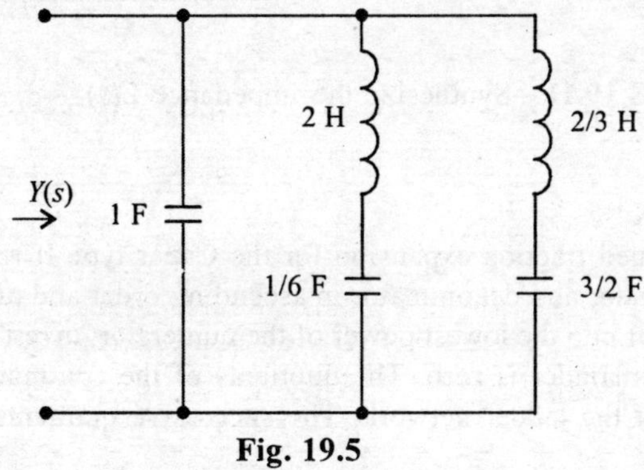

Ans.

Fig. 19.5

19.6.2 Cauer Type Realization

Networks synthesized by *continued fraction* expansion are called **Cauer type networks**. There are two types: Cauer I (series) and Cauer II (parallel).

EXAMPLE 19.10 Synthesize a Cauer type network for $Z(s) = \dfrac{2s^2+2s+1}{s(s^2+s+1)}$.

SOLUTION:
$Z(s)$ can be written as:

$$Z(s) = \frac{1}{s} + \frac{s+1}{s^2+s+1}$$

Continued fraction for the second term in the RHS is:

$$\begin{array}{r} s+1\ \big)\ s^2+s+1\ (\ s \\ \underline{s^2+s\quad} \\ 1 \end{array}$$

Thus,
$$Z(s) = \frac{1}{s} + \frac{1}{s + 1/(s+1)}$$

The realization of $Z(s)$ is shown in Fig. 19.6.

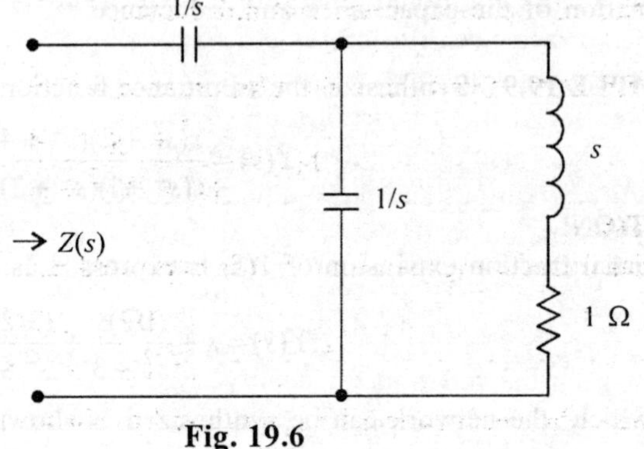

Fig. 19.6

EXAMPLE 19.11 Synthesize the impedance $Z(s) = \dfrac{(s^2+1)(s^2+3)}{s(s^2+2)}$ by Cauer type II realization.

SOLUTION:
The continued fraction expansion for the Cauer type II realization consists in arranging both numerator and denominator in ascending order and dividing the lowest power of the denominator into the lowest power of the numerator; invert the remainder and divide again until the remainder is zero. The quotients of the continued fraction expansion give the elements of the ladder network. The successive quotients alternate between Z and Y.

$$\begin{array}{l} 2s + s^3\ \big)\ 3 + 4s^2 + s^4\ (\ 3/(2s) \\ \qquad\qquad\ \underline{3 + 3/2s^2\qquad\quad} \\ \qquad\qquad\qquad 5/2s^2 + s^4\ )\ 2s + s^3\ (\ 4/(5s) \\ \qquad\qquad\qquad\qquad\qquad\qquad \underline{2s + 4/5s^3\qquad} \\ \qquad\qquad\qquad\qquad\qquad\qquad\quad 1/5s^3\ )\ 5/2s^2 + s^4\ (\ 25/(2s) \\ \qquad\qquad\qquad\qquad\qquad\qquad\qquad\qquad\quad \underline{5/2s^2\qquad\quad} \\ \qquad\qquad\qquad\qquad\qquad\qquad\qquad\qquad\qquad s^4\ )\ 1/5s^3\ (\ 1/5s \\ \qquad\qquad\qquad\qquad\qquad\qquad\qquad\qquad\qquad\qquad \underline{1/5s^3} \end{array}$$

The synthesized network is shown in Fig. 19.7.

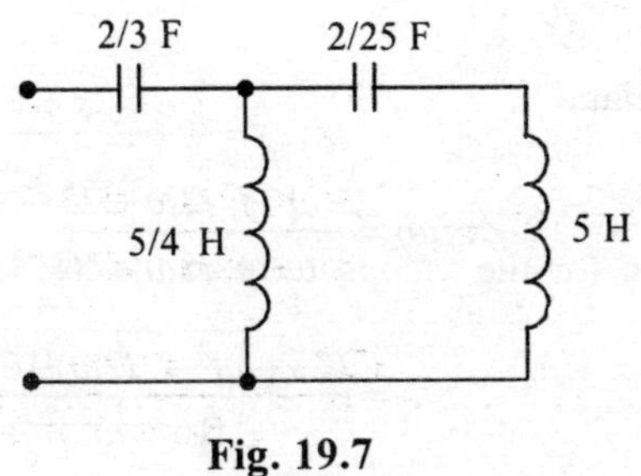

Ans.

Fig. 19.7

SUMMARY

The basic concepts in network synthesis are given in this chapter. Network synthesis is based on a class of rational functions of *s* which satisfy certain conditions; such functions are called **positive real functions**. All *RLC* networks yield positive real functions; also all positive real functions are realizable as *RLC* networks. Test for positive real functions and the properties of these functions are considered in some detail. The concept of Hurwitz polynomials is also highlighted. This is followed by a brief consideration of elementary relizability conditions and the synthesis of one-port networks. The chapter is meant only as an introduction to the very vast topic of network synthesis.

Solved Problems

P19.1 Check the function for positive realness:

$$Z(s) = \frac{s+1}{s^2+2}$$

SOLUTION:

The poles are $s = \pm j\sqrt{2}$

Zeros $s = -1$

On the *j*-axis,

$$Z(j\omega) = \frac{j\omega+1}{-\omega^2+2}$$

Re $Z(j\omega)$ is positive only for $\omega^2 < 2$. While condition 1 is satisfied, condition 2 is *not* satisfied. Hence $Z(s)$ is not prf. **Ans.**

P19.2 Test for positive realness for the following $Z(s)$:

$$Z(s) = \frac{s^2+2s+25}{s^2+5s+16}$$

SOLUTION:

The poles are $s = -5/2 \pm \frac{j}{2}\sqrt{39}$

Zeros $s = -1 \pm \frac{j}{2}\sqrt{96}$

All in the LHS plane.

$$Z(j\omega) = \frac{-\omega^2 + j2\omega + 25}{-\omega^2 + j5\omega + 16}$$

$$= \frac{(25 - \omega^2 + j2\omega)(16 - \omega^2 - j5\omega)}{(16 - \omega^2)^2 + 25\omega^2}$$

The denominator is an even function.

Real part of the numerator $= 400 - 31\omega^2 + \omega^4$

$$= (\omega^2 - 31/2)^2 + \left(400 - \frac{31^2}{4}\right)$$

Re $Z(j\omega)$ is therefore positive for all values of ω. Thus, the conditions of Section 19.3 are satisfied and $Z(s)$ is prf. **Ans.**

P19.3 Test whether the polynomial $P(s)$ is Hurwitz.

$$P(s) = s^4 + s^3 + 5s^2 + 3s + 4$$

SOLUTION:

$$m/n = \frac{s^4 + 5s^2 + 4}{s^3 + 3s} = \frac{(s^2 + 1)(s^2 + 4)}{s(s^2 + 3)}$$

Poles $s = 0, \quad s = \pm j\sqrt{3}$

Zeros $s = \pm j1, \pm j2$

Then

$$Z(j\omega) = \frac{\omega^4 - 5\omega^2 + 4}{-j\omega^3 + 3j\omega} = \frac{-j(\omega^4 - 5\omega^2 + 4)}{-\omega^3 + 3\omega}$$

It follows Re $Z(j\omega) = 0$ and m/n is therefore prf and by Theorem II, $P(s)$ is Hurwitz. **Ans.**

P19.4 Test whether $Z(s) = \dfrac{s^2 + 1}{s(s^2 + 2)}$ is LC.

SOLUTION:

Expand $Z(s)$ in partial fractions and show that each of the residues is positive.

$$\frac{Z(s)}{s} = \frac{s^2 + 1}{s^2(s^2 + 2)}$$

$$= \frac{A}{x} + \frac{B}{x + 2} \qquad \text{where } x = s^2,$$

Now, A = 1/2 and B = 1/2.

$$\frac{Z(s)}{s} = \frac{1/2}{s^2} + \frac{1/2}{s^2+2}$$

or
$$Z(s) = \frac{1}{2s} + \frac{s}{2(s^2+2)}$$

which is of the form of Eq. (19.6). The residues are positive and hence $Z(s)$ is prf and may be realized by LC network of Fig. P19.1.

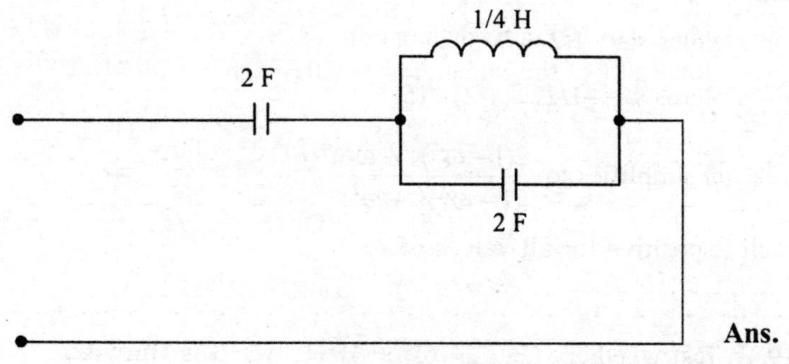

Ans.

Fig. P19.1

P19.5 Test whether the function $Z(s) = \dfrac{k(s^2+1)(s^2+9)}{s(s^2+4)}$ is LC.

SOLUTION:

The poles are $s = 0$, $s = \pm j2$ and zeros are $s = \pm j$, $s = \pm j3$. $Z(s)$ is the ratio of odd to even function. The poles and zeros are simple and lie on the $j\omega$ axis. The poles and zeros interlace on the $j\omega$ axis. The highest powers of numerator and denominator differ by unity; the lowest powers also differ by unity. At the origin, there is a pole and at infinity there is a zero. Hence properties 1 to 5 are satisfied and thus $Z(s)$ is LC. **Ans.**

Exercises

E19.1 Prove that if $Z_1(s)$ and $Z_2(s)$ are both positive real then

$$Z(s) = \frac{Z_1(s)Z_2(s)}{Z_1(s)+Z_2(s)}$$ must be positive real.

E19.2 Check the function for positive realness:

$$Z(s) = \frac{s+4}{s^2+2s+1}$$

[**Hint:** Poles $s = -1, -1$
Zeros $s = -4$

$$Z(j\omega) = \frac{j\omega + 4}{-\omega^2 + j2\omega + 1}$$

which can be shown to be positive only when $\omega^2 < 2$. Hence $Z(s)$ is not prf.

E19.3 Check the function for positive realness:

$$Z(s) = \frac{2s^2 + s + 2}{s^2 + s + 1}$$

[Hint:

Poles $s = -1/2 \pm j\sqrt{3}/2$

Zeros $s = -1/2 \pm (1/2)\sqrt{15}$

Re $Z(j\omega)$ simplifies to $\dfrac{2(1-\omega^2)^2 + \omega^2}{(1-\omega^2)^2 + \omega^2}$

which is positive for all values of ω.

[*Ans:* $Z(s)$ is prf]

E19.4 Test whether $s^5 + s^4 + 6s^3 + 4s^2 + 8s + 3$ is Hurwitz.

E19.5 Test whether the function $Z(s) = \dfrac{s^5 + 4s^3 + 5s}{3s^4 + 6s^2}$ is *LC*.

[Hint: Check property 2 for *LC* impedance.]

[*Ans:* Not *LC*.]

Matrices and Determinants

Appendix A

A matrix is a rectangular array of scalars, arranged in m rows and n columns and is denoted by

$$A = [a_{pq}] \qquad 1 \le p \le m \qquad 1 \le q \le n$$

The order or dimension of A is given by $m \times n$. If the reader has any difficulty with the double suffix notation, remember **r**oman **c**atholic will come to his rescue.

A.1 TYPES OF MATRICES

- A square matrix has the same number of rows and columns, $m = n$.
- A row matrix or row vector has one row.
- A column matrix or column vector has one column.
- A diagonal matrix is square with elements off its main diagonal all zero, $a_{pq} = 0$ for $p \neq q$.
- A matrix is triangular if $a_{pq} = 0$ for $p \ge q$ (upper triangular) or if $a_{pq} = 0$ for $p \le 0$ (lower triangular).
- The unit matrix (also called identity matrix) denoted by U, is a diagonal matrix whose elements are 1, $a_{pp} = 1$.
- The null (or zero) matrix 0 has all its elements zero.
- The trace of a square matrix is the sum of its diagonal elements.

$$\text{tr } A = \sum a_{pp}$$

- Two matrices are equal if they have the same order and their corresponding elements are equal. That is $A = B$ implies $a_{pq} = b_{pq}$.
- The transpose of A denoted by A^T is formed by interchanging rows and columns.

- A real matrix satisfies the equation $A = A^*$ where A^* is the complex conjugate of every element in A.
- An imaginary matrix satisfies the equation $A = -A^*$.
- A symmetric matrix satisfies the equation $A = A^T$.
- A skew symmetric matrix satisfies the equation $A = A^{T*}$.

A.2 ADDITION, SUBTRACTION AND MULTIPLICATION

- Matrices can be added (or subtracted) only if they have the same order. The operation is performed element by element. That is

$$A + B - C = [a_{pq} + b_{pq} - c_{pq}]$$
$$A - A = 0, \text{ the null matrix.}$$

- The commutative and associative laws hold good, that is

$$A + B = B + A$$
$$A + (B + C) = (A + B) + C$$

- A matrix when multiplied by a scalar α gives a matrix with all elements multiplied by α.

  The product of two matrices AB is defined only if the number of columns in A equals the number of rows in B. The multiplication operation is illustrated in the following example:

$$A = \begin{bmatrix} 1 & 2 \\ 8 & 9 \end{bmatrix} \qquad B = \begin{bmatrix} 1 & 2 \\ 3 & 1 \end{bmatrix}$$

$$AB = \begin{bmatrix} 1\times1+2\times3 & 1\times2+2\times1 \\ 8\times1+9\times3 & 8\times2+9\times1 \end{bmatrix} = \begin{bmatrix} 7 & 4 \\ 35 & 25 \end{bmatrix}$$

  The product BA is found to be $BA = \begin{bmatrix} 17 & 20 \\ 11 & 15 \end{bmatrix}$

  In general, commutative law of multiplication does not hold as indicated above. It is therefore important to state clearly which is the first factor in the product.

- Distributive and associative laws are valid for multiplication.

$$A(B + C) = AB + AC$$
$$(AB)C = A(BC)$$
$$AU = UA = A$$

  where the order of U is chosen to be compatible with A.

- Positive integer powers of a square matrix are defined by $A^2 = AA$, $A^3 = A^2 A$, ..., Also A^0 is U.
- The transpose of a product is the product of the individual transposed matrices in reverse order.

$$(AB)^T = B^T A^T$$

A.3 INVERSION

The inverse of a square matrix A is denoted by A^{-1} and is defined as:

$$AA^{-1} = A^{-1}A = U$$

A matrix is non-singular if its inverse exists, singular otherwise.

Before considering inversion of matrices, the following properties of determinants may be noted:

- The determinant of a square matrix can be determined by the familiar method in terms of the products of the elements in any row or column and the corresponding cofactor (signed minor).
- The determinant of a product of square matrices is equal to the product of their determinants.

$$\det(AB) = \det A \det B$$

- The determinant of A is the same as that of A^T.
- If one column vector in A is multiplied by a scalar α, the value of the determinant is multiplied by α.
- The determinant of A changes sign when any two column vectors in A are interchanged.
- The determinant of A remains unchanged if a column vector is multiplied by α and then added to another column vector.

A.4 ADJOINT MATRIX OF A MATRIX

The adjoint matrix of A is a matrix whose elements are the cofactors of A^T. A four step rule may be used for inversion.

- Find the value Δ of the determinant of A.
- Form A^T.
- Replace each element of A^T by its cofactor and obtain adjoint A(adj).
- Divide all elements of adj A by Δ.

$$A^{-1} = \frac{\operatorname{adj} A}{\det A} \tag{A.1}$$

EXAMPLE A.1 Invert

$$A = \begin{bmatrix} a_{11} & a_{12} \\ a_{21} & a_{22} \end{bmatrix}$$

SOLUTION:

$$\Delta = a_{11}a_{22} - a_{21}a_{12}$$

Then

$$A^T = \begin{bmatrix} a_{11} & a_{21} \\ a_{12} & a_{22} \end{bmatrix}$$

and $$\text{adj } A = \begin{bmatrix} a_{22} & -a_{12} \\ -a_{21} & a_{11} \end{bmatrix}$$

Thus,

$$A^{-1} = \begin{bmatrix} \dfrac{a_{22}}{\Delta} & \dfrac{-a_{12}}{\Delta} \\ \dfrac{-a_{21}}{\Delta} & \dfrac{a_{11}}{\Delta} \end{bmatrix} \quad \textbf{Ans.}$$

Check $AA^{-1} = U$

- Note that the inverse of a square matrix exists provided its determinant is non-zero. This inverse is given by Eq. (A.1).
- The inverse of a product of matrices is the product of the individual inverse matrices taken in reverse order.

$$(AB)^{-1} = B^{-1}A^{-1}$$

provided each of the inverses exists.

A.5 RANK AND LINEAR DEPENDENCE

A matrix of order $(m \times n)$ is of **rank** r if it contains at least one $(r \times r)$ non-zero minor while all possible minors of higher order are zero. The rank r cannot exceed m or n, whichever is smaller but it may be less.

Reference [13] may be made to find the relation between rank of a matrix and the maximum number of linearly independent rows and columns.

A.6 ELEMENTARY MATRIX TRANSFORMATIONS

Two matrices are said to be equivalent if one is obtained from the other by a sequence of elementary transformations listed below:

- Interchange of two rows.
- Multiplication of a row by a constant.
- Addition of a multiplied row to another row.

The solution of simultaneous equations can be expedited by using elementary transformations on the augmented coefficient matrix.

A.7 LINEAR SYSTEMS

The different methods available for the solution of simultaneous equations are:

- Gauss elimination method
- Matrix inversion method and
- Cramer's method.

Complex Variables

Appendix B

The theory of complex variables is fundamental to the subject of electric circuits. There are elementary problems for which the knowledge of college algebra and calculus is sufficient but there is a large class of problems which require a knowledge of the theory of complex analytic functions.

B.1 COMPLEX NUMBERS

The motivation for the introduction of complex numbers is the fact that there are equations such as

$$x^2 + 1 = 0$$

which are *not* satisfied by any real number. A complex number is an ordered pair of real numbers subject to the operation defined below. It is usually written in the form:

$$z = x + jy$$

where x and y are real numbers; x is called the real part of the complex number z and y is called the imaginary part of z.

$$\text{Re } z = x \quad \text{and} \quad \text{Im } z = y$$

Complex numbers can be represented as points in the Cartesian complex plane or Argand diagram. The x-axis is called the **real axis** and y-axis is the **imaginary axis**.

B.2 EQUALITY OF COMPLEX NUMBERS

If $z_1 = x_1 + jy_1$ and $z_2 = x_2 + jy_2$

are defined to be equal only if their real parts are equal and their imaginary parts are equal, that is

$$z_1 = z_2 \text{ if and only if } x_1 = x_2 \text{ and } y_1 = y_2$$

B.3 ADDITION AND SUBTRACTION OF COMPLEX NUMBERS

$$z_1 + z_2 = (x_1 + x_2) + j(y_1 + y_2)$$
$$z_1 - z_2 = (x_1 - x_2) + j(y_1 - y_2)$$

See Fig. B.1.

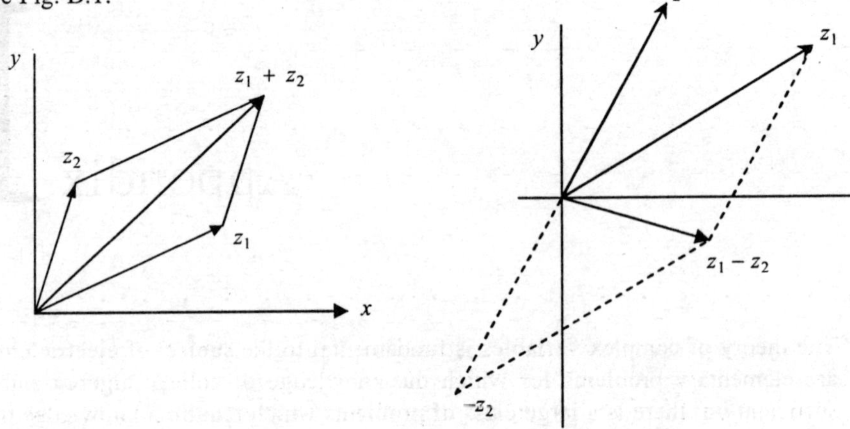

Fig. B.1

B.4 MULTIPLICATION OF COMPLEX NUMBERS

$$\begin{aligned} z_1 z_2 &= (x_1 + jy_1)(x_2 + jy_2) \\ &= (x_1 x_2 - y_1 y_2) + j(x_1 y_2 + x_2 y_1) \end{aligned}$$

replacing j^2 by -1.

B.5 DERIVATIVE, ANALYTIC FUNCTIONS

If x and y are real variables then $z = x + jy$ is called a **complex variable**. If each value of z in some region of the complex s plane is assigned a single value of a complex variable w then w is said to be a function of z defined in that region and written as $w = f(z)$. The set of complex numbers which $f(z)$ can assume as z varies in that region, is called the **range** of values of the function w.

If u and v are the real and imaginary parts of w then

$$f = u(x,y) + jv(x,y)$$

EXAMPLE If $f(z) = z^2 + z$, determine the value of $f(z)$ at $z = 1 + j3$

$$u(x,y) = \text{Re } f(z) = x^2 - y^2 + x$$

and
$$v(x,y) = \text{Im } f(z) = 2xy + y$$

at
$$z = 1 + j3$$

w has the value $(1+ j3)^2 + (1 + j3) = -7 + j9$

B.6 ANALYTICITY

A function $f(z)$ is said to be differentiable at a point $z = z_0$, if the limit

$$f'(z_0) = \lim_{\Delta z \to 0} \frac{f(z_0 + \Delta z) - f(z_0)}{\Delta z}$$

exists. This limit is called the **derivative** of $f(z)$ at the point $z = z_0$.

A function $f(z)$ is said to be *analytic* at a point $z = z_0$, if it has a derivative at every point in the neighbourhood of z_0. The function is analytic in a domain D if it is analytic at every point in D.

Complex analysis is concerned exclusively with analytic functions. The theory of analytic functions is indeed a fascinating study; at the same time most useful for several practical purposes.

B.7 COMPLEX INTEGRAL

The fundamental result is Cauchy's integral theorem from which Cauchy's integral formula follows. Integration by means of residues will be briefly considered.

B.7.1 Cauchy's Integral Theorem

If $f(z)$ is analytic in a simply connected bounded domain then for every simple closed path C in D

$$\int_C f(z)\,dz = 0$$

The proof is rather involved but can be found in Reference (10).

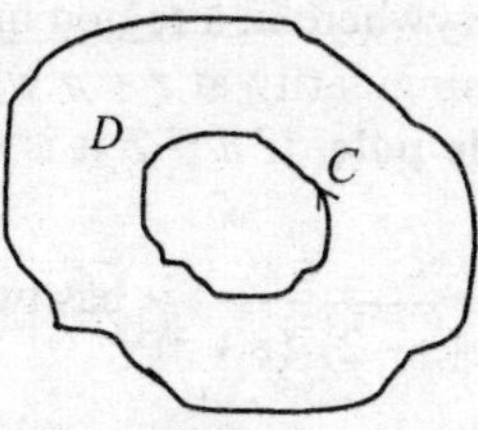

Fig. B.2

A simple closed path, such as C in Fig. B.2 is called a **contour** and integrals along such a path are called **contour integrals**.

B.7.2 Cauchy's Integral Formula

Let $f(z)$ be analytic in a simply connected domain D. Then for any point a in D and any simple closed path C in D which encloses a,

$$\int_C \frac{f(z)}{z-a}\,dz = 2\pi j f(a)$$

the integration being taken in the counter clockwise sense.

For proof see Reference (10).

B.8 TAYLOR'S SERIES

Let $f(z)$ be analytic inside and on a circle having its centre at $z = a$ then for all point z in the circle $f(z)$ can be represented by the Taylor series given by

$$f(z) = f(a) + \frac{f'(a)}{1!}(z - a) + \frac{f''(a)}{2!}(z - a)^2 + \cdots$$

EXAMPLE B.1

$$\frac{1}{1-z} = \sum_{n=0}^{\infty} z^n = 1 + z + z^2 + \cdots \qquad |z| < 1$$

B.9 SINGULAR POINTS

A singular point of a function $f(z)$ is a value of z at which $f(z)$ fails to be analytic. If $f(z)$ is analytic everywhere except at an interior point $z = a$, then a is an isolated singularity of $f(z)$.

EXAMPLE B.2 If $f(z) = \dfrac{1}{(z-3)^2}$ then $z = 3$ is an isolated singularity of $f(z)$.

B.10 POLES

If
$$f(z) = \frac{\phi(z)}{(z-a)^n}, \qquad \phi(a) \neq 0$$

where $\phi(z)$ is analytic everywhere in a region including $z = a$ and if n is a positive integer then $f(z)$ has an isolated singularity at $z = a$ which is called a **pole** of order n. If $n = 1$, the pole is called a **simple pole**, if $n = 2$ it is called a **double pole** and so on.

EXAMPLE B.3 $f(z) = \dfrac{z}{(z-2)^2(z+3)}$ has two singularities—a double pole at $z = 2$ and a simple pole at $z = -3$.

EXAMPLE B.4 $f(z) = \dfrac{2z-1}{z^2+9} = \dfrac{2z-1}{(z+j3)(z-j3)}$ has two simple poles at $z = \pm j3$.

B.11 LAURENT'S SERIES

If $f(z)$ has a pole of order n at $z = a$ but is analytic at every other point inside and on a circle C with centre at a then $(z - a)^n f(z)$ is analytic at all points inside and on C, and has a Taylor series about $z = a$ so that

$$f(z) = \frac{a_{-n}}{(z-a)^n} + \frac{a_{-n+1}}{(z-a)^{n-1}} + \cdots + \frac{a_{-1}}{z-a} + a_0 + a_1(z-a) + a_2(z-a)^2 + \cdots \qquad \text{(B.1)}$$

The part consisting of powers of $(z - a)$ is called the **analytic part** and the remainder consisting of the inverse powers of $(z - a)$ is called the **principle part**.

B.12 RESIDUE

The coefficient a_1 of $1/(z - a)$ in Eq. (B.1) is called the **residue** of $f(z)$ at pole a. For simple poles, it can be shown that

$$a_{-1} = \frac{1}{2\pi j} \int_C f(z)\, dz$$

and, therefore,

$$\int_C f(z)\, dz = 2\pi j\, a_{-1}$$

which is the **residue theorem.**

There are various methods for determining the residues which can then be used for evaluating contour integrals. More generally, if $f(z)$ is analytic within and on the boundary of a region R except at a finite number of poles $a, b, c, \ldots,$ within D having residues $a_{-1}, b_{-1}, \ldots,$ respectively then

$$\int_C f(z)\, dz = 2\pi j\, (a_{-1} + b_{-1} + \cdots)$$

That is the integral of $f(z)$ is $2\pi j$ times the sum of the residues of $f(z)$ at the poles enclosed by D.

B.12.1 Use of Residue Theorem in finding Inverse Laplace Transforms

If $F(s) = Lf(t)$ then $L^{-1} F(s)$ is given by

$$f(t) = \frac{1}{2\pi j} \int_{\sigma - j\infty}^{\sigma + j\infty} e^{st} F(s)\, ds \qquad t > 0 \tag{B.2}$$

and $f(t) = 0$ for $t < 0$.

This result is called the **complex inversion formula** or **Bromwich integral formula** which provides a direct means of obtaining the inverse Laplace transform of a function $F(s)$. The integration is to be performed along a line $s = \sigma$ in the complex plane where $s = x + jy$. The real number σ is chosen such that $s = \sigma$ lies to the right of all the singularities. In practice, the integral in Eq. (B.2) is evaluated by considering the contour integral $\dfrac{1}{2\pi j} \int_C e^{st}\, F(s)\, ds$ where C is the contour of Fig. B.3.

The integral around the arc approaches zero as $R \to \infty$ then by the residue theorem $f(t)$ = Sum of residues of $e^{st}\, F(s)$ at poles of $F(s)$.

EXAMPLE B.5 Evaluate $L^{-1}\, 1/(s - 2)$ by using complex inversion formula.

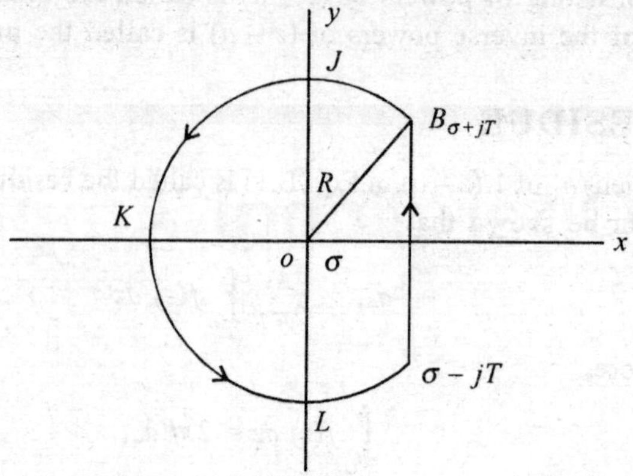

Fig. B.3

SOLUTION:

Note that $1/(s-2)$ satisfies the requirements of a Bromwich contour.
The residue of $e^{st}/(s-2)$ at pole $s = 2$ is given by

$$\lim_{s\to 2}(s-2)\frac{e^{st}}{(s-2)} = e^{2t}$$

$$L^{-1}\frac{1}{s-2} = \text{Sum of residues of } e^{st} F(s) = e^{2t} \quad \textbf{Ans.}$$

EXAMPLE B.6 Evaluate $L^{-1}\dfrac{1}{(s+2)(s-4)^2}$ by using the method of residues.

SOLUTION:

$$f(t) = \frac{1}{2\pi j}\int_C \frac{e^{st}}{(s+2)(s-4)^2}\,ds$$

$$= \text{Sum of the residues of } \frac{e^{st}}{(s+2)(s-4)^2} \text{ at poles } s = -2 \text{ and } s = 4$$

Residue at $s = -2$ is:

$$\lim_{s\to -2}\frac{(s+2)e^{st}}{(s+2)(s-4)^2} = \frac{e^{-2t}}{36}$$

Residue at double pole $s = 4$ is:

$$\lim_{s\to 4}\frac{d}{ds}\left[\frac{(s-4)^2 e^{st}}{(s+2)(s-4)^2}\right] = \frac{6te^{4t} - e^{4t}}{36}$$

Finally, Required inverse = Sum of residues = $\dfrac{e^{-2t}}{36} + \dfrac{te^{4t}}{6} - \dfrac{e^{4t}}{36}$ **Ans.**

Operational Amplifier

Appendix C

C.1 BASIC OPERATIONAL AMPLIFIER

An operational amplifier is an amplifier with a very high open-loop voltage gain. The amplifier is connected with appropriate networks of electronic components in order to perform a particular mathematical function on the input signals, that is, the output voltage is related to the input by a specific mathematical relationship and, therefore, the amplifier is dubbed the operational amplifier (op-amp). The earliest use of these amplifiers was in analogue computers. Analogue computers found wide spread applications and with the advent of solid state devices the size and cost of op-amps decreased with the result that they found their way into every type of electronic equipment.

Gain: The op-amp is an exceptionally good voltage amplifier. Typically the open-loop gain (at low frequencies) is from 10^5 to 10^9 and the output is ±10 V maximum. For an open-loop gain of 1 million, the full-scale input will be 10 μV. The extremely high input impedance results in negligible input current.

The following assumptions are usually made for ideal op-amps.

- Open-loop dc gain ∞.
- Input impedance ∞.
- Voltage appearing at the input terminals negligible, that is, input terminal is virtually at zero volt with respect to system ground. The input terminal potential is termed as virtual ground.
- Output impedance = 0.

C.2 OP-AMP SYMBOLS

Figure C.1 shows the symbol for an op-amp with differential input and a single output.

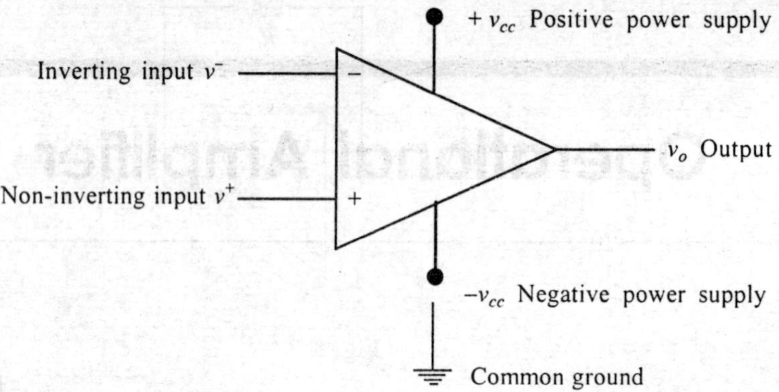

Fig. C.1 Circuit symbol for an op-amp.

A positive voltage applied to the + input will result in a positive voltage at the output. Frequently this input is not used. The inverting input causes a reversal of polarity of the signal at the output. When only one input is shown, it is the inverting input. The DIP package (Refer Fig. C.2) is shown for the μA741 op-amp.

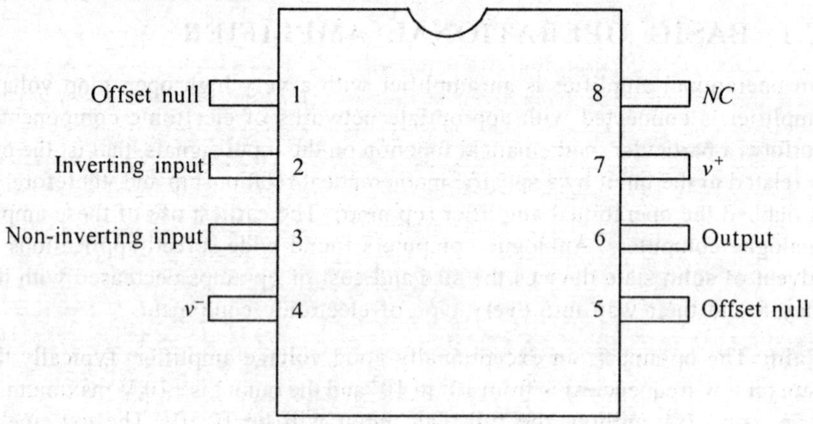

Fig. C.2 8 lead DIP package.

C.3 CLOSED LOOP CHARACTERISTICS

The op-amp is seldom used in its open loop configuration. Connecting external components at the input and output terminals and between input and output completely changes the amplifier characteristics. The use of the op-amp thus modified with external components is termed as **closed loop operation**.

C.4 INVERTING INPUT

In an inverting circuit, the input signal is connected through R_1 to the inverting terminal of the op-amp; the output terminal is connected back through a feedback resistor R_f to the inverting terminal. The non-inverting terminal is grounded as shown in Fig. C.3.

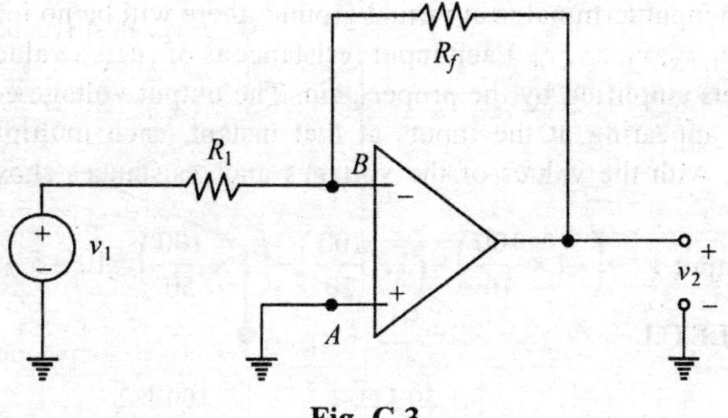

Fig. C.3

The gain v_2/v_1 is obtained by applying KCL to the currents at node B. Thus,

$$\frac{v_1}{R_1} + \frac{v_2}{R_f} = 0 \qquad \text{(C.1)}$$

and closed loop gain is:

$$\frac{v_2}{v_1} = \frac{-R_f}{R_1} \qquad \text{(C.2)}$$

Note that the closed loop gain depends only on the ratio R_f/R_1 (of the order of 100). Thus, the negative feedback reduces the system gain considerably.

Input impedance:

$$\frac{v_1}{i_1} = R_1$$

Input impedance is reduced from near infinity to the value of input resistance.

C.5 SUMMING CIRCUIT

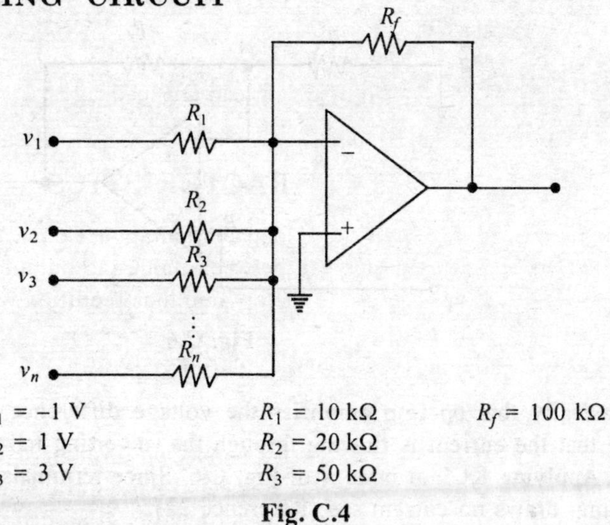

Fig. C.4

Since the input terminal is at virtual ground, there will be no interaction between different circuits $v_1, v_2, v_3, \ldots, v_n$. Each input resistance is of such a value as to make that particular signal gets amplified by the proper gain. The output voltage equals algebraic sum of the voltages appearing at the inputs at that instant, each multiplied by its own gain. For example, with the values of the voltages and resistances shown in Fig. C.4

$$\text{Output} = -\left(-1\times\frac{100}{10}\right)-\left(1\times\frac{100}{20}\right)-\left(3\times\frac{100}{50}\right)=10-5-6=-1\text{ V}$$

EXAMPLE C.1

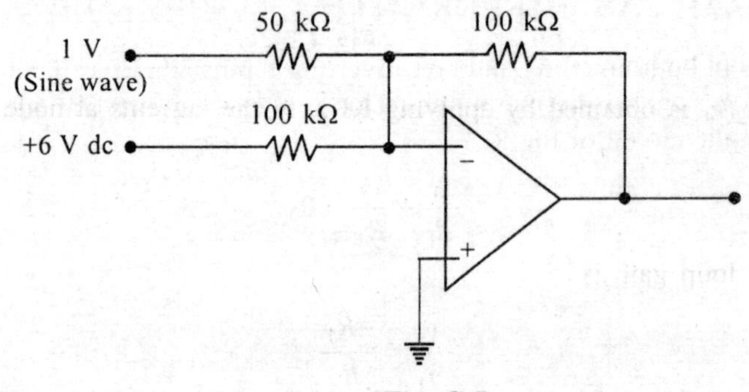

Fig. C.5

Output = 2 V (peak to peak) riding on a –6 V dc level.

C.6 USE OF THE OP-AMP WITH NON-INVERTING INPUT

The non-inverting terminal of the op-amp is seldom used with feedback from the output. The general use of the non-inverting input is with a voltage source connected directly to it plus some feedback to the inverting input, with the inverting input resistor connected to system ground as shown in Fig. C.6.

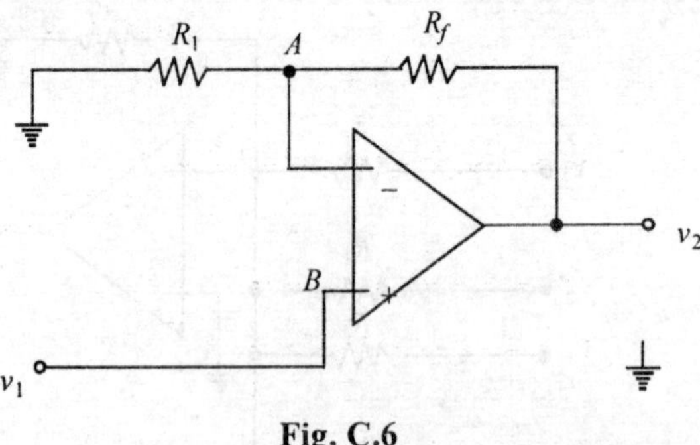

Fig. C.6

Basically this op-amp amplifies the voltage difference between the two inputs. Assume that the current is flowing through the (inverting terminal) input and feedback resistor. Applying KCL at node A of Fig. C.6. Since terminals A and B are both v_1 and the op-amp draws no current see (Reference 22).

$$\frac{v_1}{R_1} + \frac{v_1 - v_2}{R_f} = 0$$

or
$$\frac{v_2}{v_1} = 1 + \frac{R_f}{R_1}$$

Thus, closed loop gain is determined by the ratio of the resistors. If R_1 is removed (infinity) and if R_f is short circuited (zero), the gain becomes 1— it is called **unity gain voltage follower**.

C.7 OP-AMP AS DIFFERENTIAL AMPLIFIER

Op-amp having both inverting and non-inverting inputs amplifies the difference between the voltages present on the two inputs. This is a case of differential amplification. For example, see the circuit of Fig. C.7.

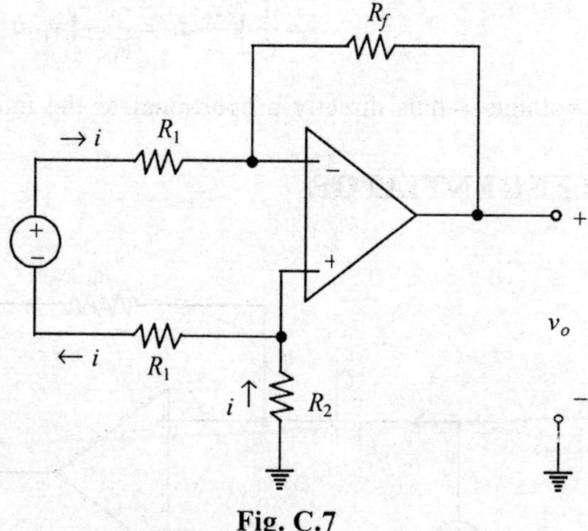

Fig. C.7

C.8 OP-AMP INTEGRATOR

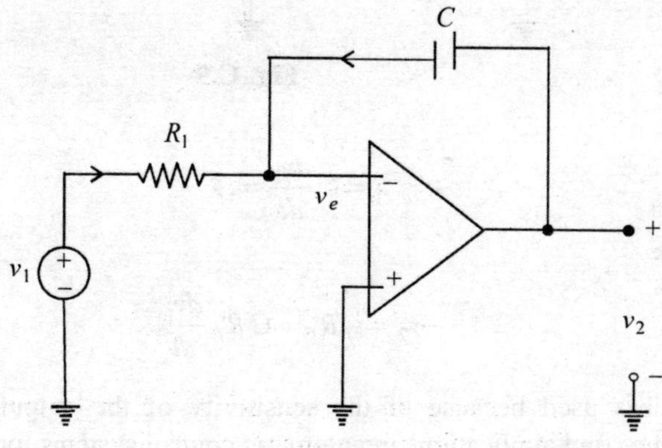

Fig. C.8

With a capacitor in the feedback loop, the op-amp can be used as an integrator. As long as the capacitor is not fully charged, it acts as a dc path and does not change the characteristics of an op-amp with feedback loop.

As a result,

$$i_f = i_1 = \frac{v_1}{R_1}$$

Because the inverting input is virtually grounded ($v_a \ll v_1$), the output voltage is given by v_2 = negative voltage across the capacitor C.

Hence

$$-v_2 = v_c = \frac{Q_c}{C} = \frac{1}{C}\int i_f \, dt$$

$$= \frac{1}{C}\int \frac{v_1}{R_1} dt = \frac{1}{R_1 C}\int v_1 \, dt$$

The output voltage is thus directly proportional to the integrated input voltage v_1.

C.9 DIFFERENTIATOR

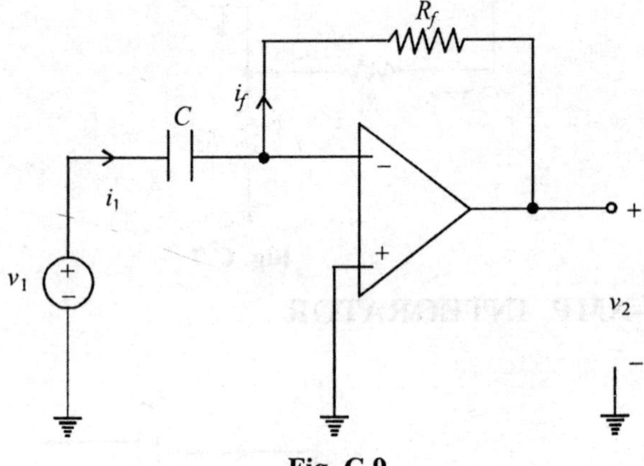

Fig. C.9

$$i_1 = C\frac{dv_1}{dt} = i_f$$

Hence

$$-v_2 = i_f R_f = C R_f \frac{dv_1}{dt}$$

These are less used because of the sensitivity of the output to the electrical noise. However, they find applications in automatic control systems for derivative control, noise amplifiers, filters, function generators and trigger circuits.

C.10 ANALOGUE COMPUTERS

Basically analogue computer consists of numerous op-amps. The input and the output terminals are brought out to the front console. Figure C.10 shows an analogue computer schematic diagram.

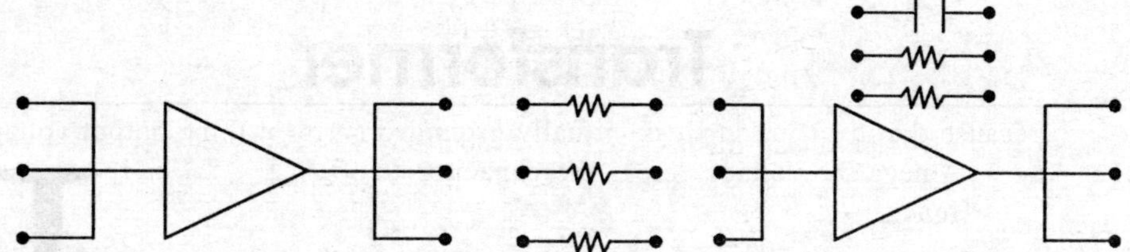

(a) Basic analogue computer op-amp schematic

(b) Complete analogue computer op-amp schematic

Fig. C.10

Coupled Circuits and the Transformer

Appendix D

The concept of mutual inductance has been introduced in Section 3.4. Two circuits which have a mutual inductance of M henry are shown in Fig. D.1.

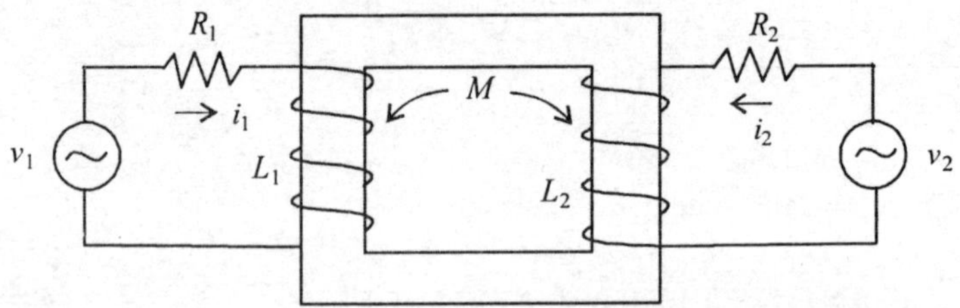

Fig. D.1

Equations for the two circuits are:

Time domain | *Frequency domain*

$$R_1 i_1 + L_1 \frac{di_1}{dt} \pm M \frac{di_2}{dt} = v_1 \qquad (R_1 + j\omega L_1)I_1 \pm j\omega M I_2 = v_1$$

$$R_2 i_2 + L_2 \frac{di_2}{dt} \pm M \frac{di_1}{dt} = v_2 \qquad (R_2 + j\omega L_2)I_2 \pm j\omega M I_1 = v_1$$

Note:
- Because of the coupling, each equation contain a term which depends on current change in the other circuit.
- The polarity of the mutually induced emf depends on the way in which the coils are wound in relation to the reference direction of coil current. There is however a simple and convenient alternative method for keeping track of the polarity of the induced emf. This is the **dot notation** or convention in which coils are marked with dots at the terminals which are instantaneously of the same polarity.

To assign the dots to a pair of coils, a dot is placed on one terminal of coil 1 in Fig. D.2. When current I_1 enters the dotted end of coil 1, flux ϕ_1 leaves the top of the coil. This flux links with coil 2 and induces an emf in it, which in turn causes I_2 to flow in coil 2. Lenz's law requires that I_2 produces a flux ϕ_2 which opposes ϕ_1. To produce ϕ_2 in the direction shown, I_2 must flow out of the bottom terminal of coil 2; this terminal is marked with a dot.

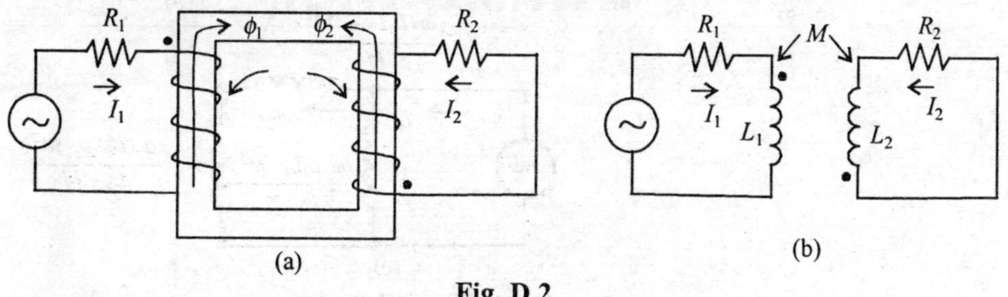

Fig. D.2

The dot convention can now be stated as:

If the current in one coil enters by a dotted terminal while the current in the other coil leaves by a dotted terminal, the sign of the M-term will be opposite to the sign on the L-terms.

The mesh equations for the circuits in Fig. D.2 may be written as:

$$(R_1 + j\omega L_1)I_1 - j\omega M I_2 = V_1$$
$$-j\omega M I_1 + (R_2 + j\omega L_2)I_2 = 0$$

With the instantaneous polarity of the coupled coils given by the dots, the representation of the core and the winding sense is not needed. The coupled coils are conveniently represented by the schematic diagram of Fig. D.2(b).

D.1 CONDUCTIVELY COUPLED EQUIVALENT CIRCUITS

It is possible to convert the magnetically coupled circuit in Fig. D.3(a) into an equivalent conductively coupled circuit. The mesh equations for the two circuits in Fig. D.3(a) are:

$$V_1 = j\omega L_1 I_1 - j\omega M I_2 = j\omega[(L_1 - M) + M]I_1 - j\omega M I_2$$
$$0 = -j\omega M I_1 + j\omega L_2 I_2 = -j\omega M I_1 + j\omega\,[(L_2 - M) + M]I_2$$

The T-equivalent conductively coupled circuit corresponding to the second expression in each equation is shown in Fig. D.3(b). The π-equivalent circuit is given in Fig. D.3(c).

If the connections to one of the coils is reversed, the dot is placed on the opposite end of coil 2 and the mutually induced emf terms become $j\omega M I_1$ and $j\omega M I_2$ in the above expressions. This has the effect in Fig. D.3(b) of making the inductances in the top branches as $(L_1 + M)$ and $(L_2 + M)$; and that of the common vertical branch $-M$.

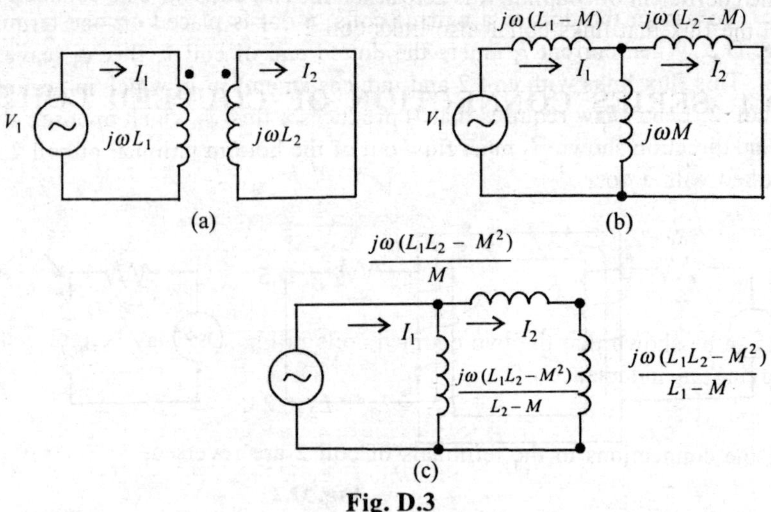

Fig. D.3

D.2 THE IDEAL TWO-WINDING TRANSFORMER

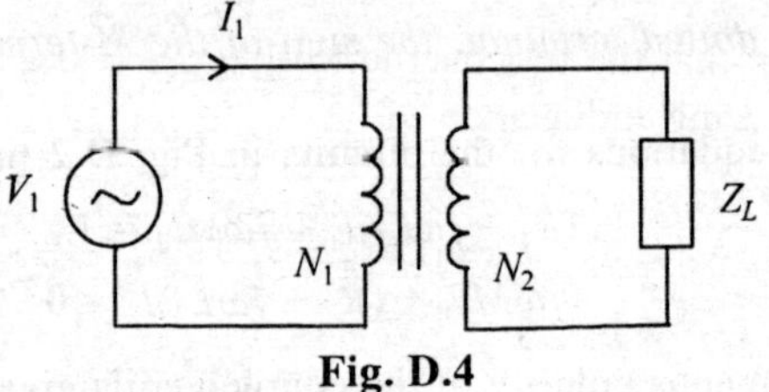

Fig. D.4

An ideal transformer (as shown in Fig. D.4) is a lossless transformer with a coupling coefficient of unity. This transformer has ampere turn balance between the windings and has same volts per turn on the windings.

$$I_1N_1 = I_2N_2, \quad V_1/N_1 = V_2/N_2$$

Combining the two equations,

$$V_1/V_2 = N_1/N_2 = I_2/I_1$$

The power supplied to primary winding is equal to the power transferred to secondary winding,

$$V_1I_1 \cos \phi = V_2I_2 \cos \phi$$

where $\cos \phi$ is the load power factor.

If an impedance Z_2 is connected to a winding of N_2 turns on an ideal transformer, the impedance Z_{2e} seen when looking into a winding of N_1 turns on the transformer is:

$$Z_{2e} = Z_2(N_1/N_2)^2$$

D.3 MUTUAL INDUCTANCE IN TERMS OF SELF INDUCTANCE (See Section 3.4)

$$M = K\sqrt{L_1L_2}$$

$$0 \le K \le 1$$

The coefficient of coupling K is zero when the two coils have no common flux; $K = 1$ when all the flux that links coil 1 also links coil 2.

D.4 SERIES CONNECTION OF COUPLED COILS

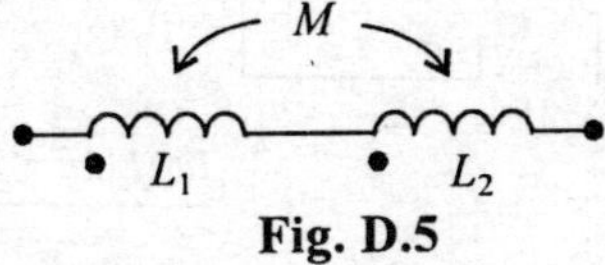

Fig. D.5

It can be shown that the two coupled coils in Fig. D.5 may be replaced by a single coil having an inductance of

$$L_{12} = L_1 + L_2 + 2M$$

If the connections to the terminals of coil 2 are reversed,

$$L_{12} = L_1 + L_2 - 2M$$

D.5 PARALLEL CONNECTION OF COUPLED COILS

It can be shown that the two magnetically coupled coils in Fig. D.6 may be replaced by a single coil having inductance of

$$L_{12} = \frac{L_1 L_2 - M^2}{L_1 + L_2 - 2M}$$

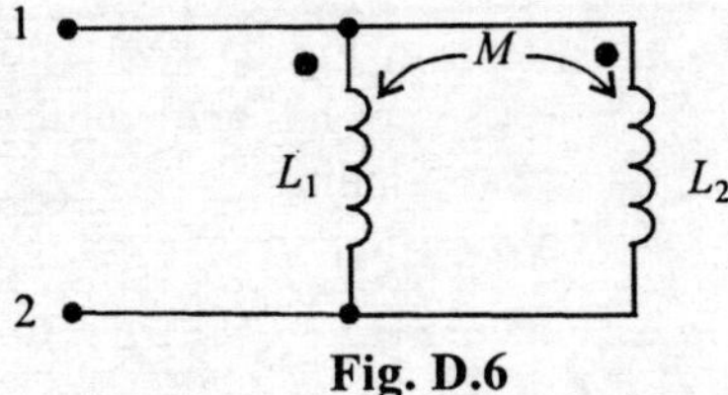

Fig. D.6

If the magnetic polarity of coil 2 is reversed, show that

$$L_{12} = \frac{L_1 L_2 - M^2}{L_1 + L_2 + 2M}$$

D.6 ENERGY CALCULATION

The energy stored in a pair of coupled coils can be shown to be

$$W = \frac{1}{2} L_1 I_1^2 + \frac{1}{2} L_2 I_2^2 + I_1 I_2 M$$

Review Problems

1. Find the current in a conductor through which 10×10^{20} electrons pass in 4 s. [**Hint:** Electronic charge = 1.6×10^{-19} C]

[*Ans:* Current I = 40 A]

2. Determine the charge that requires 1.5×10^3 J energy to be moved from infinity to a point having 15 V potential.

[*Ans:* Charge $Q = \dfrac{\text{Energy}}{\text{Potential}} = 100$ C]

3. A resistor draws a current i = 16 sin ωt at a voltage v = 100 sin ωt. Calculate the average power developed in the resistor. Hence determine the energy consumed by the resistor per cycle of current wave.
[**Hint:** Instantaneous power $p = vi$]

[*Ans:* Average power P = 800 W, Energy = $800 \times \dfrac{2\pi}{\omega}$ J]

4. A high voltage direct current transmission line between two stations A and B is operating at 400 kV and carries 1000 A as shown in Fig. 1.

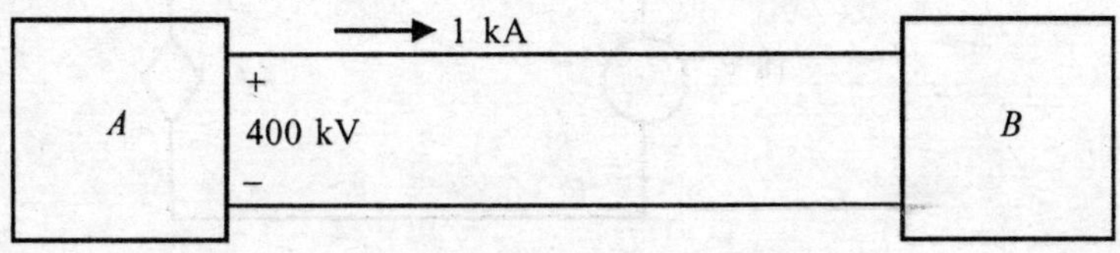

Fig. 1

[**Hint:** Note that the reference for I is in the direction of the voltage drop across the line]

[*Ans:* A is delivering 400 MW and B is absorbing 400 MW]

5. Determine the dimensions of
(a) Magnetic flux
(b) Permeability
(c) Resistance
(d) Capacitance
[**Hint:** See Table 1.2]

6. Check the dimensional validity of the expression:

$$V = \frac{B^2 r^2 e}{2m}$$

where V is the accelerating potential required to give an electron of mass m and charge e, a path of radius of curvature r in a magnetic field of flux density B.

7. Form a circuit of effective resistance 8/3 Ω using five 1 Ω resistors in a suitable series-parallel combination.

8. Three resistors of value 4 Ω, 8 Ω, 16 Ω are connected in parallel to a 14 A source. Find (a) the current passing through the 8 Ω resistor, (b) the power dissipated by the 4 Ω resistor. What is the source voltage?

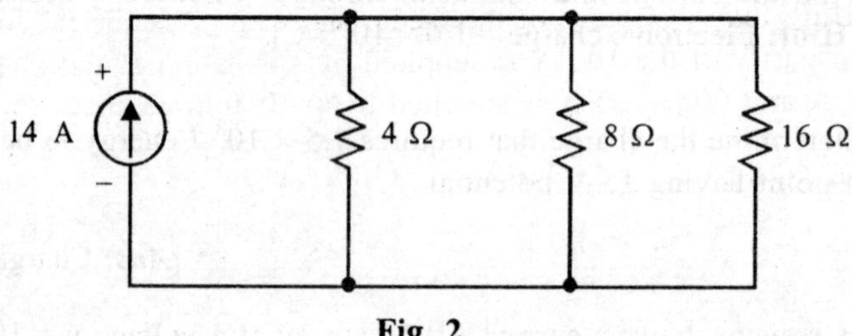

Fig. 2

[**Hint:** In parallel circuits, it is best to work with conductances (rather than resistances)]
[***Ans:*** $i_8 = 4$ A, $P_4 = i_4^2 \times 4 = 256$ W, Source voltage = 32 V]

9. In the circuit shown in Fig. 3, determine the power supplied by the 18 V source and that by the 4I-dependent voltage source.

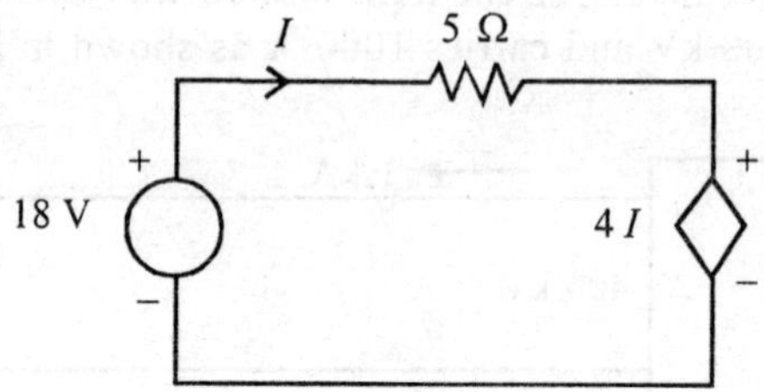

Fig. 3

[**Hint:** By Ohm's law $18 = 5I + 4I$ or $I = 2$ A]
[***Ans:*** Power supplied by 18 volt source = 18 × 2 = 36 W
Power supplied by the dependent source = −2 × (4 × 2) = −16 W
Note the negative sign for the current. The dependent source is thus absorbing power.]

10. A distribution system with a feeder supplied at both ends at voltage V as shown in Fig. 4. The distance between the houses is 100 m and that between the end houses and the substations P and Q is 200 m. Determine the current supplied by each of these substations.

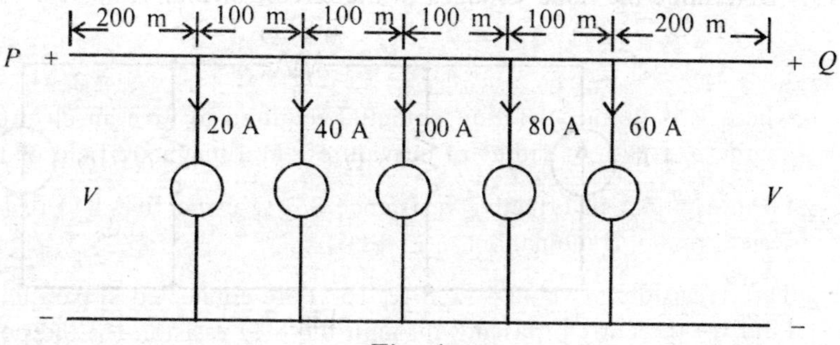

Fig. 4

[**Hint:** Take resistance of the feeder as r Ω/m. Assume the load current 20 A, 40 A and part x of the 100 A is supplied by substation P and the load current of 60 A, 80 A and $(100 - x)$ A is supplied by Q. It follows that

$(20 + 40 + x)\ 200 \times 2r + (40 + x)\ 100 \times 2r + x \times 100 \times 2r = (60 + 80 + 100 - x)$
$200 \times 2r + (80 + 100 - x)\ 100 \times 2r + (100 - x)\ 100 \times 2r$

which gives $x = 75$ A

Thus, the total current supplied by $P = 135$ A, and current supplied by $Q = 165$ A.

Check: The total current supplied by P and Q is $20 + 40 + 100 + 80 + 60 = 300$ A]

11. Calculate the current I in Fig. 5.

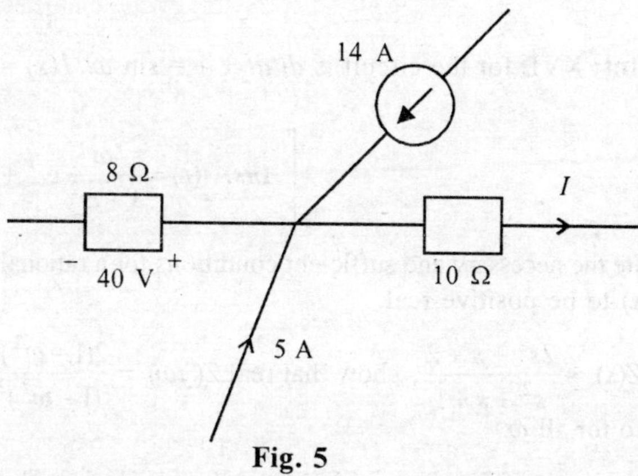

Fig. 5

[*Ans:* 15 A]

12. In the circuit shown in Fig. 6, if the power dissipated in the unknown resistor R is 8 W, find the value of R.

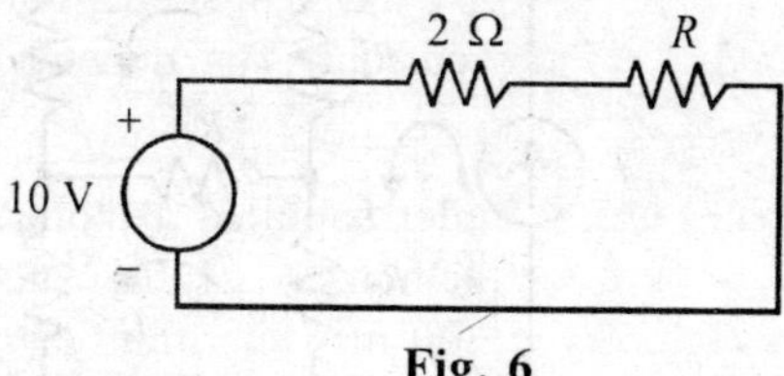

Fig. 6

[*Ans:* 8 Ω or 0.5 Ω]

13. Determine the node voltages in the circuit of Fig. 7.

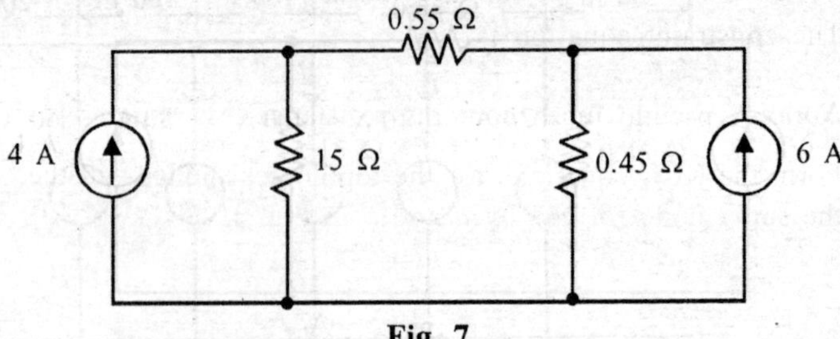

Fig. 7

[*Ans:* $V_1 = 6$ V, $V_2 = 10$ V]

14. A circuit with resistance 1 Ω and inductance 1 H connected in series is shown in Fig. 8, find the response of the circuit to an input voltage $e = \sin \omega t$.

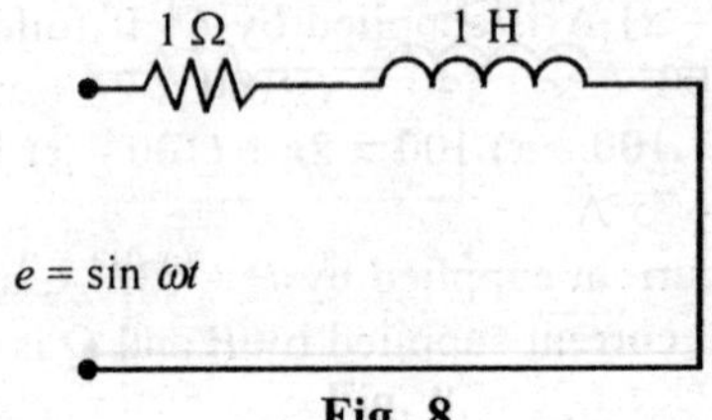

Fig. 8

[**Hint:** KVL for the circuit is $di/dt + 1 = \sin \omega t$, $I(s) = \dfrac{\omega}{(s+1)(s^2+\omega^2)}$]

$$\left[\textit{Ans: } i(t) = \frac{\omega}{1+\omega^2} e^{-t} + \frac{1}{\sqrt{1+\omega^2}} \sin(\omega t - \tan^{-1} \omega/1)\right]$$

15. State the necessary and sufficient conditions for a rational function with real coefficients $F(s)$ to be positive real.

16. If $Z(s) = \dfrac{2s^2+s+2}{s^2+s+1}$, show that real $Z(j\omega) = \dfrac{2(1-\omega^2)^2+\omega^2}{(1-\omega^2)^2+\omega^2}$, which is greater than zero for all ω.

17. The circuit of Fig. 9 consists of two topological meshes and one pseudo mesh (with an ideal current source). Write down the mesh currents I_1, I_2, I_3 as well as the constraint equation.

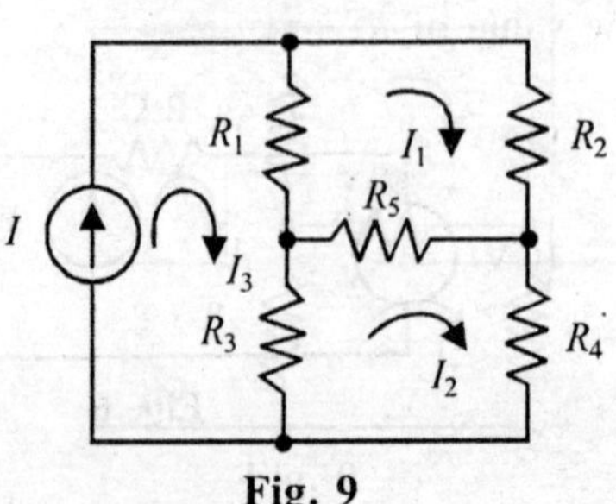

Fig. 9

[**Ans:** $R_2I_1 + R_5(I_1 - I_2) + R_1(I_1 - I_3) = 0$
$R_4I_2 + R_3(I_2 - I_3) + R_5(I_2 - I_1) = 0$

The constraint equation is:

$$I_3 = I$$

Note: A pseudo mesh contains resistance-less sources.]

18. Form the KCL equations for the topological nodes and the constraint equation for the super node formed by a and b in Fig. 10.

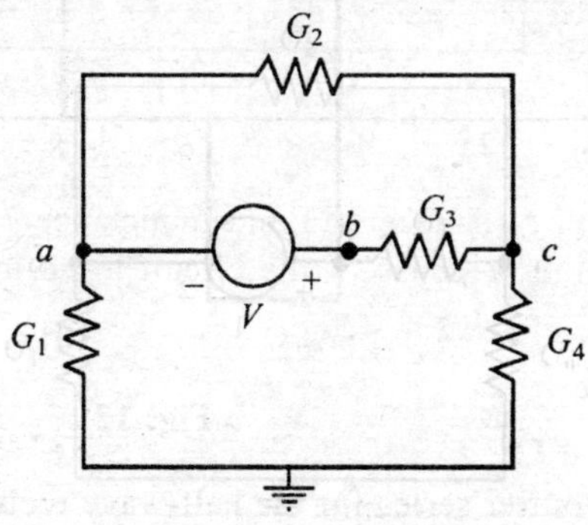

Fig. 10

[**Ans:** For node a $(v_b - v_c)\,G_3 + (v_a - v_c)\,G_2 + v_aG_1 = 0$
For node c $(v_c - v_b)\,G_3 + (v_c - v_a)\,G_2 + v_cG_4 = 0$

The constraint equation is:

$$v_b - v_a = V$$

Note: A super node is a collection of nodes within a network forming a subnetwork.]

19. Check whether $s^3 + 3s^2 + s + 5$ is a Hurwitz polynomial.

[**Ans:** $F(s)$ is not Hurwitz]

20. Find the inverse Laplace transform of $F(s) = \dfrac{1}{s(s+3)^2}$

[**Ans:** $f(t) = 1/9 - \frac{1}{9}te^{-3t} - \frac{1}{9}e^{-3t}$]

21. Determine (a) the Thevenin and (b) the Norton equivalent of the circuit connected between terminals A and B of Fig. 11(a).

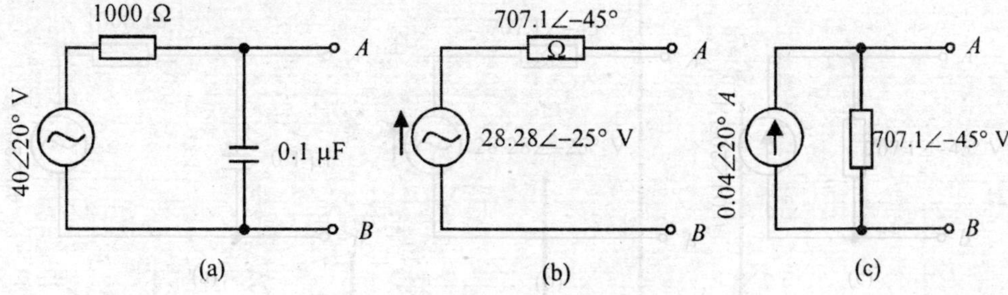

Fig. 11

[**Ans:** See Fig. 11(b) and Fig. 11(c)]

22. For the waveform in Fig. 12, evaluate the Fourier components a_0, a_1 and b_1.

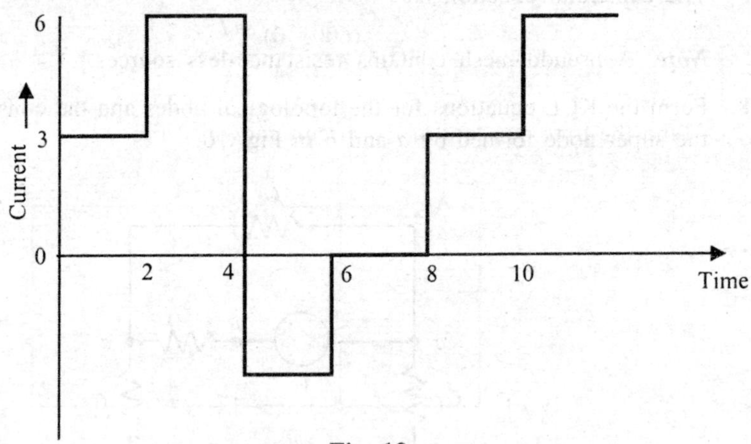

Fig. 12

23. Deduce the Fourier series for the half-wave rectified sine wave in Fig. 13 and draw the line spectra for the wave.

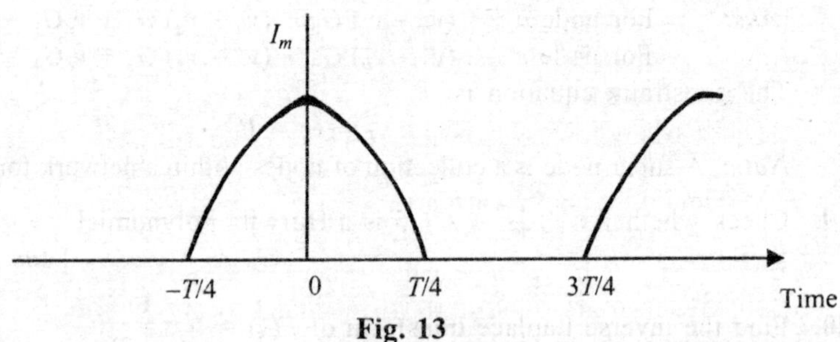

Fig. 13

[***Ans:*** $f(t) = \dfrac{I_m}{\pi}$ $(1 + \pi/2 \cos \omega t + (2/3) \cos 2\omega t - (2/15) \cos 4\omega t + (2/35) \cos 6\omega t)$]

The magnitude spectrum is shown in Fig. 13(a).

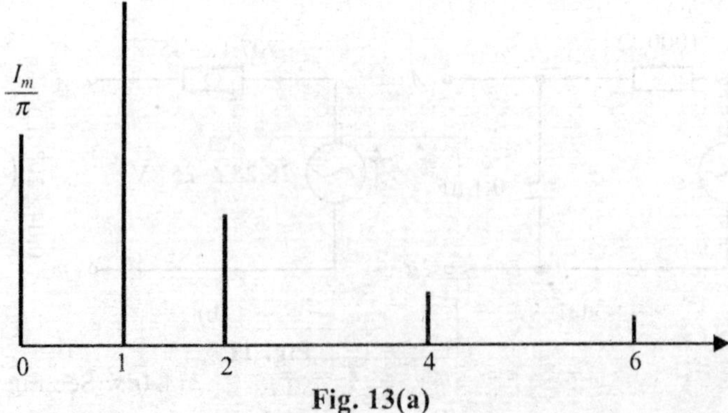

Fig. 13(a)

24. A 5.4 μF capacitance initially charged to 600 V is discharged through a 100 kΩ resistance. Find the capacitor voltage at 0.25 s after the capacitor starts to discharge.
[**Hint:** Time constant = $100 \times 10^{-3} \times 5.4 \times 10^{-6} = 0.54$ s

$$v(t) = \frac{(600 - 0)e^{-t}}{0.54}$$

[*Ans:* $v(0.25) = 378$ V]

25. Find the time constant of the circuit shown in Fig. 14.

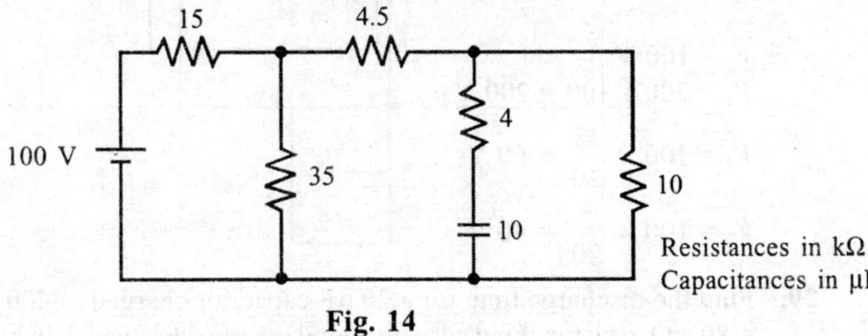

Fig. 14

[**Hint:** Time constant = $R_{Th}C$]

[*Ans:* 0.10 s]

26. How long does it take a 10 μF capacitor charged to 300 V to discharge through a 10 MΩ resistor. Assume that the discharge is completed after 5 time constants. Find the maximum discharge current and the time at which it occurs.
[**Hint:** Discharge time = $5 \times TC = 5 \times 10 \times 10^6 \times 10 \times 10^{-6} = 500$ s]
The discharge current is maximum at $t = 0$ and

Maximum discharge current $i_0 = \dfrac{300}{10 \times 10^6} = 30 \mu\text{A}$]

27. An *RC* timer has a switch that when closed connects in series a 500 V source, a 10 MΩ resistance and a 16 μF uncharged capacitance. Find the time between the closing and the opening of the switch if the capacitor charges to 20 V during this time.
[**Hint:** $TC = 16 \times 10^{-6} \times 10 \times 10^6 = 160$]
The capacitor voltage equation is:
$v = (500 - 500)e^{-t/160}$

For $v = 20$ V $t = 160 \times \ln \dfrac{500}{480} = 6.53$ s]

28. Find the voltage across each capacitor in the circuit shown in Fig. 15.

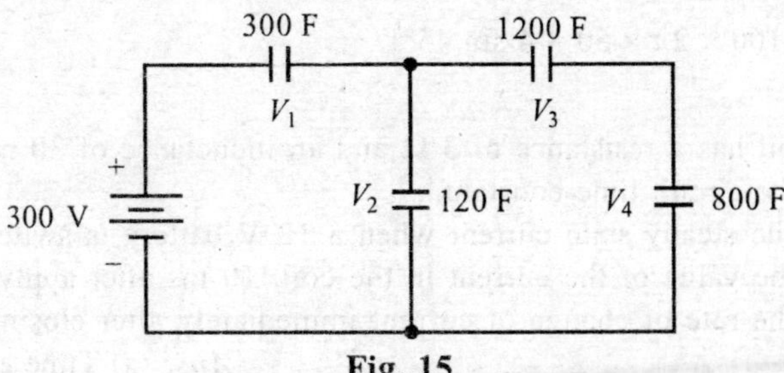

Fig. 15

[**Hint:** The circuit may be reduced as follows:

300 F

300 V +
−

600 F

$V_2 = 100$ V
$V_1 = 300 - 100 = 200$ V
$$V_4 = 100 \times \frac{12}{20} = 60 \text{ V}$$
$$V_3 = 100 \times \frac{8}{20} = 40 \text{ V}$$

29. Find the discharge time for a 20 μF capacitor charged to 400 V to discharge through a 80 kΩ resistor. Find also the total energy dissipated in the resistor.
[**Hint:** $TC = 1.6$, Discharge time $\cong 1.6 \times 5 = 8$ s, Energy $= (1/2)CV^2 = 1.6$ J]

30. Find the TC of the circuit shown in Fig. 16.

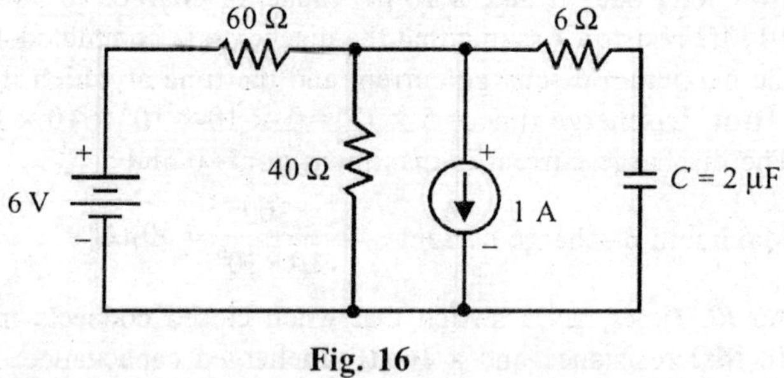

Fig. 16

[**Hint:** R_{Th} at terminals of C is $\frac{60 \times 40}{100} + 6 = 30$ Ω and $TC = 60$ μs]

31. A 100 V, 50 Hz source energises a load of 25 ∠45° Ω. Determine the capacitance to be connected in parallel to produce an overall power factor of unity.

[**Hint:** $I_L = \frac{100}{25} = 4\text{A}$,
$C \times 100 \times 2\pi \times 50 = 4 \sin 45°$]

[*Ans:* $C = 90$ μF]

32. A coil has a resistance of 3 Ω and an inductance of 30 mH. Calculate:
(a) the circuit time-constant.
(b) the steady state current when a 12 V battery is switched across the coil.
(c) the value of the current in the coil, 10 ms after applying the battery.
(d) the rate of change of current immediately after closing the switch.
[*Ans:* (a) Time constant $T = L/R = 10$ ms

(b) Steady state current = V/R = 12/3 = 4 A

(c) $i = \frac{V}{R}(1 - e^{-t/T})$

At time $t = 10 \times 10^{-3}$ $i = 4 \times 0.632 = 2.528$ A

(d) $\frac{di}{dt} = \frac{V}{R}\frac{R}{L}e^{-Rt/L}$

$$\left.\frac{di}{dt}\right|_{t=0} = \frac{12}{30 \times 10^{-3}} = 400 \text{ A/s}$$

33. A 50 kΩ resistance and a 1 μF capacitance are connected in series with a battery as shown in Fig. 17. The voltage across the capacitor terminals immediately before applying the battery was 20 V assisting the voltage of the battery. Calculate:
(a) the circuit time-constant.
(b) the time after $t = 0$, when the voltages across the capacitor and the resistor are the same.

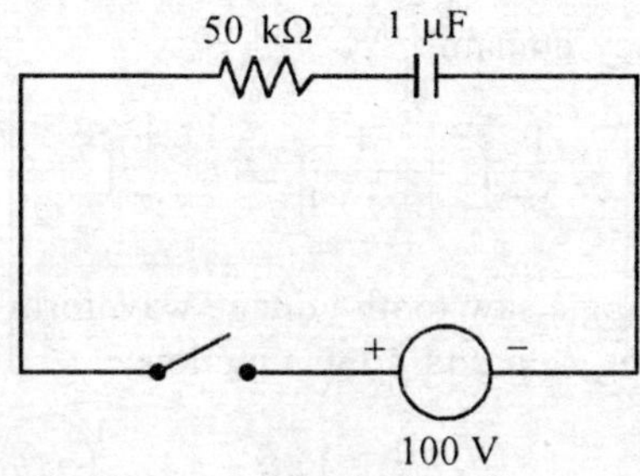

Fig. 17

[*Ans:* (a) Time constant = $CR = 1 \times 10^{-6} \times 50 \times 10^{3}$ = 50 ms

(b) If $V_R = V_c = \frac{100}{2}$ at t_1,

$$50 = 100 - [100 - (-20)]\, e^{-t/50}$$

Solving,

$$t_1 = 43.8 \text{ ms}]$$

34. The voltage transform of a network is:

$$\frac{V_o}{V_i} = G(s) = \frac{16(s+2)}{s^2 + 12s + 32}$$

Determine the transient response for a step input of 1 V to the network.

[*Ans:* $V_o(s) = \frac{1}{s}\frac{16(s+2)}{(s^2 + 12s + 32)}$

$$= \frac{1}{s} + \frac{2}{(s+4)} - \frac{3}{(s+8)}$$

and $v_o = 1 + 2e^{-4t} - 3e^{-8t}$]

35. A current of $i = 5 + 1.4 \cos(t + 200) + 0.5 \cos 3t$ flows in a circuit when the voltage $v = 10 + \cos(t + 45) + 18 \cos(3t - 10)$ is applied to it. Calculate the total power consumed.

$$\left[\textit{Ans: } P = 10 \times 5 + \frac{6}{\sqrt{2}} \times \frac{1.4}{\sqrt{2}} \cos 25 + \frac{1.8}{\sqrt{2}} \times \frac{0.5}{\sqrt{2}} \cos(-100) = 54.25 \text{ W}\right]$$

36. In Problem 35, calculate the overall power factor of the circuit.

[*Ans:* Current in the circuit $I = (52 + 1.42 + 0.52)/2 = 4.74$ A and $V = (102 + 62 + 1.82)/2 = 21.45$ V

$$\text{Overall power factor} = \frac{\text{Total power}}{VI} = 0.53\,]$$

37. Find the energy spectral density function of $f(t) = 4e^{-t}u(t)$ and check the result by Parseval's theorem.

[*Ans:* In the time domain $\int_0^{\infty} (4e^{-t})^2\, dt = 8$ J

In the frequency domain,

$$\frac{1}{2\pi}\int_{-\infty}^{\infty} \left|\frac{4}{j\omega + 1}\right|^2 d\omega = \frac{1}{\pi}\int_0^{\infty} \frac{16}{\omega^2 + 1} d\omega = 8 \text{ J}]$$

38. Synthesize a single saw tooth voltage waveform shown in Fig. 18, by superposition of the following step and ramp functions:

$$\frac{V}{t_0} r(t), \quad Vu(t - t_0), \quad \frac{V}{t_0} r(t - t_0).$$

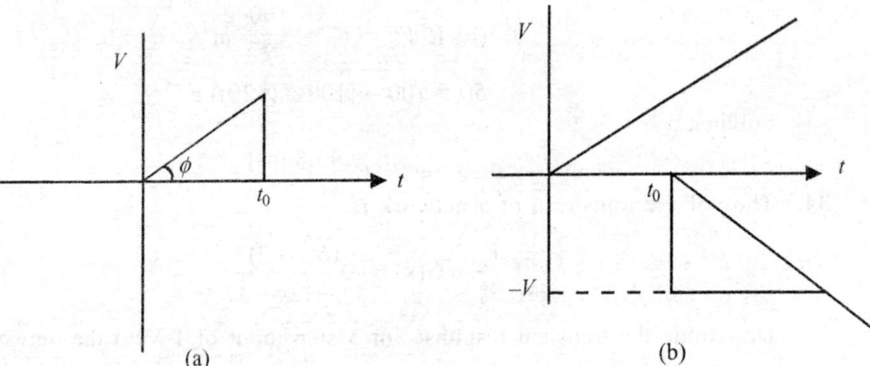

Fig. 18

39. The transfer impedance between branches *ad* and *ce* is given as $30 \angle 40°$ at 500 Hz. Calculate the current in the 9 Ω resistor if the generator voltage is $50 \angle 10°$ V at 500 Hz. (Refer Fig. 19)

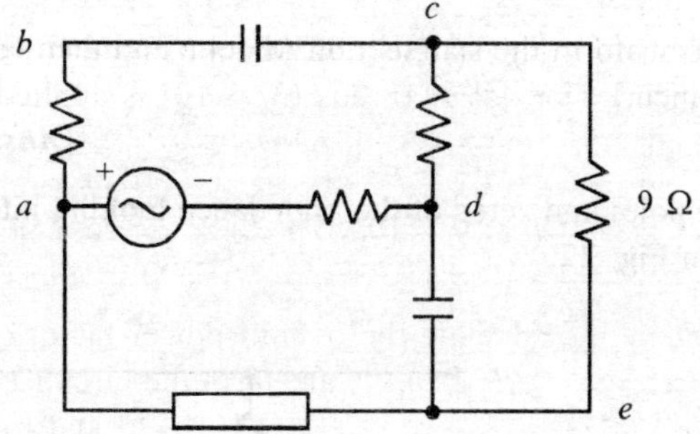

Fig. 19

[*Ans:* $Z_1 = \frac{V_{in}}{I_{out}}$, $30\angle 40° = \frac{50\angle 10°}{I_{out}}$, $I_{out} = 1.66\angle -30°$ A]

40. The Thevenin equivalent voltage and the equivalent impedance of a circuit section are $60\angle 30°$ V and $20\angle 30°$ Ω, respectively. Determine the Norton equivalent.
[**Hint:** $Z_N = Z_{Th} = 20\angle 30°$

$I_N = V_{Th}/Z_N = 3\angle 0°$ A]

41. Determine the current I in the circuit of Fig. 20.

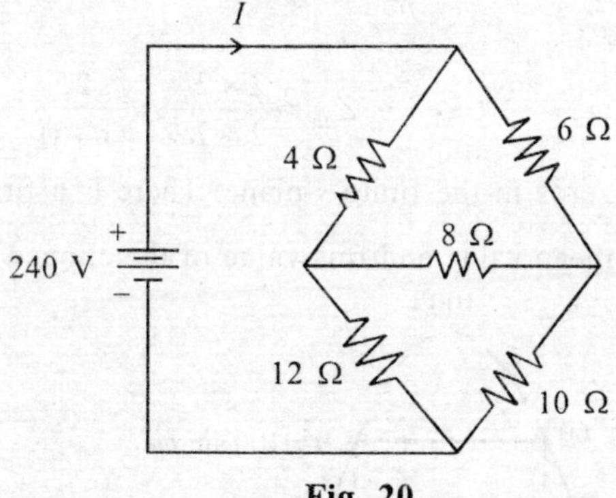

Fig. 20

[**Hint:** Use star delta transformation]

[*Ans:* $I = 30.25$ A]

42. A balanced star connection of $3\angle 20°$ Ω per branch is in parallel with a delta section of $1.5\angle 30°$ Ω per branch. Determine the equivalent delta section that can replace the two paralleled sections. See Fig. 21.

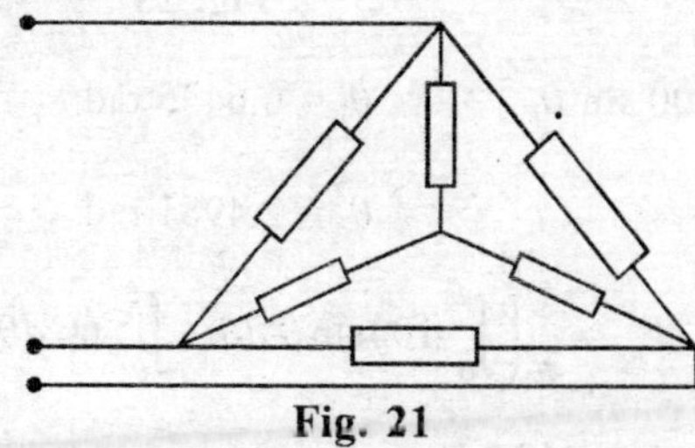

Fig. 21

[**Hint:** Transform the star section to delta and then take the parallel combination for each branch.]

[*Ans:* $1.295\angle(37.18°)$ Ω per branch]

43. Find the poles and zeros of the impedance looking into the terminals *ab* of the circuit shown in Fig. 22.

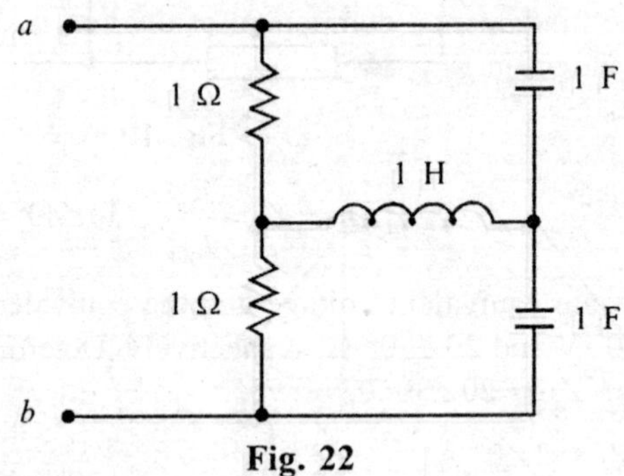

Fig. 22

[**Hint:** For a balanced bridged circuit there is no current in the detector branch. Hence

$$Z_{ab} = \frac{2 \times 2/s}{2 + 2/s} = \frac{2}{(s+1)}$$

There are no zeros in the finite *s* plane. There is a first order pole at $s = -1$]

44. Calculate the mean value and rms value of the clipped sine wave shown in Fig. 23.

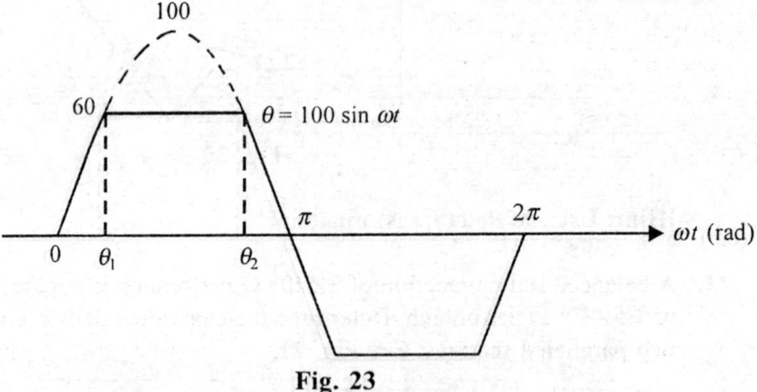

Fig. 23

[**Hint:** $60 = 100 \sin \theta_1$ $\qquad \theta_1 = 0.6435$ rad

$$\theta_2 = 2.4981 \text{ rad}$$

$$V_{av} = \frac{1}{\pi}\left(\int_0^{\theta_1} 100 \sin\theta \, d\theta + \int_{\theta_1}^{\theta_2} 60 \, d\theta + \int_{\theta_2}^{\pi} 100 \sin\theta \, d\theta\right)$$

$$= 48.2 \text{ V}$$

$$V_{\text{rms}} = \frac{1}{\pi}\left(\int_0^{\theta_1} (100 \sin\theta)^2 \, d\theta + \int_{\theta_1}^{\theta_2} 60^2 \, d\theta + \int_{\theta_2}^{\pi} (100 \sin\theta)^2 \, d\theta\right)$$

$$= 51.4 \text{ V}$$

45. Calculate the mean value of the trapezoid in Fig. 23.

[***Ans:*** $V_{av} = 47.7$ V]

46. Derive the fundamental component of the Fourier series for the trapezoidal voltage function shown in Fig. 24.

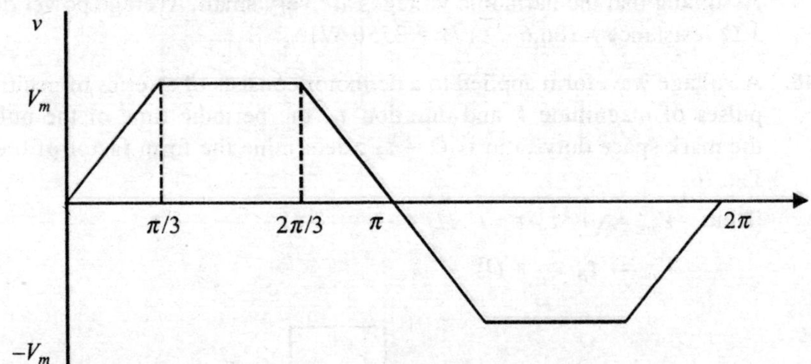

Fig. 24

[**Hint:** Note that the function is odd and has both half and quarter wave symmetry and therefore $a_k = 0$ for all values of k. Also $b_k = 0$ for even values of k. For $k = 1$,

$$b_1 = \frac{12V_m}{\pi^2} \sin \pi/3$$

and the fundamental component of the Fourier series is $V_g = \dfrac{12V_m}{\pi^2} \sin \pi/3 \sin \omega_0 t$.

If $V_m = 82.25$ and $\omega_0 = 1$ then $V_g = 86.6 \sin t$]

47. The trapezoidal voltage function in Problem 46 is applied to the circuit as shown in Fig. 25. Estimate the average power delivered to the resistor by the fundamental component of the voltage.

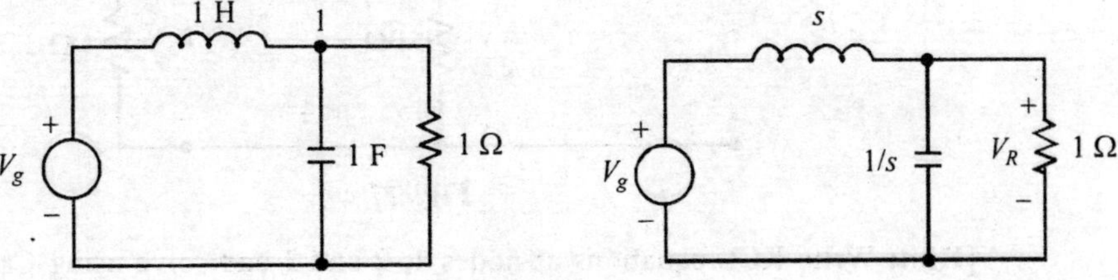

Fig. 25

[**Hint:** KCL at node 1 yields $\dfrac{V_R - V_g}{s} + V_R s + \dfrac{V_R}{I} = 0$

The transfer function

$$H(s) = \frac{V_R}{V_g} = \frac{1}{s^2 + s + 1}$$

From Problem 46, $V_g = 86.6 \sin t$

$H(j1) = 1/j$

$V_R = 86.6 \sin (t - 90)$

Assuming that the harmonic voltages are very small, Average power delivered to the 1 Ω resistance $\cong (86.6/\sqrt{2})^2/1 = 3750$ W]

48. A voltage waveform applied to a dc motor consists of a series of positive rectangular pulses of magnitude V and duration τ_d, the periodic time of the pulse being τ. If the mark space duty ratio is $D = \tau_d/\tau$ determine the form factor of the waveform in Fig. 26.

[**Hint:** $V_{\text{rms}} = \sqrt{V^2 \tau_d/\tau} = V.\sqrt{D}$

$V_{av} = V\tau_d/\tau = V.D$]

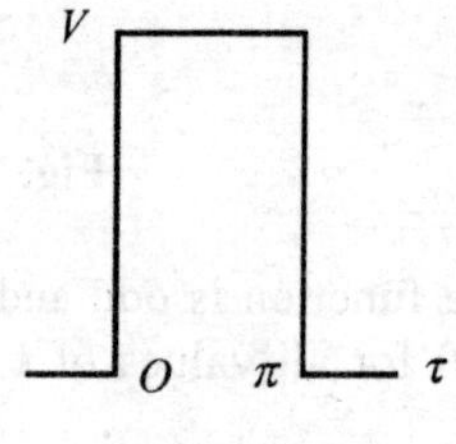

Fig. 26

[*Ans:* Form factor = $V_{\text{rms}}/V_{av} = 1/\sqrt{D}$]

49. Determine Z_{12} for the circuit of Fig. 27.

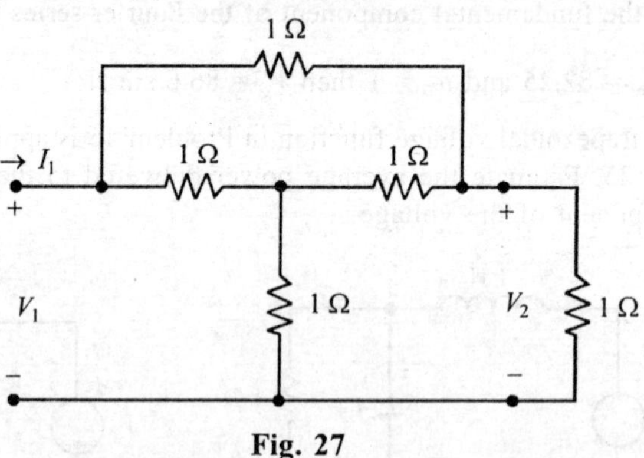

Fig. 27

[**Hint:** Write KCL equations at nodes 1, 2 and 3 and solve using Cramer's rule,

$$Z_{12} = \frac{V_2}{I_1} = \frac{1}{2}\ \Omega$$

50. Find the ratio $Y_{11}(s) = I(s)/V_1(s)$ for the circuit of Fig. 28.

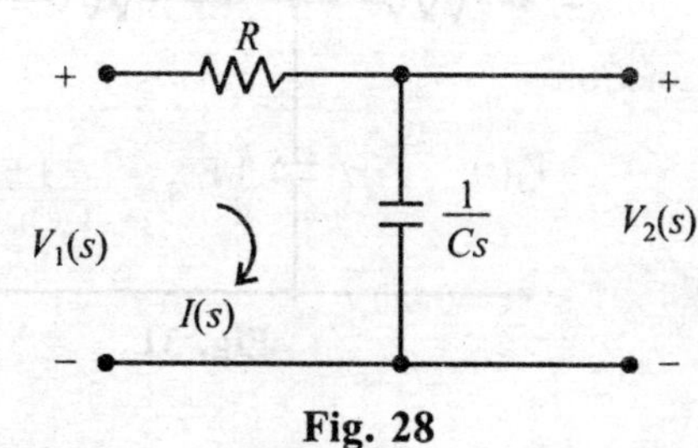

Fig. 28

$$\left[\textit{Ans: } Y_{11}(s) = \frac{1}{R}\frac{s}{(s + 1/RC)}\right]$$

51. Determine $Z(s)$ for the circuit of Fig. 29.

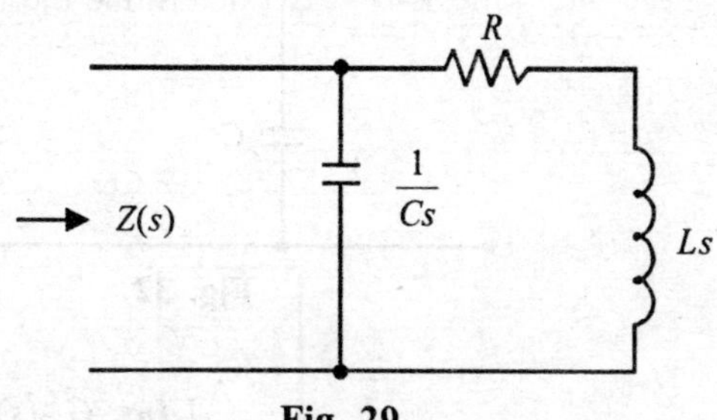

Fig. 29

$$\left[\textit{Ans: } Z(s) = \frac{1}{C}\frac{R + Ls}{(s^2 + Rs/L + 1/LC)}\right]$$

52. Find the ratio $G_{12}(s)$ for the circuit shown in Fig. 30.

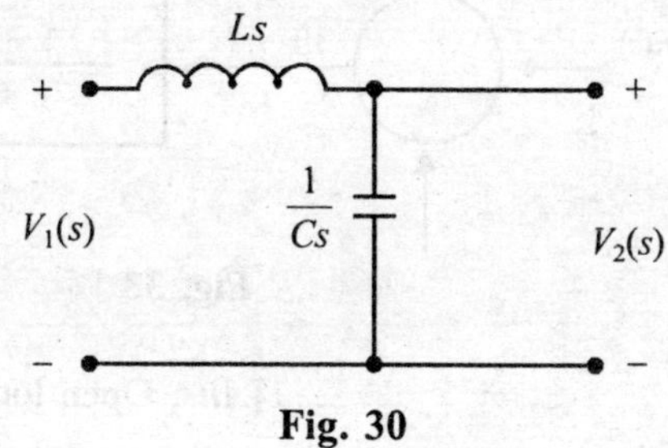

Fig. 30

$$\left[\textit{Ans: } G_{12}(s) = \frac{1/LC}{s^2 + 1/LC}\right]$$

53. Determine the ratio $\alpha_{12}(s) = I_2(s)/I_1(s)$ for the circuit of Problem 49.

[*Ans:* $\alpha_{12}(s) = -1/2$]

54. Find $G_{12}(s)$ for the circuit of Fig. 31.

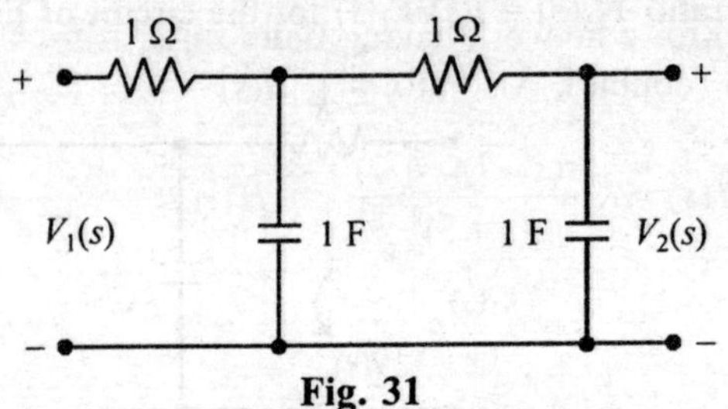

Fig. 31

[*Ans:* $G_{12}(s) = \dfrac{1}{s^2 + 3s + 1}$]

55. Find $G_{12}(s)$ for the circuit of Fig. 32.

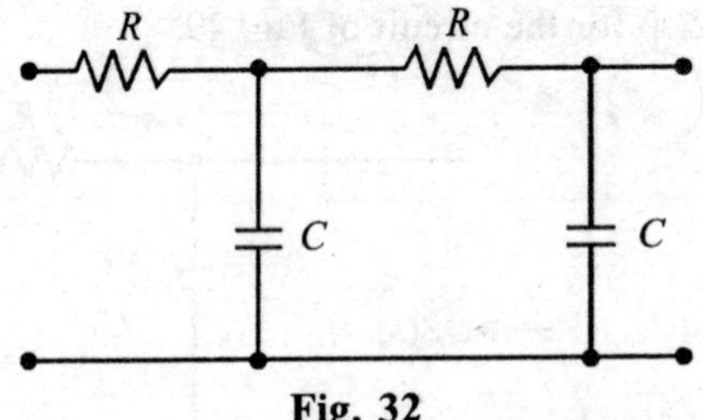

Fig. 32

[*Ans.* $G_{12}(s) = \dfrac{1}{\tau^2 s^2 + 3\tau s + 1}$ where $RC = \tau$]

56. A simple unity feedback system is illustrated in Fig. 33. Show that as k(positive) increases, the transient response decays more quickly.

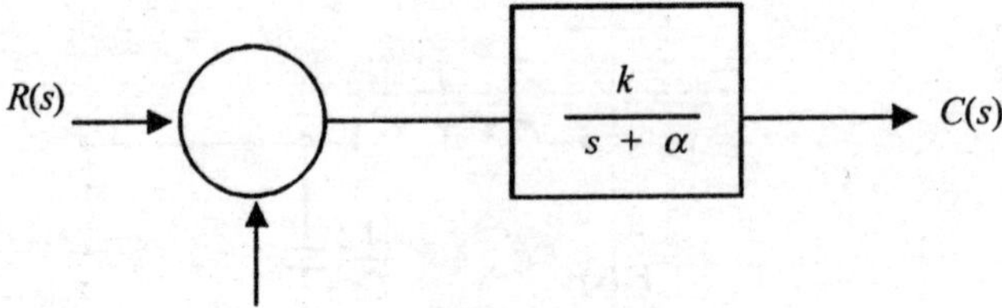

Fig. 33

[*Ans.* Open loop transfer function $G(s) = \dfrac{k}{s + \alpha}$

Unity impulse response is $C(s) = \dfrac{k}{s + \alpha}$ with unity feedback $C(s) = \dfrac{1}{s + \alpha + k}$

[Taking inverse transform

$$c(t) = ke^{-\alpha t} \text{ (Open loop)}$$

$$= ke^{-(\alpha + k)t} \text{ (Feedback)}$$

Note that as k increases transient response decays more quickly.]

57. Find $H(s)$ for a network having finite zeros at $s = 0$, $s = -1$, $s = -2$ and finite poles at $s = -3$ (double). Also at $s = 1$, $H(s) = 3$.

$$\left[\textbf{Hint: } H(s) = K\frac{s(s+1)(s+2)}{(s+3)^2} \qquad H(1) = \frac{K\times 2\times 3}{42}\right]$$

$$\left[\textit{Ans: } H(s) = \frac{8s(s+1)(s+2)}{(s+3)^2}\right]$$

58. Solve the circuit illustrated in Fig. 34.

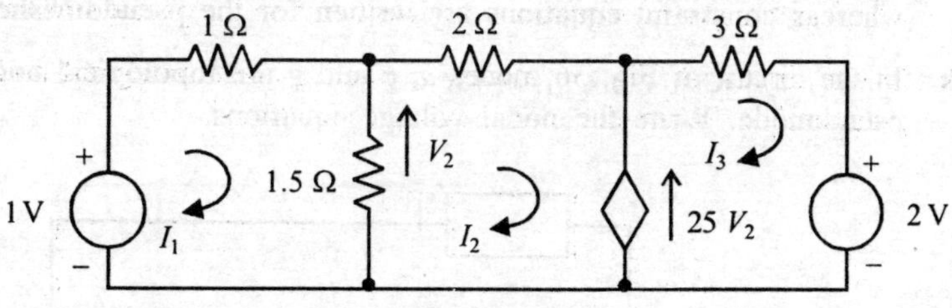

Fig. 34

[**Hint:** Applying KVL to the three meshes:

$$1 = 2.5I_1 - 1.5I_2$$
$$-2.5V_2 = -1.5I_1 + 3.5I_2$$
$$2.5V_2 - 2 = 3I_3$$

also $V_2 = 1.5\,(I_1 - I_2)$

Solve these equations.]

[*Ans:* $I_1 = 1.09$ A, $I_2 = 1.16$ A, $I_3 = -1.47$ A]

59. Solve the circuit shown in Fig. 35 (with two topological meshes 1 and 2 and one pseudomesh 3).

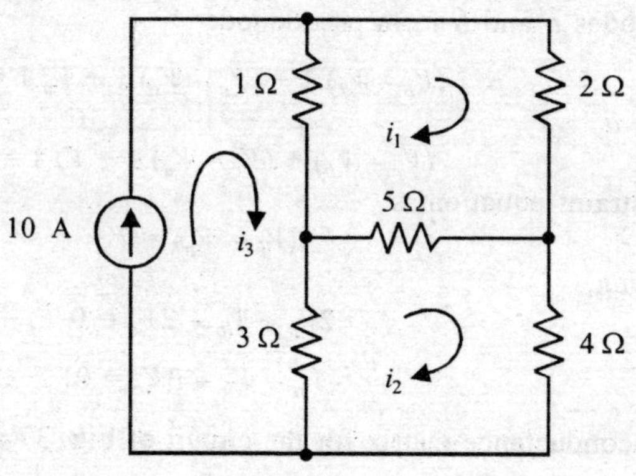

Fig. 35

[**Hint:** KVL equations for the topological meshes are:

$$20I_1 + 50(I_1 - I_2) + 10\,(I_1 - I_3) = 0$$
$$40I_2 + 30(I_2 - I_3) + 50(I_2 - I_1) = 0$$

The constraint equation for the pseudomesh 2 is

$$I_3 = 10 \text{ A}$$

Thus

$$7I_1 - 5I_2 = 10$$
$$-5I_1 + 12I_2 = 30$$

Solving

$$I_1 = 4.57 \text{ A}$$
$$I_2 = 4.41 \text{ A}]$$

Note: A resistance-less current source may create an apparent or pseudomesh in a circuit. In mesh analysis, KVL equations are written for the topological meshes, whereas constraint equations are written for the pseudomeshes.

60. In the circuit of Fig. 36, nodes *a*, *c* and *g* are topological nodes and node *b* is a pseudonode. Write the nodal voltage equations.

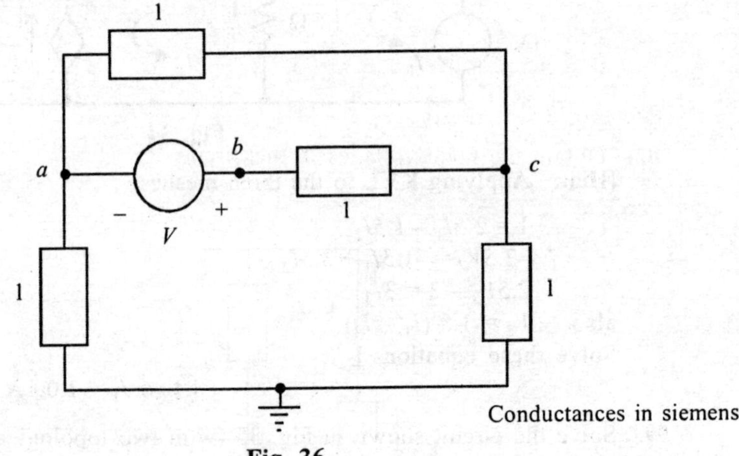

Fig. 36

[**Hint:** Nodes *a* and *b* form pseudonode

Then $$(V_b - V_c).1 + (V_a - V_c).1 + V_a.1 = 0$$

For node *a*,

$$(V_c - V_b) + (V_c - V_a).1 + V_c.1 = 0$$

The constraint equation is:

$$(V_b - V_a) = V$$

Rearranging,

$$2V_a + V_b - 2V_c = 0$$
$$-V_a - V_b - 3V_c = 0]$$

61. Find the conductance matrix for the circuit of Fig. 37 and solve for node voltages.

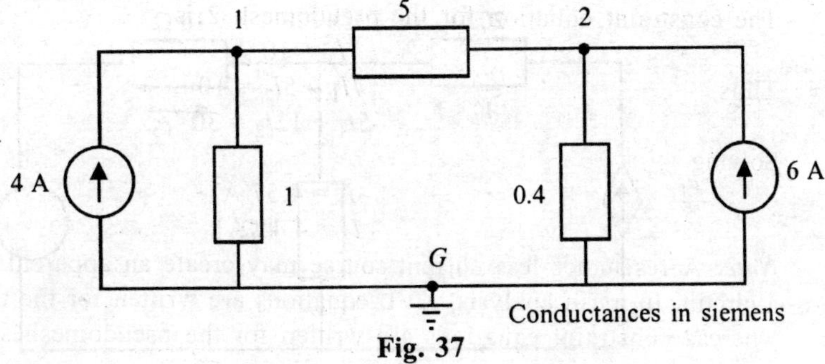

Fig. 37

$$Ans: \begin{bmatrix} 4 \\ 6 \end{bmatrix} = \begin{bmatrix} 1.5 & -0.5 \\ -0.5 & 0.9 \end{bmatrix} \begin{bmatrix} V_1 \\ V_2 \end{bmatrix}$$

$$V = \begin{bmatrix} 6 \\ 10 \end{bmatrix}$$

62. Obtain the node voltages in the circuit of Fig. 38.

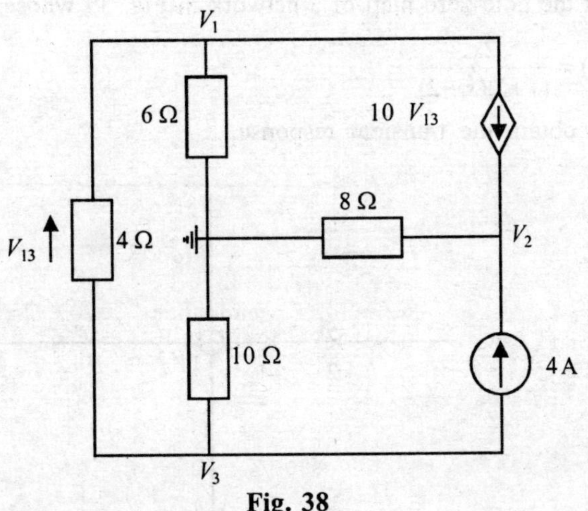

Fig. 38

[**Hint:** KCL at node 1 $10V_{13} + 6V_1 + 4(V_1 - V_3) = 0$

node 2 $8V_2 - 10V_{13} - 4 = 0$

node 3 $2 + 5V_3 + 4(V_3 - V_1) = 0$

Also $V_{13} = V_1 - V_3$]

[*Ans:* $V_1 = -0.25$ V, $V_2 = 0.63$ V, $V_3 = -0.36$ V]

63. Determine the node pair voltages V_A and V_B of the circuit shown in Fig. 39.

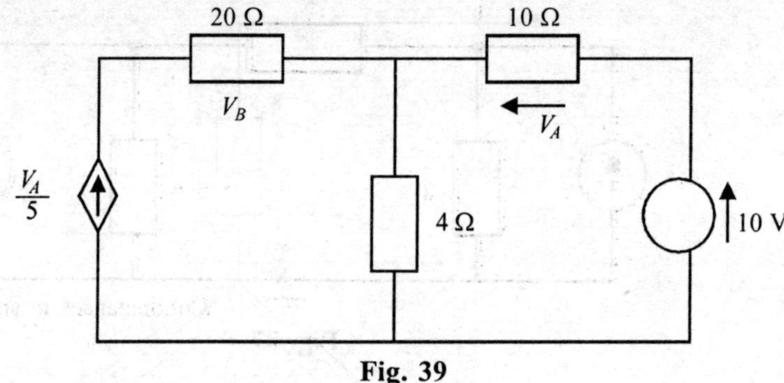

Fig. 39

[**Hint:** $\frac{V_A}{5} = \frac{V_B}{20}$,

$\frac{V_B}{20} = \frac{V_A}{10} + \frac{10 + V_A}{4}$]

[*Ans:* $V_A = -16.67$ V, $V_B = -66.67$ V]

64. Plot the pole-zero map of a network in Fig. 39 whose transfer function is:

$$I(s) = \frac{2s}{(s+1)(s+2)}$$

and obtain the transient response.

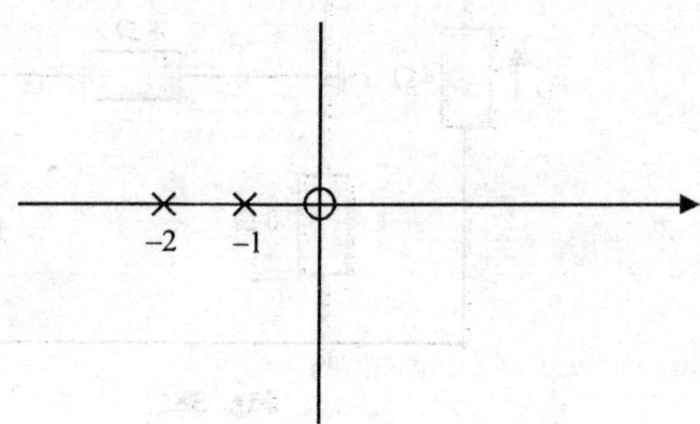

[*Ans:* $i(t) = -2e^{-t} + 4e^{-2t}$]

65. A symmetrical three-phase 400 V source supplies an unbalanced load shown in Fig. 40. Determine the line currents and power taken by the load.

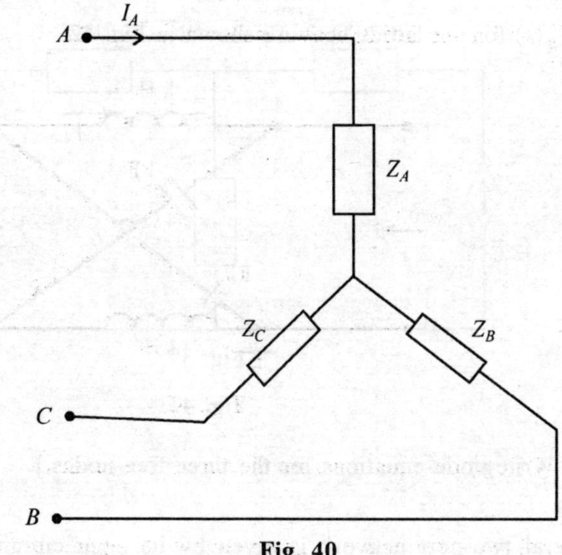

Fig. 40

[**Hint:** Equivalent delta impedances are:

$Z_{BC} = 41\angle 0°\ \Omega$, $Z_{CA} = 41\angle -30°\ \Omega$, $Z_{AB} = 41\angle 30°\ \Omega$

Phase currents in the equivalent delta are:

$V_{AB} = 400 \angle 0°$, $V_{BC} = 400 \angle -120°$, $V_{CA} = 41 \angle 120°$)

$I_{AB} = 4.75 \angle -30°$, $I_{BC} = 9.75 \angle -120°$, $I_{CA} = 9.75 \angle 150°$]

Line currents

$I_A = I_{AB} - I_{CA} = 19.49 \angle -30°$ A

$I_B = 13.75 \angle -105°$ A

$I_C = 13.75 \angle 105°$ A

Total power taken by the load $= I_A^2 R_A + I_B^2 R_B + I_C^2 R_C = 10.62$ kW]

66. Find the response to a current $i(t) = \delta(t)$ of the network shown in Fig. 41.

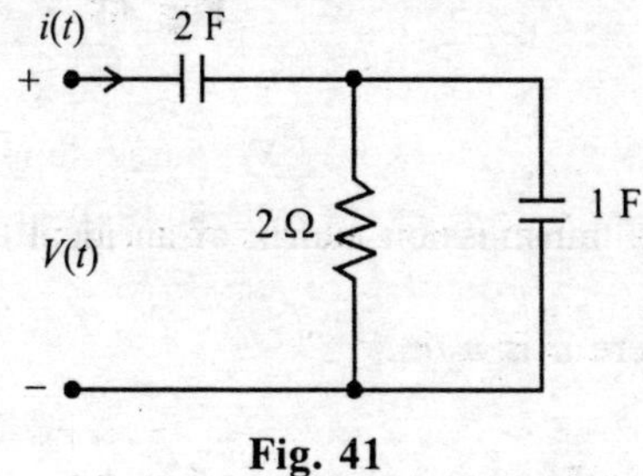

Fig. 41

[*Ans:* $i(t) = \left[\frac{1}{2} + e^{(-1/2t)} u(t)\right]$

67. Find $Z_{dp}(s)$ for the lattice network shown in Fig. 42.

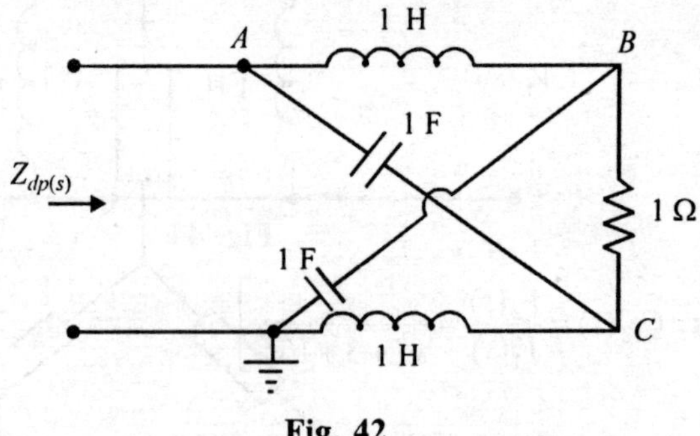

Fig. 42

[**Hint:** Write node equations for the three free nodes.]

[***Ans:*** $Z_{dp}(s) = 1$]

68. A general two-port network is given by its open circuit impedance parameters. A load impedance Z_L. is connected to port 2. Find the driving point impedance at port 1.
[**Hint:** See section 11.2.9]

[***Ans:*** $Z_{dp} = z_{11} - \dfrac{z_{12}z_{21}}{z_{22} + Z_L}$]

69. Find the natural frequencies of the network shown in Fig. 43.

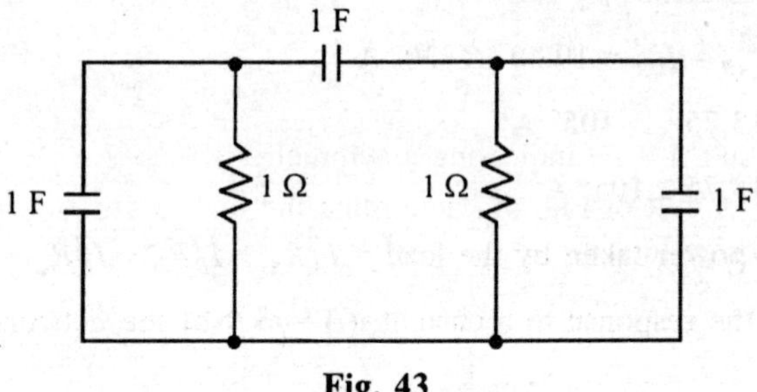

Fig. 43

[***Ans:*** $s_1 = -1/3$, $s_2 = -1$]

70. Show that the transmission matrix of an ideal transformer is given by

$\begin{bmatrix} 1/a & 0 \\ 0 & a \end{bmatrix}$ where a is n_2/n_1.

71. Find $G_{12}(s)$ for the circuit shown in Fig. 44.

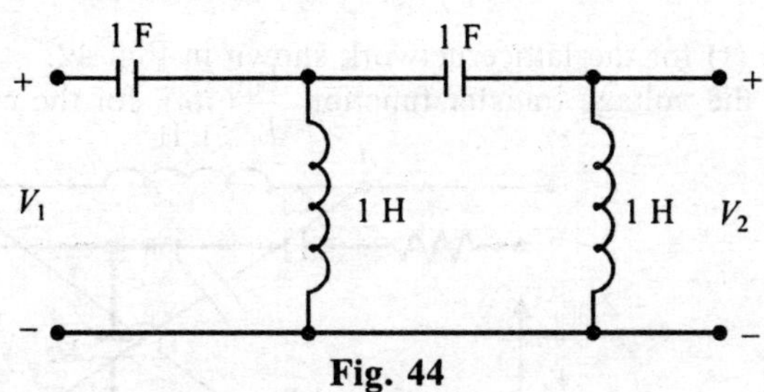

Fig. 44

$$\left[\textbf{Hint: } G_{12}(s) = \frac{V_2(s)}{V_1(s)} = \frac{s^2}{s^2 + 3 + 1/s^2}\right]$$

$$\left[\textbf{\textit{Ans:}}\ G_{12}(s) = \frac{s^4}{s^4 + 3s^2 + 1}\right]$$

72. Consider the chain connection shown in dotted lines in Fig. 45. Find the chain parameters of the overall two-port network.

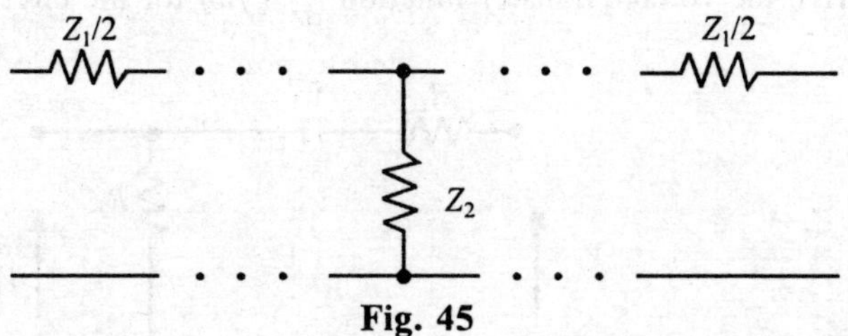

Fig. 45

$$\left[\textbf{Hint: } \begin{bmatrix} 1 & Z_1/2 \\ 0 & 1 \end{bmatrix} \begin{bmatrix} 1 & 0 \\ 1/Z_2 & 1 \end{bmatrix} \begin{bmatrix} 1 & Z_1/2 \\ 0 & 1 \end{bmatrix}\right]$$

$$\left[\textbf{\textit{Ans:}} = \begin{bmatrix} 1 + Z_1/2Z_2 & Z_1(1 + Z_1/4Z_2) \\ 1/Z_2 & 1 + Z_1/2Z_2 \end{bmatrix}\right]$$

Note that $A = D$, indicating a symmetric network.

73. In the circuit of Fig. 46, determine the number of edges, number of nodes and the number of loops.

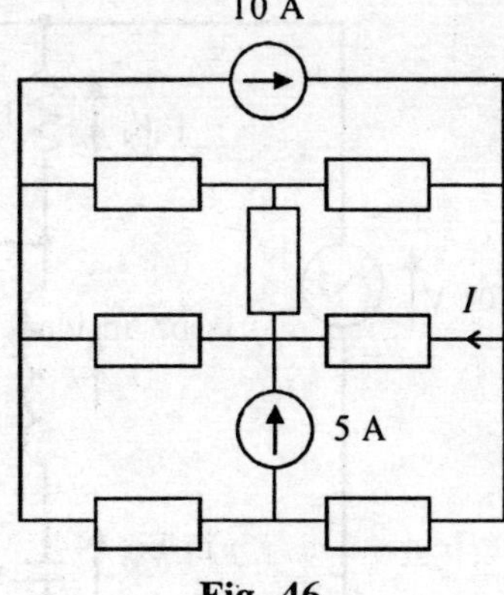

Fig. 46

[***Ans:*** Edges = 9, Nodes = 5, Loops = 5]

74. Derive the voltage transfer function $\frac{V_0}{V_1}(j\omega)$ for the circuit in Fig. 47.

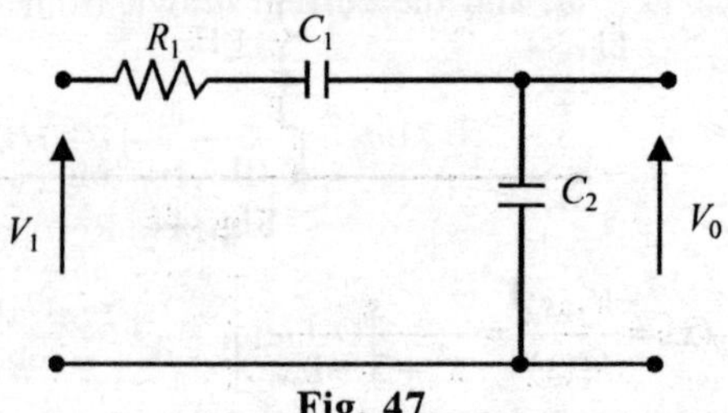

Fig. 47

[*Ans:* $\frac{C}{1+j\omega T}$ where $C = \frac{C_1}{C_1 + C_2}$, $T = \frac{R_1 C_1 C_2}{C_1 + C_2}$]

75. Derive the voltage transfer function $\frac{V_0}{V_1}(j\omega)$ for the circuit in Fig. 48.

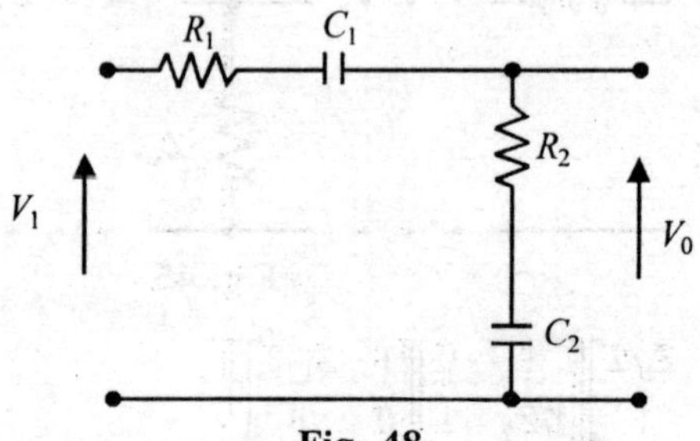

Fig. 48

$$\left[\textit{Ans:} \ \frac{C(1+j\omega T_2)}{1+j\omega T} \text{ where } C = \frac{C_1}{C_1 + C_2},\ T_1 = \frac{C_1 C_2 (R_1 + R_2)}{C_1 + C_2},\ T_2 = R_2 C_2 \right]$$

76. For the circuit in Fig. 49, calculate the resonant frequency, the Q-factor at resonance and the current drawn by the circuit.

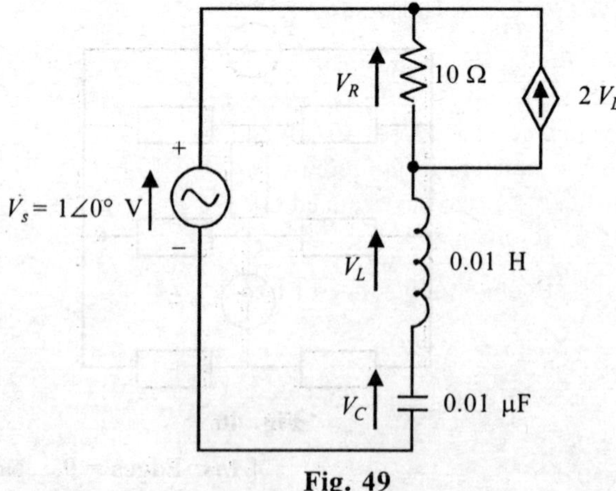

Fig. 49

[**Hint:** Voltage across the circuit $1 + j0 = 10\,(I + 2V_L) + j0.01\,\omega I + \dfrac{I}{j0.01\times 10^{-6}\omega}$
where $V_L = j0.01\omega I$
At resonance $\omega = \omega_0$ and the current drawn from the supply is in phase with the supply voltage, hence

$$0.21\omega_0 - \frac{1}{0.01\times 10^{-6}\omega_0} = 0$$

[*Ans:* Resonant frequency ω_0 = 21821 rad/s

[Q-factor = $Q_0 = \dfrac{\omega_0 L}{R} = 21.8$, Current = 0.1 + j0 A]

77. Calculate the resonant frequency of the circuit in Fig. 50, the Q factor at resonance, the bandwidth, the upper and lower cut off frequencies and the current in the circuit at resonance.

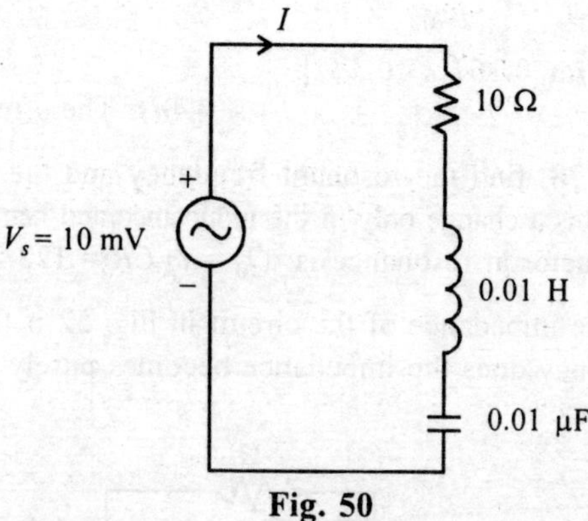

Fig. 50

[**Hint:** Resonant frequency $\omega_0 = 1/\sqrt{LC}$ = 100000 rad/s

Q-factor at resonance $Q_0 = \dfrac{1}{R}\sqrt{L/C}$ = 100

Bandwidth $BW = \dfrac{\omega_0}{Q_0}$ = 1000 rad/s

Lower cut off frequency = $\omega_0 - BW/2$ = 99500 rad/s
Upper cut off frequency = $\omega_0 + BW/2$ = 100500 rad/s
Current at resonance $I = V/R = 10^{-3}$ A
Note that the parameters in the circuit of Fig. 49 and Fig. 50 are the same but the voltage dependent source has reduced the resonant frequency and thereby reducing the Q-factor at resonance.

78. Determine the effective inductance of the voltage source at resonance of the circuit in Fig. 51.

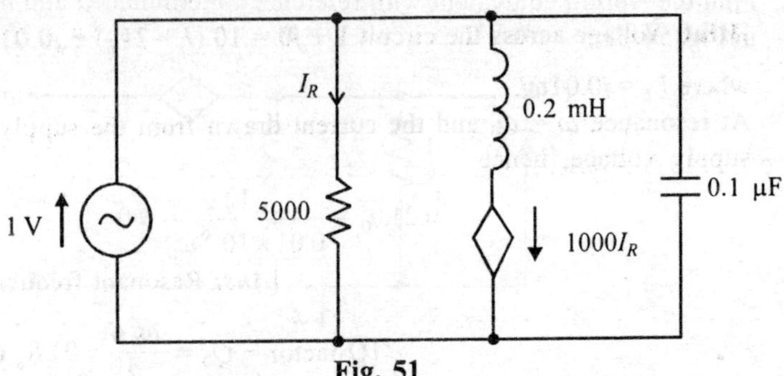

Fig. 51

[**Hint:** The current in the inductive branch at resonance is:

$$\frac{V_s}{Z_L} = \frac{1 + 1000 I_R}{j\omega_0 L} = \frac{6000}{j\omega_0}$$

Hence $Z_L = j\omega_0\, 0.167 \times 10^{-3}\ \Omega$]

[*Ans:* The effective inductance is 0.167 mH]

79. In Problem 78, find the resonant frequency and the Q-factor at resonance.
[**Hint:** There is a change only in the inductance and hence $\omega_0 = 1/\sqrt{LC} = 244925$ rad/s and the Q-factor at resonance is: $Q_0 = \omega_0 CR = 122.5$]

80. Calculate the impedance of the circuit in Fig. 52 if the frequency is 2000 rad/s. At what frequency does the impedance becomes purely resistive. What is the value of this resistance?

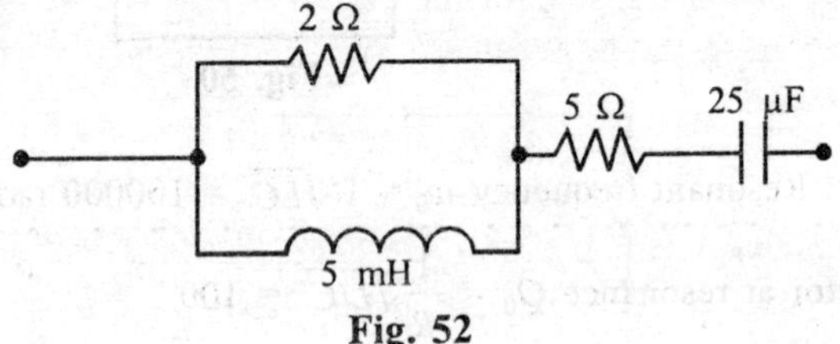

Fig. 52

[*Ans:* Impedance = $9 - j12\ \Omega$, Resonant frequency = 1000 rad/s, Impedance at resonance = 15 Ω]

81. Find the Thevenin equivalent with reference to terminals a and b in the circuit shown in Fig. 53.

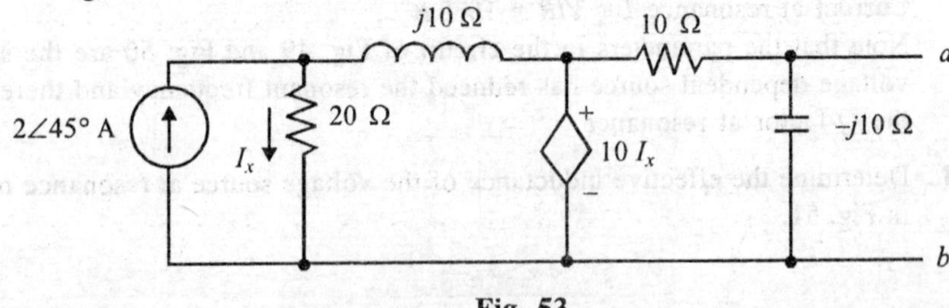

Fig. 53

[*Ans:* $V_{Th} = V_{ab} = 0.5\ \angle -45°$V, $Z_{Th} = 5 - j5\ \Omega$]

82. Find the Norton equivalent with reference to terminals a and b in the circuit shown in Fig. 54.

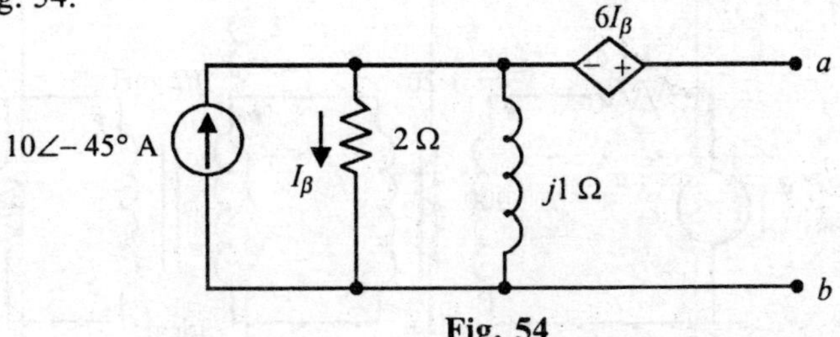

Fig. 54

[*Ans:* $I_N = 10\ \angle{-45°}$ A, $Z_N = 1.6 + j3.2\ \Omega$]

83. Find the voltage V_s(rms) in the circuit shown in Fig. 55 if loads L_1 and L_2 absorb 15 kVA at 0.6 pf lagging and 6 kVA at 0.8 leading.

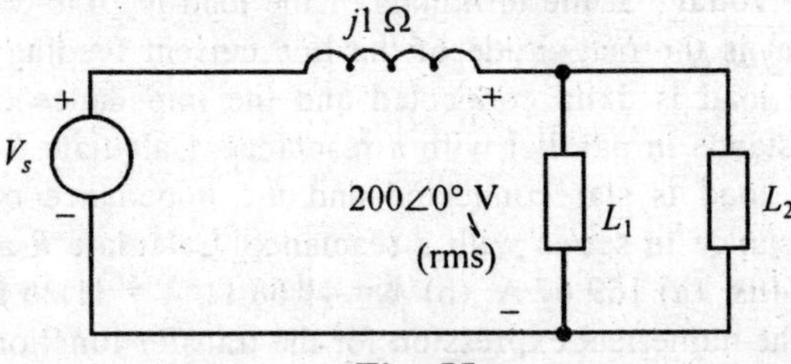

Fig. 55

[*Ans:* $251.6\ \angle159°$ V]

84. Find the value of the impedance to be connected across terminals ab for maximum average power transfer for the circuit shown in Fig. 56. What is the average power transferred to the impedance?

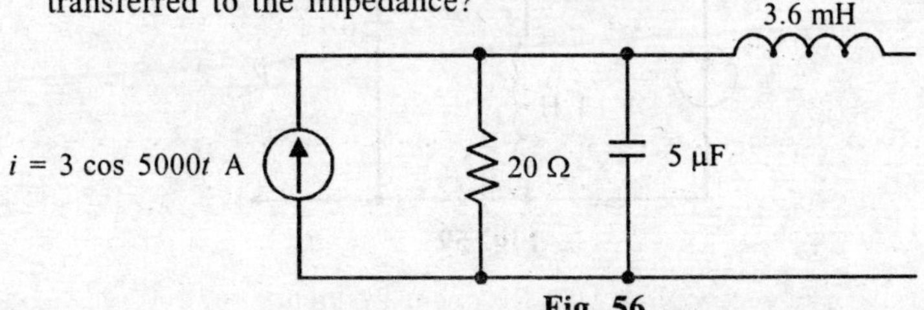

Fig. 56

[*Ans:* Load impedance = $20 - j10\ \Omega$, Average power transferred to the impedance = 18 W]

85. The magnetically coupled coils in the circuit of Fig. 57 carry initial currents of 6 A and 4 A. Find the initial energy stored in the circuit.

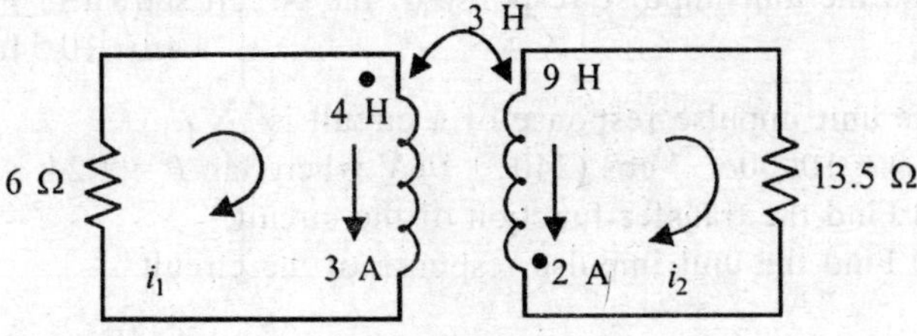

Fig. 57

[*Ans:* 18 J]

86. Find the average power delivered to the 10 Ω resistor in the circuit shown in Fig. 58.

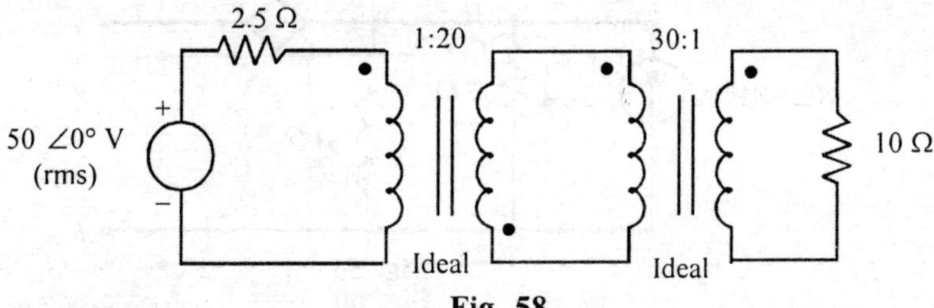

Fig. 58

[*Ans:* 90 W]

87. The complex power associated with each phase of a balanced load is 144 + *j*192 kVA. The line voltage at the terminals of the load is 2450 V.

(a) What is the magnitude of the line current feeding the load?

(b) The load is delta connected and the impedance of each phase consists of a resistance in parallel with a reactance. Calculate R and X.

(c) The load is star connected and the impedance of each phase consists of a resistance in series with a reactance. Calculate R and X.

[*Ans:* (a) 169.67 A; (b) R = 41.68 Ω, X = 31.26 Ω; (c) R = 5 Ω, X = 6.67 Ω]

88. Derive the numerical expression for the transfer function V_o/I_g for the circuit shown in Fig. 59 and give the numerical value of each pole and zero of $H(s)$.

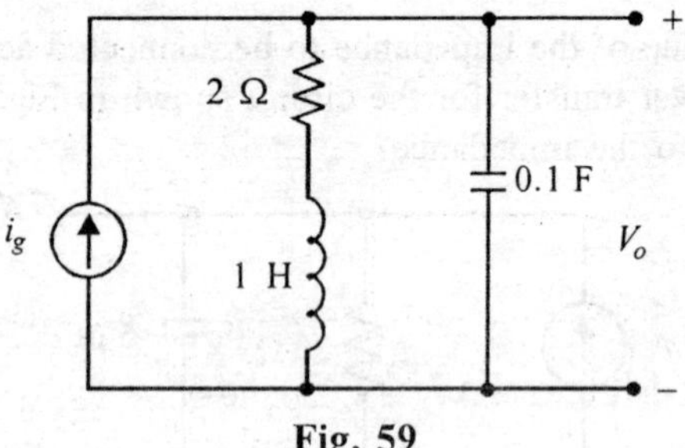

Fig. 59

[*Ans:* $H(s) = 10(s + 2)/(s^2 + 2s + 10)$ Poles $s_1 = -1 + j\,3$, $s_2 = -1 - j\,3$ Zero $s = -2$]

89. Find the unit step response of the circuit shown in Fig. 59.

[*Ans:* $[2 + 3.33e^{-t} \cos (3t + 126.87°)]\, u(t)$ V]

90. Find the unit impulse response of the circuit shown in Fig. 59.

[*Ans:* $10.54e^{-t} \cos(3t - 18.43°)\, u(t)$ V]

91. The unit impulse response of a circuit is:
$v_o(t) = 10000e^{-70t} \cos (240t + \theta)$ V where $\tan \theta = 7/24$.
(a) Find the transfer function of the circuit.
(b) Find the unit impulse response of the circuit.

[*Ans:* (*a*) $\dfrac{9600s}{s^2 + 140s + 62500}$; (b) $40e^{-70t} \sin 240t$ V]

92. A rectangular voltage pulse $v_i = [u(t) - u(t-1)]$ V is applied to the circuit shown in Fig. 60. Use the convolution integral to find v_o.

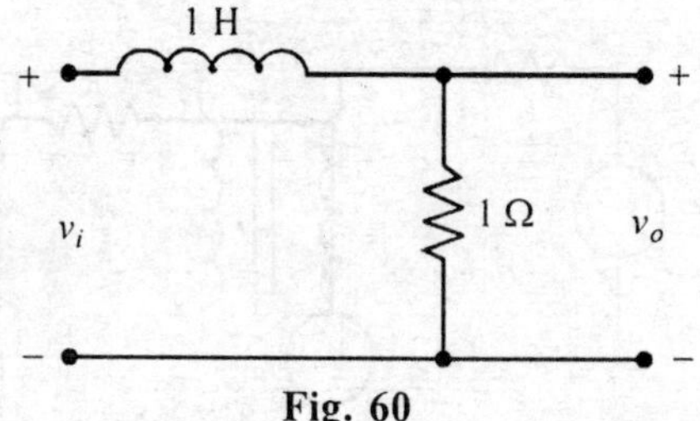

Fig. 60

[*Ans:* $v_o = 1 - e^{-t}$ V, $0 \le t \le 1$;
$v_o = (e - 1)e^{-t}$ V, $1 \le t \le \infty$]

93. In Problem 92, interchange the inductor and resistor and again use convolution to find v_o.

[*Ans:* $v_o = e^{-t}$ V, $0 < t < 1$;
$v_o = (1 - e)e^{-t}$ V, $1 < t \le \infty$]

94. Find the poles and zeros of the impedance seen looking into the terminals *ab* of the circuit shown in Fig. 61.

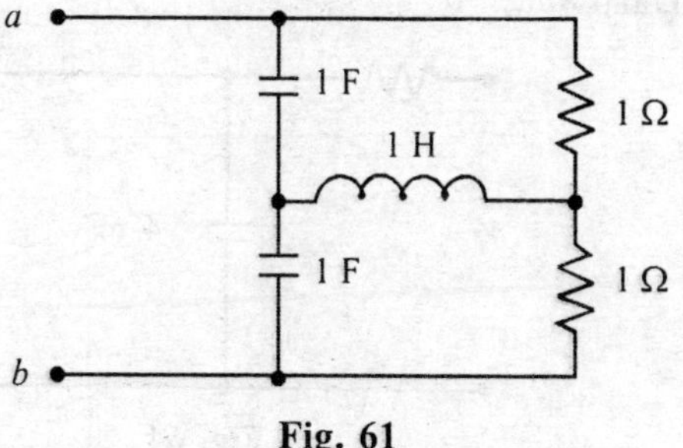

Fig. 61

[**Hint:** Note that the circuit is a balanced bridge circuit.]

[*Ans:* Poles $s = 1$ and no zeros]

95. Show that after V_oC_{eq} coulombs are transferred from C_1 to C_2 in the circuit shown in Fig. 62, the voltage across each capacitor is $C_1V_o/(C_1 + C_2)$.

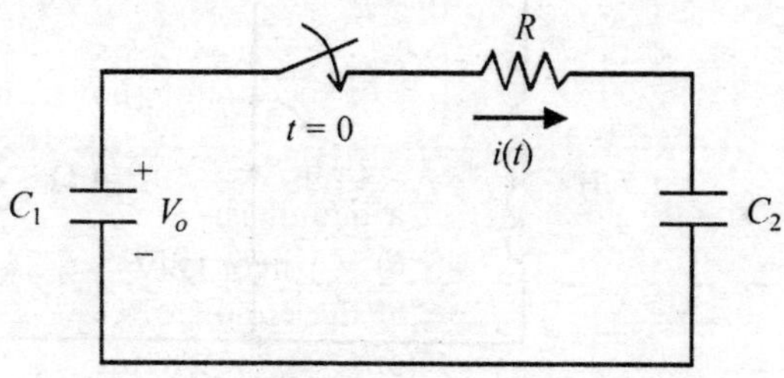

Fig. 62

[**Hint:** Use the conservation of charge principle.]

96. Design the component values for the series *RLC* band reject filter shown in Fig. 63 so that the centre frequency is 4 kHz and the quality factor is 5. Use a 500 nF capacitor.

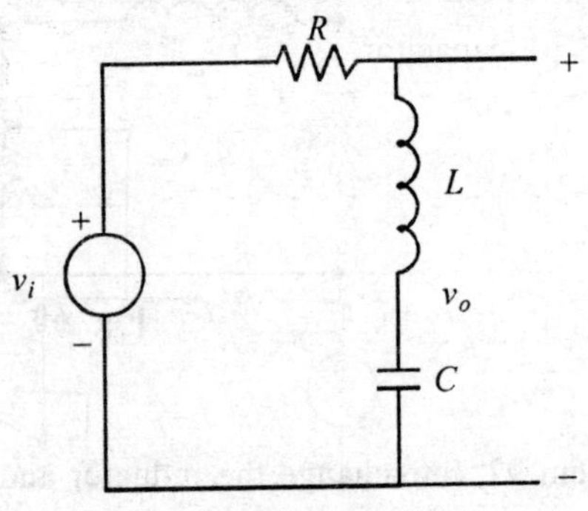

Fig. 63

[*Ans:* $L = 3.17$ mH, $R = 14.92\ \Omega$]

97. (a) Find the cut off frequency of the low pass filter shown in Fig. 64.
(b) Calculate $H(j\omega)$ at ω_c, $0.2\omega_c$ and $8\omega_c$.
(c) If $v_i = 480 \cos \omega t$ mV, write the steady-state expression for v_o when $\omega = \omega_c$, $0.2\ \omega_c$ and $8\omega_c$.

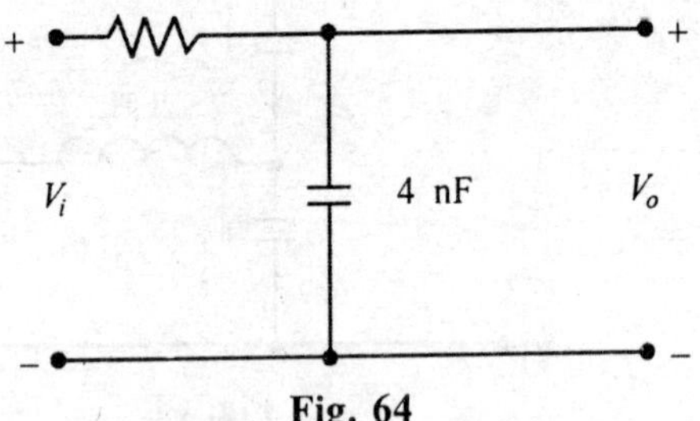

Fig. 64

98. (a) Determine the expression for $v_o(t)$ after the switch in the network in Fig. 65 is opened at $t = 0$; (b) Repeat part (a) when the inductor is replaced by a 1/6 mF capacitor.

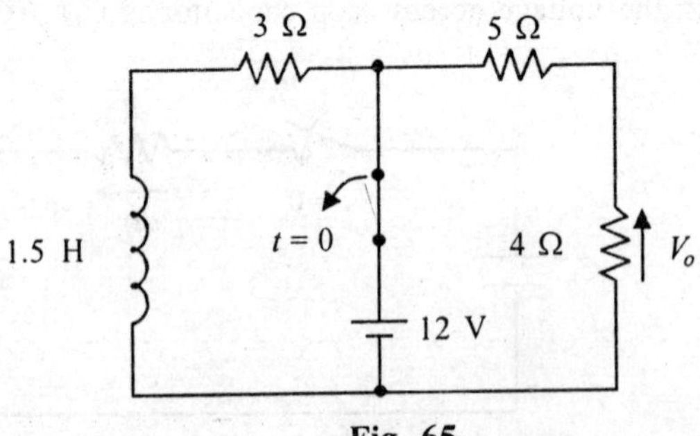

Fig. 65

[*Ans:* (a) $v_o(t) = -16e^{-8t}$ V, (b) $v_o(t) = 4e^{-500t}$ V]

99. Compute the resistor values needed for the low-pass filter circuit in Fig. 66 to have the transfer function

$$H(s) = \frac{-20000}{s + 5000}$$

Use a 5 μF capacitor.

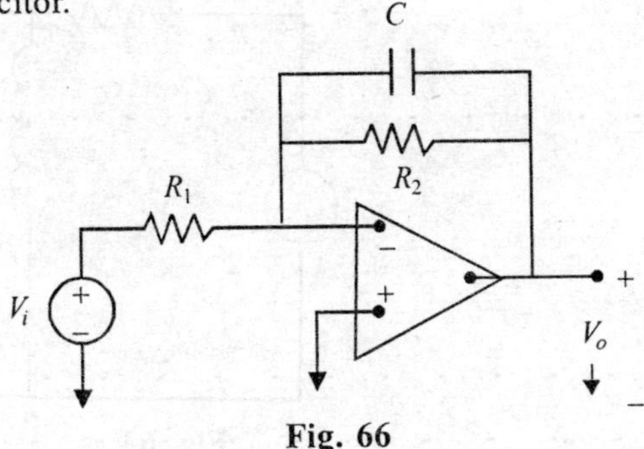

Fig. 66

[*Ans:* R_1 = 10 Ω, R_2 = 40 Ω]

100. For the circuit in Fig. 67, find the values of R_1 and R_2 that yields a second-order prototype Butterworth high-pass filter.

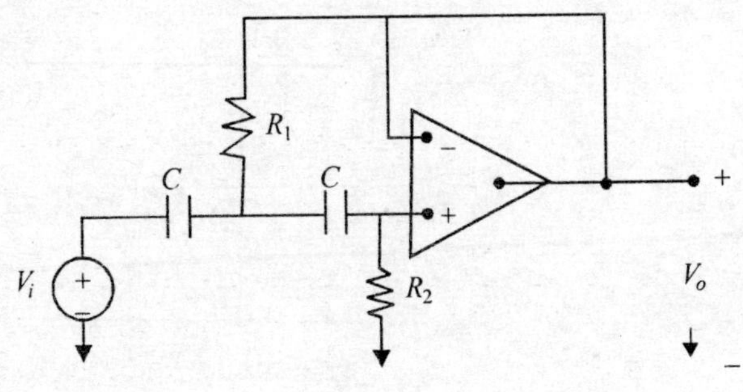

Fig. 67

[*Ans:* R_1 = 0.707 Ω, R_2 = 1.41 Ω]

101. Find the average power in a 15 Ω resistor if the current is:
$i = 10 \sin \omega t + 5 \sin 3\omega t + 2 \sin 5\, \omega t$ A

[*Ans:* 967.5 W]

References

1. Williams, G., *An Introduction to Electrical Circuit Theory*, Macmillan, ELBS, 1973.
2. Franklin, Kuo, F., *Network Analysis and Synthesis,* Wiley, 1966.
3. Nilson, J.W. and Riedel, A. Susan, *Electric Circuits*, Prentice Hall, New Jersey, 2001.
4. Mix, F. Dwight, and Schmidt, Neil M., *Circuit Analysis for Engineers,* John Wiley, 1985.
5. Karni, Shlomo, *Analysis of Electrical Networks*, John Wiley, 1986.
6. O'Malley, John, *Basic Circuit Analysis,* McGraw-Hill, 1992.
7. Edminister, Joseph, *Electric Circuits,* McGraw-Hill, 1987.
8. Van Valkenburg, M.E., *Network Analysis,* Prentice Hall, 1996.
9. Morris, Noel M., and Senior, Frank W., *Electric Circuits,* Macmillan, 1993.
10. Kreyzig, *Advanced Engineering Mathematics,* John Wiley, 1972.
11. Spencer, A.J.M., et al., *Engineering Maths,* Vol I, Van Nostrand Reinhold, London, 1978.
12. Spiegel, Murray R., *Laplace Transforms,* McGraw-Hill, 1987.
13. Pipes, L.A. and Harvill, L.R., *Applied Mathematics for Engineers and Physicists,* McGraw-Hill, 1971.
14. Mithal, G.K., *Network Analysis,* Khanna, 1976.
15. Nilson, J.W., *Introduction to PSpice Manual,* Prentice Hall, 2000.
16. Rashid, Muhammad H., *Spice for Circuits and Electronics using PSpice,* Prentice-Hall of India, 1995.
17. Nassar, Syed A., *Electric Circuits,* Book I & II, McGraw-Hill, 1989.
18. Nagrath, J. and Gopal, M., *Control Systems Engineering,* Affiliated East-West Press, 1989.
19. Sinha, Umesh, *Circuit Theory,* Sathya Prakash, 2001.
20. Charles A., et al., *Basic Circuit Theory,* McGraw-Hill, 1969.
21. Punagin, R.P., *Electrical Circuit Analysis,* Interline, 1994.
22. Bouwen, A.J., *Digital Instrumentation,* Tata McGraw-Hill, 1997.
23. Coughlin, Robert F. and Driscoll, Frederick F., *Operational Amplifiers and Linear Integrated Circuits,* 6th ed., Prentice-Hall of India, 2003.

Index